Luminescence in Electrochemistry

Fabien Miomandre · Pierre Audebert
Editors

Luminescence in Electrochemistry

Applications in Analytical Chemistry, Physics and Biology

Springer

Editors
Fabien Miomandre
PPSM
Ecole Normale Superieure de Cachan
Cachan
France

Pierre Audebert
PPSM
Ecole Normale Superieure de Cachan
Cachan
France

ISBN 978-3-319-49135-6 ISBN 978-3-319-49137-0 (eBook)
DOI 10.1007/978-3-319-49137-0

Library of Congress Control Number: 2016956820

Printed on acid-free paper

This Springer imprint is published by Springer Nature
The registered company is Springer International Publishing AG
The registered company address is: Gewerbestrasse 11, 6330 Cham, Switzerland

Foreword

When our team, in collaboration with korean friends, started some while ago (in 2004) to work on electrofluorochromism (EF), we were unaware at which point this field was virgin, like many others closely related and now fully emerging. After some pioneering works in the 1990s by several renowned researchers like A.J. Bard in the US, R. Compton in UK, C. Amatore and E. Levillain in France among others, it seems that a fresh momentum in the interplay of luminescence and electrochemistry appeared with new targets like the detection of a few chemical events or the design of new high contrast emitting displays.

As strange as it may seem, at the time we started to be interested in the crossover of electrochemical and photophysical properties, only electrochromism (EC, the most ancient related research area) and electrochemiluminescence (ECL) along with some of its applications in biosensing, had already been really developed. Undoubtedly the field of 'luminescence and electrochemistry coupling' had not revealed all its potentialities at that time and the last twelve years confirmed that matter of fact with several significant new contributions, where both fields were concerned and sometimes intimately intricate.

Therefore, we were both delighted and honored when the renowned Springer edition house contacted us asking whether we were willing to serve as guest editors of a book closely related to this ensemble of topics. This is how the concept of "Luminescence in Electrochemistry" started on. After thorough discussions, we were finally able to identify, besides the well-established domains as aforementioned, several emerging ones, like the design of new molecules and materials or the investigation of nanometer size systems where electrochemistry and light could bring real added values to the knowledge. The combination of all these fields finally led to the organization of this book.

This organization lets three different sections appear. The first one (chapters Emission Spectroelectrochemistry: Cell Design and Setup, In Situ Spectroelectrochemical Fluorescence Microscopy for Visualizing Interfacial Structure and Dynamics in Self-assembled Monolayers, Electrochemically Modulated Luminescence in Nanophotonic

Structures) highlights the development of new instrumental set-up allowing to investigate the coupling of electrochemistry and luminescence : emission spectroelectrochemistry which was at first designed with the model of its absorption counterpart progressively evolved to let electrochemically coupled luminescence microscopy techniques occupy the front of the stage, because their outstanding sensitivities enable to lower the detection threshold to monolayers or nearly single molecules at the electrode interface. The second section (chapters Electrochemically Monitored Photoluminescence of Conjugated Polymers, Electrofluorochromic Devices with Organic Dyes and Conjugated Polymers, Control of Emission and Coloration in Electrochemical Systems and Its Applications) deals with electrofluorochromism (EF), that is the control of the photogenerated light through the electrode potential by changing the redox state of a molecule or a polymer film. This field opens the way to new systems where absorption, reflection and emission can be smartly combined to lead to high contrast, night and day operable and more efficient low cost displays. The last section of the book (chapters Theoretical Insights in ECL, Applications of Electrogenerated Chemiluminescence in Analytical Chemistry, Electrochemically Driven Luminescence in Organometallic and Inorganic Systems, Light-Emitting Electrochemical Cells) summarizes the latest results in the related fields of ECL and light electrochemical cells (LEC). We have taken care to give a large place to ECL, with chapters covering at the same time the theoretical aspects, as well as the experimental ones and their applications especially in the medical diagnostic. Finally the principles and potentialities of LEC, an electrochemical version of LED, are described in the last chapter.

Thus it appears that electrochemistry and luminescence can be combined in different ways, either by using the former to generate the latter as in ECL or LEC, or simply to tune it like in EF. The issues in all these fields rely on various aspects, typically from the molecules to the device. This requires the skills or organic chemists to design new molecular systems, but also of physicists to develop more and more informative techniques allowing pushing the detection to the limit both in the time and space domains. Thus, some parts of the book will be more devoted to chemistry (e.g. chapters Electrofluorochromic Devices with Organic Dyes and Conjugated Polymers, Electrochemically Driven Luminescence in Organometallic and Inorganic Systems) while others will be more concerned with the physics (chapter In Situ Spectroelectrochemical Fluorescence Microscopy for Visualizing Interfacial Structure and Dynamics in Self-assembled Monolayers) or the devices (chapters Control of Emission and Coloration in Electrochemical Systems and Its Applications, Light-Emitting Electrochemical Cells) thus covering the main aspects of the topic. In the image of its applications coming from electro-optics (light emitting devices, displays …) to biology (biosensors, diagnostic, mechanisms involved in biochemical phenomena …), the scope is actually very wide and thus requires crossing knowledge of several disciplines (electronics, optics, spectroscopy, chemistry, biology, hydrodynamics …). It is also a key point of this book to bring together contributions from experts with various backgrounds.

This brief introduction would of course not be complete without warmly thanking all our friends and colleagues that have kindly accepted to contribute to this book. Not only their works are brilliant, but also they are essential to the usefulness and future success of this book. Let alone one or two contributors had declined, and this book would already have lost of its consistence and completeness. We are really indebted towards them for their huge efforts in giving the best of themselves to make sometimes complicated issues accessible to non-specialists while keeping the scientific content at a very high level.

To conclude we hope that this compilation, beyond highlighting the potentiality and versatility of coupling 'luminescence and electrochemistry', will stimulate worldwide interest for this field. We would like that this book stimulates further research everywhere, and thus contributes to the emergence of this field, although it is already in progress. While the theoretical concepts are more or less now established, we are certainly not aware of all the possible applications. Our wish is that upon reading this book, people develop new ideas, thus contributing to the further expansion of this fascinating research domain, and why not even further, that this expansion leads to everyday life products and/or devices with high technological impact.

Thanks to all the contributors and enjoy your reading.

Contents

Emission Spectroelectrochemistry: Cell Design and Setup

Olivier Alévêque and Eric Levillain

Abstract This chapter reviews the most recent developments in the fluorescence spectroelectrochemistry, coupled detection of fluorescence and electrochemical signals. It focuses on the instrumental development in fluorescence spectroelectrochemistry and recent coupling of electrochemical techniques with fluorescence microscopy. The first part is dedicated to conventional fluorescence spectroelectrochemistry cells and the second one to the electrochemistry and fluorescence microscopy coupling.

1 Introduction and Chapter Scope

Spectroelectrochemistry (SEC) combines both electrochemical and spectroscopic experiments and affords the investigation and identification of electroactive species under different redox states or products arising from redox reactions [1–13]. The strength of this technique relies on generating spectroscopic information about in situ electrogenerated species. The coupling can be done, for example, with electronic absorption (UV/VIS/NIR absorption), light emission and scattering (fluorescence), vibrational modes and frequencies (IR and Raman), magnetic resonance (NMR) and electron spin resonance (ESR).

Combined results from electrical and optical responses allow detailed insights in underlying mechanisms and more precise studies than each process taken separately.

The typical application fields that involve spectroelectrochemistry are bioelectrochemistry, redox polymer processes, molecular electrochemistry, molecular switches and organometallic reactions among others.

Historically, time-resolved spectroelectrochemistry was introduced in the 1990s to study molecules in solution. Over the years, it became a complementary analytical technique to conventional electrochemical methods. The electrochemical

O. Alévêque · E. Levillain (✉)
Laboratoire MOLTECH Anjou—Université d'Angers/CNRS UMR6200,
2, Boulevard Lavoisier, 49045 ANGERS Cedex, France
e-mail: eric.levillain@univ-angers.fr

F. Miomandre and P. Audebert (eds.), *Luminescence in Electrochemistry*,
DOI 10.1007/978-3-319-49137-0_1

measurement (i.e, cyclic voltammetry and/or chronoamperometry) was first coupled to multichannel UV–visible–NIR spectrophotometry [14–18] and then adapted to micro-Raman [19, 20], rapid scan infrared [21–23], fluorescence [24, 25], ESR [26, 27], EXAFS [28], etc. Over the past 15 years or so, the improvement in electronic devices and the lowering cost of CCD detectors have extended the spectroelectrochemical analytical range to modified electrodes (i.e., mono-, multilayers or conducting polymers).

Since there are a lot of potential analytical applications, especially in biochemistry [4, 29–32], where in situ dual detection of electrochemical and fluorescence signals can lead to very sensitive and selective biosensors, this chapter reviews the most recent cell design and setup developments in the fields of fluorescence spectroelectrochemistry (F-SEC), coupled detection of fluorescence and electrochemical signals of molecules and materials exhibiting electrofluorochromic properties.

The first part (Part 2) focuses on the conventional F-SEC, whereas the second one is dedicated to the coupling of electrochemistry and fluorescence microscopy (Part 3). It is noteworthy that this chapter is largely inspired by the excellent minireview of Pierre Audebert and Fabien Miomandre, published in Chemical Science in 2013 [33].

2 Conventional Fluorescence Spectroelectrochemistry Cells

2.1 *Fluorescence Spectroelectrochemical Cells*

The general principle of the fluorescence spectroelectrochemical technique is to visualize in situ, simultaneously, the electrical (current) and the spectral (fluorescence) responses of a molecular system close to the surface of a working electrode surface subjected to both a linear or staircase potential modulation and an incident monochromatic light [14, 34].

Despite a seemingly simple principle, the F-SEC has received less attention than its absorption counterpart [i.e., absorption spectroelectrochemistry (A-SEC)] for investigating molecular systems. While sharing a common principle, this is essentially due to the difficulties to design specific dedicated cells.

Actually, this technique requires that the immediate vicinity of the working electrode or the electrode surface itself must be in interaction with the excitation light. This also implies that the optical path of the light beams must not be disturbed by the elements present in the cell, such as counter and reference electrodes for example. Finally, the cell has to be fully compatible with the ideal 90° detection angle, between the excitation and the emission light beams, required for luminescence measurements in order to limit the quantity of excitation light reaching the detector [12].

Over years, cells have been designed according to technologies available at that time and to the specific experimental conditions. This chapter lists the most used and the most effective solutions selected among the most promising [35–38].

2.1.1 Optically Transparent Thin-Layer Electrochemical (OTTLE) Cells

The first setup designed to generate interpretable results was "optically transparent thin-layer electrochemical" (OTTLE) cells, which have been recurrently used for A-SEC [39]. The later presents the advantages to work with small volume and with a high $(\text{surface/volume})_{\text{exposed}}$ ratio, thus achieving rapid electrolysis of the bulk solution confined in a pathway of ca 250 μm. However, cells cannot make real-time-resolved spectroelectrochemistry due to the presence of a non-negligible diffusion layer.

In practice, this type of cells was built with working electrodes based on vapor-deposited metallic films [40], transparent conducting oxide films [40, 41] or presented small holes in its surface [42] such as minigrid electrodes [2, 43–47]. In addition, the PVD technique has the advantages of producing reproducible metallic surfaces with high crystallographic quality, low roughness, low electrical resistance and with a defined geometry [15]. With this technique, optically transparent electrodes are prepared with ease.

The experimental configuration is set with a cell positioned with a 45° angle versus both excitation and emission beams. This configuration gives, in the case of thin layers (≈250 μm), a maximum emission intensity combined with minimal interference regarding the excitation radiation and satisfies the 90° detection angle between the excitation and the emission light beams [12, 40]. A schematic of possible cell configuration is depicted in Fig. 1.

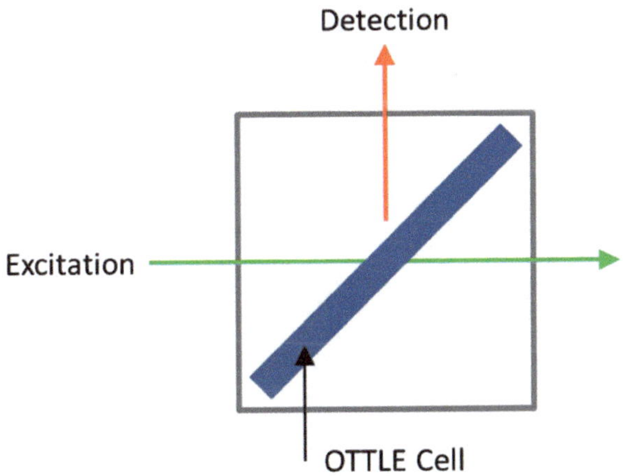

Fig. 1 Schematic representation of a spectroelectrochemical experiment using OTTLE cell with minigrid or vapor-deposited metal film working electrode

Despite results in agreement with the potential of F-SEC in terms of sensitivity, the fluorescence detection is mainly limited by the light scattering from the front face of the cell leading to a low signal-to-noise ratio and to poor reproducibility.

2.1.2 Long Optical Path Electrochemical Cells

To overcome limitations encountered with OTTLE cells, e.g., short pathway inducing low signal intensities, "long optical path electrochemical" (LOPE) cells based on gold resinate film electrode [48] or reticulated vitreous carbon electrode (RVC) [49–51] were designed (Fig. 2). Their configuration permitted detection of the emitted light at 90° and avoided scattering effect from the faces of the cell. Unfortunately, the equilibration time is longer than the one required for OTTLE cells and shows the pseudo-thin-layer nature of the long optical path electrochemical cells. However, bunch of experiments can be easily performed with this type of cells, particularly for stable electroactive species.

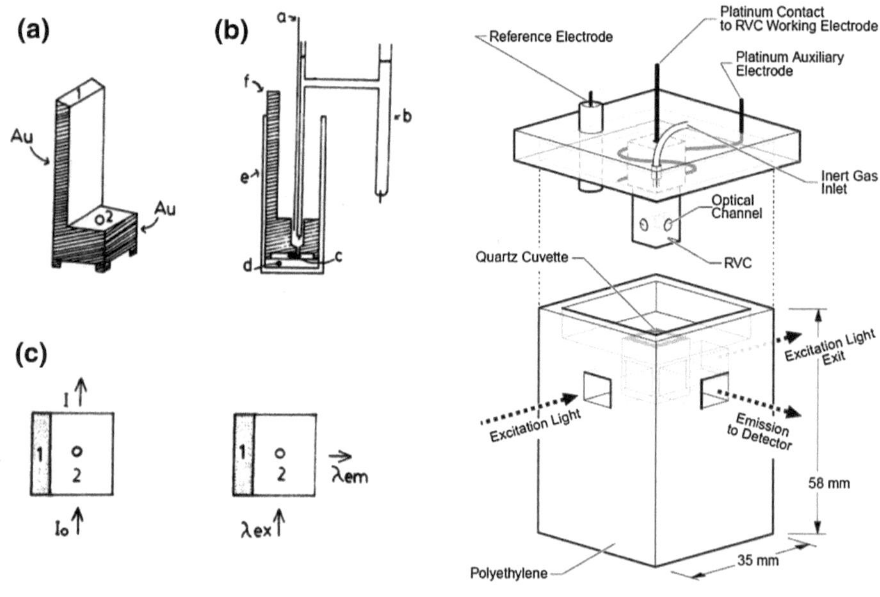

Fig. 2 *Left* **a** Gold-coated Teflon working electrode insert, areas *1* and *2* as in *C*. **b** Completed cell diagram showing (*a*) Pt auxiliary electrode, (*b*) SCE reference electrode. (*c*) Radiation path, (*d*) Teflon bottom spacer, (*e*) quartz cuvette and (*f*) gold-coated Teflon working electrode. **c** Top view of the cell showing radiation paths in absorption and fluorescence experiments, areas *1* and *2* as in *A* (from Ref. [48]). *Right* Design of a two-body spectrofluoroelectrochemical cell with right angle detection (from Ref. [51])

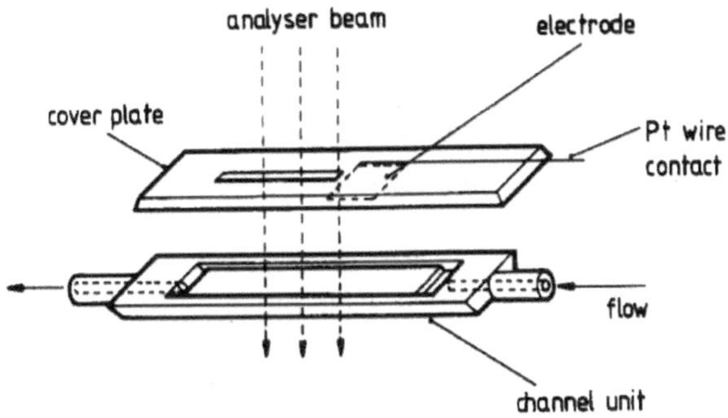

Fig. 3 The channel electrode luminescence cell from Ref. [52]

2.1.3 Hydrodynamic Voltammetry

Following on from cells mainly limited to stable species, OTTLE cells have been redesigned by Compton et al. to work as flow cells (Fig. 3), in hydrodynamic voltammetry experiments, especially for complex electrochemical processes [52–54]. Hydrodynamic voltammetry techniques using such cells facilitate the analysis of fluorescent solution-phase electrogenerated species, by modeling, from a proposed model, their spatial and temporal distributions in the flow cell, and finally correlate the fluorescence intensity to the electrode current.

2.1.4 Variable Thickness Thin-Layer Cells

In the beginning of 2000s, a new evolution of thin-layer electrochemical cells, able to operate at variable thicknesses, was designed. A first version [55], dedicated to UV–Vis and fluorescence spectroscopies, was proposed by Yu et al. and is composed of a working electrode disk inserted normally to the emission detector, in a Teflon body up to an experimental chamber. The excitation light and the emission detector were placed at 90° to each other. This type of flow cell is illustrated in Fig. 4.

A second version, proposed by Levillain et al. [14, 25] and inspired by the works of Salbeck [56] and Wertz [44], can be used in different configurations due to its

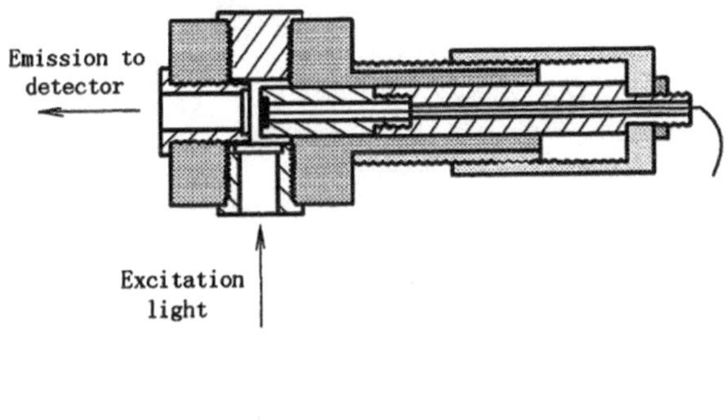

Fig. 4 Schematic diagram of the thin-layer spectroelectrochemical cell from Ref. [55]

wide versatility and dedicated to UV–Vis, IR or fluorescence spectroscopies. Regarding F-SEC, the excitation light makes a 30, 45 or 60° angle with the electrode surface and the emission light is recorded normal to the surface. The path length can range from few micrometers (thin-layer conditions) to few millimeters (diffusion layer conditions). This type of spectroelectrochemical cell is highlighted in Fig. 5. These two examples show that the transparent electrodes are replaced by commonly used non-transparent electrodes such as gold, vitreous carbon, platinum. Recently, Levillain et al. have proposed to record simultaneously both the excitation and emission lights, with a zero angle configuration versus both excitation and detection directions [15]. They can observe two concomitant processes, namely the extinction of fluorescence and the inevitable increasing of the absorbance at the wavelength of the excitation light. This setup allows performing a fluorescence and absorption spectroelectrochemistry without tuning the configuration of the bench.

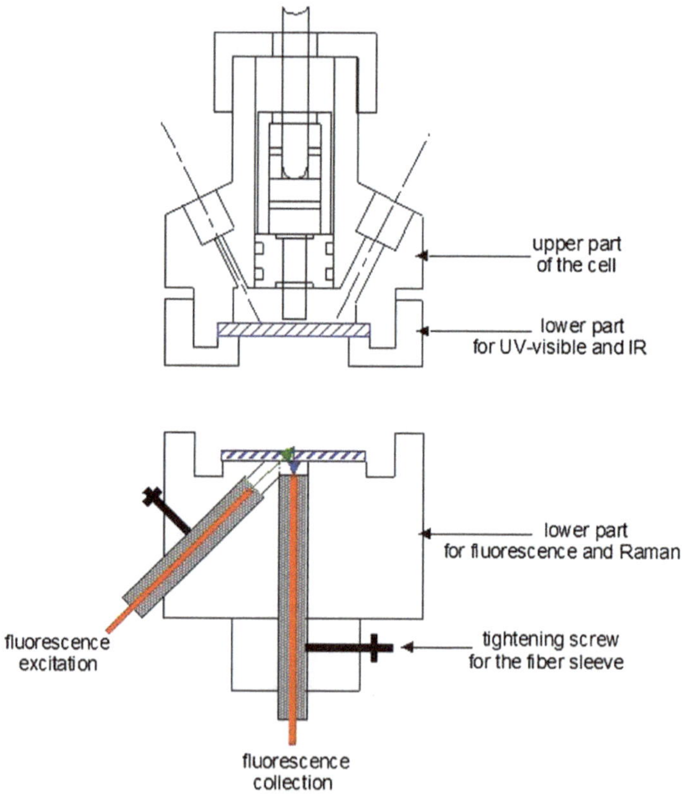

Fig. 5 Schematic view of the spectroelectrochemical cell from Ref. [25]

2.2 Data Processing

The improvement in cameras (i.e, very sensitive and stable with low noise and large dynamics), sources (i.e, stability vs time), optical fibers and 3D printing technologies enables now the development of very efficient time-resolved spectroelectrochemical measurement bench capable of:

- An accurate monitoring of the evolution of the spectroscopic signature as a function of an electrical perturbation such as a potential step or linear scan.
- A recording of the emission light with a nonzero or a zero angle versus both the excitation and detection directions.
- A probing of very low-intensity signals at high signal-to-noise ratio.

A-SEC and F-SEC:

Since the fluorescence intensity (I_{Fluo}) depends on the absorbance (A) (Table 1), F-SEC is closely linked to A-SEC. Setup capable of carrying out these two techniques in parallel emerges in literature and allows an overall view of the optical and fluorescent properties under potential. This approach offers more flexibility and gives a comprehensive overview of emission and absorption processes (Fig. 6).

Voltafluorogram and voltabsorptogram:

Even though it is well established since a while that derivative cyclic voltabsorptogram (DCVA = dA/dt versus potential) is an efficient tool for characterizing optical properties under potential, the pioneering work dedicated to derivative cyclic voltafluorograms (DCVF) was only published in 2004 [25]. The high performance of cameras allows now calculating the derivative of the fluorescence (or absorbance) signal from a three-dimensional representation [potential (or time), wavelength, signal] and then extracting the DCVF (or DCVA) at any potential (or time) (Fig. 6). In this regard, it is worth reminding here that a good signal-to-noise ratio is a prerequisite to extract, with or without smoothing via 1D or 2D digital filters, the derivative of the fluorescence versus potential (or time).

Simultaneous recording of excitation and emission lights:

The change of the redox state of fluorescent species occurring during a spectroelectrochemical experiment in thin-layer conditions typically leads to two concomitant processes, namely the fluorescence quenching and the unavoidable

Table 1 Usual relationships between fluorescence and absorbance

	Solution	Modified electrode
Absorbance	$A = \varepsilon \ell C$ with ε, $mol^{-1}\ L\ cm^{-1}$ ℓ, cm C, $mol\ L^{-1}$	$A = 1000\ \varepsilon\ \Gamma$ with ε, $mol^{-1}\ L\ cm^{-1}$ Γ, $mol\ cm^{-2}$ (Γ is the surface coverage)
Fluorescence	$I_{Fluo} = k\phi I_{Exc}[1 - e^{10}]$ with k, instrument constant ϕ, fluorescence quantum yield I_{Exc}, intensity of excitation light	

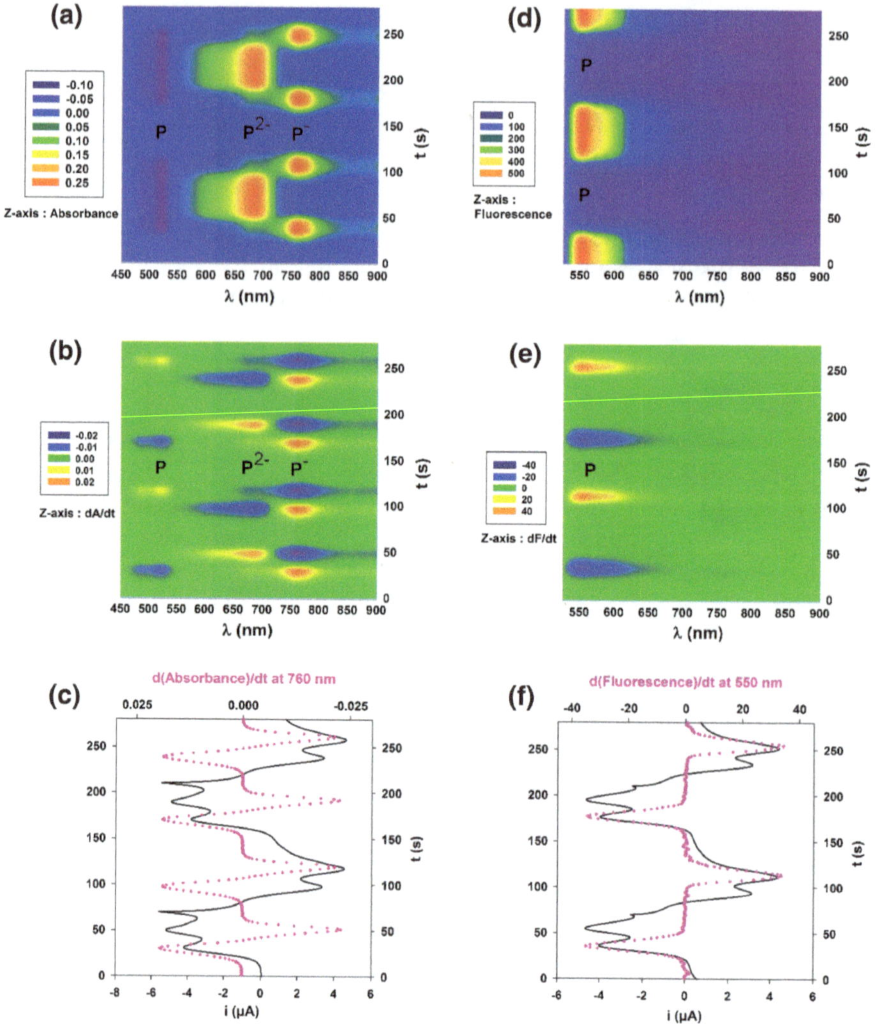

Fig. 6 Absorbance and F-SEC of 5.10^{-4} M perylenediimide (*P*) on Pt electrode in 0.1 M $TBAPF_6$ in CH_2Cl_2 in TLCV (∼50 μm). **a** 3D representation: X-axis = wavelength, Y-axis = time and Z-axis = absorbance. **b** 3D DCVA representation: Z-axis = d*A*/d*t*. **c** Unfolded CV versus time at 5 mV s^{-1} and absorbance at 550 nm extracted from (**b**). **d** 3D representation: Z-axis = fluorescence intensity. Excitation at 495 nm and detection by reflection on Au electrode between 380 and 980 nm. **e** 3D DCVF representation: Z-axis = d*F*/d*t*. (***F***) CV at 10 mV s^{-1} and fluorescence intensity at 550 nm extracted from (**b**). Note that these results have not been published because these experiments are the same as those described in ref. [25] but performed with the setup of Refs. [15, 57]

increasing of the absorbance at the wavelength of the excitation light ($\lambda_{excitation}$). The monitoring of both the excitation and the emission processes affords the opportunity of improving the measurement chain. The best setup is a simultaneous recording of the excitation and emission lights with a zero angle versus excitation and detection directions (Fig. 6). Moreover, it is noteworthy that, with this setup, no configuration change is necessary to perform A-SEC experiments.

Monitoring very low signals:

Since the frantic development of nanoscale materials requires very sensitive characterizations, monitoring very low signals by spectroelectrochemistry rapidly becomes a challenge.

To illustrate, consider the "worst-case" nanomaterial of all: the self-assembled monolayer (SAM) with a surface coverage typically close to 10^{-10} mol cm^{-2}.

A simple numerical calculation from Beer's law shows that the absorbance on a monolayer is, by nature, very low (close to 0.001 with $\Gamma = 10^{-10}$ mol cm^{-2} and $\varepsilon = 10{,}000$ M^{-1} cm^{-1}). However, a recent work has demonstrated that it is possible to perform a visible–NIR time-resolved spectroelectrochemistry on SAMs with a high molar attenuation coefficient chromophore, through the study of the oxidation of a 5,5′-disubstituted-2,2′-bithiophene immobilized on Au substrate [15].

For a fluorescence emission, we are facing the same problem. A numerical calculation confirms that, at the best, $I_{\mathrm{Fluo}} = 10^{-3} I_{\mathrm{Exc}}$ with $\Gamma = 10^{-10}$ mol cm^{-2}, $\varepsilon = 10{,}000$ M^{-1} cm^{-1}, $\phi = 100\%$ and $k = 1$ (i.e, a fluorophore with a high fluorescence quantum yields is required). Because the molecular luminescence on SAM is, in agreement with the classical energy transfer theory, quenched by a metallic substrate, a F-SEC may not be performed on SAM. However, it is possible to reach a SAM situation under thin-layer conditions. Figure 7 provides evidence of an emission spectroelectrochemistry of a perylenediimide (i.e, $\varepsilon = 50{,}000$ M^{-1} cm^{-1} and $\phi = 100\%$) in solution under drastic conditions (i.e, a 50-μm thin layer at 10^{-4} M, corresponding to a 5×10^{-10} mol cm^{-2} surface coverage).

All these results illustrate the benefits provided by the latest technological breakthroughs and offer promising prospects to probe nanomaterials by absorption and emission spectroelectrochemistry.

3 Electrochemistry and Fluorescence Microscopy Coupling

The coupling of electrochemical techniques with fluorescence microscopy is very recent compared to the conventional F-SEC previously described in the precedent part.

The development of the electrochemistry and fluorescence microscopy coupling (TIRF and confocal configuration) is mainly due to the very high sensitivity of optical detection involved in the fluorescence microscopy benches that can be pushed to the single-molecule detection. In most cases, the electrochemical technique is only devoted to tune the redox state of the fluorescent molecules.

One of the most important challenges relies in detecting single-molecular fluorescence under electrochemical conditions [58].

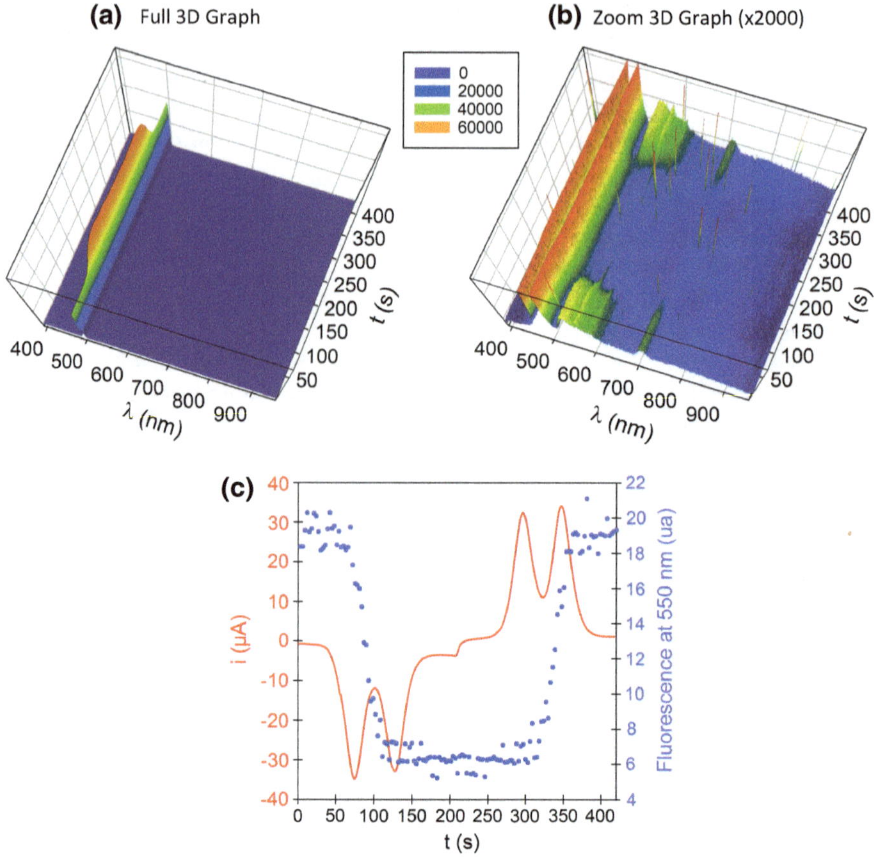

Fig. 7 F-SEC of 10^{-4} M perylenediimide on Pt electrode in 0.1 M $TBAPF_6$ in CH_2Cl_2 characterized in a thin layer close to 50 µm, corresponding to a SAM with a surface coverage close to 5×10^{-10} mol cm^{-2}. Excitation at 495 nm and detection by reflection on Au electrode between 380 and 980 nm. **a** 3D representation: X-axis ≐ wavelength, Y-axis = time and Z-axis = fluorescence intensity. **b** Zoom of 3D representation (raw data without smoothing via a 2D filter). As expected, the intensity ratio i_{Fluo}/i_{Exc} (15/60,000) is very small 2.5×10^{-4}. **c** CV at 5 mV s^{-1} and fluorescence intensity at 550 nm extracted from (**b**) versus frames. As expected, the first reduction of the perylene moiety turns the fluorescence off. Note that these results have not been published because these experiments are the same as those described in Ref. [25] but performed with the setup of Ref. [15]

3.1 *Fluorescence Detection*

The laser-induced fluorescence spectroscopy technique is typically involved for fluorescence detection. The latter is mainly composed of three different experimental setups when coupled to a microscope: (1) classical fluorescence microscopy (epifluorescence), (2) confocal fluorescence microscopy and (3) total internal

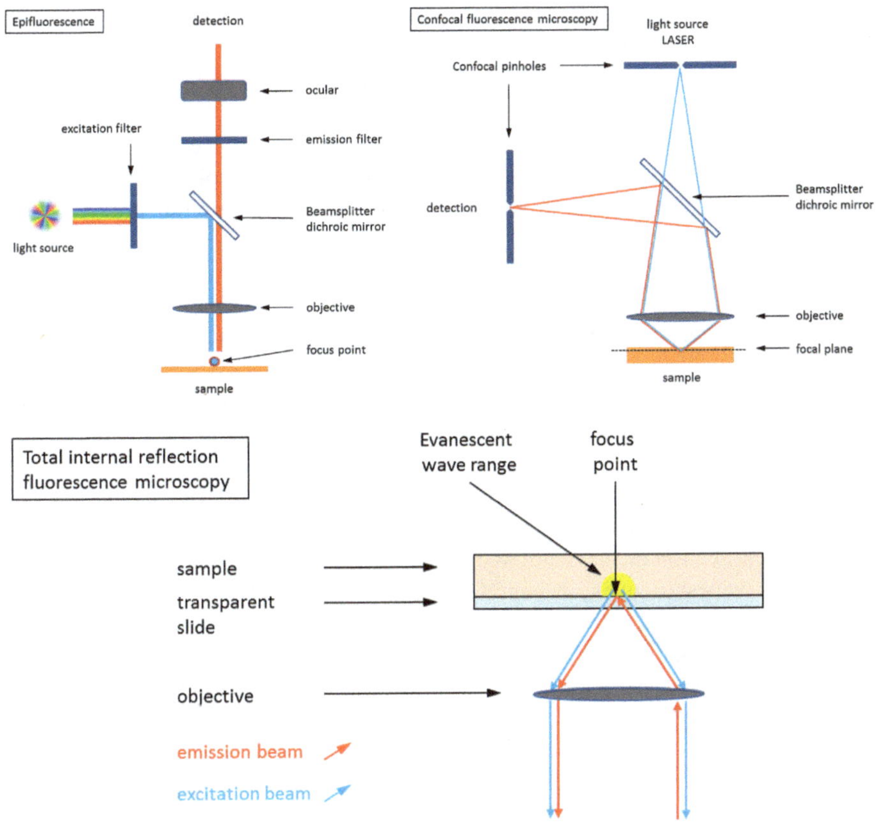

Fig. 8 Schematic configurations of classical fluorescence microscopy (epifluorescence—*top left*), confocal fluorescence microscopy (*top right*) and total internal reflection fluorescence microscopy (TIRFM—*bottom*)

reflection fluorescence microscopy (TIRFM). Schematics of these three configurations are depicted in Fig. 8.

3.1.1 Epifluorescence

In this configuration, excitation of the fluorophore and detection of the emitted fluorescence are performed through the same light path. The monochromatic excitation light goes through the microscope lens to be focused on the sample. Fluorescence is collected in the same direction as the excitation (i.e, through the objective), normal to the sample, and separated from it in the detection line by dichroic mirrors and filters.

3.1.2 Confocal Fluorescence Microscopy

Confocal microscopy increases the optical resolution (axial and lateral) and contrast of classical microscopes by adding a spatial pinhole placed at the confocal plane of the lens in order to eliminate light coming out of the focus plane. The fluorescence is thus spatially probed and is only coming from the focal plane. It enables the reconstruction of three-dimensional images from those obtained at different focal planes.

3.1.3 Total Internal Reflection Fluorescence Microscopy (TIRFM)

A total internal reflection fluorescence microscope (TIRFM) is a type of microscope in which only a restricted region of the sample can be observed near the glass–solution interface.

To reach these special conditions, the excitation light hits the sample with an incident angle greater than the critical angle determined by refraction laws. In this case, a total reflection of the incident light occurs at the glass–solution interface, and an induced evanescent wave is generated at the interface. The evanescent intensity decays exponentially from the interface and thus penetrates into the sample to a depth of approximately a few hundreds of nanometers. This very thin probed thickness leads to a very good axial resolution in the perpendicular direction to the interface.

3.2 *Coupling Setup*

The most promising spectroelectrochemistry cells are used in combination with total internal reflection fluorescence microscopy; consequently, the non-exhaustive examples of setups listed below are mostly based on this configuration.

The first example of fluorescence microscopy coupled to a three-electrode electrochemical, reported by Audebert and Miomandre [24, 47, 59], was dedicated to the investigation of a controlled potential redox switch of organic fluorophores (Fig. 9). The three-electrode cell was composed of a working electrode made with platinum thin layer coated on a microscope glass slide. In order to produce valuable information about quenching mechanisms during a redox state change event, the lifetime variations along with the fluorescence intensity have been measured via the implementation of a pulsed laser to the microscope (Fig. 10). Confocal fluorescence microscopy and total internal reflection fluorescence microscopy were adapted to the electrochemical cell to probe the fluorescence emission.

A similar bench (Fig. 11) was proposed by Ackerman et al. in a confocal fluorescence microscopy configuration [60, 61]. The spatial/temporal resolution of this technique combined with a nanomolar concentration of active molecule permit to detect single events as "bursts" in the modulated signal.

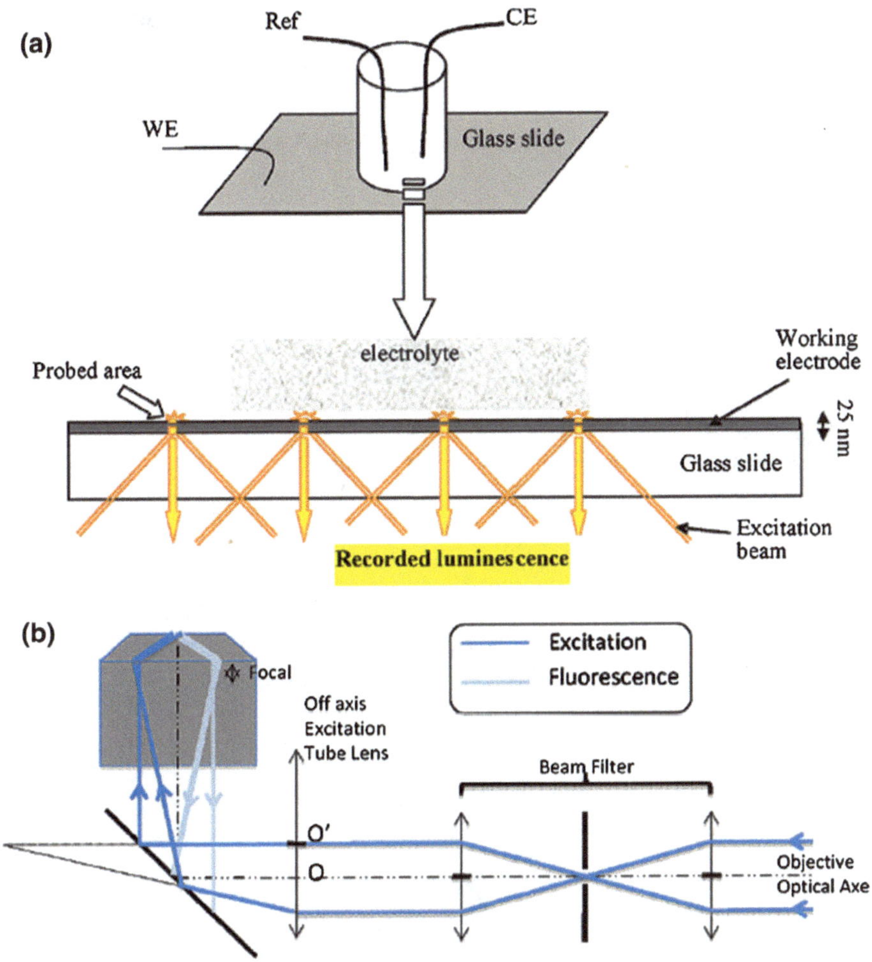

Fig. 9 **a** Scheme of the electrochemical cell (*top*) and platinum-coated microscope slide used for TIRF measurements (*bottom*) and **b** optical path for the excitation and emitted beams in the epifluorescence setup. From Ref. [59]

This study shows that it is a priori possible with this method to identify the fluorescence of single molecules. Thus, molecular electrochemical characteristics such as standard potentials and electron transfer kinetics may be determined with a similar method. This was accomplished with specific molecules designed to allow energy transfer to be modulated by the electrode potential as for example when the redox active fluorophore was adsorbed on nanoparticles [61, 62], when the redox center was connected to a fluorescent molecule as a protein [63], or when the active molecule is a part of a polymer with high molecular weight [58].

Voltafluorograms show that the fluorescence intensity varies with the potential with the same trend as the electrochemical current [62–64]. Compared to other

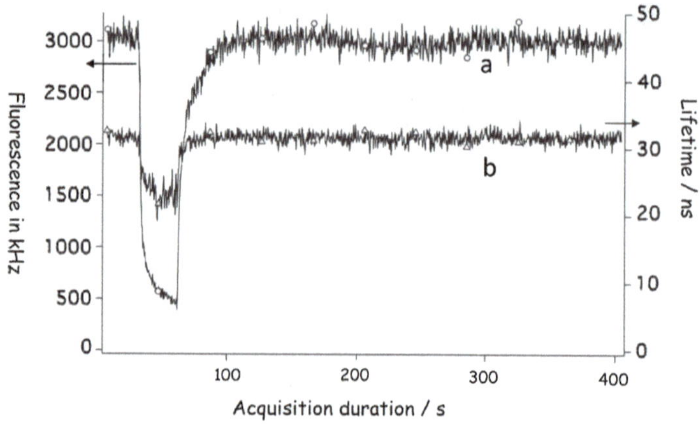

Fig. 10 Evolution of the fluorescence lifetime of tetrazine dye (*right scale*) along with the fluorescence intensity (*left scale*) for a double-step potential between 0 and −0.85 V. From Ref. [59]

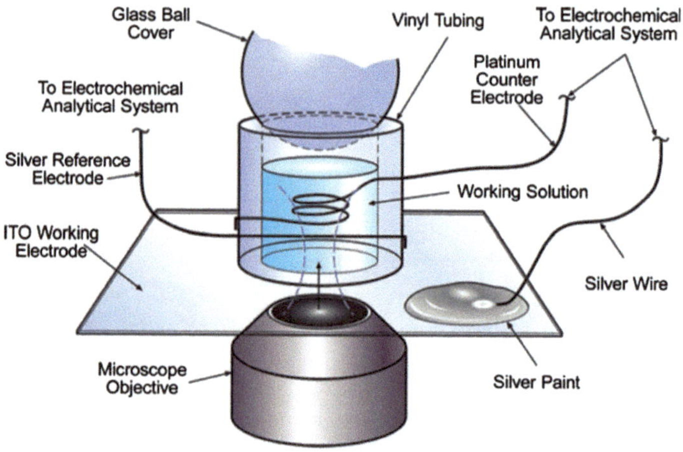

Fig. 11 Schematic of an electrochemical cell coupled with scanning confocal fluorescence microscopy from Ref. [60]

techniques, the spatial resolution of fluorescence microscopy (image of few microns) ranks this technique among the most promising for the future.

Recently, thermodynamics and kinetics of electron transfer of single immobilized particles have been determined based on single-molecule spectroelectrochemistry (SMS-EC) [58], and this new technique reported by Palacios and co-workers is based on total internal reflection fluorescence microscopy.

Last but not least, another promising method consists of using microelectrodes combined with total internal reflection fluorescence microscopy. Indeed, Amatore et al. [64] have reported the real-time observation of fluorescence and amperometric

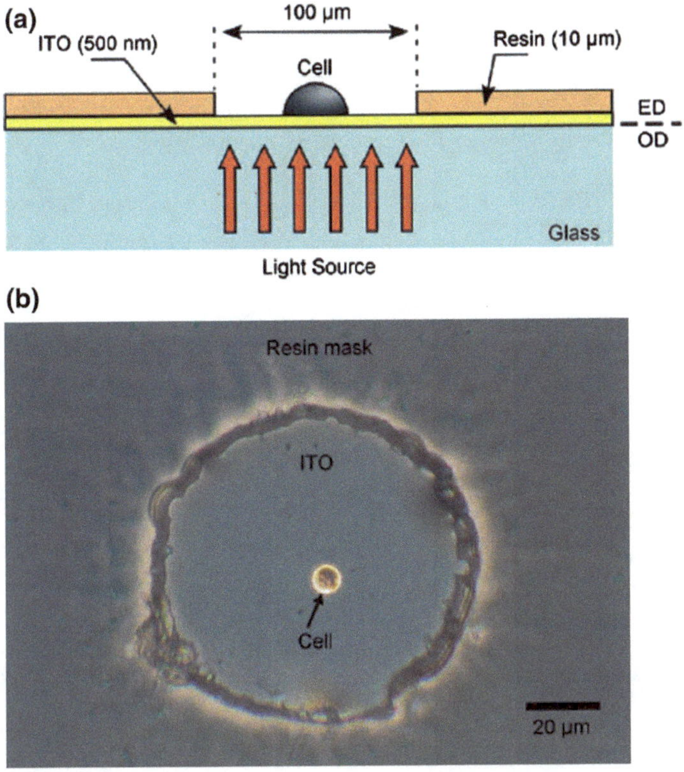

Fig. 12 **a** Schematic vertical cross section of an ITO microelectrode. The electrochemical detection (*ED*) is performed on the ITO surface (*upper part* of the device), while the optical detection (*OD*) is obtained through the glass and ITO (*lower part*). **b** Photograph (inverted microscopy) of a chromaffin cell adhering to an ITO microelectrode (100 mm diameter); the insulating resin appears as the dark gray external area. From Ref. [64]

measurements of exocytosis events at single biologic cell on transparent indium tin oxide (ITO) microelectrodes (Fig. 12). Similar investigations were done using microfluidic configuration [65].

4 Conclusion

Over the last decade, the benefits provided by the latest technological breakthroughs have improved the instrumental setup dealing with electrofluorochromism, i.e, electrochemical monitoring of the fluorescence properties.

Nowadays, the technology facilities are no longer a hindrance to the development of emission spectroelectrochemistry and offer promising prospects in forthcoming research on fluorescent molecules, materials and devices. In the reasonably near future, significant improvements in the miniaturization of the devices should emerge since it is now possible to follow very low absorption and emission events, up to the detection of a single molecule. For example, in analytical and bioanalytical chemistry, many analytes are redox active and could be detected and mapped with a high sensitivity through fluorescence via nanosensors.

Acknowledgement The authors thank Flavy Alévêque and Dr. Clément Cabanetos for their critical reading of the manuscript.

References

1. Blubaugh, E.A., Yacynych, A.M., Heineman, W.R.: Thin-layer spectroelectrochemistry for monitoring kinetics of electrogenerated species. Anal. Chem. **51**(4), 561–565 (1979). doi:10.1021/ac50040a026
2. Kaim, W., Fiedler, J.: Spectroelectrochemistry: the best of two worlds. Chem. Soc. Rev. **38** (12), 3373–3382 (2009). doi:10.1039/b504286k
3. Dunsch, L.: Recent advances in in situ multi-spectroelectrochemistry. J. Solid State Electrochem. **15**(7–8), 1631–1646 (2011). doi:10.1007/s10008-011-1453-1
4. Melin, F., Hellwig, P.: Recent advances in the electrochemistry and spectroelectrochemistry of membrane proteins. Bio. Chem. **394**(5), 593–609 (2013). doi:10.1515/hsz-2012-0344
5. Pruiksma, R., McCreery, R.L.: Observation of electrochemical concentration profiles by absorption spectroelectrochemistry. Anal. Chem. **51**(13), 2253–2257 (1979). doi:10.1021/ac50049a045
6. Skully, J.P., McCreery, R.L.: Glancing incidence external reflection spectroelectrochemistry with a continuum source. Anal. Chem. **52**(12), 1885–1889 (1980). doi:10.1021/ac50062a025
7. Robinson, R.S., McCreery, R.L.: Absorption spectroelectrochemistry with microelectrodes. Anal. Chem. **53**(7), 997–1001 (1981). doi:10.1021/ac00230a017
8. Brewster, J.D., Anderson, J.L.: Fiber optic thin-layer spectroelectrochemistry with long optical-path. Anal. Chem. **54**(14), 2560–2566 (1982). doi:10.1021/ac00251a035
9. Robinson, R.S., McCurdy, C.W., McCreery, R.L.: Microsecond spectroelectrochemistry by external reflection from cylindrical microelectrodes. Anal. Chem. **54**(13), 2356–2361 (1982). doi:10.1021/ac00250a049
10. Bard, A.J., Faulkner, L.R., Brisset, J.L.: Electrochimie: principes, méthodes et applications. Masson (1983)
11. Bard, A.J., Faulkner, L.R.: Electrochemical methods: fundamentals and applications. Wiley (2000)
12. Zoski, C.G.: Handbook of electrochemistry. Elsevier (2007)
13. Kaim, W., Klein, A.: Spectroelectrochemistry. Roy. Soc. Chem. (2008)
14. Gaillard, F., Levillain, E.: Visible time-resolved spectroelectrochemistry—application to study of the reduction of sulfur (S-8) in dimethylformamide. J. Electroanal. Chem. **398**(1–2), 77–87 (1995). doi:10.1016/0022-0728(95)04144-1
15. Aleveque, O., Levillain, E., Sanguinet, L.: Spectroelectrochemistry on electroactive self-assembled monolayers: cyclic voltammetry coupled to spectrophotometry. Electrochem. Commun. **51**, 108–112 (2015). doi:10.1016/j.elecom.2014.12.014
16. Neudeck, A., Dunsch, L.: Cyclic voltammetry at microstructured electrodes. J. Electroanal. Chem. **370**(1–2), 17–32 (1994). doi:10.1016/0022-0728(93)03206-5
17. Neudeck, A., Dunsch, L.: Microstructured electrode materials in UV-visible spectroelectrochemistry. J. Electroanal. Chem. **386**(1–2), 135–148 (1995). doi:10.1016/0022-0728(95)03824-z

18. Niu, J.J., Dong, S.J.: Transmission spectroelectrochemistry. Rev. Anal. Chem. **15**(1–2), 1–171 (1996)
19. Frank, O., Dresselhaus, M.S., Kalbac, M.: Raman spectroscopy and in situ Raman spectroelectrochemistry of isotopically engineered graphene systems. Acc. Chem. Res. **48** (1), 111–118 (2015). doi:10.1021/ar500384p
20. Gaillard, F., Levillain, E., Dhamelincourt, M.C., Dhamelincourt, P., Lelieur, J.P.: Polysulphides in dimethylformamide: a micro-Raman spectroelectrochemical study. J. Raman Spectrosc. **28**(7), 511–517 (1997). doi:10.1002/(sici)1097-4555(199707)28:7<511:aid-jrs119>3.3.co;2-w
21. Bellec, V., De Backer, M.G., Levillain, E., Sauvage, F.X., Sombret, B., Wartelle, C.: In situ time-resolved FTIR spectroelectrochemistry: study of the reduction of TCNQ. Electrochem. Commun. **3**(9), 483–488 (2001). doi:10.1016/s1388-2481(01)00158-8
22. Wartelle, C., Viruela, P.M., Viruela, R., Orti, E., Sauvage, F.X., Levillain, E., Le Derf, F., Salle, M.: A study by spectroelectrochemical FTIR and density functional theory calculations of the reversible complexing ability of an electroactive tetrathiafulvalene crown. J. Phys. Chem. A **109**(6), 1188–1195 (2005). doi:10.1021/jp045397y
23. Ashley, K., Pons, S.: Infrared spectroelectrochemistry. Chem. Rev. **88**(4), 673–695 (1988). doi:10.1021/cr00086a006
24. Miomandre, F., Meallet-Renault, R., Vachon, J.-J., Pansu, R.B., Audebert, P.: Fluorescence microscopy coupled to electrochemistry: a powerful tool for the controlled electrochemical switch of fluorescent molecules. Chem. Commun. **16**, 1913–1915 (2008). doi:10.1039/b718899d
25. Dias, M., Hudhomme, P., Levillain, E., Perrin, L., Sahin, Y., Sauvage, F.X., Wartelle, C.: Electrochemistry coupled to fluorescence spectroscopy: a new versatile approach. Electrochem. Commun. **6**(3), 325–330 (2004). doi:10.1016/j.elecom.2004.01.010
26. Petr, A., Dunsch, L., Neudeck, A.: In situ UV-vis ESR spectroelectrochemistry. J. Electroanal. Chem. **412**(1 2), 153–158 (1996). doi:10.1016/0022-0728(96)04582-2
27. Rapta, P., Neudeck, A., Petr, A., Dunsch, L.: In situ EPR/UV-VIS spectroelectrochemistry of polypyrrole redox cycling. J. Chem. Soc. Faraday Trans. **94**(24), 3625–3630 (1998). doi:10.1039/a806423g
28. Sharpe, L.R., Heineman, W.R., Elder, R.C.: EXAFS spectroelectrochemistry. Chem. Rev. **90** (5), 705–722 (1990). doi:10.1021/cr00103a002
29. Dong, S.J., Niu, J.J., Cotton, T.M.: Ultraviolet-visible spectroelectrochemistry of redox proteins. Biochem. Spectrosc. **246**, 701–732 (1995)
30. Taboy, C.H., Bonaventura, C., Crumbliss, A.L.: Anaerobic oxidations of myoglobin and hemoglobin by spectroelectrochemistry. Redox Cell Biol. Genet. Pt B **353**, 187–209 (2002)
31. Murgida, D., Hildebrandt, P.: Surface-enhanced vibrational spectroelectrochemistry: Electric-field effects on redox and redox-coupled processes of heme proteins. Sur. Enhanced Raman Scattering Phy. Appl. **103**, 313–334 (2006)
32. Best, S.P.: Spectroelectrochemistry of hydrogenase enzymes and related compounds. Coord. Chem. Rev. **249**(15–16), 1536–1554 (2005). doi:10.1016/j.ccr.2005.01.012
33. Audebert, P., Miomandre, F.: Electrofluorochromism: from molecular systems to set-up and display. Chem. Sci. **4**(2), 575–584 (2013). doi:10.1039/c2sc21503a
34. Heineman, W.R.: Spectroelectrochemistry: the combination of optical and electrochemical techniques. J. Chem. Educ. **60**(4), 305 (1983). doi:10.1021/ed060p305
35. Kim, Y., Kim, E., Clavier, G., Audebert, P.: New tetrazine-based fluoroelectrochromic window; modulation of the fluorescence through applied potential. Chem. Commun. **34**, 3612–3614 (2006). doi:10.1039/b608312a
36. Schroll, C.A., Chatterjee, S., Heineman, W.R., Bryan, S.A.: Thin-layer spectroelectrochemistry on an Aqueous Microdrop. Electroanalysis **24**(5), 1065–1070 (2012). doi:10.1002/elan.201100711
37. Voicescu, M., Rother, D., Bardischewsky, F., Friedrich, C.G., Hellwig, P.: A combined fluorescence spectroscopic and electrochemical approach for the study of thioredoxins. Biochemistry **50**(1), 17–24 (2011). doi:10.1021/bi1013112

38. Brisendine, J.M., Mutter, A.C., Cerda, J.F., Koder, R.L.: A three-dimensional printed cell for rapid, low-volume spectroelectrochemistry. Anal. Biochem. **439**(1), 1–3 (2013). doi:10.1016/j.ab.2013.03.036
39. Kuwana, T., Darlington, R.K., Leedy, D.W.: Electrochemical studies using conducting glass indicator electrodes. Anal. Chem. **36**(10), 2023–2025 (1964). doi:10.1021/ac60216a003
40. Yildiz, A., Kissinger, P.T., Reilley, C.N.: Evaluation of an improved thin-layer electrode. Anal. Chem. **40**(7), 1018–1024 (1968). doi:10.1021/ac60263a012
41. Wilson, R.A., Pinyayev, T.S., Membreno, N., Heineman, W.R.: Rapid prototyped optically transparent thin-layer electrode holder for spectroelectrochemistry in bench-top spectrophotometers. Electroanalysis **22**(19), 2162–2166 (2010). doi:10.1002/elan.201000267
42. Compton, R.G., Fisher, A.C., Wellington, R.G.: A thin-layer electrode cell for fluorescence measurements on electrogenerated intermediates. Electroanalysis **3**(1), 27–29 (1991). doi:10.1002/elan.1140030105
43. Simone, M.J., Heineman, W.R., Kreishman, G.P.: Preliminary spectrofluoroelectrochemical studies indicate a possible conformational change in horse heart cytochrome c upon reduction. J. Colloid Interface Sci. **86**(2), 295–298 (1982). doi:10.1016/0021-9797(82)90075-3
44. McLeod, C.W., West, T.S.: Spectroelectrochemistry of morphine and related alkaloids and their investigation by fluorescence in a gold micromesh cell. Analyst **107**(1270), 1–11 (1982). doi:10.1039/AN9820700001
45. Jones, E.T.T., Faulkner, L.R.: Luminescence spectroelectrochemistry in thin layer cells. J. Electroanal. Chem. Interfacial Electrochem. **179**(1), 53–64 (1984). doi:10.1016/S0022-0728(84)80274-0
46. Heineman, W.R., Norris, B.J., Goelz, J.F.: Measurement of enzyme E.deg'. values by optically transparent thin layer electrochemical cells. Anal. Chem. **47**(1), 79–84 (1975). doi:10.1021/ac60351a001
47. Miomandre, F., Allain, C., Clavier, G., Audibert, J.-F., Pansu, R.B., Audebert, P., Hartl, F.: Coupling thin layer electrochemistry with epifluorescence microscopy: An expedient way of investigating electrofluorochromism of organic dyes. Electrochem. Commun. **13**(6), 574–577 (2011). doi:10.1016/j.elecom.2011.03.013
48. Simone, M.J., Heineman, W.R., Kreishman, G.P.: Long optical path electrochemical cell for absorption or fluorescence spectrometers. Anal. Chem. **54**(13), 2382–2384 (1982). doi:10.1021/ac00250a058
49. Lee, Y.F., Kirchhoff, J.R.: Design and characterization of a spectroelectrochemistry cell for absorption and luminescence measurements. Anal. Chem. **65**(23), 3430–3434 (1993). doi:10.1021/ac00071a016
50. Lee, Y.F., Kirchhoff, J.R.: Absorption and luminescence spectroelectrochemical characterization of a highly luminescent rhenium(II) complex. J. Am. Chem. Soc. **116**(8), 3599–3600 (1994). doi:10.1021/ja00087a056
51. Kirchhoff, J.R.: Luminescence spectroelectrochemistry. Curr. Sep. **16**(1), 11–14 (1997)
52. Compton, R.G., Fisher, A.C., Wellington, R.G., Winkler, J.: Spectrofluorometric hydrodynamic voltammetry—theory and practice. J. Phys. Chem. **96**(20), 8153–8157 (1992). doi:10.1021/j100199a061
53. Compton, R.G., Wellington, R.G.: Spectrofluorometric hydrodynamic voltammetry—the investigation of electrode-reaction mechanisms. J. Phys. Chem. **98**(1), 270–273 (1994). doi:10.1021/j100052a045
54. Compton, R.G., Winkler, J., Riley, D.J., Bearpark, S.D.: Spectrofluorometric hydrodynamic voltammetry—investigation of reactions at solid/liquid interfaces. J. Phys. Chem. **98**(27), 6818–6825 (1994). doi:10.1021/j100078a026
55. Yu, J.-S., Yang, C., Fang, H.-Q.: Variable thickness thin-layer cell for electrochemistry and in situ UV–VIS absorption, luminescence and surface-enhanced Raman spectroelectrochemistry. Anal. Chim. Acta **420**(1), 45–55 (2000). doi:10.1016/S0003-2670(00)01005-9
56. Salbeck, J.: An electrochemical cell for simultaneous electrochemical and spectroelectrochemical measurements under semi-infinite diffusion conditions and thin-layer conditions. J. Electroanal. Chem. **340**(1–2), 169–195 (1992). doi:10.1016/0022-0728(92)80297-H

57. Bkhach, S., Le Duc, Y., Aleveque, O., Gautier, C., Hudhomme, P., Levillain, E.: Highly stable perylenediimide based self-assembled monolayers studied by spectroelectrochemistry. Chem. ElectroChem. **3**(6), 887–891 (2016). doi:10.1002/celc.201600034
58. Palacios, R.E., Fan, F.R.F., Bard, A.J., Barbara, P.F.: Single-molecule spectroelectrochemistry (SMS-EC). J. Am. Chem. Soc. **128**(28), 9028–9029 (2006). doi:10.1021/ja062848e
59. Miomandre, F., Lepicier, E., Munteanu, S., Galangau, O., Audibert, J.F., Meallet-Renault, R., Audebert, P., Pansu, R.B.: Electrochemical monitoring of the fluorescence emission of tetrazine and bodipy dyes using total internal reflection fluorescence microscopy coupled to electrochemistry. Acs Appl. Mater. Interfaces **3**(3), 690–696 (2011). doi:10.1021/am100980u
60. Lei, C., Hu, D., Ackerman, E.J.: Single-molecule fluorescence spectroelectrochemistry of cresyl violet. Chem. Commun. **43**, 5490–5492 (2008). doi:10.1039/b812161c
61. Lei, C., Hu, D., Ackerman, E.: Clay nanoparticle-supported single-molecule fluorescence spectroelectrochemistry. Nano Lett. **9**(2), 655–658 (2009). doi:10.1021/nl802998e
62. Galland, C., Ghosh, Y., Steinbruck, A., Sykora, M., Hollingsworth, J.A., Klimov, V.I., Htoon, H.: Two types of luminescence blinking revealed by spectroelectrochemistry of single quantum dots. Nature **479**(7372):203–207 (2011). doi:http://www.nature.com/nature/journal/v479/n7372/abs/nature10569.html#supplementary-information
63. Salverda, J.M., Patil, A.V., Mizzon, G., Kuznetsova, S., Zauner, G., Akkilic, N., Canters, G. W., Davis, J.J., Heering, H.A., Aartsma, T.J.: Fluorescent cyclic voltammetry of immobilized azurin: direct observation of thermodynamic and kinetic heterogeneity. Angew. Chem. Int. Ed. **49**(33), 5776–5779 (2010). doi:10.1002/anie.201001298
64. Amatore, C., Arbault, S., Chen, Y., Crozatier, C., Lemaitre, F., Verchier, Y.: Coupling of electrochemistry and fluorescence microscopy at indium tin oxide microelectrodes for the analysis of single exocytotic events. Angewandte Chem. Int. Ed. **45**(24), 4000–4003 (2006). doi:10.1002/anie.200600510
65. Shi, B.-X., Wang, Y., Lam, T.-L., Huang, W.-H., Zhang, K., Leung, Y.-C., Chan, H.L.W.: Release monitoring of single cells on a microfluidic device coupled with fluorescence microscopy and electrochemistry. Biomicrofluidics **4**(4), 043009 (2010). doi:10.1063/1.3491470

In Situ Spectroelectrochemical Fluorescence Microscopy for Visualizing Interfacial Structure and Dynamics in Self-assembled Monolayers

Jannu Casanova-Moreno, Zhinan Landis Yu, Jonathan Massey-Allard, Brian Ditchburn, Jeff F. Young and Dan Bizzotto

Abstract In situ analysis of electrochemical interfaces modified with molecular adsorbates using fluorescence microscopy is outlined. The fluorescence intensity from the fluorophore-modified adsorbate is strongly quenched when the separation of the fluorophore from the metal electrode surface is decreased below 200 nm. The theory describing this important characteristic is outlined with emphasis on the lifetime and far-field intensity of the fluorophore as a function of the separation from the metal. A number of examples are given in which fluorescence microscopy is used to study surfaces modified with the self-assembled monolayers (SAMs) composed of either alkylthiols, peptides, or DNA. The ability to interrogate both the lateral and axial distributions of the adsorbed monolayers within the micron scale optical resolutions is highlighted. The influence of the electrode potential (or charge) on the fluorescence images is shown for the reductive or oxidative removal of the adsorbate. The preparation of modified electrode surfaces is also reviewed, illustrating the influence of surface crystallography on the resulting surface modification or thiol exchange processes. Preliminary results of a DNA SAM studied using 2-photon fluorescence lifetime imaging microscopy are presented, demonstrating the measurement of lifetime distributions and its correspondence with the theory. In situ spectroelectrochemical fluorescence microscopy is thus shown to be useful in studying the electrochemical interface in terms of its homogeneity of modification, the structure in the axial direction away from the electrode surface and the influence of charge (or potential) on the dynamics of the interface.

J. Casanova-Moreno · Z.L. Yu · J. Massey-Allard · J.F. Young · D. Bizzotto (✉)
Advanced Materials and Process Engineering Laboratory,
The University of British Columbia, Vancouver, Canada
e-mail: bizzotto@chem.ubc.ca

J. Casanova-Moreno · Z.L. Yu · B. Ditchburn · D. Bizzotto
Department of Chemistry, The University of British Columbia, Vancouver, Canada

J. Massey-Allard · J.F. Young
Department of Physics and Astronomy,
The University of British Columbia, Vancouver, Canada

F. Miomandre and P. Audebert (eds.), *Luminescence in Electrochemistry*,
DOI 10.1007/978-3-319-49137-0_2

1 Introduction

The analysis of a modified electrode surface in electrolyte and the quantification of the influence of the interfacial electric field on the electrode properties have been addressed by many combined optical and electrochemical methods. The detailed analysis of interfacial reactivity, mechanisms, and kinetics can be realized with many of the hyphenated methods that have been developed over the past decades starting with electroreflectance methods [1–4]. Measuring in situ UV–Vis reflectance on solid substrates and on Hg has been demonstrated and coupled with changes in a redox active chromophoric adsorbate-like methylene blue [4–8]. The influence of the electrode potential on the interfacial properties of adsorbed molecules can be studied through the use of chemically specific analyses such as IR absorption which has enabled significant advances in understanding small molecule redox pathways [9]. Molecular electrosorption was probed with a signal enhancement achieved through the use of structured electrode surfaces (e.g., SEIRAS) [10, 11]. Other in situ FTIR approaches have used PMIRRAS to investigate the electrode surface, providing structural and orientation information as a function of potential [12–18]. Electrochemistry has been also coupled with methods sensitive to refractive index changes at the metal surface, such as ellipsometry and surface plasmon resonance [19–23]. A conceptually similar approach using neutrons has also been demonstrated [14, 24, 25]. In the majority of these challenging measurements, achieving acceptable signal to noise requires measuring from a large area which results in the surface being considered uniformly active and the average changes measured are interpreted with this inherent assumption. Most often non-spectroscopic-based approaches, not covered in this chapter, have been used to characterize the modified electrode surface. For example, the presence of defects in SAMs has been evidenced using Faradaic redox species either dissolved in the electrolyte or modifying one of the components of the SAM.

These buried electrochemical interfaces can be analyzed at the atomic scale using scanning probe microscopy such as AFM and STM. Numerous examples of high resolution in situ study of electrochemical processes at the submicron length scales have been demonstrated, many of them for electrode surfaces modified by physisorption or chemisorption [15, 26–36]. These molecular scale studies provide a detailed picture of how the surface is modified, the distribution of the adsorbates, the registration with the surface, and molecular environment of the adsorbates. However, high-resolution images are achieved over a limited field of view and there is a need for larger-scale studies of inhomogeneities in the surface modification. These methods may also have difficulty with analysis of more complex multicomponent monolayers or adsorbates that extend further than a few nanometers from the electrode surface typical of multilayer biosensor surfaces. Achieving both non-averaging measurements coupled with larger-scale molecular characterization of the surface uniformity or variation in surface monolayer characteristics requires many methods. In situ optical microscopy can study the surface down to the optical resolution (~ 1 μm). There are many examples of Raman microscopy being used to study the electrochemical interface, relying on surface enhancement to create measurable Raman signals [37–43]. This requirement has not allowed Raman microscopy to be easily used for many surfaces since SERS requires a roughened surface and is limited to coinage metals. Laser confocal microscopy with interference contrast has been reported to achieve an axial (z) resolution that can resolve atomic layers of gold during its dissolution [44].

Spectroelectrochemical fluorescence measurements have also been applied to the modified electrode interface containing a fluorophore, or for studying the interfacial pH-induced changes in fluorescence from a fluorophore near the electrode, or to monitor the removal of fluorophores from the SAM-covered electrode surface [5, 45–52]. In addition, fluorescence microscopy has been used to study the electrochemical interface with the adsorbate suitably modified with a fluorophore. Successful measurements require specific conditions to limit the background signal from solution, and an understanding of quenching due to the metal surface. Here, a review of in situ spectroelectrochemical fluorescence microscopy imaging (herein abbreviated iSEFMI) will be presented. This chapter focuses on the gold electrode interface and is organized in sections corresponding to the different adsorbates studied. Examples are provided to demonstrate that variations of this method are useful in understanding the adsorbate, the substrate, and the surface modification per se. Additionally, electrochemically driven dynamic processes of the modified surface are described. This surface specific method is dependent on the interaction between the fluorophore in ground and excited states and the metal surface.

2 Metal-Mediated Fluorescence Quenching

The key to the characterization of the electrochemical interfaces considered in this chapter is to understand the different metal-mediated decay processes of fluorophores near metal surfaces. To do so, we first examine fluorescence in a homogeneous dielectric, introducing the idea of the photonic density of states. The concept of density of states is then used to explain the impact of a planar metal surface on a nearby fluorophore's lifetime and integrated spectral intensity in the far field.

2.1 Fluorescence and Photonic Density of States

Fluorescence occurs when an emitter, such as an atom or molecule, already in an excited state of energy E_j decays down to a lower-energy state E_i by generating a photon of energy $E_j - E_i = \hbar\omega_{ji}$. This is an example of spontaneous emission, which is fundamentally governed by quantum mechanics. An important aspect of spontaneous emission, which is not entirely obvious at first glance, is that it is intrinsically influenced by the emitter's environment. More specifically, the rate of spontaneous emission of a certain transition, γ_{ij}, is derived using a famous result of quantum mechanics called Fermi's golden rule [53]:

$$\gamma_{ij} = \frac{\pi\omega_{ij}}{3\hbar\varepsilon_0}\left|\mu_{ij}\right|^2\rho(\omega_{ij}) \tag{1}$$

Fermi's golden rule states that the transition probability per unit time between the emitter's excited and lower states is proportional to the square of the transition dipole moment, $\left|\mu_{ij}\right|^2$, and the density of photonic states at the transition frequency, $\rho(\omega_{ij})$. The μ_{ij} is simply a number with units of a dipole moment, and its value is determined by the overlap and symmetries of the excited and lower states' wavefunctions, which will only be perturbed by the environment if the emitter is very close to inhomogeneities in the medium (on a scale of the characteristic size of the wavefunction, i.e., angstroms). The latter relates to photon states, which are the harmonic solutions of the source-free Maxwell's equations for some dielectric environment defined by a local dielectric function $\varepsilon(\mathbf{r},\omega)$. In other words, these are solutions that exist without external excitation. The spatial dependence of $\varepsilon(\mathbf{r},\omega)$ defines the geometry of the dielectric system (e.g., a metal/vacuum half-space), and the frequency dependence indicates that the medium may be dispersive.

The density of photon states quantifies the ability of the environment to support electromagnetic modes. In an infinite homogeneous lossless space characterized by a dielectric constant ε, plane electromagnetic waves with spatiotemporal solutions $\mathbf{E}(\mathbf{r},t) = \mathrm{Re}\{\mathbf{E}_0(\mathbf{r},\omega)\exp[\mathrm{i}(\mathbf{k}\cdot\mathbf{r} - \omega t)]\}$ represent one complete set of such photon states if one includes all (three-dimensional) $\mathbf{k}$ wavevectors and two orthogonal polarizations of the electric field vector. In such a uniform medium, the wavevectors have a dispersion relation of the form $|\mathbf{k}| = k = \sqrt{\varepsilon}\omega/c$. Physically, each state has a wavelength $\lambda = 2\pi/k$, and a direction of propagation defined by the direction of the vector $\mathbf{k}$. For a given wavelength, there are thus many states propagating in all different directions. Between k and $k + \mathrm{d}k$, the total number of photon states is $\rho(k)\mathrm{d}k = k^2/(c\pi^2)\mathrm{d}k$ [54]. Using the dispersion relation for a uniform dielectric, this means that $\rho(\omega) = \omega^2\varepsilon/(c^3\pi^2)$. The decay rate in a lossless homogeneous space is then:

$$\gamma_{ij}^0 = \frac{\omega_{ij}^3\left|\mu_{ij}\right|^2\sqrt{\varepsilon}}{3\pi\hbar\varepsilon_0 c^3} \tag{2}$$

The ω^2 dependence of the photon density of states associated with plane waves in a homogeneous dielectric is both the same everywhere in space and easy to derive analytically. For almost any non-trivial dielectric environment, the density of photon states will change based on the location $\mathbf{r}_d$ of the emitter and is much more difficult to obtain. For the type of problems considered in this chapter, it is in fact only the variation of γ_{ij} with $\mathbf{r}_d$ that is of interest. Since $\rho(\omega_{ij})$ is the only quantity in Eq. (1) that depends on $\mathbf{r}_d$, it is the only quantity that has to be calculated. To generalize the determination of $\rho(\omega_{ij})$ to a *local* density of states (LDOS), $\rho(\mathbf{r}_d, \omega_{ij})$, one uses the fact that the power electromagnetically coupled out of a *classical* point dipole located at $\mathbf{r}_d$ driven harmonically at frequency ω is also proportional to $\rho(\mathbf{r}_d, \omega)$. Thus, the LDOS, and therefore the spatial dependence of γ_{ij}, can be determined using a classical electromagnetic dipole emission calculation.

Specifically, a general way to determine $\rho(\mathbf{r}_d, \omega)$ in an arbitrary dielectric environment is to calculate $\mathbf{E}_d(\mathbf{r}_d)$, the electric field generated by a classical point dipole forced to oscillate harmonically with amplitude $\mathbf{p}_0$ at frequency ω at $\mathbf{r}_d$ [53]:

$$\rho(\mathbf{r}_d, \omega) = \frac{6\varepsilon_0}{\pi\omega|\mathbf{p}_0|^2} \mathrm{Im}\{\mathbf{p}_0^* \cdot \mathbf{E}_d(\mathbf{r}_d)\} \tag{3}$$

The power coupled out to the field by the dipole is simply the product of the velocity of the moving charge, $\mathrm{d}\mathbf{r}/\mathrm{d}t$, and the force $q\mathbf{E}_d(\mathbf{r}_d)$ on the oscillating charge q due to the electric field. Using the fact that the dipole moment is $\mathbf{p} = q\mathbf{r}$, the time-averaged power coupled to the field by the dipole oscillating harmonically with amplitude $\mathbf{p}_0$ is [53]:

$$P = \frac{\omega}{2} \mathrm{Im}\{\mathbf{p}_0^* \cdot \mathbf{E}_d(\mathbf{r}_d)\} \tag{4}$$

In terms of the LDOS, the power coupled to the electromagnetic field by a forced harmonic dipole oscillation is therefore $P = \frac{\pi\omega^2}{12\varepsilon_0} |\mathbf{p}_0|^2 \rho(\omega_{ij}, \mathbf{r}_d)$, which establishes the fact that both the spontaneous emission rate of a quantum emitter with resonant frequency ω, and the power radiated by a classical dipole forced to radiate harmonically at ω, are proportional to the LDOS at ω, $\rho(\mathbf{r}_d, \omega)$. Comparing to Eq. (1), we see that within factors of unity, the electromagnetic decay rate of a quantum two-level system is analogous to the rate that power is dissipated (in units of photon energy) by a classical, harmonically driven dipole with strength $|\mathbf{p}_0|^2 = \hbar\omega|\mu_{ij}|^2/4$ [55]. In this classical dipole analogy, the quadrature or imaginary component of the self-field (i.e., $\mathbf{E}_d(\mathbf{r}_d)$) extracts power from the excited state, with the total amount of energy emitted equal to $\hbar\omega_{ij}$. Practically, this has a profound implication for our interpretation of spontaneous emission since we can now relate what is intrinsically a quantum mechanical problem described by Fermi's golden rule, to something that we can think about classically: the electric field of a dipole.

In general, the total decay rate of a quantum emitter in an arbitrary dielectric environment, γ, can be written in terms of the total decay rate in a reference homogeneous system, γ^0, and the additional decay rate induced by scattering from dielectric inhomogeneities with respect to the homogeneous reference material, γ^m:

$$\gamma = \gamma^0 + \gamma^m \tag{5}$$

The decay rate in a reference homogeneous system is generally given as: $\gamma^0 = \gamma_r^0 + \gamma_{nr}^0$, where γ_r^0 is given by Eq. (2), and γ_{nr}^0 is the internal non-radiative decay rate within the quantum emitter. Introducing the intrinsic quantum efficiency of the emitter as $q = \gamma_r^0/\gamma^0$, and using the expression for γ_r^0 in Eq. (2), we can write the *normalized* decay rate as:

$$\frac{\gamma}{\gamma^0} = 1 + q\frac{\pi^2 c^3}{\omega^2\sqrt{\varepsilon_1}}\rho_m(\omega, \mathbf{r}_d) \tag{6}$$

where ε_1 is the dielectric constant of the homogeneous reference system and $\rho_m(\omega, \mathbf{r}_d)$ is the density of states introduced by the inhomogeneities. This quantity is related to the field of a harmonic dipole scattered back to the dipole's position due to the inhomogeneities introduced in the otherwise homogeneous dielectric environment.

In summary, fluorescence is a spontaneous emission process, which is not an intrinsic property of the emitter, but depends on the environment. This dependence is quantified by the local density of photonic states at the emitter's location. In a homogeneous lossless space, these states take the form of electromagnetic plane waves. Despite the fact that spontaneous emission is a quantum mechanical process, we showed that we can calculate the decay rate of an emitter in an arbitrary medium by simply finding the electric field generated by a classical harmonic dipole, at its location. In the next sections, we will describe how the photon states that exist in the more complex geometry of a metal half-space affect the power dissipated by a nearby dipole and thus how this will affect the fluorescence decay rate of a quantum emitter at that location.

2.2 Fluorescence Near a Metal Surface

Consider, as shown in Fig. 1, a dipole at a certain distance z_0 along the z-axis above a metal surface in a dielectric half-space characterized by a real dielectric constant ε_1. The additional power dissipated by the dipole due to the metal surface will be due to the reflected field, at the location of the dipole. For the geometry illustrated in Fig. 1, we can therefore find $\rho_m(\omega, \mathbf{r}_d)$, the LDOS due to the metal, by solving for $\mathbf{E}_{\text{ref}}(z_0)$, the reflected dipole field at z_0. Using Eqs. (3) and (6), the normalized decay rate can be written as:

$$\frac{\gamma}{\gamma^0} = 1 + q\frac{6\pi\varepsilon_0 c^3}{\sqrt{\varepsilon_1}|p|^2\omega_{ij}^3}\mathrm{Im}\{\mathbf{p}^* \cdot \mathbf{E}_{ref}(z_0)\} \tag{7}$$

For reasons that will be made clear in the next section, it is useful to write $\mathbf{E}_{ref}(z_0)$ as an expansion in terms of $s = k_\rho/k_1$, where k_ρ is the component of the electric field wavevector in the plane of the interface and k_1 its magnitude in the medium above the metal [53, 56]:

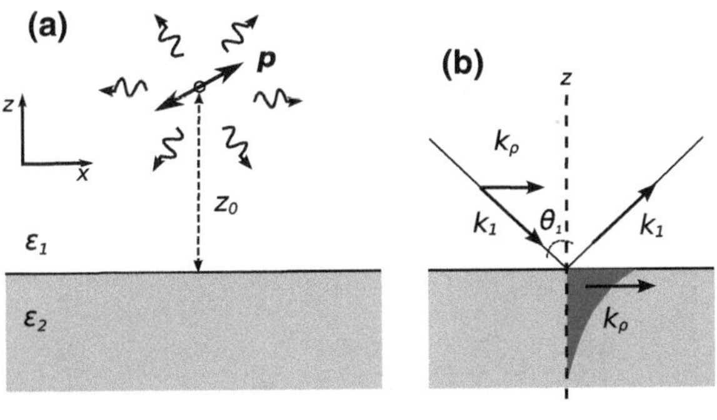

Fig. 1 **a** Point dipole with arbitrary orientation and dipole moment **p** at a distance z_0 above an interface in the x–y plane separating a lossless medium with dielectric constant $\varepsilon_1(\omega)$ for $z > 0$ and a metallic substrate characterized by a complex dielectric constant $\varepsilon_2(\omega)$ for $z < 0$. **b** The source dipole field in this geometry can be decomposed in terms of photon states with well defined in-plane wavevectors k_ρ. This figure illustrates the case of a certain downward propagating plane wave with component $k_\rho < k_1$ incident on the interface, for which both a reflected plane wave and a rapidly attenuated transmitted wave are generated by the metal. There are also evanescent components (not shown) of the source dipole field with $k_\rho > k_1$ that can excite polarization in the metal, which produces evanescently decaying electric fields in both the metal and the top half-space

$$\mathbf{E}_{\text{ref}}(z_0) = \frac{\mathrm{i}k_1^3}{8\pi\varepsilon_0\varepsilon_1}\int_0^\infty \frac{s}{s_z}\left\{p_\parallel \hat{p}_\parallel\left[r_s - s_z^2 r_p\right]\mathrm{e}^{2ik_1 s_z z_0} + 2p_\perp \hat{p}_\perp s^2 r_p \mathrm{e}^{2ik_1 s_z z_0}\right\}\mathrm{d}s \tag{8}$$

In the above expression, $s_z = \sqrt{1-s^2}$ and r_s and r_p are the Fresnel reflection coefficient for s- and p-polarized light, respectively, evaluated as a function of s. The dipole moment components along the z-axis and in the plane of the interface are $p_\perp$ and $p_\parallel$, respectively. Obtaining the analytical expression for the reflected dipole field from a metal surface is a problem that has been treated extensively in the literature [53, 56, 57] and will not be discussed further.

Figure 2 shows the lifetime calculated from Eqs. (7) and (8) for three point dipoles oscillating at different frequencies as a function of the distance between the dipole and the surface of a gold half-space. The oscillations at relatively large separations can be easily understood from the partial constructive and destructive interference of radiative components (essentially a superposition of plane waves) of the original dipole field that are reflected back to the dipole. The oscillations clearly damp out at larger separation, due to the fact that the reflected dipole field amplitude decreases as the dipole/metal separation increases. The dramatic drop-off in the lifetime at distances less than ~100 nm is due to more subtle effects involving the coupling of the dipole near-field with non-plane wavelike contributions to the LDOS near the metal surface.

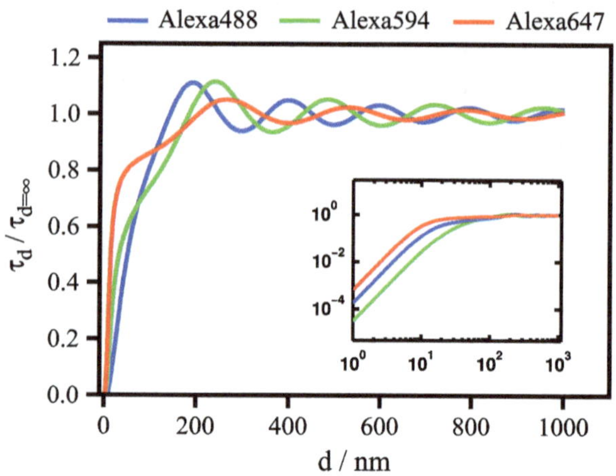

Fig. 2 Normalized lifetime for various AlexaFluor dye molecules in water ($\varepsilon_1 = 1.33^2$) above a gold substrate calculated using isotropic point dipoles at the dye's peak emission frequencies (as given in Table 1). The dyes are AlexaFluor647 (*red*), AlexaFluor594 (*blue*) and AlexaFluor488 (*green*). *Inset* shows the same data on a log–log plot, highlighting the clear $1/d^3$ dependence for the decay rate emerging at short metal/emitter separations

2.3 Near-Field Decay Mechanisms and Quenching

Due to the planar symmetry of our geometry, the photonic states of the environment are naturally defined in terms of a two-dimensional wavevector parallel to the surface, $\mathbf{k}_\rho$. The behavior of the field in the z direction is determined by the z component of the wavevector, $k_{z1} = \sqrt{k_1^2 - k_\rho^2}$. Using the dispersion relation for a uniform dielectric medium, this quantity, for a given $\mathbf{k}_\rho$ and frequency ω, can be written as:

$$k_{z1} = \sqrt{\varepsilon_1 \omega^2 / c^2 - k_\rho^2} \tag{9}$$

As per Eqs. (3) and (8), the LDOS can be calculated from the dipole field and thus be expressed as an expansion in terms of the in-plane wavevector. Components of that superposition with $k_\rho < k_1$ have a real k_{z1} and so contribute to the field that propagates away from the interface in the positive z direction, to the "far field." Components of the field generated by the dipole with $k_\rho > k_1$ have an imaginary k_{z1} and thus describe the dipole's near field. Similarly, components of the reflected field with $k_\rho > k_1$ are bound to the metal surface and do not propagate to the far field. The total LDOS at any one location above the interface is the sum of all available photon states with a certain in-plane wavevector at that location. In Fig. 3, we plot $\mathrm{d}\rho(\omega)/\mathrm{d}s$ versus $s = k_\rho/k_1$ for several different locations above the metal surface.

For distances further than ≈200 nm from the metal, the LDOS is completely dominated by photon states similar to the propagating plane waves characteristic of

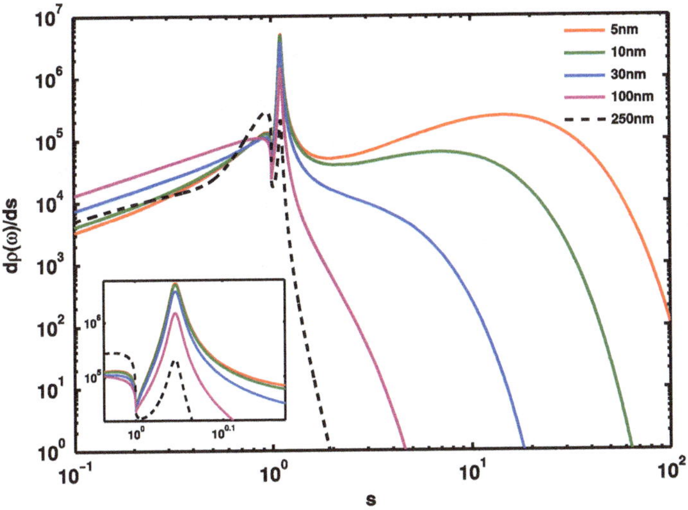

Fig. 3 Available photon states as a function of normalized in-plane wavevector $s = k_\rho/k_1$ calculated for a z-polarized dipole in water ($\varepsilon_1 = 1.33^2$) at 619 nm (corresponding to AlexaFluor594 peak emission). The different curves are for different gold substrate/dipole separations. The *inset* details the peak in the LDOS at $s \simeq 1.1$ corresponding to the SPP states that exist at the gold/water interface at this particular frequency

an infinite lossless homogeneous medium, and as per the above discussion, they must correspond to the part of the LDOS with $k_\rho/k_1 < 1$ in Fig. 3. However, these original plane wave states are modified by the interface and can be roughly thought of as the superposition of a plane wave incident from the top half-space, and a reflected wave from the metal surface. The associated interference explains the oscillations in the lifetime apparent in that regime.

For $k_\rho/k_1 > 1$, we begin to consider photon states that are evanescent in nature (they only exist near the metal surface) and can thus only be excited when the dipole is near the surface. These are the states that are responsible for the initial drop-off in the lifetime of an emitter as it approaches the metal interface. In the transition region below ≈200 nm, the states associated with the peak in the LDOS at $k_\rho/k_1 \sim 1$ are particularly important. These are called surface plasmon polaritons (SPPs) and are surface wave solutions of Maxwell's equations for the specific geometry of a metal/dielectric interface. They are surface waves because their k_z wavevectors are imaginary (they are evanescent) in both the metal and the dielectric half-space above it. SPPs have a well-defined dispersion relation given by [53]:

$$k_\rho^2 = \frac{\varepsilon_1 \varepsilon_2}{\varepsilon_1 + \varepsilon_2} \frac{\omega^2}{c^2} \tag{10}$$

This equation implies that SPPs always have an in-plane wavevector greater than that of a plane wave propagating freely in the medium above it. The increased momentum of these modes is linked to the fact that the electromagnetic wave propagating along the surface also has to "drag" along the surface charges of the metal [53]. Excitation of these modes is thus not possible by the radiative components of the dipole field; only evanescent components of the dipole field that possess a high enough in-plane wavevector can excite such states. SPPs can be resonantly excited for an in-plane wavevector component of the dipole field that matches the plasmon wavevector given by Eq. (10), which accounts for the peak at $s \simeq 1.1$ in Fig. 3.

In summary, due to their evanescent nature, SPPs only contribute to the LDOS near the interface. As such, coupling to these modes is only possible when the dipole is in the vicinity of the interface (less than ≈200 nm away). It is also important to note that any power coupled into the SPPs remains attached to the interface (eventually decaying due to absorption in the metal) and thus cannot be detected in the upper half-space by traditional means.

The SPP modes dominate the LDOS, and hence the spontaneous emission rate, for a range of near-interface distances down to ≈20 nm. Below that separation, the divergent (non-propagating) near-field components of the dipole field start to penetrate the metal and transfer power directly via ohmic heating (inducing electron currents that scatter and heat the metal). Since these near fields decay as $\sim 1/r^3$, the sharp increase in decay rate at short metal/dipole separations also follows $\sim 1/d^3$ (as shown via the lifetime in the inset of Fig. 2) [58, 59]. There is no simple interpretation of how these processes manifest themselves in the LDOS of the metal–dielectric interface, but they do contribute via the broad peak in the LDOS at large in-plane wavevectors that dominates the total LDOS for small separations.

In this chapter, the focus is on the metal/emitter separation regime that can be achieved using DNA adsorbates on gold, which is typically for distances greater than ≈3 nm. For shorter separations, the exact manner in which the near field interacts with the electrons in the metal cannot be accurately described using the local (wavevector-independent) dielectric function we have limited ourselves to considering, and more complicated analysis is required to capture the physics at play [56].

2.4 Measuring Fluorescence in the Far Field

Although fluorescence lifetime measurements of molecules are incredibly powerful in terms of directly probing the LDOS of an arbitrary environment, they are often very challenging to realize for metal surfaces modified with fluorophores under electrochemical control. A much more convenient experiment, which can often be done in situ in an epifluorescence setup, is the measurement of the continuous wave (CW) fluorescence spectrum, or its integrated spectral intensity, in the far field.

Before discussing the modeling of CW fluorescence in complex dielectric environments, it is important to comment on how the discussion up to this point in terms of a single quantum mechanical transition actually relates to the spontaneous emission of a complex dye molecule. The emission spectrum actually comprises a superposition of radiation emitted by the thermalized excited state, to a variety of empty vibrational and rotational levels of the ground state. The single transition model can work reasonably well to describe the impact of the dielectric environment on the overall lifetime of the excited state of real molecules if to some degree of approximation, each of these transitions is similarly influenced by roughly the same LDOS, and each transition has roughly the same non-radiative dephasing rate.

When considering how the CW fluorescence *strength* is impacted by proximity to the metal surface, care must be taken to properly include both the absorption and emission processes, since both are influenced by the dielectric environment. Absorption is typically via higher energy transitions than those associated with the spontaneous emission that occurs after subsequent thermalization of the photoexcited electrons, and so, there is no coherence between emission and absorption processes. If we assumed that all of the absorption took place via a single transition between ground state k and excited state l, the total power radiated into the upper half-space due to cyclic absorption via the k to l transition followed by spontaneous emission from the j to i transition would be:

$$P^r_{\text{out}} = N_{\text{ex}} \gamma^r_{ij} \hbar \omega_{ij}. \tag{11}$$

where N_{ex} is the steady-state probability of the molecule being in the j excited state and γ^r_{ij} is the decay rate of that transition via radiation away from the substrate. We can determine N_{ex} by balancing the total input power with the total output power. The latter is given by the total decay rate of the emitter: $P_{\text{out}} = N_{\text{ex}} \gamma_{ij} \hbar \omega_{ij}$. The input power depends on the intensity of the electric field exciting the emitter, $I(\mathbf{r}_d) = \frac{1}{2} c \varepsilon_0 \sqrt{\varepsilon_1} |\mathbf{E}(\mathbf{r}_d)|^2$, and the absorption cross section of the k to l transition at the excitation frequency:

$$P_{\text{in}} = \sigma_{\text{abs,kl}} I(\mathbf{r}_d) \tag{12}$$

If one neglects the power loss associated with the Stokes shift between absorption and emission $(\omega_{\text{kl}} - \omega_{ij})$, the steady-state excited state probability is:

$$N_{\text{ex}} = \frac{\sigma_{\text{abs,kl}} I(\mathbf{r}_d)}{\gamma_{ij} \hbar \omega_{ij}} \tag{13}$$

The steady-state power radiated into the upper half-space from the j to i transition would then be:

$$P^r_{\text{out}} = \sigma_{\text{abs,kl}} I(\mathbf{r}_d) q_{a,ij} \tag{14}$$

where we introduced the apparent quantum yield as $q_{a,ij} = \gamma^r_{ij} / \gamma_{ij}$.

Accurate modeling of the CW absorption and emission kinetics would require summing over all the pairs of states that absorb at the excitation frequency, and separately, calculating the net lifetime of the excited state by summing over all of its possible decay transitions (independent of the collection bandwidth). This would be a complex numerical task for any given molecule, but given some simplifying assumptions we can state that the total steady-state power radiated into the upper half-space from fluorescence is:

$$P^{r}_{\mathrm{out},f} \propto |\mathbf{E}(\mathbf{r}_d)|^2 q_a(\omega_{\mathrm{em}}) \tag{15}$$

where the apparent quantum yield, q_a, is evaluated at the peak emission frequency, ω_{em}, of the molecule. The total decay rate in the denominator of the expression for q_a is calculated as discussed extensively above (cf. Eq. (7)), while γ^r_{ij} requires a calculation of the dipole power radiated away from the interface. Knowing the total dipole field in our geometry, this can be readily achieved with Poynting's theorem. The result is once again (cf. Eq. (8)) expressed in terms of an expansion in terms of $s = k_\rho/k_1$, the normalized in-plane wavevector [59, 60]:

$$\begin{aligned}\gamma^r_{ij}/\gamma^0_{ij} = {} & q\frac{p_\parallel^2}{|p|^2}\left[\frac{1}{2} + \frac{3}{8}\int_0^1 \left[\frac{s}{s_z}|r_s|^2 + ss_z|r_p|^2\right]\mathrm{d}s\right. \\ & \left. + \frac{3}{4}\int_0^1 \mathrm{Re}\left\{\left[\frac{s}{s_z}r_s - ss_z r_p\right]\mathrm{e}^{2ik_1 s_z z_0}\mathrm{d}s\right\}\right] \\ & + q\frac{p_\perp^2}{|p|^2}\left[\frac{1}{2} + \frac{3}{4}\int_0^1 \frac{s^3}{s_z}|r_p|^2\mathrm{d}s + \frac{3}{2}\int_0^1 \mathrm{Re}\left\{\frac{s^3}{s_z}r_p \mathrm{e}^{2ik_1 s_z z_0}\mathrm{d}s\right\}\right]\end{aligned} \tag{16}$$

The above expression shows that the decay rates via radiation for all dipole orientations is solely due to photon states with $k_\rho < k_1$ that are propagating away from the dipole into the upper half-space. The first terms for both dipole orientations relates to the decay rate into plane wave modes radiating directly away from the dipole into the upper half-space. In the absence of the metal interface, we can see that γ^r_{ij} yields precisely half the radiative homogeneous medium decay rate, $q\gamma^0_{ij}$, as it must. The second terms relate to the contributions of plane waves reflected away from the interface, while the third terms result from the interference of the primary and reflected dipole fields. It is important to note that Eq. (16) cannot be obtained from a simple calculation of the non-evanescent photonic states contribution to the LDOS since those also include the downward propagating plane waves that are not reflected back into the upper half-space but are instead rapidly attenuated in the lossy metal substrate. As shown in Fig. 4, the apparent quantum yield is sensitive to both the emitter's distance from the metal and its frequency. The oscillations at large distances can be explained by the same partial interference effect that affected the lifetime.

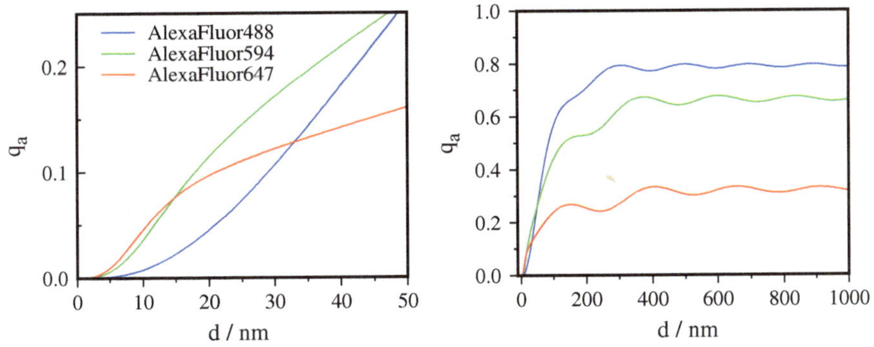

Fig. 4 Apparent quantum yield (q_a) calculated for various Alexa dye molecules in water as a function of emitter/gold separation for regions close to and far from the gold surface (*left* and *right* figures, respectively). Peak emission frequency used in the calculation is shown in Table 1

In an actual experiment, the integrated fluorescence intensity will depend on the numerical aperture of the objective used. Since the contribution of a particular plane wave to the dipole far field has an in-plane wavevector that can be expressed as $k_\rho = k_1 \sin\theta_1$, where the angle θ_1 is between the z-axis and the wavevector $\mathbf{k}_1$ (see Fig. 1), the integration in Eq. (16) can be trivially carried out as a function of θ_1 [53]. The apparent quantum efficiency can thus be restricted to only include radiation modes that propagate within the collection angle of the objective given by its numerical aperture: $\mathrm{NA} = n \sin\theta_1$ (where n is the refractive index of the medium in contact with the objective). A small NA objective will essentially crop out parts of the far-field radiation pattern occurring at shallow angles, which can result in dramatic oscillations of the apparent quantum yield when the emitter is far from the metal, as shown in Fig. 5.

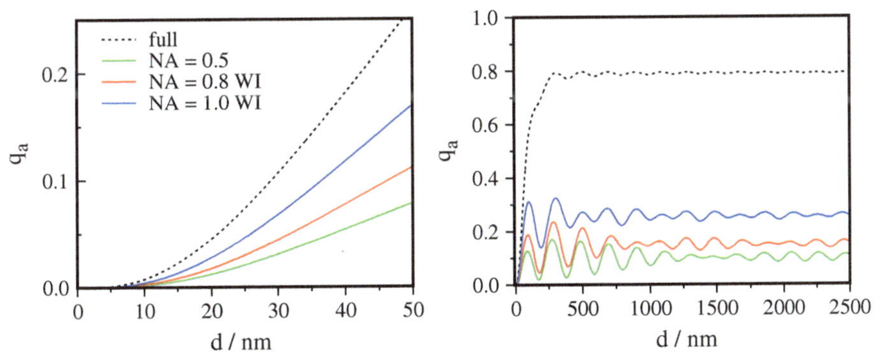

Fig. 5 Impact of the collection angle of objectives with different numerical apertures (NA) on the apparent quantum yield (q_a) of Alexa488 in water as a function of emitter/gold separation for regions close to and far from the gold surface (*left* and *right* figures, respectively). High NA are realized using water immersion (WI) objectives

3 In Situ Spectroelectrochemical Fluorescence Microscopy Imaging (iSEFMI) Methodology

Luminescence-based spectroelectrochemistry evolved from electroreflectance measurements through the use of a similar experimental arrangement. To measure luminescence, the emission is acquired normal to the electrode surface plane with the illumination coming into the surface at an angle [5, 45, 61]. Fluorescence-based spectroelectrochemical measurements were used to investigate the changes in the interfacial chemical environment [46, 51, 62] and to study the electric field at the surface [52, 63]. The fluorescence-based in situ spectroelectrochemical technique based on intensity measurements from the whole electrode surface was typically used to study molecules adsorbed onto the surface and the influence due to the applied potential [48–50, 61]. Implementation of the spectroelectrochemical measurement using a fluorescence microscope enabled laterally resolved microscopic measurements [64]. Fluorescence images from electrode surfaces can be acquired in several configurations defined by the type of microscope used (upright or inverted). In our laboratory, the inverted epifluorescence microscope has been preferred due to its versatility of the variety of electrode shapes that can be studied (e.g., hanging meniscus). The basic setup is shown in Fig. 6. A spectroelectrochemical glass cell with a thin glass window as the optical port is positioned atop of the inverted fluorescence microscope. This cell houses all the electrodes placed in such a way that the bottom of the working electrode is imaged through the optical window and the fluorescence is detected using a PMT or camera. Electrochemical experiments

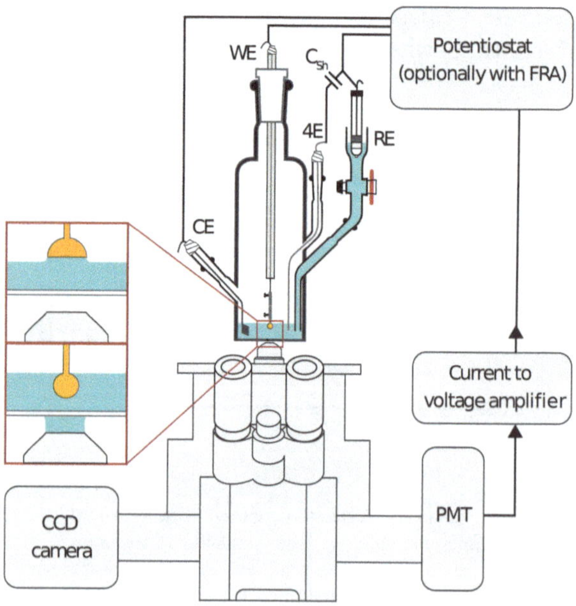

Fig. 6 Schematic of the spectroelectrochemical setup designed for use in an inverted epifluorescence microscope. Two types of detectors are shown. Also, two types of electrode geometries are shown for hanging meniscus (*top*) and immersed Au bead electrode (*bottom*) with a dry objective and water immersion objective, respectively

such as cyclic voltammetry (CV), differential capacitance and impedance spectroscopy can then be performed while simultaneously recording the fluorescence response as images or as the average intensity. Note that the attainment of spatial resolution comes at a price of decreasing the temporal resolution of the measurements. While a PMT rise time is a few nanoseconds, the use of a camera requires exposure times on the order of seconds for the typically low-intensity levels that are emitted by an adsorbed monolayer. A number of improvements have been realized since the last review of this method [65] with the evolution of a few key components of the microscopy system, substantially improving the quality data acquired. Here we present a summary of the most important changes.

3.1 Spectroelectrochemical Cell Designs

Presently, the spectroelectrochemical cells which are suitable for in situ fluorescence microscopy for use in the inverted epifluorescence arrangement are custom made by the University of British Columbia glassblowing shop. The details of how to produce these cells, in particular the procedure used to seal the thin optical windows, are presented in Appendix.

Originally, the cell design included a 25-mm-diameter, 0.17-mm-thick optical window (cover slip) sealed onto a cylindrical port protruding into the center of the cell large enough for the objective (Fig. 7a, c) [65]. This enabled the electrode surface to be at the focal plane even in hanging meniscus configuration while maintaining a large volume of electrolyte solution. Additionally, it facilitates positioning of the counter electrode, connection to the reference electrode via a salt bridge and bubblers in positions which did not block the optical access to the electrode surface. This approach worked well for objectives with long working distances (>7 mm). The 50× dry objective has a long working distance with a NA of 0.5 which limited the illumination and collection efficiency. However, most higher magnification (40× and higher) objectives have a working distance of less than 2 mm, making the placement of the objective and electrode dangerously close to the window risking breakage.

A modified cell design was developed in which the window is positioned at the bottom of a cylindrical cell (Fig. 7b, d). The distance between the objective and the optical window can be accurately monitored permitting the use of high NA/small working distance objectives. In addition, the window can be more easily replaced in the case of breakage. This arrangement also enables easy objective changes, addition of the immersion water, and translating the cell to image different regions on the electrode. The trade-off is that working with these high NA objectives is only possible using a small volume of electrolyte. The low volume of electrolyte also increases the risk of significant volume reduction due to evaporation by the inert gas used to blanket the electrolyte keeping oxygen out of the cell. The use of an external bubbler is recommended so that the connection to the reference electrode via a salt bridge is not compromised due to evaporation. However, care must be taken to

Fig. 7 Pictures of the two spectroelectrochemical cell arrangements used highlighting the position of the objective and the optical window

ensure that excessive moisture does not condense on the stem of the electrode, which can affect the electrochemical measurements.

The use of thin optical windows that have the same typical thickness as a coverglass slip (0.17 mm thick) was found to warp during the heating process necessary to seal the window to the cell. Increasing the optical window thickness to 0.25 mm resulted in significantly reduced warping enabling the use of larger window diameters (up to 50 mm). The edges were not useful for imaging because of the distortion caused by sealing to the body of the cell. These larger diameter windows result in cells that have a larger undistorted optically flat region in the center providing excellent image quality.

3.2 Digital Imaging System

The spectroelectrochemical cell is placed on the stage of an inverted epifluorescence microscope equipped with a UV–Vis light source, appropriate optical filters, and dichroic mirrors arranged in filter cube sets, objectives, and either a CCD or PMT detection method. The electrode surface in the cell is brought into focus taking care that the objective does not touch the window. This requires the electrode to be carefully positioned in the cell a small distance away from the optical window. The modified electrodes investigated are composed of monolayers that are only nm thick, which leads to strongly quenched fluorescence. A significant amount of excitation light reflected from the gold surface also bleeds through the dichroic mirror and barrier filters resulting in a high non-fluorescent background. These two effects combined decrease the signal-to-noise ratio. Since fluorescence from adsorbed monolayers is difficult to measure, potential modulation can be used to induce changes in the fluorescence intensity, a standard method in spectroelectrochemical studies. When images need to be acquired, modulation takes the form of potential steps.

In addition, the use of a cooled electron multiplying (EM) CCD camera was found to significantly increase the signal over cooled CCD cameras. This is particularly useful in the study of monolayers where the number of dye molecules is limited, especially when examining small electrode areas, for mixed SAMs or for short exposure times needed to measure fast processes.

Imaging an electrode surface in an operating electrochemical environment requires the use of long working distance objectives. Our first attempts used dry (metallographic) low NA ($\leqq 0.5$) objectives which limited measurement efficiency. The presence of many interfaces with large differences in refractive index (RI) (air/glass, glass/electrolyte) also limited the efficiency of the imaging system. Water immersion objectives are used to decrease the RI gradients in the optical path by replacing the air between the objective and the glass window with water. This change increased the image quality and the signal-to-noise ratio (Fig. 8). These objectives had shorter working distance (5 mm) but had a larger NA (0.8) which, along with the removal of large gradients in RI, resulted in a more efficient illumination and collection of emitted photons. This is demonstrated in Fig. 8, where a much clearer, higher-contrast, and better resolved image of the interface is obtained with a reduced background signal, which is particularly important in fluorescence measurements. We found that this improvement is dramatic for the higher magnification objectives (40×) but much less so for the 10× water immersion objectives. These changes can be rationalized through the influence of NA on resolution, which, in the case of diffraction-limited wide-field epifluorescence images, is given by the Airy radius, defined as [66].

$$r_{\mathrm{Airy}} = \frac{0.61\lambda_o}{NA_{\mathrm{obj}}}$$

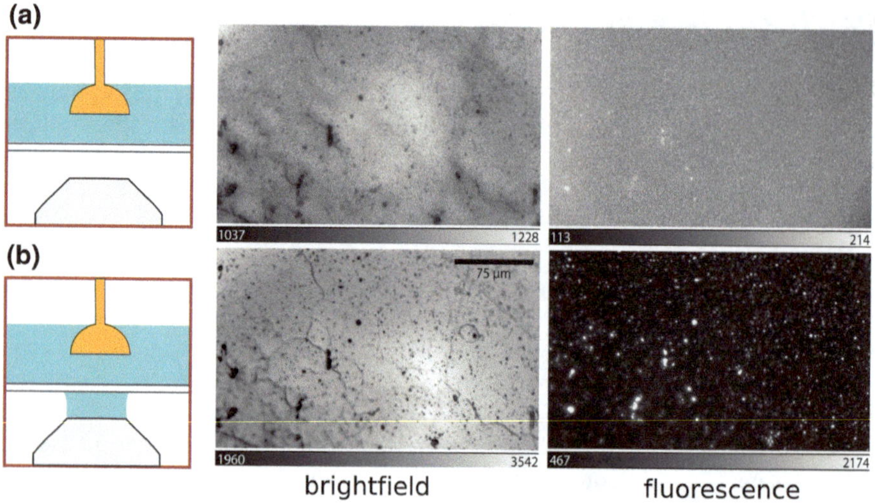

Fig. 8 Comparison of brightfield and fluorescence images of a MCH/DNA layer deposited on a polished gold electrode, using a 50× (NA = 0.5) dry objective and a 40× (NA = 0.8) water immersion objective. Bars under the images show the minimum and maximum intensity values used in the grayscale images. Adapted with permission from [167]

where r_{Airy} is the smallest distance between two points in focus that can be resolved, $\mathrm{NA}_{\mathrm{obj}}$ is the numerical aperture of the objective, and λ_o is the wavelength in vacuum. For example, using an excitation of 450 nm, and an objective with a NA of 0.5, $r_{\mathrm{Airy}} = 0.55\ \mu\mathrm{m}$. The resolution in the direction normal to the electrode (axial) is given by:

$$z_{\min} = \frac{2\lambda_o \eta}{(NA_{\mathrm{obj}})^2}$$

where η is the refractive index of the electrolyte. For the same wavelength and objective in electrolyte ($\eta = 1.33$), $z_{\min} = 4.8\ \mu\mathrm{m}$.

These in situ measurements required coordination of the application of electrochemical potentials and the measurement of capacitance or current with the collection of the fluorescence images. This was accomplished with a LabView program. Typically, the fluorescence image is acquired during the measurement of the differential capacitance at a fixed frequency using a lock-in amplifier. The typical duration of the potential steps necessitates capacitance measurements at 200 Hz, a frequency higher than normal to ensure a quick response from the lock-in amplifier, as the phase lock is lost after a large potential step. Care must be taken that the time needed to transfer the image data is sufficient so as not to run into timing problems (Fig. 9). For this, digital triggers are used to coordinate the imaging and electrochemical measurements. Since the fluorophores used may photobleach, the use of a timed shutter on the light source is also advocated.

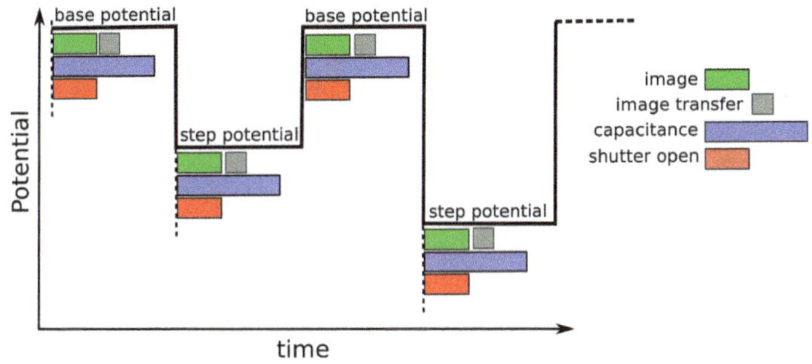

Fig. 9 Typical timing for image collection, data transfer, capacitance measurement, and shutter opening/closing for a potential step program using a constant base potential and a variable step potential

Turning the illumination off during the electrochemical measurements and data transfer minimizes the photobleaching.

3.3 *In Situ Fluorescence Lifetime Imaging (FLIM)*

An important characteristic of the fluorophore is its excited state lifetime [67–70]. Fluorescence lifetime imaging is a well-established method that is used to study biological materials [69–72] and is particularly useful in correlating the different species or environments that are present in a sample, since fluorophore lifetimes are sensitive to the local environment. While the more common intensity measurements are a convolution of fluorophore concentration and relative decay rates, lifetime measurements are independent of the amount of fluorophore. Therefore, this technique is suitable to isolate the effects of quenching by environmental factors such as O_2, energy transfer, or in the case that pertains to us, the separation from a metal surface. Imaging an electrode surface modified with a fluorescent SAM is also limited by this effect. In this case, the intensity is a convolution of the surface density of fluorophores and their average separation from the electrode surface which influences the quenching efficiency.

Interpretation of intensity-based images must consider these parameters, but it can be difficult to control or know any one of them a priori. Preparing a well-characterized monolayer with the fluorophore a known distance from the electrode surface would enable a direct relationship between coverage and the intensity to be measured. In the same vein, changes in intensity for a fixed coverage could be used as a measure of the changes in the separation between the electrode and the adsorbed layer with potential or other perturbations. A complete characterization of the modified interface and the influence of potential would require lifetime imaging which would distinguish between low intensity due to quenching or due to low coverage. In the high-intensity regions, lifetime may also distinguish between bound molecules (e.g., covalently attached to the surface) and those which are adsorbed onto the modified electrode surface via non-specific interactions. These typically exist further from the surface, thereby having a lifetime similar to an unquenched or free fluorophore.

FLIM systems rely on measuring the decay in the fluorescence which typically occurs in the nanosecond timescale. Therefore, FLIM measurements are most commonly associated with confocal microscopes which use pulsed laser excitation. Confocal microscopes have better lateral and axial resolution than the wide-field fluorescence imaging systems [66] in large part due to the use of pinholes which restricts the sample volume excited and ensures that fluorescence is only collected from the optical plane which is in focus. The volume under investigation is raster scanned across the sample, and the image is created by measuring the fluorescence intensity at each specified point volume on the sample. The density of these points is controlled by the scanning speed and measurement frequency, but the ultimate resolution is still defined by the optics and the sample medium [66, 73] as in wide-field fluorescence imaging. Eliminating the out-of-focus fluorescence signal increases the image contrast which effectively improves the lateral resolution by $\sqrt{2}$ [66]. Using a time-correlated single-photon counting (TCSPC) measurement of the fluorescence emission in conjunction with a pulsed laser allows for fluorescence lifetime to be recorded for each predefined pixel in the image. The TCSPC method bins the photons based on their delayed arrival time with respect to the laser pulse.

Two-photon microscopy is another method for high-resolution imaging which uses long wavelength photons from a high-power femtosecond pulsed laser [71, 74]. Creating an excited state requires the near simultaneous absorption of two photons that each has approximately half the energy needed for absorption [75]. This requirement results in an absorption rate that is dependent on the square of the excitation intensity. The spatial and temporal overlap required for the two-photon absorption event is only satisfied in the small focal volume of the excitation laser where the photon density is the highest, resulting in localized excitation. Fluorescence from the out-of-focus regions does not exist since the excitation volume is the same as the collection which allows for more sensitive non-descanned detection of the fluorescence eliminating the need for pinholes which are used in confocal systems. In addition, since the fluorescence is blueshifted from the long wavelength excitation, it will be measured on a low background region of the spectrum. It is important to note that although heuristics exist to extrapolate the main features of a two-photon absorption spectrum from a corresponding single-photon spectrum, two-photon absorption is fundamentally governed by different quantum mechanical selection rules than its single-photon counterpart [76].

In this work, we adapt our approach for wide-field epifluorescence microscopy for use in an inverted two-photon confocal laser scanning microscope (Zeiss LSM-510 MP with a Chameleon XR (Coherent) Ti–sapphire pulsed laser (minimum pulse width of 140 fs with tunable output from 710 to 980 nm) setup using water immersion objectives (40× W Plan-Apochromat NA = 1.0). A schematic is

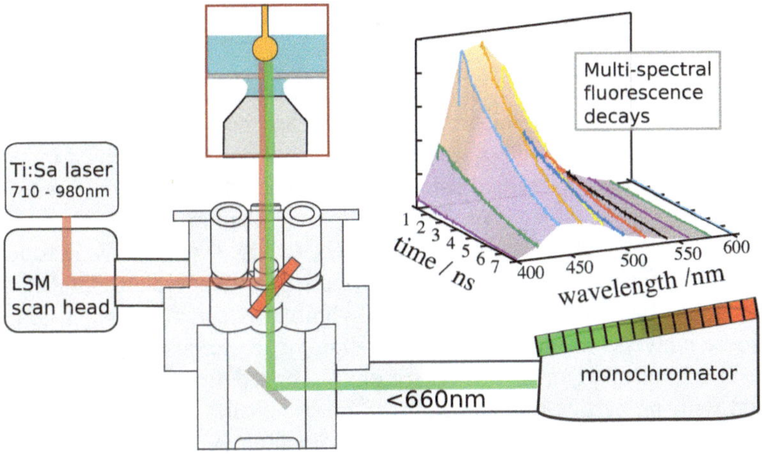

Fig. 10 A schematic of the two-photon confocal fluorescence lifetime imaging (FLIM) setup used for spectroelectrochemical studies. The fluorescence collected from the electrode surface is spectrally resolved into 16 detector channels that are time resolved

shown in Fig. 10. A non-descanned detection approach was used. Since the fluorescence photon flux is low, efficient photon collection is necessary in order to perform a reliable fit to the time-resolved fluorescence signal to extract the lifetime. This requires high NA objectives and accumulation over many image scans (minutes to hours). Photobleaching of the fluorophore must also be considered and balanced with respect to fluorescence photon accumulation. At times, this is a difficult balancing act.

The TCSPC system used here (Becker and Hickle SPC-830) spectrally resolves the photons (using a 125-mm spectrograph (Oriel model 77,400) dispersing a 200-nm spectral region into 16 channels, 12.5-nm bandwidth per channel) which enables the measurement of many luminescence processes for the fluorophores used. This can be particularly useful in FRET or other molecular energy transfer measurements between two fluorophores. Furthermore, the use of spectrally resolved lifetime measurements ensures that the correct/expected wavelength for the fluorescence is measured. The reflection of the excitation from the electrode surface is collected and must be filtered before reaching the detectors. As demonstrated in Sect. 2, the excited state lifetime is dependent on the distance the fluorophore is from the electrode surface. With short lifetimes, as expected for the quenched fluorophore, convolution of the emission with the instrument response (e.g., excitation pulse width, detector bandwidth) must be taken into account in the analysis. This can be done by measuring the instrument response function (IRF) from the second-harmonic signal of a sample like urea crystals. It is important to ensure as similar an optical path as possible.

4 Study of Alkylthiol SAMs Using iSEFMI

Modification of electrode surfaces has been accomplished by many methods, classified as physi- and chemisorption, where Langmuir–Blodgett deposition and self-assembled thiol monolayers are two examples. Interfaces modified by either of these methods have been studied using iSEFMI. The advantages and limitations of the fluorescence imaging method will be demonstrated by exploring fluorophore-labeled self-assembled monolayers (SAMs). Originally introduced in the 1940s [77], the SAM approach is often used since noble metals like Au, Ag, Cu, Pd, and Pt can be coated with layers of alkylthiolated compounds [78–83]. These chemisorbed layers offer a modification strategy to purposely functionalize the distal end of the alkyl molecule. Therefore, so-called functional surfaces can be designed with properties that are different from the ones of the bulk substrate or adsorbate. Examples of these modifications include redox active groups [84], molecular [85] and biomolecular [86] recognition elements, as well as terminal groups suitable for further modification (e.g., carboxylic-amine bonds and click chemistry). Several reviews have been published discussing the generalities of self-assembled monolayers [81, 87–90]. Many of the substrates for thiol-based self-assembly are traditionally employed as electrode materials; therefore, electro-analytical techniques were used to analyze the SAMs, as reviewed by Finklea [91]. This section highlights the use of iSEFMI for characterizing SAMs in which at least one of the adsorbates possesses a fluorescent moiety at their distal end (Fig. 11).

This in situ spectroelectrochemical technique based on fluorescence takes advantage of the quenching of molecular fluorescence when near (in the nm range) a metallic surface (as discussed in Sect. 2) [59, 92]. Potential-induced changes in the separation of the fluorophore from the metal can be monitored by measuring fluorescence intensity. These molecules are rarely longer than 20 carbon atoms and thus form layers that are at most 3 nm thick [78]. Consequently, the fluorescence of these SAMs is strongly quenched. In order to easily observe fluorescence, they must be desorbed from the electrode surface. This can be achieved by controlling the potential of the metallic substrate (E) to favor reductive or oxidative desorption processes.

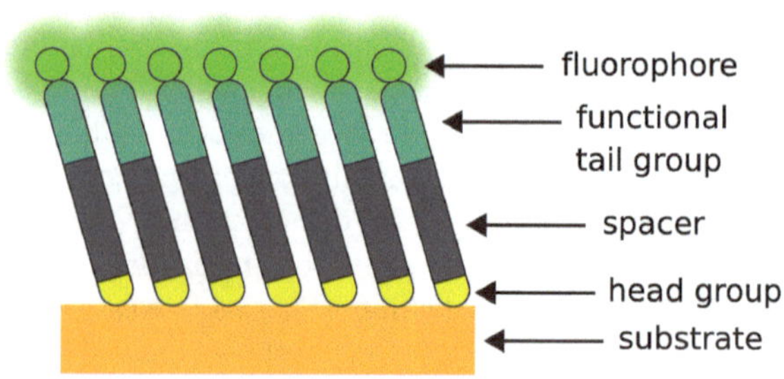

Fig. 11 Schematic of the structure of a fluorescently labeled self-assembled monolayer

4.1 Reductive Desorption

Applying potentials of approximately −0.9 to −1.2 V versus a saturated calomel electrode (SCE) to a gold substrate covered with an alkylthiol SAM will reduce the sulfur–metal bond. Without its headgroup covalently tethered to the metal, the monolayer is displaced from the surface [93–98], as thiolates

$$\mathrm{AuSR} + e^- \rightarrow \mathrm{Au(0)} + \mathrm{RS}^- \tag{17}$$

This process is usually carried out in basic media to circumvent the overlap with hydrogen evolution. During this reaction, the gold atom is reduced, while the sulfur remains in the −2 oxidation state. Equation (17) can be rewritten more formally to emphasize the effect that the solvent and the surface have on this displacement equilibrium [99]:

$$\mathrm{RS}_{(\mathrm{surf},\mathrm{M}(hkl))} + x\mathrm{H_2O}_{(\mathrm{aq})} + \mathrm{e}^-_{(\mathrm{M}(hkl))} \rightleftharpoons \mathrm{RS}^-_{(\mathrm{aq})} + x\mathrm{H_2O}_{(\mathrm{surf},\mathrm{M}(hkl))} \tag{18}$$

where the subindices "surf" and "aq" denote the species at the metallic surface with a (*hkl*) orientation and in solution, respectively. This process has been studied for a large variety of thiolated compounds [99–104]. The most common application of reductive desorption of SAMs is the estimation of the surface coverage by integration of the charge under the cathodic peak [105] although it has been shown that this method overestimates the amount of thiol in the surface by as much as 20% [106, 107]. Moreover, reductive desorption has also been employed to regenerate clean surfaces [108–110] and in systems where controlled release is desired [95, 111, 112]. Although most often the reductive potential is applied to the electrode by means of a potentiostat, reductive conditions can also be achieved by adding a reducing agent to the solution [113].

Selective reductive desorption. The desorption of alkylthiol SAMs has been studied in our laboratory using iSEFMI. An alkyl thiol labeled with a BODIPY fluorophore (HS-C10-BODIPY as shown in Fig. 12a) was used to form a SAM that would fluoresce when desorbed. Initial experiments on a multicrystalline Au bead electrode showed that a small increase in the fluorescence could be observed when the potential is stepped to ~−0.6 V versus SCE). The fluorescent signal, however, becomes much more intense (1000×) at potentials more negative than −1.2 V versus SCE. This has been interpreted as the thiolates—resulting from the desorption—separating significantly from the electrode surface [114]. In agreement with this explanation, differential capacitance measurements show a noticeable increase at $E \leq -1.2$ V versus SCE. Capacitance increases when the low dielectric constant alkyl layer is substituted by electrolyte which possesses a higher dielectric constant. Despite some readsorption upon the positive sweep back to potentials where the layer is stable, most of the thiol was lost to the solution. The desorption process removes the thiol specifically from certain parts of the electrode surface depending on potential and surface crystallography (discussed below). These

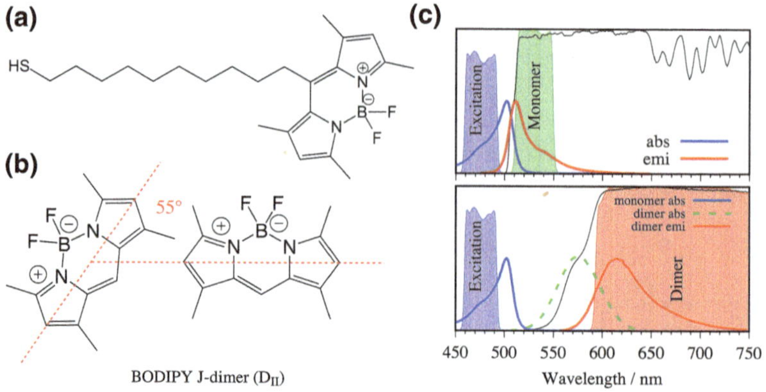

Fig. 12 **a** Structure of the BODIPY-labeled decanethiol, **b** the arrangement of the BODIPY dimer DII, **c** the absorption and emission spectra for BODIPY monomer (*top*) and dimer (*bottom*) from [168], and the wavelength ranges of the excitation, dichroic, and emission filters used in the fluorescence measurements for the BODIPY monomer (*top*) and dimer (*bottom*)

regions remain free of thiol when cycled to more negative potentials in the subsequent second and third cycles. The exclusive removal of the SAM from specific regions on the electrode surface is clearly observed in Fig. 13 by comparing the fluorescence images. Complete desorption occurs after a scan to −1.4 V versus SCE where the capacitance of the electrode resembled that of a Au electrode without a SAM.

During the desorption experiments described above, it was noted that the thiolate would leave the electrode surface as a diffuse fluorescent plume which would displace over time occasionally moving laterally across the surface. The study of this displacement on the multicrystalline bead electrodes is difficult since the relatively large domains of different surface crystallography are next to each other. As a consequence, those thiols that desorb first can move over top of the neighboring crystalline faces for which desorption may start. These multiple sources of thiolate for a single sampled volume complicate the potential dependent fluorescence changes of the modified electrode.

Desorption from microelectrodes. To study the movement of thiolates after desorption, it is more convenient to use electrodes whose dimensions are on the micrometer scale. The use of patterned microband electrodes to study the dynamics of the thiol release through fluorescence microscopy was first reported by Ghaly et al. [47]. To allow for more accurate control of independent variables influencing the movement of desorbed thiolates such as desorption potential, tilt angle of the electrode surface, and distance to the counter electrode, we chose to work with microdisc electrodes embedded in glass pipettes [115]. Due to the fabrication method employed, domains with different crystallography (if present) are so small that the reductive fluorescence signal occurs as one spatially homogeneous signal. The small size of the electroactive area (25 μm in diameter) also results in a behavior similar to a point source, which facilitates studying the fate of the

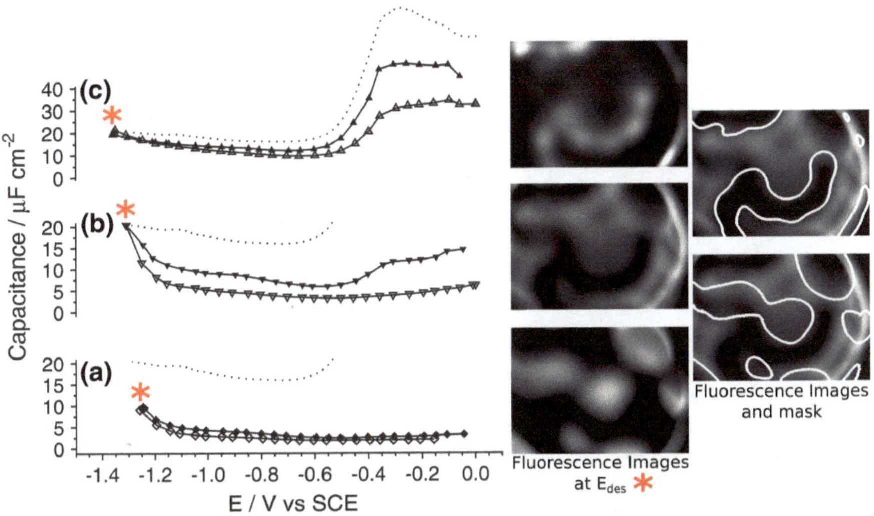

Fig. 13 Capacitance (*left*) of a multicrystalline Au bead covered with a HS-C10-BODIPY self-assembled monolayer. The potential was sequentially stepped in −50 mV intervals (*closed symbols*) until a potential limit of **a** −1.25, **b** −1.3, and **c** −1.35 versus SCE. The potential was stepped in +50 mV intervals (*open symbols*) until the initial potential (0 V) was reached. Fluorescence images taken at the negative potential limit are shown in addition to the outlines of the features from potential scans (**a**) and (**c**) overlayed on the image for potential scan (**b**). The electrolyte was composed of 0.1 M NaOH in ultrapure (>18 MΩ) water. The *dotted lines* are the capacitance of the clean gold electrode in 0.1 M NaOH. Adapted with permission from [114]. Copyright (2004) American Chemical Society

desorbed molecules. Furthermore, this setup permits the more accurate control of independent variables influencing the movement of desorbed thiolates such as desorption potential, tilt angle of the electrode surface, and distance to the counter electrode. With this system, it was possible to observe three different forces determining the velocity of the desorbed thiolate: (i) an isotropic diffusion of the thiolate molecules, (ii) a small electrophoretic attraction toward the positively charged counter electrode (Fig. 14a), and (iii) a strong buoyant force caused by hydrogen evolution at the negative potentials used (Fig. 14b). The knowledge of such a combination of forces is useful in designing systems where control is desired over the movement of SAM desorption products.

Reductive desorption and surface crystallography. As demonstrated above, the selective removal of the SAM from certain areas of the electrode was explained as a consequence of the different surface crystallography [104] which is a result of the differences in the potentials of zero charge [116]. Additionally, the atomic arrangement at the substrate surface can influence the substrate–adsorbate and adsorbate–adsorbate interactions [89, 99]. In situ fluorescence microscopy enabled a direct observation of the influence of surface crystallinity on the desorption potential. The SAM-covered gold bead electrode used by Shepherd et al. displayed an increase in the fluorescence only at selected regions at a characteristic negative limit potentials (Fig. 13). This bead electrode was fabricated by melting the end of a

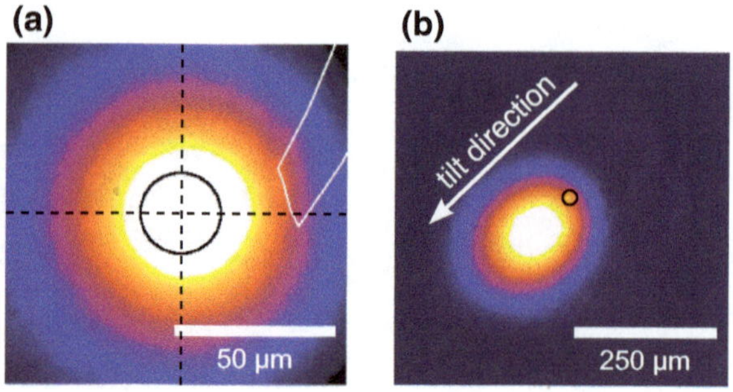

Fig. 14 Fluorescence images of the reductive desorption of a HS-C10-BODIPY SAM deposited on microelectrodes by immersion in a 1 mM solution of the alkylthiol in 1:1 MeOH:$CHCl_3$ and later desorbed by decreasing the potential in 25 mV steps. The *black circle* represents the electrode metal surface. **a** Desorption performed with the counter electrode in close proximity to the working electrode. The position of the counter electrode is delineated by the *white line*. The *dotted lines* are visual aids. **b** Desorption performed with the electrode surface tilted 3° with respect to the horizontal showing the movement of the desorbed thiol. Adapted with permission from [115]. Copyright (2013) American Chemical Society

gold wire [117] using a butane torch and was multicrystalline in nature. By aligning the brightfield and fluorescence images, a facetted area with a (111) plane crystallography was identified. The flat facet enabled electron backscattered diffraction (EBSD) measurements, which confirmed the crystallography of that area (within 10°). However, other areas were more curved and could not be unambiguously assigned by EBSD.

This limitation was recently overcome by the fabrication of single-crystal bead electrodes. This was achieved by using ultrapure gold (>99.999%), performing the process in a area protected from air currents and slowing the melting and cooling rates. While only the (111) facets are flat enough for EBSD analysis, the surface crystallography of other regions on the bead can be deduced by their relative positions with respect to the (111) facets. The order of the observed desorption potentials for the low index surfaces (Fig. 15c) is in excellent agreement with reported values from purely electrochemical experiments using polished single crystals [99]. Performing a potential desorption sweep to negative desorption potentials while imaging revealed a number of symmetrically organized regions that desorbed at specific potentials (Fig. 15a, b). By choosing a fluorescence intensity threshold, a map of the desorption potentials for the gold face-centered cubic (fcc) crystallographic triangle could be obtained in one single experiment in a highly self-consistent measurement. This is advantageous since traditionally obtaining the same information would require the fabrication of individual single crystals for each desired crystalline orientation, relying on reproducible assembly and analysis conditions for all samples.

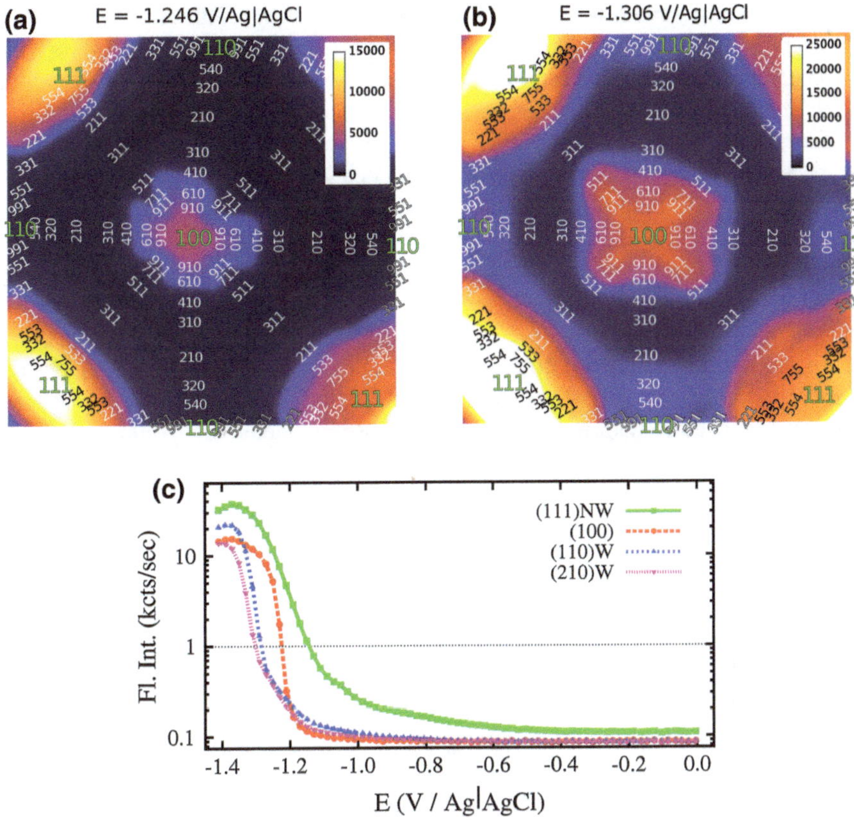

Fig. 15 Overlay of the fluorescence image and crystalline orientation map at **a** −1.246 V and **b** −1.306 V versus Ag|AgCl. **c** Fluorescence as a function of potential for selected crystalline orientations. Fluorescence images in false color. Adapted with permission from [136]. Copyright (2015) American Chemical Society

Reductive desorption and Förster resonance energy transfer (**FRET**). The spectral shift of the emission of some fluorescent probes is used to report on the change in its environment. Such is the case of the BODIPY dyes which are known to form dimers on close contact (Fig. 12). One of these dimers, usually referred to in the literature as D_{II}, is believed to be a ground state dimer and acts as a FRET acceptor with the excited state of the monomer as the donor (Förster radius = 42 Å). The resulting emission from the D_{II} dimer (simply referred to as the BODIPY dimer in the rest of the text) at 630 nm is redshifted from the monomer emission [118, 119]. Therefore, measuring the dimer emission can give an indication of the degree of aggregation of BODIPY molecules (which in turn reflects the packing of adsorbate molecules) in either the adsorbed or desorbed state.

Reductive desorption experiments where the intensities for monomer and dimer fluorescence were sequentially monitored showed that the dependence of fluorescence on potential for the two emission wavelengths is different [120]. Such an observation suggests that the dimer fluorescence does not depend on the monomer fluorescence. This can be interpreted as the dimer forming on the surface of the electrode rather than during the desorption process. Since the dimer fluorescence is observed through FRET from the monomer, it can also be inferred that the SAM is composed of a mixture of dimers and monomers in close proximity. A limitation of a standard epifluorescence setup (i.e., employing one set of filters per fluorophore studied) is that only one wavelength range can be measured at a given time. This renders any comparison sequential and relies on the similarity between the fabricated layers. An improvement to the setup could include the addition of a beamsplitter and the use of two cameras simultaneously. It is also possible to use three-channel color cameras, but they have a much lower performance than dedicated (e.g., electron multiplying) monochromatic cameras. Confocal microscopy with multiple spectrally resolved detectors would be a useful alternative.

4.2 *Oxidative Desorption of Alkylthiol SAMs*

Alkylthiol SAMs can be oxidatively desorbed through a complex reaction that occurs when applying potentials of approximately +0.5 to +0.6 V versus SCE at pH≈13. The voltammetric desorption of *n*-propanethiol SAMs from Au on mica has been proposed to proceed through a reaction that yields a sulfinic acid product [121]

$$\mathrm{AuSR} + 2\mathrm{H_2O} \rightarrow \mathrm{Au(0)} + \mathrm{RSO_2^-} + 3\mathrm{e^-} + 4\mathrm{H^+} \tag{19}$$

However, experiments performed by Yang et al. at slower scan rates on a Au (111) single-crystal electrode showed that nonane- and butanethiol SAMs require as many as 11 electrons per thiol oxidized [97]

$$\mathrm{AuSCH_2R} + 14\mathrm{OH^-} \rightarrow \mathrm{Au(0)} + \mathrm{RCO_2^-} + \mathrm{SO_4^{2-}} + 11\mathrm{e^-} + 8\mathrm{H_2O} \tag{20}$$

This reaction was supported by the appearance of CO stretching modes in subtractively normalized interfacial Fourier transform infrared spectra (SNIFTIRS). Experimental conditions precluded the definitive identification of the final sulfur species, limiting the sulfate presence to speculation.

The alkylthiols labeled with the BODIPY fluorophore described in the previous section have also been used to study the oxidative desorption of SAMs [120]. While reductive desorption occurred in a narrow potential range, its oxidative counterpart was observed over a large potential range and overlapping with the oxidation potential for gold. Furthermore, while the capacitance change precedes the fluorescence change in reductive step potential experiments, the opposite is true for the oxidative process. This suggests a different mechanism for the two processes. Reduced layers might be more stable, remaining closer to the electrode after desorption. Oxidized layers, on the other hand, might stay mainly adsorbed, while a few molecules are expelled further from the surface.

The kinetics of the desorption process also seem to vary noticeably between the two processes. This is evidenced by the time that the fluorescence signals take to reach a maximum value, which is much longer for oxidation than for the reduction process. Furthermore, significant roughening of the electrode was observed after the oxidative removal procedure, presumably due to the fact that the coupling of the thiol removal and Au oxidation processes results in a loss of Au atoms.

5 Study of Thiol SAMs Composed of BODIPY-Labeled Aib Peptides

Alkanethiol SAMs can undergo various phase transitions on substrate surfaces, which can result in non-ideal packing of the SAMs. For example, a decanethiol monolayer assembled on a Au(111) surface can undergo transitions between as many as six phases depending on the surface coverage [122]. To circumvent this issue, researchers have devoted much effort to find new adsorbate candidates with rigid conformation and favorable intermolecular interaction. One class of such competent candidates, α-aminoisobutyric acid (Aib) peptides, has attracted recent research interest [123]. Figure 16a shows the general structure of the HS-Aib_{n+1} series (where n is the number of intrachain hydrogen bonds) with the possibility of the end group being replaced by other functional groups. An important structural motif of the Aib peptide is its rigid 3_{10}-helix secondary structure induced by intramolecular C=O···H–N hydrogen bonds between residues i and $i + 3$ [124]. This helical structure in turn induces a strong dipole moment along the peptide backbone as the sum of multiple polar carbonyl bonds along the helix [125]. Orientation of the dipole moment is maintained with increasing number of Aib residues due to the accompanying increase in the number of hydrogen bonds. SAMs formed by

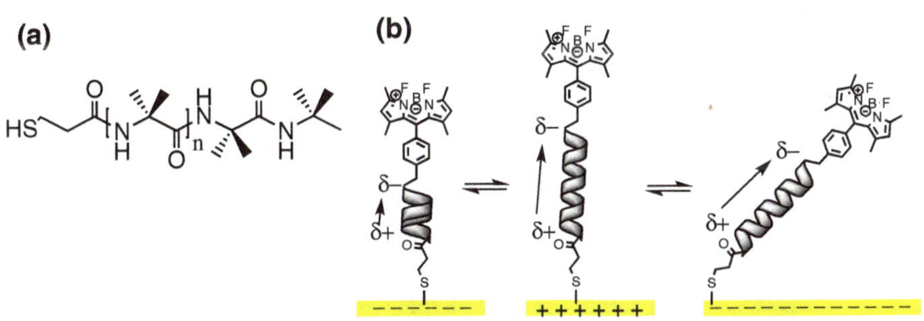

Fig. 16 **a** The primary structure of the thiolated Aib peptide series (positive pole of the dipole moment is at the thiol side). **b** An illustration of the hypothesized changes in the adsorbate due to changes in the electrode potential where a negative electrode surface can induce a compression or a tilting of the adsorbed Aib peptide SAM

thiolated Aib peptides on gold nanoclusters [126, 127] and planar gold surfaces [128] have been reported by Maran and coworkers. It was found that the peptide molecules form well-packed monolayers on both substrates. In addition to the intramolecular hydrogen bonds, a peptide molecule can form three-center hydrogen bonds with adjacent molecules in a SAM, which creates a large hydrogen bond network on the surface [126–128]. This hydrogen bond network accounts for the well-defined organization of the Aib peptide SAMs. Venanzi and coworkers also studied SAMs on gold surfaces using peptides with 3_{10}-helix Aib peptide as the core structure [123, 129–133]. These peptide SAMs were also densely and homogeneously packed. Furthermore, the dipole moment along the peptide backbone could facilitate the formation of bicomponent nano-structure through helix–helix interactions between opposite dipole moments on the surface [129]. These properties have made Aib peptide SAMs promising platforms for biosensing [128] and photovoltaic [130, 131] applications.

As mentioned in Sect. 4.1, the adsorbate–adsorbate interactions have a strong influence on the stability of the SAM which is reflected in the potential required to desorb it. Furthermore, the effect that substrate crystallography has on these interactions depends on the structure of the adsorbate [99]. Short-range hydrophobic interactions and H bonds hold the alkylthiol and Aib peptide SAMs together, respectively. The effects of this difference are shown in Fig. 17 which presents the interpolated desorption potentials for HS-C10-BODIPY and HS-Aib_4-BODIPY monolayers. Alongside, a calculation of the density of broken bonds (d_{bb}) is used in lieu of the point of zero charge of the bare Au surfaces [134, 135]. It can be seen that the Aib layer follows the d_{bb} trend more closely than the alkylthiol layer [136]. This indicates that (under the conditions employed) the differences in Au atoms arrangement at the surface have a larger effect on the stability of the alkyl chains as compared to the Aib peptide moieties.

The fluorescence of low-coverage HS-Aib_4-BODIPY peptide SAMs was found to be modulated by potential in a unique manner. Potential steps were applied to the

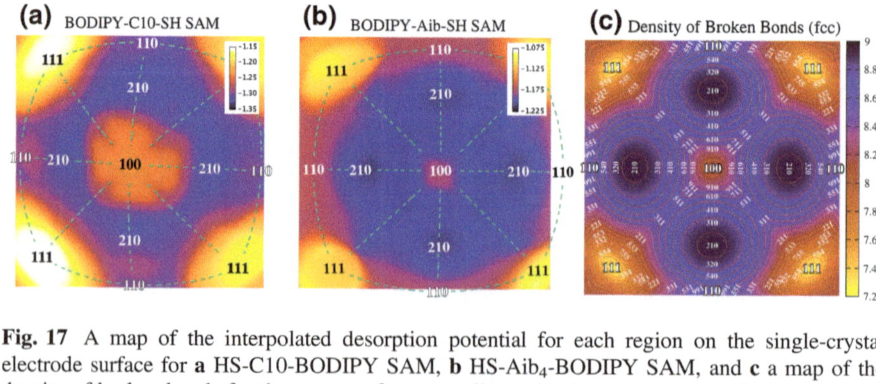

Fig. 17 A map of the interpolated desorption potential for each region on the single-crystal electrode surface for **a** HS-C10-BODIPY SAM, **b** HS-Aib_4-BODIPY SAM, and **c** a map of the density of broken bonds for the same surface crystallography shown in the experimental images. Adapted with permission from [136]. Copyright (2015) American Chemical Society

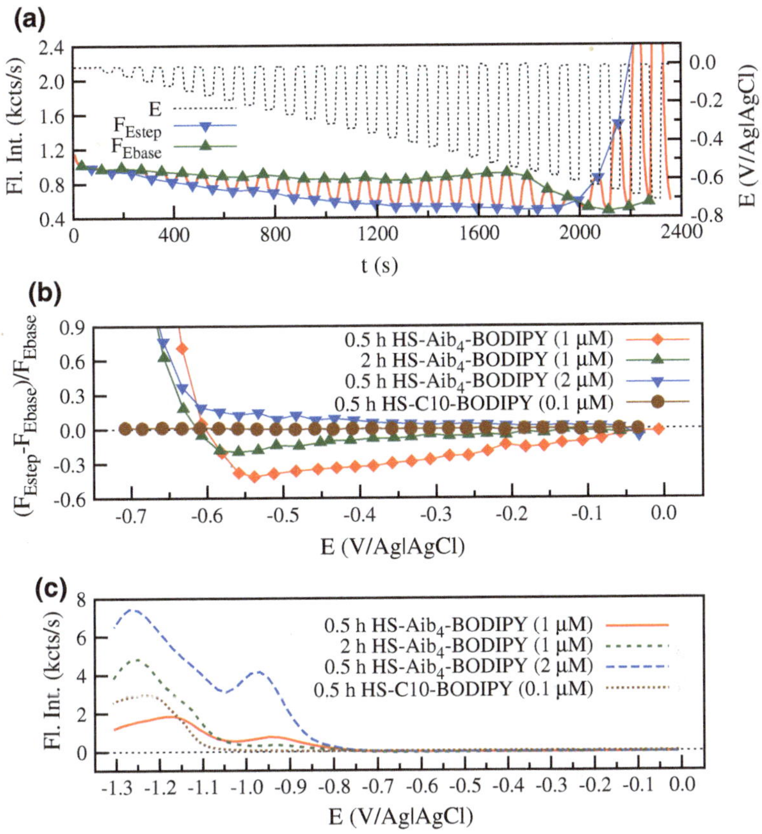

Fig. 18 **a** Profiles of the modulated potential steps and fluorescence intensity of a HS-Aib_4-BODIPY SAM. **b** The relative fluorescence changes as a function of the step potential of HS-Aib_4-BODIPY SAMs prepared under different deposition conditions and a HS-C10-BODIPY SAM. **c** The fluorescence intensity as a function of potential during desorption of the HS-Aib_4-BODIPY and HS-C10-BODIPY SAMs shown in (**b**)

modified electrode to induce a modulated change of the fluorophore–substrate distance. This set of potential steps was limited by the onset of reductive desorption. Specifically, the potential (profiled in Fig. 18a) was stepped from 0 mV (vs. Ag|AgCl) to −700 mV (vs. Ag|AgCl) in −25 mV intervals returning to the base potential (0 mV) after each step potential.

A typical low-coverage peptide SAM was prepared by immersing a gold bead electrode in a 1 μM HS-Aib_4-BODIPY ethanoic solution for 0.5 h. The modified electrode was then introduced into the spectroelectrochemical setup adjusting the orientation and position so that the (111) facet was at the bottom of the bead. Figure 18a shows the modulated fluorescence intensity profile measured from the center of the (111) facet, when applying potential steps. For step potentials less negative than −0.55 V, the fluorescence intensity is lower at the step potentials than at the base potential (0 V). This trend continues until about −0.6 V where the fluorescence intensity at step potentials increases significantly and eventually surpasses that at base potential, indicating the onset of desorption. Using the fluorescence intensity values at the step potential and at the base potential, the relative fluorescence change can be calculated for each step potential using: $(F_{\text{Estep}} - F_{\text{Ebase}})/F_{\text{Ebase}}$. The result as a function of the step potential is presented in Fig. 18b. The relative change becomes more evident (−50%) until −0.55 V and then quickly becomes positive. The decrease in intensity suggests that the SAM reorients, with the fluorophore becoming closer to the electrode surface at more negative potentials as long as no desorption occurs. The proposed mechanisms of the structural change of the peptide are shown in Fig. 16b. Note that in this set of experiments the positive pole of the dipole moment is at the thiol side. This effect is overwhelmed by desorption of the SAM at more negative potentials.

The coverage of the SAM can be increased by either extending the immersion time or increasing the concentration of the deposition solution. The SAM coverages were confirmed by an increasing fluorescence intensity during the reductive desorption measurement as shown in Fig. 18c. As shown in Fig. 18b, fluorescence modulation is strongly dependent on the SAM coverage. Extending the bead immersion time to 2 h significantly reduced the relative fluorescence change per potential step, while doubling the concentration of the deposition solution to 2 μM completely eliminated the fluorescence modulation at potentials more positive than the start of desorption.

The potential modulated response is thus a property of low-coverage Aib peptide SAMs. As a comparison, alkanethiol SAMs of similar or even lower coverage have been tested in the same way. The result of one HS-C10-BODIPY SAM is shown in both Fig. 18b, c. The HS-C10-BODIPY SAM exhibits no fluorescence intensity modulation before the start of desorption even though it has a coverage similar to the HS-Aib4-BODIPY SAMs which did show significant intensity modulation. Note that the potentials where the alkylthiol and Aib peptide SAMs start to desorb shown in Fig. 18c are comparable to what is presented in Fig. 17. Experimentally, the modulation of the Aib peptide SAM fluorescence was highly reproducible within a limited potential range (less negative than the onset of the desorption). Once the SAM experienced a potential negative enough to initiate desorption, the potential modulated behavior was no longer observed. This suggests that the Aib peptide SAM responds to the potential modulation without reduction in the Au–S bond. The proposed structural changes in the SAM, consistent with the changes in fluorescence, are shown in Fig. 16b. A more negative potential decreases the fluorescence intensity which requires the fluorophore to approach the electrode surface. The peptide has a strong dipole moment with the positive end near the thiol. The electrode potential may act to bring the SAM closer to the electrode surface by changing its tilt angle or by compressing the molecule. The peptide could tilt in response to the electrode potential similar to the DNA SAMs. Compressing the peptide is also possible since it has been reported that a 3_{10}-helix peptide, analogous to a spring, can undergo solvent polarity mediated conformation change [137]. Therefore, it is possible for potential to change the local solvent polarity, inducing a change in the peptide conformation.

6 iSEFMI Studies of Fluorescently Labeled DNA SAMs

The immobilization of DNA onto an electrode surface through self-assembly has been studied extensively for use as a biosensor [138–148]. The most common immobilization strategy uses a DNA molecule synthesized with a "linker" hydrocarbon chain (typically containing 6 carbon atoms) terminating in a thiol. To use single-stranded DNA (ssDNA) as a biorecognition element, a fundamental requirement is the change in conformation upon binding with the target molecule (complementary strand or a small molecule for the case of aptamers). The quality of the surface modification can have a large influence on the sensor performance. For example, Peterson et al. have shown that if the strand density is as high as 12×10^{12} strands/cm^2, the hybridization efficiency is reduced to ~15% compared to a value of 73% for a lower surface density (2×10^{12} strands/cm^2) [149]. This difference is explained in terms of having the space necessary for the complementary strand to access the probe strand and create a stable duplex. To limit the DNA surface density and prevent the adsorption of DNA bases to the gold surface, a small alkylthiol with an alcohol functional group (6-mercapto-1-hexanol, MCH) is used, creating a two-component SAM. It has been commonly assumed that the majority of the ssDNA are uniformly distributed on the surface. Therefore, a method capable of characterizing the surface bound biorecognition molecules (e.g., DNA) and their environment is desirable. The characterization parameter most often reported is the surface density of the ssDNA which can be determined by electrochemical methods [150] providing an average measurement of the entire surface. Its sole use implicitly assumes that the majority of the ssDNA are uniformly distributed on the surface. Therefore, a method capable of characterizing the distribution of surface bound biorecognition molecules (e.g., DNA) and their environment is desirable. Recent in situ AFM studies of these surfaces have characterized the local structure of the DNA SAMs on a Au surface in addition to the influence of the electrode potential [28, 151, 152].

In situ fluorescence microscopy has been used to better understand the structure of DNA modified electrodes for length scales greater than a few micrometers. To this end, oligonucleotides modified with a fluorophore (AlexaFluor series) on one end (3′) and a thiol at the other (5′) are used to prepare the DNA SAMs. Specifically, the modified DNA adsorbate used for all the measurements discussed is (from 5′ to 3′) HS-C6-CTG-TAT-TGA-GTT-GTA-TCG-TGT-GGT-GTA-TTT-AlexaFluor. The fluorescent label was either AlexaFluor488, AlexaFluor594, or AlexaFluor647. The preparation method most often found in the literature consists of immersing the clean Au surface into a 1–10 μM solution of thiolated DNA for several hours followed by immersion in a 1 mM solution of MCH for 1–2 h [149, 150] (DNA/MCH SAM).

Using this procedure, initial microscopy images showed heterogeneous layers in which regions of high-intensity ("hot spots") were observed and interpreted as DNA aggregates [153]. If the gold surface was exposed to MCH first and DNA second (MCH/DNA SAM), a significant reduction in the number of "hot spots" was observed, along with lower average coverage. Lower heterogeneity by this procedure has also been reported through in situ AFM [152].

Subsequently, we found that these intense hot-spot regions (presumably DNA aggregates) can be significantly reduced by careful preparation of the electrode surface (e.g., ensuring a smooth surface free of defects, completely removing previously deposited SAMs, avoiding disrupting the surface during the deposition procedure) even though their presence is not solely correlated with imperfections in the gold surface. This is particularly true for surfaces with a very high coverage of ssDNA, although even in the absence of hot spots, large-scale non-uniform fluorescence is still observed (Fig. 19b, c). When attempting to create low-coverage surfaces via the MCH/DNA thiol exchange process (Fig. 19a), another type of non-uniform coverage heterogeneity was observed. First, there were notable differences in intensity for different crystalline orientations of the substrate (Fig. 19d). By employing the single-crystal bead electrode described in Sect. 4.1, the thiol exchange (displacement of MCH with thiolated DNA) was found to favor certain crystallographic orientations (in the sequence 311 > 100, region around 111 > 111 > 110 > 210). Furthermore, a closer view of the surface shows that even within regions of the same crystallinity a clear irregular structure could still be seen (Fig. 19e) consisting of regions with higher fluorescence intensity and regions virtually devoid of any fluorophore.

The adsorption of thiol modified DNA is considered to occur specifically through the thiol Au interaction, but physisorption of the nitrogenous bases onto Au or adsorption onto the modified gold surface is also possible. This non-specifically

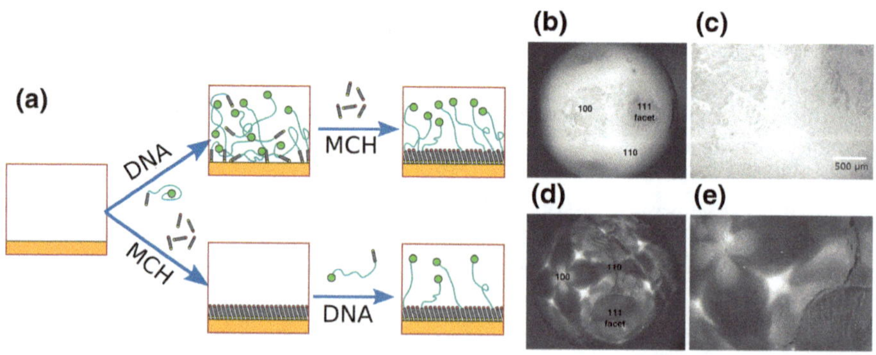

Fig. 19 Two methods used to modify gold bead surfaces with a mixed monolayer of MCH and single-stranded DNA labeled with AlexaFluor488. **a** Schematic of the two methods used to make the DNA/MCH (*top*) and MCH/DNA (*bottom*) SAMs. The resulting fluorescence images are shown for the corresponding methods: **b** and **c** DNA/MCH, **d** and **e** MCH/DNA. All images are in the same intensity scale. Adapted with permission from [167]

adsorbed DNA will also contribute to the fluorescence intensity measured and therefore must be removed. It was discovered that simply rinsing in buffer was not sufficient to remove physisorbed DNA regardless the method of preparation (DNA/MCH or MCH/DNA) [154, 155]. The desorption of these physisorbed species is evident as a fluorescent plume diffusing away from the surface at potentials not negative enough to cause reductive desorption. We speculate that this removal is facilitated by the electrostatically driven DNA reorientation described in the following section. It was possible to almost eliminate this physisorbed DNA by soaking the electrode in buffer over many days.

The "switching" DNA Construct—potential-induced reorientation of DNA SAMs. One of the most interesting examples of the coupling of fluorescently labeled SAMs and electrochemistry takes advantage of DNA's negatively charged backbone at physiological pH. As a result, DNA immobilized on an electrode surface can be manipulated to some extent by changing the charge on the electrode surface. When negatively charged, the surface will repel the DNA molecules making them "stand up" from the surface. Conversely, when a positive charge is accumulated on the surface, the DNA molecules will be attracted and "lay flat" on the surface (Fig. 20a). If the distal end of the DNA is labeled with a fluorophore, modulation of the fluorescence intensity is observed with changes in the electrode potential (Fig. 20b), as a consequence of the separation-dependent fluorescence quenching by the metal.

Rant et al. have devoted significant efforts in understanding this system from both the theoretical [156–159] and experimental [160–162] point of view. Square and sinusoidal potential perturbations have been employed so as to use these systems as biosensor platforms. A change in the amplitude of the fluorescence response was observed when ssDNA binds to its complementary DNA strand [161]

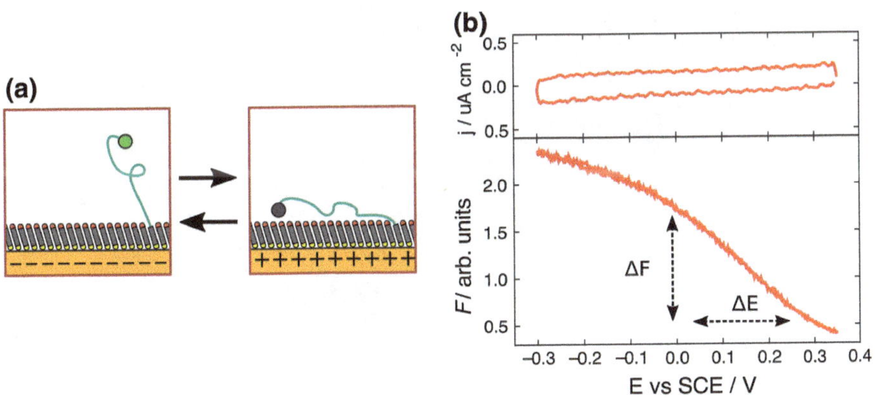

Fig. 20 Changes in DNA SAM orientation due to changes in the charge of the metal surface. **a** Schematic of potential-induced DNA reorientation, **b** CV (100 mV s^{-1}) of the MCH/ssDNA SAM and simultaneous fluorescence intensity measurements. The current has been smoothed with a Savitzky–Golay filter. The potential and fluorescence intensity ranges employed in the FRA experiments (Fig. 22) are shown with *dashed arrows*. Adapted from [155]

or when coupled to a protein [162]. Hybridization increases the stiffness of the DNA strand, and therefore, it can extend further in response to the electric field as compared to the ssDNA, which results in a larger amplitude of the modulated fluorescence signal. Furthermore, the highest frequency at which the DNA responds to the external potential perturbation depends on the effective hydrodynamic radius of the molecule, making it a useful tool to measure the diameter of attached proteins [157, 158, 162]. The largest amplitudes are achieved at low surface coverage DNA SAMs since the reorientation requires space for the movement to occur unimpeded. In fact, the high-coverage layers created by the DNA/MCH procedure do not change orientation with the electrode potential. The low-coverage MCH/DNA approach on the other hand results in large changes in the fluorescence. The relative change in fluorescence due to a potential perturbation is shown in Fig. 21b for different surface crystallographies on the single-crystal Au bead electrode. The orientation change is reported as a maximum change in fluorescence (ΔF_{max}) normalized by the fluorescence at the positive potential (F_{dc}). A maximum in $\left(\frac{\Delta F}{F_{\text{dc}}}\right)$ is observed in regions with low fluorescence (Fig. 21a) directly indicating the DNA packing density in those regions.

The rate at which the reorientation occurs can also be determined for these different regions. Since the imaging process with a CCD camera is slow, the analysis presented in Fig. 21b can be considered as measured at low frequency (<1 Hz). Frequency response analysis (FRA) can be used to measure the compliance of the DNA reorientation with respect to the applied potential perturbation. FRA can be accomplished using the same approach as in electrochemical

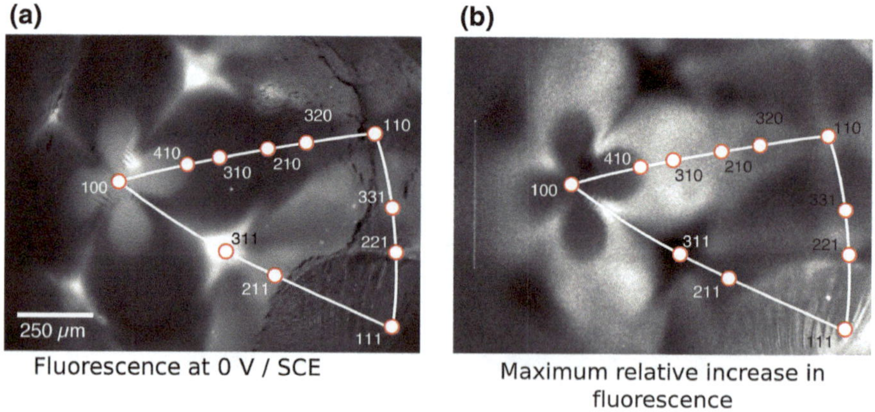

Fig. 21 Fluorescence images of **a** MCH/DNA on a single-crystal gold bead electrode at 0 V versus SCE with the surface crystallography overlay, **b** the normalized change in fluorescence $\left(\frac{\Delta F}{F_{dc}}\right)$ due to potential modulation from +0.35 to −0.4 V versus SCE. Note that the fluorescence image (**a**) is the result of merging a collection of images at different focal planes providing an extended depth of field [169, 170], while (**b**) is acquired at a single focal plane. Adapted with permission from [167]

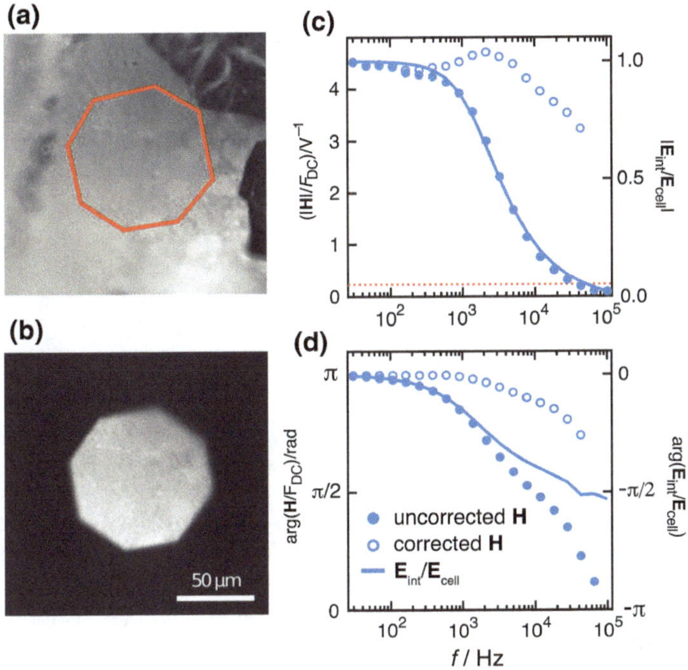

Fig. 22 Frequency response analysis for a MCH/DNA SAM. **a** The fluorescence image of the modified electrode surface, **b** the region of the electrode surface analyzed using a 90-μm aperture. The magnitude (**c**) and phase (**d**) of the FRA transfer function before and after correction for the electrochemical time constant along with the fraction of the applied cell potential realized at the interface (E_{int}/E_{cell}). The electrolyte was composed of 10 mM KNO_3 + 10 mM Tris. E_{dc} = +150 mV; ac amplitudes of 200 mV *p–p* and 5 mV rms were used for the fluorescence and electrochemical measurements, respectively. (**a**) was adapted from [154] Copyright (2014), with permission from Elsevier. (**c**) and (**d**) adapted with permission from [155]. Copyright (2015) American Chemical Society

impedance spectroscopy (EIS). In this case, an area of interest is first found using the CCD images. The detector is switched to a photomultiplier tube, and the illumination aperture is decreased, encompassing the area of interest (Fig. 22a, b). In this way, the response from ~90 μm diameter areas can be compared. The change in fluorescence intensity magnitude and phase with the frequency of the sinusoidal potential perturbation was measured and a transfer function calculated. The transfer function (H), similar in many respects to admittance, is shown in Fig. 22 (magnitude and phase in c and d, respectively) for this interface. While the amplitude varied between regions with the expected inverse dependency on DNA coverage, the frequency response presented only slight differences. However, it was found that the charging time of the electrode strongly influenced the observed response and a method for correction was developed [154]. Measuring EIS in the same spectroelectrochemical cell enables deconvolution of the electrochemical time constant from the measured fluorescence response. This method allows the measurement of the inherent DNA reorientation kinetics driven by potential. The change in this frequency response signals a binding event and can be used as a transduction motif for measurement.

FLIM of dsDNA SAMs on Au Fluorescence lifetimes are representative of the fluorophore and its environment. As described in Sect. 2, the fluorescence lifetime of a labeled SAM depends on variables such as fluorophore–metal separation, the metal, and the intrinsic characteristics of the fluorophore (e.g., lifetime, quantum yield, excitation/emission energies). Here we show preliminary results that examine the influence of the fluorophore characteristics on the decrease in its lifetime for SAMs that are prepared so that the fluorophore is a fixed distance from the surface. This is realized through the use of double-stranded DNA (dsDNA) SAMs that are prepared with high-coverage (DNA/MCH procedure). The 30-base pair DNA sequence is the same as used in the previous section. The total estimated length of this construct is 11 nm, assuming that the alkylthiol-tethered dsDNA sequence is normal to the electrode surface. The high-coverage SAM is closely packed such that no significant potential-induced reorientation is observed. The long persistence length of dsDNA (25–55 nm [156, 163, 164]) makes this assumption quite reasonable. Lifetime measurements were taken on three electrodes, each modified by SAMs that have the same DNA sequence with the same alkylthiol modification on the 5′ end. These three different layers were prepared with one of the three fluorophores (listed in Table 1) modifying the 3′ end.

The electrode surface was modified with HS-C6-dsDNA-AlexaFluor which was prepared by heating a solution that contains HS-ssDNA-AlexaFluor and corresponding complementary strand to about 90 °C and then slowly cooling it down to room temperature over 1 h. The HS-ssDNA-AlexaFluor and the complementary strand were mixed with 1:2 ratio in $pH = 7.5 \pm 0.05$ Tris 10 mM and NaCl 100 mM solution. DNA SAMs were prepared by immersing a cleaned gold bead electrode in 1 μM HS-dsDNA-AlexaFluor in an immobilization buffer (IB) of $pH = 7.5 \pm 0.05$ Tris 10 mM, NaCl 100 mM, and $MgCl_2$ solution for 18 h followed by 1 mM MCH in IB for 1 h. As reported previously [154, 155], it is important to remove any non-specific adsorption of fluorescent species before further fluorescence measurements as described previously. The dsDNA SAM was prepared on Au bead electrodes that contain Au(111) facets.

For the FLIM measurement, the dsDNA/MCH modified electrode was placed into a spectroelectrochemical cell that contains a buffer composed of $pH = 7.5 \pm 0.05$

Table 1 Characteristics of the AlexaFluor fluorophores used in the lifetime imaging work. λ_{ex}^{max}, λ_{em}^{max}, ε^{max}, and τ are the maximum excitation wavelength, maximum emission wavelength, maximum extinction coefficient and lifetime in homogeneous medium, respectively, as reported in [171]. τ_{expt} is the experimentally measured lifetime of the ssDNA–dye in solution

Fluorophore	λ_{ex}^{max} (nm)	λ_{em}^{max} (nm)	ε^{max} (M^{-1} cm^{-1})	τ (ns)	τ_{expt} (ns)
AlexaFluor488	495	519	73,000	4.1	3.45
AlexaFluor594	590	617	92,000	3.6	3.75
AlexaFluor647	650	668	270,000	1.0	1.03

10 mM Tris and 10 mM KNO_3 prepared in Milli-Q water. The electrolyte was purged with Ar to remove dissolved oxygen. The microscope setup described in Sect. 3.3 was used for the in situ two-photon measurements. The available laser wavelength range (710–980 nm) limited the options for excitation of the fluorophores used since the two-photon absorption process is larger in energy than the S1 excitation peak maximum and would presumably excite the molecules into the S2 manifold which would quickly relax to the S1 state from which the fluorescence lifetimes are measured [75]. It is not clear whether two-photon absorptivities are similar to linear absorption [76, 165], so an optimum was used which balances the excitation efficiency and minimizes photobleaching.

The lifetime analysis can be performed on all the photons collected in a specific wavelength range, providing an average lifetime. In many cases, the assumption of a single lifetime may not be accurate and analysis using multi-exponential analysis may be needed to accurately describe the measured decay rates. These fitting decisions are facilitated by a number of programs that are available for FLIM analysis. This lifetime analysis can also be performed for each pixel in the image that has sufficient signal. Binning pixels together can improve the fitting at the cost of a loss in lateral resolution.

Using the AlexaFluor647-labeled dsDNA as an example, the decay curves (within an emission wavelength range of 670–710 nm) for the fluorescently tagged DNA in electrolyte (purged of O_2) and from the surface species are shown in Fig. 23 in addition to the IRF. The decay of the fluorescence intensity (I) of the fluorophore in solution is fit with a stretched exponential [166] after correcting for the IRF using

$$I = Z + A_1\left(-\exp\left(\frac{t}{\tau_k}\right)\right)^{\frac{1}{h}}$$

where Z is the offset, A_1 is the intensity, τ_k is the characteristic decay time, and h is a measure of the width of the lifetime distribution (ideally equal to 1) with the mean lifetime calculated using $\tau = h \cdot \tau_k \cdot \Gamma(h)$ [166] where $\Gamma()$ is the gamma function. This approach is typically used for fluorophores that do not completely relax within the measurement time. The dsDNA SAM is fit with a bi-exponential decay

$$I = Z + A_1 - \exp\left(\frac{t}{\tau_1}\right) + A_2 - \exp\left(\frac{t}{\tau_2}\right)$$

revealing the presence of two environments (with lifetimes of τ_1 and τ_2) on the gold surface. The difference between the experimentally measured data and fitted curves shows no systematic deviations, indicative of a well-defined system (as shown in Fig. 23b). The constants resulting from the fit are given in Table 2.

It is clear that the dsDNA-AlexaFluor647 lifetime is strongly decreased near the gold surface. Interestingly, the bi-exponential analysis shows the presence of a strongly quenched component (τ_2) and a unquenched component (τ_1) which are 90

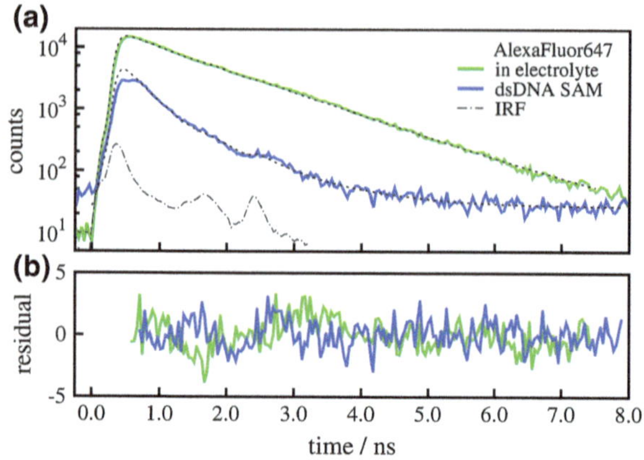

Fig. 23 **a** Fluorescence decay of HS-C6-DNA-AlexaFluor647 in electrolyte and bound to the electrode surface, excited with 720 nm. The experimental data (*solid lines*), the fitted decay (*dashed*), and IRF (*dash-dot*) are shown. **b** The residual from the fitting procedure

and 10% of the total signal recorded, respectively. More informative are the maps of the lifetime and the absolute (A_i) and relative $\left(\frac{A_i}{\sum_i A_i}\right)$ intensities of these two components when assembled onto the gold electrode.

The wide-field fluorescence images and the total photon count distribution across the image measured using FLIM for the dsDNA-AlexaFluor647 SAM are presented in Fig. 24a. The Au(111) facet is clearly visible in the wide-field CW fluorescence image as an intense oval region. The square region was further analyzed using the higher magnification water immersion objective. The edge of the (111) facet and a locally low-intensity fluorescence region are seen demonstrating the kind of non-uniformity that can exist on the DNA SAM surface. The FLIM photon count image reproduces the CW fluorescence image, with most photons coming from the intense region at the bottom of the image. Performing a bi-exponential fitting of the pixels in the image required binning of the data from neighboring pixels (25 pixel $\times$ 25 pixel binning was used) to achieve sufficient photon counts to ensure the fit results are significant. This binning results in distortion in the edges of the images. Images showing the distribution of the short and long lifetime components, their raw intensity (A_i), and fractional composition $\left(\frac{A_i}{\sum_i A_i}\right)$ are given in Fig. 24b.

The lifetime measurements of the two components shown in Table 2 are reflected in the images, with the majority of the AlexaFluor647 exhibiting a short lifetime which is evenly distributed across the electrode surface as a proportion of the total signal. The fitting results for the top part of the images are less certain due to low photon counts, presumably due to this region not being entirely in focus. The largest concentration of these short lifetime fluorophores is at the bottom of the image, correlated with the photon count distribution. The value of the longer

Table 2 Lifetime fitting results for dsDNA-AlexaFluor647 in electrolyte and on the Au surface modified with a DNA SAM. The wavelength used for the two-photon excitation is indicated in brackets

AlexaFluor647 (720 nm)	Component 1			Component 2		
	A_1	τ_1 (ns)	h	A_2	τ_2 (ns)	χ^2
Solution	$1.96 \pm 0.01 \times 10^4$	1.03 ± 0.01	1.08 ± 0.01	–	–	1.61
SAM	$7.39 \pm 0.80 \times 10^2$ (10%)	1.01 ± 0.04	–	$6.00 \pm 0.16 \times 10^3$ (90%)	0.32 ± 0.01	1.32

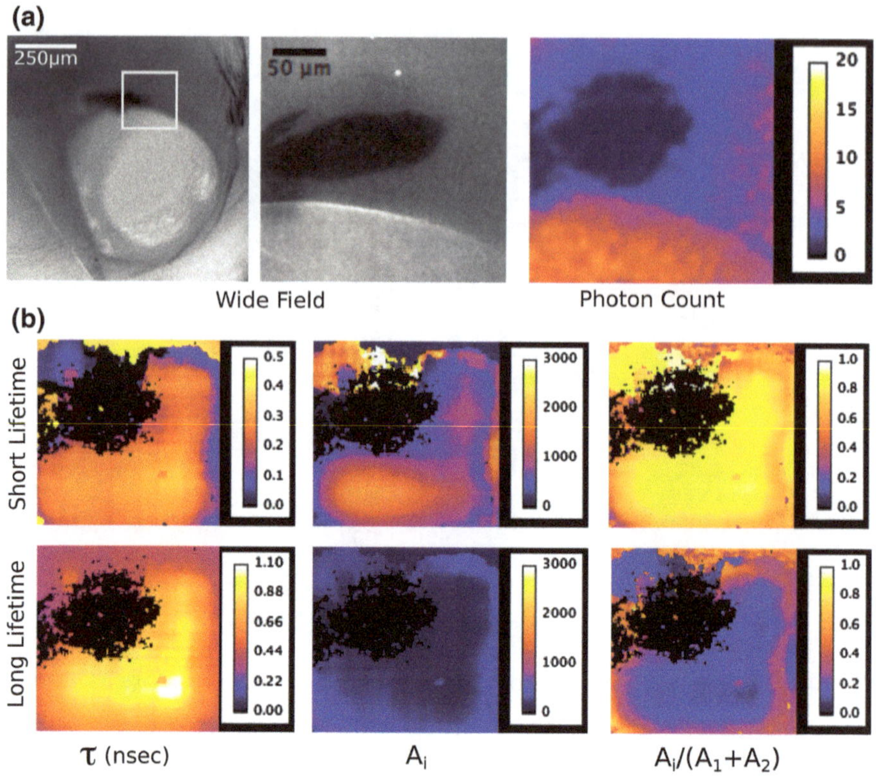

Fig. 24 FLIM analysis of the dsDNA-AlexaFluor647/MCH Au surface. **a** Wide-field fluorescence of the electrode surface showing the region analyzed, a zoomed in view, and the FLIM photon count from the same region, **b** the analyzed results from the two-photon FLIM measurements showing the lifetime distribution, the intensity, and fractional intensity for both short and long lifetime components. The *black region* in the images was not analyzed due to a lack of detected fluorescence photons

lifetime component varies across the electrode surface, with the longest lifetimes seen at the bottom right part of the image. These long lifetime fluorophores are evenly distributed across the interface comprising roughly 20% of the signal.

A comparison of the distributions of the lifetimes measured for the solution phase and surface bound species is given in Fig. 25. The solution species has a well-defined lifetime, as does the short lifetime component (which is assigned to the SAM). The longer lifetime component has a broad range of lifetimes, from 0.4 to 1.1 ns suggesting that not all the HS-C6-DNA-AlexaFluor647 is bonded to the surface in a manner expected for a thiol SAM, but rather adsorbed and positioned at various distances from the electrode surface resulting in the lifetime distribution. This appears to be evidence of non-specific adsorption, and its prevalence across the surface indicates that our efforts in removing this non-specifically adsorbed dsDNA were not sufficient to achieve a single lifetime distribution. This conclusion

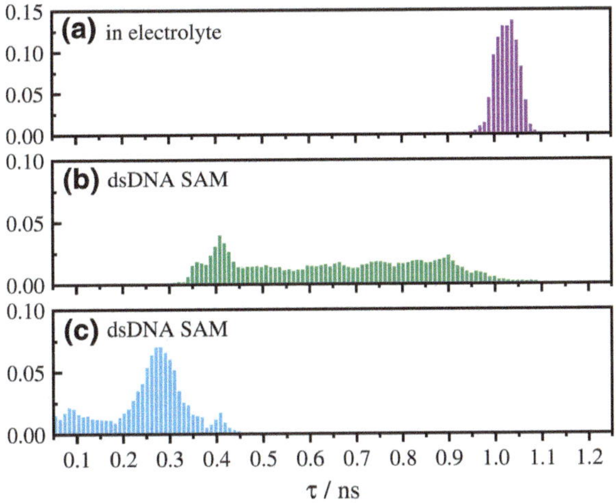

Fig. 25 The lifetime distributions measured from the FLIM analysis of **a** ssDNA-AlexaFluor647 in electrolyte, **b** long, and **c** short lifetime components of the SAM modified electrode

is really only possible by FLIM and would not be easily observed with wide-field fluorescence or through measurements that only provide an average for the surface.

Comparing FLIM lifetimes of AlexaFluor-labeled dsDNA SAMs with Theory. The lifetimes of dsDNA SAMs labeled with either AlexaFluor488 or AlexaFluor594 were also measured following the same procedure as for AlexaFluor647. These were compared with the lifetimes of the dsDNA in solution. The results from the data analysis are given in Table 3. The lifetimes of labeled dsDNA SAMs were best fit with bi-exponential decays using all the photons collected in the image around 520 and 620 nm, respectively. The AlexaFluor488 sample had little of the adsorbate in a long lifetime environment, while the AlexaFluor594 was more similar to the AlexaFluor647 SAM. The longer lifetime component was not a single value, but distributed over 1.25–1.75 ns for both AlexaFluor488 and AlexaFluor594. Lifetime measurements are able to distinguish these two components that are characteristic of the non-specifically adsorbed molecules and those that are bound to the surface. Comparing the short lifetime component with the theoretical lifetimes calculated (as discussed in Sect. 2) as a

Table 3 Lifetime fitting results for AlexaFluor488 and AlexaFluor594 in electrolyte and on the Au surface labeling the dsDNA SAM. The wavelength used for the two-photon excitation is indicated in brackets

		τ_1 (ns)	h	τ_2 (ns)
AlexaFluor488 (720 nm)	Solution	3.47 ± 0.03	1.00 ± 0.01	–
	SAM	1.36 ± 0.01	–	0.184 ± 0.001
AlexaFluor594 (780 nm)	Solution	3.75 ± 0.03	1.00 ± 0.02	–
	SAM	1.56 ± 0.01	–	0.304 ± 0.003

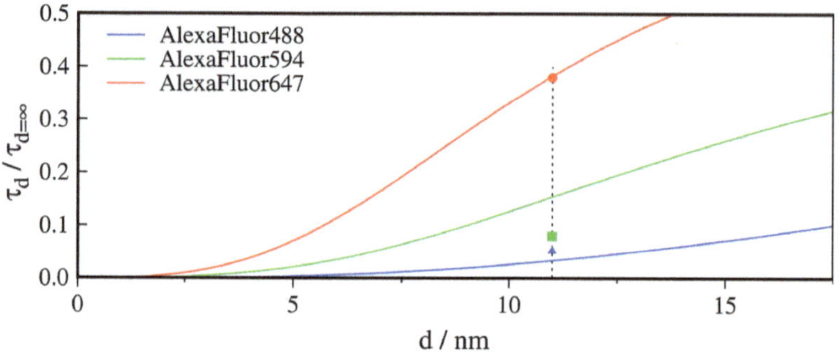

Fig. 26 Comparing the theoretical decrease in lifetime $\left(\text{as } \frac{\tau_d}{\tau_{d=\infty}}\right)$ for a dsDNA/MCH SAM on gold (*lines*) with the experimentally measured short lifetime values (*points*) for the three different fluorophore-labeled DNA SAMs. The metal/emitter distance used in the calculation is of 11 nm, corresponding to the maximum assumed length of the dsDNA

function of distance from the electrode surface shows a satisfying agreement for most of the fluorophores (Fig. 26). The separation from the electrode was assumed to correspond to the total length of the dsDNA SAM ($\sim$11 nm), and the lifetime data are shown as a fraction of the lifetime in solution $\left(\frac{\tau_{SAM}}{\tau_{solution}}\right)$. The AlexaFluor488 lifetime in solution was found to be smaller than that in the literature, so our experimental data were used for this comparison.

The theory and experiment agree for measurements on the dsDNA-AlexaFluor647. The lifetime for AlexaFluor594 and AlexaFluor488 is not as close. The measured lifetime for the dsDNA-AlexaFluor488 SAM is close to the minimum that can be measured after deconvoluting the IRF, and so we are probably over-estimating the ratio. The agreement is not ideal for AlexaFluor594 with the experimental value much smaller than the theoretical prediction. This suggests that the fluorophore may not be as far from the surface as estimated, or that an additional quenching process is active. Further work is required to determine the source of this discrepancy. This multiple fluorophore approach using dsDNA to control the separation provides insight into the influence of the wavelength range on the quenching efficiency of the fluorophores. Overall, given a well-defined surface structure with a highly controlled and characterized adsorbate, the distance of a fluorophore tethered to a dsDNA strand can be positioned accurately from the electrode surface so as to test the theoretical predictions of fluorescence quenching near a metal surface. The dsDNA SAM approach to controlling the nanometric construction of the interface if carefully created would enable measurements up to a maximum separation determined by the stiffness of the dsDNA strand, which could be as long as 50 nm [163].

7 Conclusions and Outlook

The electrochemical interface represents a complex challenge for characterization and for investigating the influence of potential on the interfacial behavior. The modification of these surfaces using molecular-based thiol self-assembly now underlies many related technologies and requires a variety of methods for characterizing the assembly, uniformity, and molecular environment. Labeling these adsorbates with a fluorophore allows for fluorescence microscopy to become a useful method for such study. We have shown that the fluorescence intensity and lifetimes can be used to quantify the behavior of the adsorbed molecules during potential perturbation. Investigating the homogeneity of the surface coating and the influence of surface crystallography on the resulting modification was demonstrated. The kinetics of charge-induced reorientation of DNA adsorbed onto the surface was measured using fluorescence microscopy coupled with frequency response analysis. Preliminary results of lifetime imaging measurements on DNA modified electrodes using three different fluorophores were compared to theoretical calculations, showing close agreement for some fluorophores. Theory also provided an indication of the influence of the measurement conditions, specifically the numerical aperture of the objective, on the magnitude of the fluorescence intensity when the fluorophore is separated from the surface within the first 200 nm. Lifetime imaging was able to reveal two lifetime components in the DNA SAMs, strongly suggesting the presence of non-specifically adsorbed molecules and informing on the local molecular environment.

Further developments hold the promise of characterizing more complex interfacial architectures such as multilayer structures formed by sequential chemical modification. These interfaces will require laterally resolved studies of uniformity as well as a measure of structure in the direction normal to the surface. Advances in method development will enable the use of fluorescence energy transfer between acceptors and donors in multicomponent monolayers and the use of confocal microscopy to map these molecular environments. A wide variety of advanced fluorescence methods developed for biological studies can be adapted for use in the study of modified electrochemical interfaces with consideration of the metal-mediated quenching. Therefore, many opportunities exist for expanding this methodology to address specific questions about the influence of potential on the modified electrode surface.

Acknowledgements The authors would like to express their gratitude to the Advanced Materials and Processing Engineering Laboratory (AMPEL), the Mechanical and Electronic Shops in the Department of Chemistry (UBC), and Laboratory for Advanced Spectroscopy and Imaging Research (LASIR) for continued support of the development of this methodology. We would also like to acknowledge the assistance by Dr. S. Kamal in using and adapting the two-photon FLIM microscope for use with the spectroelectrochemical cell. The HS-C10-BODIPY and HS-Aib4-BODIPY molecules were gratefully provided by Prof. M. Workentin (Western University) and Prof. F. Maran (University of Padova), respectively. We would also like to thank Prof C. Buess-Herman and Dr. T. Doneux of the Université Libre de Bruxelles for very helpful discussions and collaboration on method development. Funding for this work was provided by

NSERC (Canada) through the Discovery Grant and RTI programs. JCM benefited from a scholarship by CONACYT (Mexico) for his graduate studies.

Appendix

Glassblowing instruction for making the spectroelectrochemical cells. The process developed in the UBC chemistry glass shop to create the cells involves a borofloat window of the appropriate diameter and the tubing to match. The addition of ports and the location of the ports are added as requested. Most of the cells manufactured have 4 ports around the main or top joint, with a further 3–4 on the cell body. The ports on the cell body were straight through or ring seals with stems and stopcocks to direct material to the window at the bottom of the cell.

The first step is the manufacture of the cell bodies with the main or top joint (standard taper 14/23 socket). These have been made primarily from 38-mm borosilicate tubing as this diameter matches the 1.5″ windows. Cells have been made from (1″) 25.5 mm to (2″) 50 mm with (1.5″) 38 mm proving to be the most resilient. The glass joints are sealed to the tubing with an effort to maintain a shoulder for the positioning of future ports. The main body of the cell will then be pulled to a point leaving tubing to the length requested. This step is usually repeated several times to maintain a "stock" of cell bodies in preparation for sealing the windows onto the bottom of the tubing.

The bottom of the main bodies of the cells are cut using a wet saw and then washed with tap water to remove any excess grit. The cell bodies are then ready for polishing using diamond pads. Beginning with 200 grit, the bottom of the cell is polished and then rinsed with tap water. Next diamond pads with 400 grit, 500 grit, 800 grit, and finally 1500 grit were used. This is done in an effort to allow the window to seal to the cell body using less heat.

Sealing the windows onto the cell body is done with the use of a glass lathe, with the cell body in a holder using the 14/23 joint at the head stock and a 30-mm carbon rod held in the tail stock. A window is placed between the cell body and the carbon rod. The window is then sandwiched between the carbon and glass cell body and set to slowly rotate with a hand torch gently heating the bottom of the cell body and window. The torch tip used on the national hand torch is size one, which allows for a very sharp, tight fire. The fire needs to be quickly adjustable from a gassy soft fire to a strong sharp oxygen rich fire as the window and cell body warm up and are ready to be sealed.

Using a carbon paddle to ensure that the window remains in contact with the cell body, the hand torch fire is sharpened and sealing the window is started. Once a light tacking of the window occurs with the cell body, the carbon rod is withdrawn a short distance from the window face. If the carbon rod remains too long on the window, the window will be distorted and conform to the carbon rod.

Fusing the window to the cell body can begin with special regard for the wash of the torch flame. Any incidental heat can and will melt the window which can be heard flexing from the heat at this point in the procedure. The polishing done to the bottom of the cell body facilitates a quick seal between the window and cell body.

Once the window is attached, the cell is placed in a hot oven to anneal. This is usually the time that stress can cause the window to shatter; every effort needs to be taken to protect the new seal from sharp changes in temperature. These seals can be so sensitive that the barometric pressure affects the likelihood of survival; this is a speculative explanation for windows failing to survive the experience of being sealed onto the cell.

Once the windows have been annealed to 565 °C, they are very resilient to temperature changes. This allows the ports to be sealed on the cell usually beginning from the top ports around the main joint and then down the cell body to the window. The only special care that needs to be taken is to protect the window from blunt force.

References

1. Kolb, D.M.: UV-visible reflectance spectroscopy. In: R.J. Gale (ed.) Spectroelectrochemistry. Plenum Press (1988). URL http://www.worldcat.org/title/spectroelectrochemistry/oclc/18739315 (Chapter 4)
2. McIntyre, J.D.E.: Specular reflection spectroscopy of the electrode-solution interphase. In: Muller, R.H. (ed.) Advances in Electrochemistry and Electrochemical Engineering, pp. 61–166. Wiley-Inerscience, New York (1973)
3. Sagara, T.: UV-visible reflectance spectroscopy of thin organic films at electrode surfaces. In: Advances in Electrochemical Science and Engineering, vol. 9, pp. 47–95. Wiley-VCH Verlag GmbH, Weinheim, Germany (2006). doi:10.1002/9783527616817.ch2. URL http://doi.wiley.com/10.1002/9783527616817.ch2
4. Sagara, T., Igarashi, S., Sato, H., Niki, K.: Voltammetric application of electromodulated electroreflection absorption spectroscopy: electroreflectance voltammetry as an in situ spectroelectrochemical technique. Langmuir **7**(5), 1005–1012 (1991). doi:10.1021/la00053a032
5. Nagatani, H., Sagara, T.: Potential-modulation spectroscopy at solid/liquid and liquid/liquid interfaces. Anal. Sci. **23**(9), 1041–1048 (2007). doi:10.2116/analsci.23.1041
6. Sagara, T., Hiasa, H., Nakashima, N.: An electroreflectance study at the bottom-surface of a mercury drop electrode placed on an underlying nafion film. Chem. Lett. **8**, 783–784 (1998)
7. Sagara, T., Kawamura, H., Nakashima, N.: Electrode reaction of methylene blue at an alkanethiol-modified gold electrode as characterized by electroreflectance spectroscopy. Langmuir (1996). doi:10.1021/la951530p
8. Sagara, T., Zamlynny, V., Bizzotto, D., McAlees, A., McCrindle, R., Lipkowski, J.: Spectroelectrochemical investigations of the spreading of 4-pentadecyl pyridine onto the Au (111) electrode. Israel J. Chem. **37**(2–3), 197–211 (2013). doi:10.1002/ijch.199700024. URL http://doi.wiley.com/10.1002/ijch.199700024
9. Sun, S.G., Christensen, P.A., Wieckowski, A.: In-situ Spectroscopic Studies of Adsorption at the Electrode and Electrocatalysis. Elsevier Science (2011)
10. Rosendahl, S.M., Danger, B.R., Vivek, J.P., Burgess, I.J.: Surface enhanced infrared absorption spectroscopy studies of DMAP adsorption on gold surfaces. Langmuir **25**(4), 2241–2247 (2009). doi:10.1021/la803404u

11. Uchida, T., Osawa, M., Lipkowski, J.: SEIRAS studies of water structure at the gold electrode surface in the presence of supported lipid bilayer. J. Electroanal. Chem. **716**, 112–119 (2014). doi:10.1016/j.jelechem.2013.10.015. URL http://linkinghub.elsevier.com/retrieve/pii/S1572665713004645
12. Hoon-Khosla, M., Fawcett, W.R., Chen, A., Lipkowski, J., Pettinger, B.: A SNIFTIRS study of the adsorption of pyridine at the Au(111) electrode-solution interface. Electrochim. Acta **45**(4), 611–621 (1999)
13. Iwasita, T., Nart, F.C.: In situ infrared spectroscopy at electrochemical interfaces [review]. Prog. Surf. Sci. **55**(4), 271–340 (1997)
14. Leitch, J.J., Collins, J., Friedrich, A.K., Stimming, U., Dutcher, J.R., Lipkowski, J.: Infrared studies of the potential controlled adsorption of sodium dodecyl sulfate at the Au(111) electrode surface. Langmuir **28**(5), 2455–2464 (2012). doi:10.1021/la204451s
15. Pensa, E., Vericat, C., Grumelli, D., Salvarezza, R.C., Park, S.H., Longo, G.S., Szleifer, I., Méndez De Leo, L.P.: New insight into the electrochemical desorption of alkanethiol SAMs on gold. Phys. Chem. Chem. Phys. **14**(35), 12,355–12,367 (2012). doi:10.1039/c2cp41291h
16. Zamlynny, V., Lipkowski, J.: Quantitative SNIFTIRS and PM IRRAS of organic molecules at electrode surfaces. In: Advances in Electrochemical Science and Engineering, pp. 315–376. Wiley-VCH Verlag GmbH (2008). doi:10.1002/9783527616817.ch9
17. Zamlynny, V., Zawisza, I., Lipkowski, J.: PM FTIRRAS studies of potential-controlled transformations of a monolayer and a bilayer of 4-pentadecylpyridine, a model surfactant, adsorbed on a Au (111) electrode surface. Langmuir **19**(1), 132–145 (2003). doi:10.1021/la026488u
18. Zawisza, I., Bin, X., Lipkowski, J.: Spectroelectrochemical studies of bilayers of phospholipids in gel and liquid state on Au(111) electrode surface. Bioelectrochemistry **63**(1–2), 137–147 (2004). doi:10.1016/j.bioelechem.2003.12.004
19. Dahlin, A.B., Dielacher, B., Rajendran, P., Sugihara, K., Sannomiya, T., Zenobi-Wong, M., Voros, J.: Electrochemical plasmonic sensors. Anal. Bioanal. Chem. **402**(5), 1773–1784 (2012). doi:10.1007/s00216-011-5404-6
20. den Engelsen, D., de Koning, B.: Ellipsometric study of organic monolayers. Part 1—condensed monolayers. J. Chem. Soc. Faraday Trans. 1: Phys. Chem. Condens. Phases **70** (0), 1603–1614 (1974). doi:10.1039/F19747001603
21. Shan, X., Patel, U., Wang, S., Iglesias, R., Tao, N.: Imaging local electrochemical current via surface plasmon resonance. Science **327**(5971), 1363–1366 (2010). doi:10.1126/science.1186476
22. Wang, Y., Shan, X., Cui, F., Li, J., Wang, S., Tao, N.: Electrochemical reactions in subfemtoliter-droplets studied with plasmonics-based electrochemical current microscopy. Anal. Chem. **87**(1), 494–498 (2015). doi:10.1021/ac5036692
23. Yu, Y., Jin, G.: Influence of electrostatic interaction on fibrinogen adsorption on gold studied by imaging ellipsometry combined with electrochemical methods. J. Colloid Interface Sci. **283**(2), 477–481 (2005). doi:10.1016/j.jcis.2004.09.021
24. Majewski, J., Smith, G.S., Burgess, I., Zamlynny, V., Szymanski, G., Lipkowski, J., Satija, S.: Neutron reflectivity studies of electric field driven structural transformations of surfactants. Appl. Phys. Mater. Sci. Process. **74**, S364–S367 (2002)
25. Zamlynny, V., Burgess, I., Szymanski, G., Lipkowski, J., Majewski, J., Smith, G., Satija, S., Ivkov, R.: Electrochemical and neutron reflectivity studies of spontaneously formed amphiphilic surfactant bilayers at the gold-solution interface. Langmuir **16**(25), 9861–9870 (2000)
26. Grubb, M., Wackerbarth, H., Wengel, J., Ulstrup, J.: Direct imaging of hexaamine-ruthenium(III) in domain boundaries in monolayers of single-stranded DNA. Langmuir **23**(3), 1410–1413 (2007). doi:10.1021/la062555z
27. Hiasa, T., Onishi, H.: Mercaptohexanol assembled on gold: FM-AFM imaging in water. Colloids Surf. Physicochem. Eng. Aspects **441**, 149–154 (2014). doi:10.1016/j.colsurfa.2013.09.002

28. Josephs, E.A., Ye, T.: A single-molecule view of conformational switching of DNA tethered to a gold electrode. J. Am. Chem. Soc. **134**(24), 10021–10030 (2012). doi:10.1021/ja3010946
29. Lei, S., Feyter, S.: STM, STS and bias-dependent imaging on organic monolayers at the solid-liquid interface. In: P. Samori (ed.) STM and AFM studies on (bio)molecular systems: unravelling the nanoworld, pp. 269–312. Springer Berlin Heidelberg (2008). doi:10.1007/128_2007_23
30. Li, M., Chen, M., Sheepwash, E., Brosseau, C.L., Li, H., Pettinger, B., Gruler, H., Lipkowski, J.: AFM studies of solid-supported lipid bilayers formed at a Au(111) electrode surface using vesicle fusion and a combination of Langmuir-Blodgett and Langmuir-Schaefer techniques. Langmuir **24**(18), 10313–10323 (2008). doi:10.1021/la800800m
31. Li, W.H., Haiss, W., Floate, S., Nichols, R.J.: In-situ infrared spectroscopic and scanning tunneling microscopy investigations of the chemisorption phases of uracil, thymine, and 3-methyl uracil on Au (111) electrodes. Langmuir **15**(14), 4875–4883 (1999). doi:10.1021/la9815594
32. Nichols, R., Haiss, W., Fernig, D., Zalinge, H., Schiffrin, D., Ulstrup, J.: In situ STM studies of immobilized biomolecules at the electrodeelectrolyte interface. In: Bioinorganic Electrochemistry, pp. 207–247. Springer Netherlands (2008). URL http://dx.doi.org/10.1007/978-1-4020-6500-2_7
33. Pignataro, B.: Advances in SPMs for Investigation and Modification of Solid-Supported Monolayers. In: M. Tomitori, B. Bhushan, H. Fuchs (eds.) Applied Scanning Probe Methods IX, pp. 55–88. Springer Berlin Heidelberg (2008). doi:10.1007/978-3-540-74083-4_3
34. Vericat, C., Andreasen, G., Vela, M.E., Martin, H., Salvarezza, R.C.: Following transformation in self-assembled alkanethiol monolayers on Au(111) by in situ scanning tunneling microscopy. J. Chem. Phys. **115**(14), 6672–6678 (2001). doi:10.1063/1.1403000
35. Wano, H., Uosaki, K.: In situ, real-time monitoring of the reductive desorption process of self-assembled monolayers of hexanethiol on Au (111) surfaces in acidic and alkaline aqueous solutions by scanning tunneling microscopy. Langmuir **17**(26), 8224–8228 (2001). doi:10.1021/la010990h
36. Xu, S., Chen, M., Cholewa, E., Szymanski, G., Lipkowski, J.: Electric-field-driven surface aggregation of a model zwitterionic surfactant. Langmuir **23**(13), 6937–6946 (2007). doi:10.1021/la0701327
37. Cialla, D., März, A., Böhme, R., Theil, F., Weber, K., Schmitt, M., Popp, J.: Surfaceenhanced Raman spectroscopy (SERS): progress and trends. Anal. Bioanal. Chem. **403**(1), 27–54 (2011). doi:10.1007/s00216-011-5631-x
38. Cortés, E., Etchegoin, P.G., Le Ru, E.C., Fainstein, A., Vela, M.E., Salvarezza, R.C.: Monitoring the electrochemistry of single molecules by surface-enhanced Raman spectroscopy. J. Am. Chem. Soc. **132**(51), 18034–18037 (2010). doi:10.1021/ja108989b
39. Itoh, T., McCreery, R.L.: In situ Raman spectroelectrochemistry of electron transfer between glassy carbon and a chemisorbed nitroazobenzene monolayer. J. Am. Chem. Soc. **124**(36), 10894–10902 (2002). doi:10.1021/ja020398u
40. Oklejas, V., Harris, J.M.: In-situ investigation of binary-component self-assembled monolayers: a SERS-based spectroelectrochemical study of the effects of monolayer composition on interfacial structure. Langmuir **19**(14), 5794–5801 (2003). doi:10.1021/la020916e
41. Tian, Z.Q., Ren, B.: Adsorption and reaction at electrochemical interfaces as probed by surface-enhanced raman spectroscopy. Ann. Rev. Phys. Chem. 55(1), 197–229 (2004). doi:10.1146/annurev.physchem.54.011002.103833
42. Van Duyne, R.P.: Applications of Raman spectroscopy in electrochemistry. J. Phys. Colloques **38**, C5–239–C5–252 (1977). doi:10.1051/jphyscol:1977531

43. Wu, D.Y., Li, J.F., Ren, B., Tian, Z.Q.: Electrochemical surface-enhanced Raman spectroscopy of nanostructures. Chem. Soc. Rev. **37**(5), 1025–1041 (2008). doi:10.1039/B707872M
44. Wen, R., Lahiri, A., Azhagurajan, M., Kobayashi, S., Itaya, K.: A new in situ optical microscope with single atomic layer resolution for observation of electrochemical dissolution of Au (111). J. Am. Chem. Soc (2010). doi:10.1021/ja106231x
45. Dias, M., Hudhomme, P., Levillain, E., Perrin, L., Sahin, Y., Sauvage, F., Wartelle, C.: Electrochemistry coupled to fluorescence spectroscopy: a new versatile approach. Electrochem. Commun. **6**(3), 325–330 (2004). doi:10.1016/j.elecom.2004.01.010
46. Engstrom, R.C., Ghaffari, S., Qu, H.: Fluorescence imaging of electrode-solution interfacial processes. Anal. Chem. **64**(21), 2525–2529 (1992). doi:10.1021/ac00045a012
47. Ghaly, T., Wildt, B.E., Searson, P.C.: Electrochemical release of fluorescently labeled thiols from patterned gold surfaces. Langmuir **26**(3), 1420–1423 (2010). doi:10.1021/la9032282
48. Li, Meuse: C., Silin, V., Gaigalas, A.K., Zhang, Y.Z.: Application of electromodulated fluorescence to the study of the dynamics of Alexa 488 fluorochrome immobilized on a gold electrode. Langmuir **16**(10), 4672–4677 (2000). doi:10.1021/la991192i
49. Li, L., Ruzgas, T., Gaigalas, A.K.: Fluorescence from Alexa 488 fluorophore immobilized on a modified gold electrode. Langmuir **15**(19), 6358–6363 (1999). doi:10.1021/la981704d
50. Miomandre, F., Allain, C., Clavier, G., Audibert, J.F., Pansu, R.B., Audebert, P., Hartl, F.: Coupling thin layer electrochemistry with epifluorescence microscopy: an expedient way of investigating electrofluorochromism of organic dyes. Electrochem. Commun. **13**(6), 574–577 (2011). doi:10.1016/j.elecom.2011.03.013
51. Plummer, S.T., Bohn, P.W.: Spatial dispersion in electrochemically generated surface composition gradients visualized with covalently bound fluorescent nanospheres. Langmuir **18**(10), 4142–4149 (2002). doi:10.1021/la011742o
52. Pope, J.M., Tan, Z., Kimbrell, S.: Measurement of electric fields at rough metal surfaces by electrochromism of fluorescent probe molecules embedded in self-assembled monolayers. J. Am. Chem. Soc. **114**(25), 10085–10086 (1992). doi:10.1021/ja00051a065
53. Novotny, L., Hecht, B.: Principles of Nano-Optics. Cambridge University Press (2006)
54. Loudon, R.: The quantum theory of light. Clarendon Press (1973)
55. Milonni, P.W.: Semiclassical and quantum electrodynamical approaches in nonrelativistic radiation theory (1976)
56. Ford, G.W., Weber, W.H.: Electromagnetic interactions of molecules with metal surfaces. Phys. Rep. **113**(4), 195–287 (1984). doi:10.1016/0370-1573(84)90098-x
57. Chance, R.R., Prock, A., Silbey, R.: Lifetime of an emitting molecule near a partially reflecting surface. J. Chem. Phys. **60**(7), 2744–2748 (1974)
58. Barnes, W.L.: Fluorescence near interfaces: the role of photonic mode density. J. Mod. Opt. **45**(4), 661–699 (1998)
59. Chance, R.R., Prock, A., Silbey, R.: Molecular fluorescence and energy transfer near interfaces. In: Advances in Chemical Physics, pp. 1–65. John Wiley & Sons, Inc. (1978). doi:10.1002/9780470142561.ch1
60. Novotny, L.: Allowed and forbidden light in near-field optics. I. A single dipolar light source. J. Opt. Soc. Am. A **14**(1), 91–104 (1997). doi:10.1364/JOSAA.14.000091
61. Bizzotto, D., Lipkowski, J.: Electrochemical and spectroscopic studies of the mechanism of monolayer and multilayer adsorption of an insoluble surfactant at the Au(111)| electrolyte interface. J. Electroanal. Chem. **409**(1–2), 33–43 (1996). doi:10.1016/0022-0728(96)04537-8
62. Chung, D.S., Alkire, R.C.: Confocal microscopy for simultaneous imaging of Cu electrodeposit morphology and adsorbate fluorescence. J. Electrochem. Soc. **144**, 1529–1536 (1997)
63. Pope, J.M., Buttry, D.A.: Measurements of the potential dependence of electric field magnitudes at an electrode using fluorescent probes in a self-assembled monolayer. J. Electroanal. Chem. 498(1–2), 75–86 (2001). doi:10.1016/S0022-0728(00)00268-0

64. Bizzotto, D., Pettinger, B.: Fluorescence imaging studies of the electrochemical adsorption/desorption of octadecanol. Langmuir **15**(23), 8309–8314 (1999). doi:10.1021/la990249y
65. Bizzotto, D., Shepherd, J.L.: EPI-fluorescence microscopy studies of potential controlled changes in adsorbed thin organic films at electrode surfaces. In: Advances in Electrochemical Science and Engineering, pp. 97–126. Wiley-VCH Verlag GmbH, Weinheim, Germany (2006). doi:10.1002/9783527616817.ch3
66. Inoué, S.: Foundations of confocal scanned imaging in light microscopy. In: Handbook of Biological Confocal Microscopy, pp. 1–19. Springer US, Boston, MA (2006). doi:10.1007/978-0-387-45524-2_1
67. Amaro, M., Sachl, R., Jurkiewicz, P., Coutinho, A., Prieto, M., Hof, M.: Time-resolved fluorescence in lipid bilayers: selected applications and advantages over steady state. Biophys. J. **107**(12), 2751–2760 (2014). doi:10.1016/j.bpj.2014.10.058
68. Bastiaens, P.I.H., Squire, A.: Fluorescence lifetime imaging microscopy: spatial resolution of biochemical processes in the cell. Trends Cell Biol. **9**(2), 48–52 (1999). doi:10.1016/S0962-8924(98)01410-X
69. Becker, W.: Fluorescence lifetime imaging—techniques and applications. J. Microsc. **247** (2), 119–136 (2012). doi:10.1111/j.1365-2818.2012.03618.x
70. Berezin, M.Y., Achilefu, S.: Fluorescence lifetime measurements and biological imaging. Chem. Rev. **110**(5), 2641–2684 (2010). doi:10.1021/cr900343z
71. Barber, P.R., Ameer-Beg, S.M., Gilbey, J., Carlin, L.M., Keppler, M., Ng, T.C., Vojnovic, B.: Multiphoton time-domain fluorescence lifetime imaging microscopy: practical application to protein-protein interactions using global analysis. J. R. Soc. Interface **6**(Suppl 1), S93–S105 (2009). doi:10.1098/rsif.2008.0451.focus
72. Stockl, M.T., Herrmann, A.: Detection of lipid domains in model and cell membranes by fluorescence lifetime imaging microscopy. Biochim. Biophys. Acta BBA Biomembr. **1798** (7), 1444–1456 (2010). doi:10.1016/j.bbamem.2009.12.015
73. Pawley, J.B.: Fundamental limits in confocal microscopy. In: Handbook of Biological Confocal Microscopy, pp. 20–42. Springer US, Boston, MA (2006). doi:10.1007/978-0-387-45524-2_2
74. Ustione, A., Piston, D.W.: A simple introduction to multiphoton microscopy. J. Microsc. **243**(3), 221–226 (2011). doi:10.1111/j.1365-2818.2011.03532.x
75. Denk, W., Piston, D.W., Webb, W.W.: Multi-photon molecular excitation in laser-scanning microscopy. In: Handbook of Biological Confocal Microscopy, pp. 535–549. Springer US, Boston, MA (2006). doi:10.1007/978-0-387-45524-2_28
76. Drobizhev, M., Makarov, N.S., Tillo, S.E., Hughes, T.E., Rebane, A.: Two-photon absorption properties of fluorescent proteins. Nat. Methods **8**(5), 393–399 (2011). doi:10.1038/nmeth.1596
77. Bigelow,W.C., Pickett, D.L., Zisman,W.A.: Oleophobic monolayers: I. Films adsorbed from solution in non-polar liquids. J. Colloid Sci. **1**(6), 513–538 (1946). doi:10.1016/0095-8522(46)90059-1
78. Bain, C.D., Troughton, E.B., Tao, Y.T., Evall, J., Whitesides, G.M., Nuzzo, R.G.: Formation of monolayer films by the spontaneous assembly of organic thiols from solution onto gold. J. Am. Chem. Soc. **111**(1), 321–335 (1989). doi:10.1021/ja00183a049
79. Folkers, J.P., Zerkowski, J.A., Laibinis, P.E., Seto, C.T., Whitesides, G.M.: Designing ordered molecular arrays in two and three dimensions. In: Supramolecular Architecture, pp. 10–23. American Chemical Society (1992). doi:10.1021/bk-1992-0499.ch002
80. Li, Z., Chang, S.C., Williams, R.S.: Self-assembly of alkanethiol molecules onto platinum and platinum oxide surfaces. Langmuir **19**(17), 6744–6749 (2003). doi:10.1021/la034245b
81. Love, J.C., Estroff, L.A., Kriebel, J.K., Nuzzo, R.G., Whitesides, G.M.: Self-assembled monolayers of Thiolates on metals as a form of nanotechnology. Chem. Rev. **105**(4), 1103–1170 (2005). doi:10.1021/cr0300789
82. Nuzzo, R.G., Allara, D.L.: Adsorption of bifunctional organic disulfides on gold surfaces. J. Am. Chem. Soc. **105**(13), 4481–4483 (1983). doi:10.1021/ja00351a063

83. Strong, L., Whitesides, G.M.: Structures of self-assembled monolayer films of organosulfur compounds adsorbed on gold single crystals: electron diffraction studies. Langmuir **4**(3), 546–558 (1988). doi:10.1021/la00081a009
84. Hickman, J.J., Ofer, D., Laibinis, P.E., Whitesides, G.M., Wrighton, M.S.: Molecular self-assembly of two-terminal, voltammetric microsensors with internal references. Science **252**(5006), 688–691 (1991). doi:10.1126/science.252.5006.688
85. Schierbaum, K.D., Weiss, T., van Veizen, E.U.T., Engbersen, J.F.J., Reinhoudt, D.N., Gopel, W.: Molecular recognition by self-assembled monolayers of cavitand receptors. Science **265**(5177), 1413–1415 (1994). doi:10.1126/science.265.5177.1413
86. Chaki, N.K., Vijayamohanan, K.: Self-assembled monolayers as a tunable platform for biosensor applications. Biosens. Bioelectron. **17**(1–2), 1–12 (2002). doi:10.1016/S0956-5663(01)00277-9
87. Akiba, U, Fujihira, M.: Preparation of self-assembled monolayers (SAMs) on Au and Ag. In: Encyclopedia of Electrochemistry. Wiley-VCH Verlag GmbH & Co. KGaA (2007). doi:10.1002/9783527610426.bard100121
88. Jamison, A.C., Chinwangso, P., Lee, T.R.: Self-assembled monolayers: the development of functional nanoscale films. In: Functional Polymer Films, pp. 151–217. Wiley-VCH Verlag GmbH & Co. KGaA (2011). doi:10.1002/9783527638482.ch5
89. Schreiber, F.: Structure and growth of self-assembling monolayers. Prog. Surf. Sci. **65**(5–8), 151–257 (2000). doi:10.1016/S0079-6816(00)00024-1
90. Schreiber, F.: Self-assembled monolayers: from 'simple' model systems to biofunctionalized interfaces. J. Phys. Condensed Matter **16**(28), R881 (2004). doi:10.1088/0953-8984/16/28/R01
91. Finklea, H.O.: Self-assembled monolayers on electrodes. In: Encyclopedia of Analytical Chemistry. John Wiley & Sons, Ltd. (2006). doi:10.1002/9780470027318.a5315
92. Lakowicz, J.R.: Radiative decay engineering 5: metal-enhanced fluorescence and plasmon emission. Anal. Biochem. **337**(2), 171–194 (2005). doi:10.1016/j.ab.2004.11.026
93. Finklea, H.O.: Electrochemistry of organized monolayers of thiols and related molecules on electrodes. Electroanal. Chem. Ser. Adv. **19**(19), 109–335 (1996)
94. Schneider, T.W., Buttry, D.A.: Electrochemical quartz crystal microbalance studies of adsorption and desorption of self-assembled monolayers of alkyl thiols on gold. J. Am. Chem. Soc. **115**(26), 12391–12397 (1993). doi:10.1021/ja00079a021
95. Sun, K., Jiang, B., Jiang, X.: Electrochemical desorption of self-assembled monolayers and its applications in surface chemistry and cell biology. J. Electroanal. Chem. **656**(1–2), 223–230 (2011). doi:10.1016/j.jelechem.2010.11.008
96. Xu, J., Li, H.L.: The chemistry of self-assembled long-chain alkanethiol monolayers on gold. J. Colloid Interface Sci. **176**(1), 138–149 (1995). doi:10.1006/jcis.1995.0017
97. Yang, D.F., Al-Maznai, H., Morin, M.: Vibrational study of the fast reductive and the slow oxidative desorptions of a nonanethiol self-assembled monolayer from a Au(111) single crystal electrode. J. Phys. Chem. B **101**(7), 1158–1166 (1997). doi:10.1021/jp962247h
98. Zhong, C.J., Porter, M.D.: Fine structure in the voltammetric desorption curves of alkanethiolate monolayers chemisorbed at gold. J. Electroanal. Chem. **425**(1–2), 147–153 (1997). doi:10.1016/S0022-0728(96)04957-1
99. Doneux, T., Steichen, M., De Rache, A., Buess-Herman, C.: Influence of the crystallographic orientation on the reductive desorption of self-assembled monolayers on gold electrodes. J. Electroanal. Chem. **649**(1–2), 164–170 (2010). doi:10.1016/j.jelechem.2010.02.032
100. Doneux, T., Nichols, R.J., Buess-Herman, C.: Dissolution kinetics of octadecanethiolate monolayers electro-adsorbed on Au(111). J. Electroanal. Chem. **621**(2), 267–276 (2008). doi:10.1016/j.jelechem.2008.01.008

101. Doneux, T., Steichen, M., Bouchta, T., Buess-Herman, C.: Mixed self-assembled monolayers of 2-mercaptobenzimidazole and 2-mercaptobenzimidazole-5-sulfonate: determination and control of the surface composition. J. Electroanal. Chem. **599**(2), 241–248 (2007). doi:10.1016/j.jelechem.2006.03.006
102. Porter, M.D., Bright, T.B., Allara, D.L., Chidsey, C.E.D.: Spontaneously organized molecular assemblies. 4. Structural characterization of n-alkyl thiol monolayers on gold by optical ellipsometry, infrared spectroscopy, and electrochemistry. J. Am. Chem. Soc. **109** (12), 3559–3568 (1987). doi:10.1021/ja00246a011
103. Rez, P.R.I., Andreu, R., Calvente, J.J., Calzado, C.J., Rez, G.L.O.P.P.E.: Electrochemical formation and electron transfer through self-assembled monolayers of 4-mercaptophenol on mercury. J. Electroanal. Chem. **582**(1–2), 179–190 (2005). doi:10.1016/j.jelechem.2005.01.035
104. Yang, D.F., Wilde, C.P., Morin, M.: Electrochemical desorption and adsorption of nonyl mercaptan at gold single crystal electrode surfaces. Langmuir **12**(26), 6570–6577 (1996). doi:10.1021/la960365q
105. Walczak, M.M., Popenoe, D.D., Deinhammer, R.S., Lamp, B.D., Chung, C., Porter, M.D.: Reductive desorption of alkanethiolate monolayers at gold: a measure of surface coverage. Langmuir **7**(11), 2687–2693 (1991). doi:10.1021/la00059a048
106. Kunze, J., Leitch, J., Schwan, A.L., Faragher, R.J., Naumann, R., Schiller, S., Knoll, W., Dutcher, J.R., Lipkowski, J.: New method to measure packing densities of self-assembled thiolipid monolayers. Langmuir **22**(12), 5509–5519 (2006). doi:10.1021/la0535274
107. Laredo, T., Leitch, J., Chen, M., Burgess, I., Dutcher, J., Lipkowski, J.: Measurement of the charge number per adsorbed molecule and packing densities of self-assembled long-chain monolayers of thiols. Langmuir **23**(11), 6205–6211 (2007)
108. Choi, S., Chae, J.: Reusable biosensors via in situ electrochemical surface regeneration in microfluidic applications. Biosens. Bioelectron. **25**(2), 527–531 (2009). doi:10.1016/j.bios.2009.08.003
109. Kim, Y.R., Kim, H.J., Lee, M.H., Kang, Y.J., Yang, Y., Kim, H., Kim, J.S.: Electrochemically programmed chemodosimeter on ultrathin platinum films. Chem. Commun. **46**(44), 8448–8450 (2010). doi:10.1039/C0CC02528C
110. Orive, A.G., Grumelli, D., Vericat, C., Ramallo-Lopez, J.M., Giovanetti, L., Benitez, G., Azcarate, J.C., Corthey, G., Fonticelli, M.H., Requejo, F.G., Creus, A.H., Salvarezza, R.C.: "Naked" gold nanoparticles supported on HOPG: melanin functionalization and catalytic activity. Nanoscale **3**(4), 1708–1716 (2011). doi:10.1039/C0NR00911C
111. Inaba, R., Khademhosseini, A., Suzuki, H., Fukuda, J.: Electrochemical desorption of self-assembled monolayers for engineering cellular tissues. Biomaterials **30**(21), 3573–3579 (2009). doi:10.1016/j.biomaterials.2009.03.045
112. Mali, P., Bhattacharjee, N., Searson, P.C.: Electrochemically programmed release of biomolecules and nanoparticles. Nano Lett. **6**(6), 1250–1253 (2006). doi:10.1021/nl0609302
113. Yuan, M., Zhan, S., Zhou, X., Liu, Y., Feng, L., Lin, Y., Zhang, Z., Hu, J.: A method for removing self-assembled monolayers on gold. Langmuir **24**(16), 8707–8710 (2008). doi:10.1021/la800287e
114. Shepherd, J.L., Kell, A., Chung, E., Sinclar, C.W., Workentin, M.S., Bizzotto, D.: Selective Reductive Desorption of a SAM-Coated Gold Electrode Revealed Using Fluorescence Microscopy. Journal of the American Chemical Society **126**(26), 8329–8335 (2004). doi:10.1021/ja0494095
115. Casanova-Moreno, J.R., Bizzotto, D.: What happens to the Thiolates created by reductively desorbing SAMs? An in situ study using fluorescence microscopy and electrochemistry. Langmuir **29**(6), 2065–2074 (2013). doi:10.1021/la305170c
116. Trasatti, S.: Structuring of the solvent at metal/solution interfaces and components of the electrode potential. J. Electroanal. Chem. Interfac. Electrochem. **150**, 1–15 (1983). doi:10.1016/S0022-0728(83)80183-1

117. Clavilier, J., Faure, R., Guinet, G., Durand, R.: Preparation of monocrystalline Pt microelectrodes and electrochemical study of the plane surfaces cut in the direction of the 111 and 110 planes. J. Electroanal. Chem. Interfacial Electrochem. **107**(1), 205–209 (1979). doi:10.1016/S0022-0728(79) 80022-4
118. Bergström, F., Mikhalyov, I., Hägglöf, P., Wortmann, R., Ny, T., Johansson, L.B.A.: Dimers of dipyrrometheneboron difluoride (BODIPY) with light spectroscopic applications in chemistry and Biology. J. Am. Chem. Soc. **124**(2), 196–204 (2002). doi:10.1021/ja010983f
119. Tleugabulova, D., Zhang, Z., Brennan, J.D.: Characterization of bodipy dimers formed in a molecularly confined environment. J. Phys. Chem. B **106**(51), 13133–13138 (2002). doi:10.1021/jp027126y
120. Musgrove, A., Kell, A., Bizzotto, D.: Fluorescence imaging of the oxidative desorption of a BODIPY-alkyl-thiol monolayer coated Au bead. Langmuir **24**(15), 7881–7888 (2008). doi:10.1021/la800233c
121. Widrig, C.A., Chung, C., Porter, M.D.: The electrochemical desorption of n-alkanethiol monolayers from polycrystalline Au and Ag electrodes. J. Electroanal. Chem. Interfac. Electrochem. **310**(1–2), 335–359 (1991). doi:10.1016/0022-0728(91)85271-P
122. Poirier, G.E.: Coverage-dependent phases and phase stability of decanethiol on Au(111). Langmuir **15**(4), 1167–1175 (1999). doi:10.1021/la981374x
123. Gatto, E., Venanzi, M.: Self-assembled monolayers formed by helical peptide building blocks: a new tool for bioinspired nanotechnology. Polym. J. **45**, 468–480 (2013). doi:10.1038/pj.2013.27
124. Toniolo, C., Crisma, M., Formaggio, F., Peggion, C., Broxterman, Q.B., Kaptein, B.: Molecular spacers for physicochemical investigations based on novel helical and extended peptide structures. Biopolymers **76**(2), 162–176 (2004). doi:10.1002/bip.10575
125. Shin, Y.G.K., Newton, M.D., Isied, S.S.: Distance dependence of electron transfer across peptides with different secondary structures: the role of peptide energetics and electronic coupling. J. Am. Chem. Soc **125**(13), 3722–3732 (2003). doi:10.1021/ja020358q
126. Fabris, L., Antonello, S., Armelao, L., Donkers, R.L., Polo, F., Toniolo, C., Maran, F.: Gold nanoclusters protected by conformationally constrained peptides. J. Am. Chem. Soc. **128**(1), 326–336 (2006). doi:10.1021/ja0560581
127. Perera, N.V., Isley, W., Flavio, M., Gascón, J.A.: Molecular modeling characterization of a conformationally constrained monolayer-protected gold cluster. J. Phys. Chem. C **114**(38), 16043–16050 (2010). doi:10.1021/jp102585n
128. Kaplan, J.M., Shang, J., Gobbo, P., Antonello, S., Armelao, L., Chatare, V., Ratner, D.M., Andrade, R.B., Maran, F.: Conformationally constrained functional peptide monolayers for the controlled display of bioactive carbohydrate ligands. Langmuir **29**, 8187–8192 (2013). doi:10.1021/la4008894
129. Gatto, E., Porchetta, A., Scarselli, M., De Crescenzi, M., Formaggio, F., Toniolo, C., Venanzi, M.: Playing with peptides: how to build a supramolecular peptide nanostructure by exploiting helix···helix macrodipole interactions. Langmuir **28**(5), 2817–2826 (2012). doi:10.1021/la204423d
130. Gatto, E., Stella, L., Formaggio, F., Toniolo, C., Lorenzelli, L., Venanzi, M.: Electroconductive and photocurrent generation properties of self-assembled monolayers formed by functionalized, conformationally-constrained peptides on gold electrodes. J. Pept. Sci. **14**(2), 184–191 (2008). doi:10.1002/psc.973
131. Gatto, E., Venanzi, M., Palleschi, A., Stella, L., Pispisa, B., Lorenzelli, L., Toniolo, C., Formaggio, F., Marletta, G.: Self-assembled peptide monolayers on interdigitated gold microelectrodes. Mater. Sci. Eng. C **27**, 1309–1312 (2007). doi:10.1016/j.msec.2006.07.013
132. Pace, G., Venanzi, M., Castrucci, P., Scarselli, M., De Crescenzi, M., Palleschi, A., Stella, L., Formaggio, F., Toniolo, C., Marletta, G.: Static and dynamic features of a helical

hexapeptide chemisorbed on a gold surface. Mater. Sci. Eng. C **26**, 918–923 (2006). doi:10.1016/j.msec.2005.09.078
133. Venanzi, M., Pace, G., Palleschi, A., Stella, L., Castrucci, P., Scarselli, M., De Crescenzi, M., Formaggio, F., Toniolo, C., Marletta, G.: Densely-packed selfassembled monolayers on gold surfaces from a conformationally constrained helical hexapeptide. Surf. Sci. **600**, 409–416 (2006). doi:10.1016/j.susc.2005.10.040
134. Hamelin, A., Martins, A.M.: Cyclic voltammetry at gold single-crystal surfaces. Part 2. Behaviour of high-index faces. J. Electroanal. Chem. **407**(1–2), 13–21 (1996). doi:10.1016/0022-0728(95)04500-7
135. de Levie, R.: A simple empirical correlation between the potential of zero charge and the density of broken bonds. J. Electroanal. Chem. Interfac. Electrochem. **280**(1), 179–183 (1990). doi:10.1016/0022-0728(90)87093-Y
136. Yu, Z.L., Casanova-Moreno, J., Guryanov, I., Maran, F., Bizzotto, D.: Influence of surface structure on single or mixed component self-assembled monolayers via in situ spectroelectrochemical fluorescence imaging of the complete stereographic triangle on a single crystal Au bead electrode. J. Am. Chem. Soc. **137**(1), 276–288 (2015). doi:10.1021/ja5104475
137. Peggion, C., Crisma, M., Toniolo, C., Formaggio, F.: A solvent-dependent peptide spring unraveled by 2D-NMR. Tetrahedron **68**(23), 4429–4433 (2012)
138. Cheng, A.K.H., Sen, D., Yu, H.Z.: Design and testing of aptamer-based electrochemical biosensors for proteins and small molecules. Bioelectrochemistry **77**(1), 1–12 (2009). doi:10.1016/j.bioelechem.2009.04.007
139. Cosnier, S.: Electrochemical biosensors. Pan stanford series on the high-tech of biotechnology. Pan stanford (2015)
140. Drummond, T.G., Hill, M.G., Barton, J.K.: Electrochemical DNA sensors. Nat. Biotechnol. **21**(10), 1192–1199 (2003). doi:10.1038/nbt873
141. Fang, Z., Soleymani, L., Pampalakis, G., Yoshimoto, M., Squire, J.A., Sargent, E.H., Kelley, S.O.: Direct profiling of cancer biomarkers in tumor tissue using a multiplexed nanostructured microelectrode integrated circuit. ACS Nano **3**(10), 3207–3213 (2009). doi:10.1021/nn900733d
142. Ferguson, B.S., Hoggarth, D.A., Maliniak, D., Ploense, K., White, R.J., Woodward, N., Hsieh, K., Bonham, A.J., Eisenstein, M., Kippin, T.E., Plaxco, K.W., Soh, H.T.: Real-time, aptamer-based tracking of circulating therapeutic agents in living animals. Sci. Transl. Med. **5**(213), 213ra165–213ra165 (2013). doi:10.1126/scitranslmed.3007095
143. Grieshaber, D., MacKenzie, R., Voros, J., Reimhult, E.: Electrochemical biosensors—sensor principles and architectures. Sensors **8**(3), 1400–1458 (2008). doi:10.3390/s8031400
144. Li, C.Z., Long, Y.T., Sutherland, T., Lee, J.S., Kraatz, H.B.: Electronic biosensors based on DNA self-assembled monolayer on gold electrodes. In: Frontiers in Biochip Technology, pp. 274–291. Springer US, Boston (2006). doi:10.1007/0-387-25585-0_17
145. Li, N., Kerman, K.: Nanomaterial-based dual detection platforms: optics meets electrochemistry. In: Nanobiosensors and Nanobioanalyses, pp. 99–120. Springer Japan, Tokyo (2015). doi:10.1007/978-4-431-55190-4_6
146. Ronkainen, N.J., Halsall, H.B., Heineman, W.R.: Electrochemical biosensors. Chem. Soc. Rev. **39**(5), 1747–1763 (2010). doi:10.1039/B714449K
147. Wang, J.: Electrochemical biosensors: towards point-of-care cancer diagnostics. Biosens. Bioelectron. 21(10), 1887–1892 (2006). doi:10.1016/j.bios.2005.10.027
148. Wang, Y., Li, C., Li, X., Li, Y., Kraatz, H.B.: Unlabeled hairpin-DNA probe for the detection of single-nucleotide mismatches by electrochemical impedance spectroscopy. Anal. Chem. **80**(6), 2255–2260 (2008). doi:10.1021/ac7024688
149. Peterson, A.W., Heaton, R.J., Georgiadis, R.M.: The effect of surface probe density on DNA hybridization. Nucl. Acids Res. **29**(24), 5163–5168 (2001). doi:10.1093/nar/29.24.5163

150. Steel, A.B., Herne, T.M., Tarlov, M.J.: Electrochemical quantitation of DNA immobilized on gold. Anal. Chem. **70**(22), 4670–4677 (1998). doi:10.1021/ac980037q
151. Josephs, E.A., Ye, T.: Electric-field dependent conformations of single DNA molecules on a model biosensor surface. Nano Lett. **12**(10), 5255–5261 (2012). doi:10.1021/nl3024356
152. Josephs, E.A., Ye, T.: Nanoscale spatial distribution of thiolated DNA on model nucleic acid sensor surfaces. ACS Nano **7**(4), 3653–3660 (2013). doi:10.1021/nn400659m
153. Murphy, J.N., Cheng, A.K.H., Yu, H.Z., Bizzotto, D.: On the nature of DNA self-assembled monolayers on Au: measuring surface heterogeneity with electrochemical in situ fluorescence microscopy. J. Am. Chem. Soc. **131**(11), 4042–4050 (2009). doi:10.1021/ja808696p
154. Casanova-Moreno, J., Bizzotto, D.: A method for determining the actual rate of orientation switching of DNA self-assembled monolayers using optical and electrochemical frequency response analysis. Anal. Chem. **87**(4), 2255–2263 (2015). doi:10.1021/ac503919a
155. Casanova-Moreno, J.R., Bizzotto, D.: Frequency response analysis of potential-modulated orientation changes of a DNA self assembled layer using spatially resolved fluorescence measurements. Electrochim. Acta **162**, 62–71 (2015). doi:10.1016/j.electacta.2014.09.037
156. Kaiser, W., Rant, U.: Conformations of end-tethered DNA molecules on gold surfaces: influences of applied electric potential, electrolyte screening, and temperature. J. Am. Chem. Soc. **132**(23), 7935–7945 (2010). doi:10.1021/ja908727d
157. Langer, A., Hampel, P.A., Kaiser, W., Knezevic, J., Welte, T., Villa, V., Maruyama, M., Svejda, M., J a hner, S., Fischer, F., Strasser, R., Rant, U.: Protein analysis by time-resolved measurements with an electro-switchable DNA chip. Nat Commun **4** (2013). doi:10.1038/ncomms3099
158. Langer, A., Kaiser, W., Svejda, M., Schwertler, P., Rant, U.: Molecular dynamics of DNA-protein conjugates on electrified surfaces: solutions to the drift-diffusion equation. J. Phys. Chem. B **118**(2), 597–607 (2014). doi:10.1021/jp410640z
159. Rant, U., Arinaga, K., Tornow, M., Kim, Y.W., Netz, R.R., Fujita, S., Yokoyama, N., Abstreiter, G.: Dissimilar kinetic behavior of electrically manipulated single- and double-stranded DNA tethered to a gold surface. Biophys. J. **90**(10), 3666–3671 (2006). doi:10.1529/biophysj.105.078857
160. Knezevic, J., Langer, A., Hampel, P.A., Kaiser, W., Strasser, R., Rant, U.: Quantitation of affinity, avidity, and binding kinetics of protein analytes with a dynamically switchable biosurface. J. Am. Chem. Soc. **134**(37), 15225–15228 (2012). doi:10.1021/ja3061276
161. Rant, U., Arinaga, K., Scherer, S., Pringsheim, E., Fujita, S., Yokoyama, N., Tornow, M., Abstreiter, G.: Switchable DNA interfaces for the highly sensitive detection of label-free DNA targets. Proc. Natl. Acad. Sci. USA **104**(44), 17,364–17,369 (2007). doi:10.1073/pnas.0703974104
162. Rant, U., Pringsheim, E., Kaiser, W., Arinaga, K., Knezevic, J., Tornow, M., Fujita, S., Yokoyama, N., Abstreiter, G.: Detection and size analysis of proteins with switchable DNA layers. Nano Lett. **9**(4), 1290–1295 (2009). doi:10.1021/nl8026789
163. Baumann, C.G., Smith, S.B., Bloomfield, V.A., Bustamante, C.: Ionic effects on the elasticity of single DNA molecules. Proc. Natl. Acad. Sci. USA **94**(12), 6185–6190 (1997)
164. Brunet, A., Tardin, C., Salomé, L., Rousseau, P., Destainville, N., Manghi, M.: Dependence of DNA persistence length on ionic strength of solutions with monovalent and divalent salts: a joint theory-experiment study. Macromolecules **48**(11), 3641–3652 (2015). doi:10.1021/acs.macromol.5b00735
165. Yellen, G., Mongeon, R.: Quantitative two-photon imaging of fluorescent biosensors. Curr. Opin. Chem. Biol. **27**, 24–30 (2015). doi:10.1016/j.cbpa.2015.05.024
166. Lee, K.C., Siegel, J., Webb, S.E., Lévêque-Fort, S., Cole, M.J., Jones, R., Dowling, K., Lever, M.J., French, P.M.: Application of the stretched exponential function to fluorescence lifetime imaging. Biophys. J. **81**(3), 1265–1274 (2001). doi:10.1016/S0006-3495(01)75784-0

167. Casanova Moreno, J.R.: Spectroelectrochemical characterization of ultrathin organic films deposited on electrode surfaces. Ph.D. thesis, University of British Columbia (2014)
168. Mikhalyov, I., Gretskaya, N., Bergström, F., Johansson, L.B.A.: Electronic ground and excited state properties of dipyrrometheneboron difluoride (BODIPY): dimers with application to biosciences. Phys. Chem. Chem. Phys. **4**(22), 5663–5670 (2002). doi:10.1039/B206357N
169. Forster, B., Van De Ville, D., Berent, J., Sage, D., Unser, M.: Complex wavelets for extended depth-of-field: A new method for the fusion of multichannel microscopy images. Microsc. Res. Tech. **65**(1–2), 33–42 (2004). doi:10.1002/jemt.20092
170. de Lausanne, E.c.P.F.e.d.e.r.: BIG Extended Depth of Field. Tech. rep. (2011). URL http://bigwww.epfl.ch/demo/edf/
171. Johnson, I.: The Molecular Probes Handbook: A Guide to Fluorescent Probes and Labeling Technologies (2010)

Electrochemically Modulated Luminescence in Nanophotonic Structures

W. Xu, L.P. Zaino and P.W. Bohn

Abstract Electrochemistry can be applied to extract fundamental information about charge transfer processes in a large number of situations. However, at ultrasmall scales, when the system of interest contains mesoscopic collections of matter, direct electrical measurements become difficult, and there is ample reason to couple electron transfer reactions to luminescence. This chapter summarizes a new class of spectroelectrochemical experiments in which a very small numbers of redox-active molecules interact with nanophotonic structures so that the molecular luminescence is strongly coupled to the electrochemical event in zero-dimensional, 0-D, and one-dimensional, 1-D, architectures. In this chapter, we will describe various nanophotonic and nanoelectronic structures, especially the zero-mode waveguide, ZMW, that are being coupled to mesoscopic samples of redox-active molecules in order to carry out ultrasensitive measurements on the nanoscale.

1 Introduction

Mesoscopic Assemblies. Mesoscopic assemblies of matter may be distinguished both operationally—as systems that are intermediate between individual molecules and the bulk—and functionally—as samples that exhibit emergent properties which cannot be described properly either by molecular descriptions or by large ensemble behavior. Understanding electron transfer processes in systems at this scale is one of the remaining challenges of electrochemical science. Small populations of molecules exhibit unique characteristics, and these can effectively be probed using well-designed and characterized nanoscale architectures. Therefore, this chapter will explore how mesoscale interfaces, junctions, and nanophotonic systems may be coupled with redox-active systems for applications such as chemical sensing,

W. Xu · L.P. Zaino · P.W. Bohn (✉)
Department of Chemistry and Biochemistry and Department of Chemical and Biomolecular Engineering, University of Notre Dame, Notre Dame, IN 46556, USA
e-mail: pbohn@nd.edu

F. Miomandre and P. Audebert (eds.), *Luminescence in Electrochemistry*,
DOI 10.1007/978-3-319-49137-0_3

energy storage, nanoelectrochemical switching, and manipulating molecular electronic structure through charge transfer and potential-controlled electron transfer.

Nanoscale Measurements. Nanostructures exhibit a number of unique properties that motivate our interest in them. (1) Mass transport to nanostructures is significantly enhanced relative to large structures, because the diffusive boundary layer is hemispherical, rather than planar [1]. (2) It is possible to access new physical phenomena when the size of the structure is commensurate with physical scaling lengths—e.g., ballistic electron transport in quantum conductors, compared to the diffusive transport observed in larger structures [2, 3], and electrokinetic fluid flow is fundamentally altered in structures with physical dimensions of the order of the Debye length [4–6]. (3) Nanostructures exhibit enhanced surface-to-volume ratios, so surface phenomena are naturally accentuated, and small volumes produce smaller background signals. (4) Small-scale structures are inherently easier to deconstruct and regenerate, facilitating the logistics of making a series of measurements over time. (5) Nanoscale structures can be built that are inherently commensurate with molecular systems, as shown for example, by single-molecule conductivity experiments [7].

As compelling as these advantages are, making electrochemical measurements at ultrasmall structures poses special challenges. (1) Nanostructures necessarily support a small total number of molecules, meaning that signals that scale with molecular population are smaller than signals produced in larger structures, and small structures can be susceptible to environmental noise. (2) Fabrication of small-scale structures can be tedious and expensive, especially to the extent that they involve expensive lithography processes. (3) The ergodic hypothesis, which undergirds much of macro-scale sampling theory [8], must be explicitly justified for small structures. (4) Interfacing with small structures must be carefully addressed, since there is a mismatch between the scales (volumes, lengths, times for transport, equilibration, etc.) characterizing the nanostructure and the leads that connect it to the external world. (5) Statistical fluctuations in chemical properties, e.g., the chemical potential, play a much more significant role than they do in larger structures. These challenges are the key issues that set the noise floor for practical measurements and, thus, define the performance limits for nanoscale devices, especially nanoscale electrochemical studies.

Low-level Current Measurements. Direct amperometric and voltammetric measurements at small current values (at or below a few fA) are fundamentally noise-limited [9]. Consider an electron transfer process that produces 10^4 electrons s^{-1}. If we could count single electrons, a fierce technical challenge even today, then the 1.6 fA current would experience a shot noise of 10^2 s^{-1}, giving a very respectable S/N ratio ~ 100. However, current measurements are typically made in transimpedance amplifiers and are ultimately limited by diffusive scattering, rather than shot noise. The same 1.6 fA current measured through a 1 MΩ load at 1 Hz bandwidth would produce a 1.6 nV signal, but it would be accompanied by 130 nV of Johnson noise, yielding a S/N ~ 0.012, a completely untenable measurement. The driving force behind most of the works discussed in this chapter derives directly, then, from the distinction that it is quite straightforward to count single photons, but nearly impossible to count single electrons under electrochemical conditions. Thus, there is a great driving force to couple electron transfer events, especially in mesoscopic chemical systems, to luminescence measurements.

2 Background

2.1 High-Sensitivity Current Measurements

Redox Cycling. A conceptually appealing approach to the problem of a small number of redox species is to simply measure each molecule many times by cycling it back and forth between two adjacent electrodes. One widely used technique based on this simple concept is scanning electrochemical microscopy (SECM). Introduced by Bard and coworkers in 1989 [10], it is a scanning probe technique widely used for studying interfacial electron transfer events [11]. By moving the SECM tip, typically an ultramicroelectrode, to and from a substrate surface, quantitative information about the interface is obtained from the current observed at the tip, which is usually recorded as a function of tip-to-substrate distance [12]. In an experiment when both tip and substrate are immersed in the same solution of electroactive species, a steady-state current is observed at the tip at a sufficiently large tip–substrate distance, because electroactive species may diffuse freely to and from the tip [13, 14]. However, as the tip is brought very close to the substrate surface, the tip current changes in ways that reflect the intrinsic local electrical and topographical character of the substrate.

A particularly interesting case occurs during positive feedback, when regeneration of species at the metal substrate results in a large flux of molecules to the tip and produces significantly amplified currents [11, 12]. This, so-called, redox cycling (RC) effect relies on repeated oxidation and reduction of a reversible redox-active molecule between the tip and the substrate, and it exhibits strong dependence on the tip–substrate distance [15–18]. As this distance decreases to ~ 10 nm, amplification factors as large as $\sim 10^7$ can be observed [17].

Redox cycling can also be achieved in nanoscale cavities containing independent electrodes. Owing to the nanoscale interelectrode distances and trapping of the molecules inside the cavity, efficient RC is achieved, making it possible to perform electrochemical detection at ultrahigh sensitivity [15]. Alternative approaches to achieve RC have used ultrasmall volume cavities in glass or silicon to produce a thin-layer cell (TLC) with electrode gaps of the order ~ 10 nm [16, 19–21]. While producing slightly lower sensitivity, these TLC geometries are more compatible with chip-based devices for integrating with microfluidic systems for solution manipulation [16, 21]. Indeed, the TLC devices are similar in geometry to nanofluidic channels.

Interdigitated electrode arrays (IDEA) and nanopore arrays with recessed planar/ring-disk electrodes (RP/RDE) are additional conventional structures that can be used to achieve the RC effect [22–25]. In the RP/RDE geometry, the interelectrode distance can be conveniently controlled by adjusting the thickness of insulating layer separating the ring and the disk, see Fig. 1d [18, 26–30]. All these RC devices also have the capability to improve the selectivity of electrochemical detection for reversible species against irreversible species [21, 24, 30]. Since the RC effect requires species to be redox reversible, irreversible species are not regenerated for recycling and therefore cannot contribute to current amplification. Among different RC geometries, higher selectivity can be achieved with TLC nanofluidic devices and nanopore electrode arrays, because the confinement effect limits the amount of species diffusing to the electrode surface, which significantly decreases the current of the irreversible species, but has little effect on RC of reversible species [21, 26].

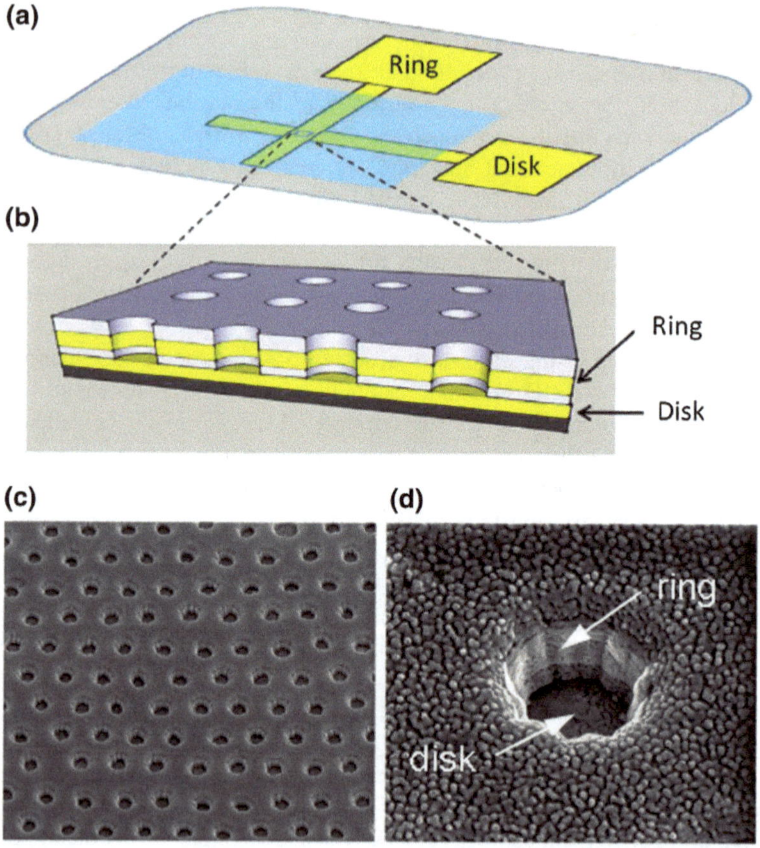

Fig. 1 **a** Schematic diagram showing the structure of a recessed ring-disk electrode array device and **b** a ring-disk geometry in the nanopores. **c** SEM image of an RRDE array at 52° tilt and **d** magnified view for a single-electrode pair at 52° tilt. Reproduced from Ref. [27]. Copyright 2013 American Chemical Society, reproduced with permission

Electron Correlation. Redox cycling is capable of addressing mesoscopic assemblies of molecules, because their diffusion back and forth between the generator and collector electrodes multiples the fundamental charge transfer event [11]. Furthermore in TLC geometries, the non-stochastic element of the diffusive motion can be monitored, if the molecular population in the structure is limited. In this case, the current can be measured separately on the two electrodes, then correlated [19, 20], because the nanofluidic geometry traps small numbers of molecules inside the channel and allows their alternative redox reactions on the two opposed planar electrodes to be measured. Using this temporal correlation strategy, single-molecule redox events can be differentiated from background noise [31]. This concept has been developed into electrochemical cross-correlation spectroscopy (ECS) by Lemay and coworkers [16]. Fluctuations in the number of reversibly redox-active molecules present in an active volume can be used to determine diffusion constants through amperometric measurements of the redox cycling current. ECS has also been employed to detect ultraslow liquid flow rates and investigate the adsorption behavior of redox molecules at the electrode surface, because it produces different behavior than freely diffusing molecules, effects which are described in more detail below [20, 32].

2.2 Coupling Electron Transfer to Luminescence

In analytical chemistry, the detection of single molecules defines the ultimate in limit of detection. Spectroscopic experiments, particularly fluorescence-based measurements, readily achieve single-molecule detection by reducing the interfering background, typically by decreasing the effective excitation and observation volumes [33]. While single-molecule detection has been possible by fluorescence spectroscopy for quite some time, extension of single-molecule measurements to electrochemistry has only developed recently, with two distinct approaches—spectroelectrochemistry and redox cycling. Utilizing redox cycling to detect single molecules relies on a high turnover rate between oxidized and reduced states to effectively measure the same molecule many times, as discussed in Sect. 2.1. Because redox cycling depends on a high molecular turnover rate, it cannot be used to study single-electron transfer events. Correlation strategies, as discussed in Sect. 2.1, also utilize a large number of cycling events for a single molecule. In contrast, spectroelectrochemical measurements can monitor single-electron transfer events and electron transfer events involving a single molecule. Single-molecule spectroelectrochemistry has been demonstrated with both fluorescence [34–37] and Raman scattering [38, 39], while electrogenerated chemiluminescence [40] and dark field scattering [41, 42] have been observed for single nanoparticle spectroelectrochemistry. Generally, fluorescence spectroscopy is used for single-molecule spectroelectrochemical experiments, due to its superior signal-to-background characteristics and relatively high signal levels (at least compared to Raman scattering), but of course, the analyte must exhibit a fluorescence change with redox state. Typically, redox-sensitive fluorescence intensity occurs because oxidization or reduction induces a change in the molecular absorbance of the fluorophore, or its quantum efficiency, or both. Thus, changes in spectroscopic behavior with changes in redox state, known collectively as electrofluorochromism, open up a variety of interesting applications and approaches to single-molecule electrochemistry by coupling electrochemical and fluorescence spectroscopy for single molecules and single enzymes [34, 36].

The first observations of single-molecule electron transfer were achieved by monitoring the redox cofactor during single-enzyme turnover [33]. The electron transfer of flavin adenine dinucleotide, FAD, when cholesterol oxidase oxidizes cholesterol, reducing FAD to non-fluorescent $FADH_2$ was observed in fluorescence microscopy by Xie and coworkers. Subsequent re-oxidation of $FADH_2$ by adventitious molecular oxygen completes the cycle. Similar strategies have been used for a number of single-enzyme investigations [43–46]. It is important to note that although these experiments involved electron transfer reactions, they were carried out in homogeneous media between molecular redox partners.

Single-molecule spectroelectrochemistry was pioneered by Bard and Barbara, who coupled total internal reflection fluorescence, TIRF, microscopy to electrochemistry of single molecules of conjugated poly(9,9-dioctylfluorene-co-benzothiadiazole), F8BT, on indium tin oxide (ITO) electrodes [34]. In these experiments, cyclic voltammetry was performed in conjunction with TIRF image acquisition, as shown in Fig. 2. Single molecule and ensemble fluorescence trajectories were acquired from image sequences, and a transition from fluorescent to non-fluorescent behavior was observed for individual molecules as the potential was increased from 0 to +1.5 V. By monitoring the fluorescence over a collection of individual molecules, the half-wave potential of single molecules was determined.

In a similar series of experiments, Chen and coworkers studied single-molecule electrocatalysis on single-walled carbon nanotubes on ITO electrodes though coupled electrochemical and fluorescence spectroscopic measurements [47]. Natively non-fluorescent resazurin was irreversibly, electrocatalytically reduced to resorufin, a fluorescent molecule, which can then be further (reversibly) reduced to non-fluorescent dihydroresorufin. Figure 3 depicts the experiment for resazurin reduction including the trajectory of a single molecule [48]. The ability to microscopically monitor many single fluorescence trajectories makes it possible to characterize individual electrocatalysis sites on the SWCNT, which in turn generates information about the heterogeneity between reaction sites at the single-molecule level.

Wide-field fluorescence microscopy has also been coupled to potential control to study the irreversible oxidation of BODIPY from a luminescent to a dark state [37]. Highly heterogeneous half-wave potential behavior was observed, which was attributed to the distribution of BODIPY in the polymer matrix in which the experiments were carried out. Single-molecule spectroelectrochemistry has also been carried out in films sandwiched between ITO electrodes by Jia and coworkers [49–51].

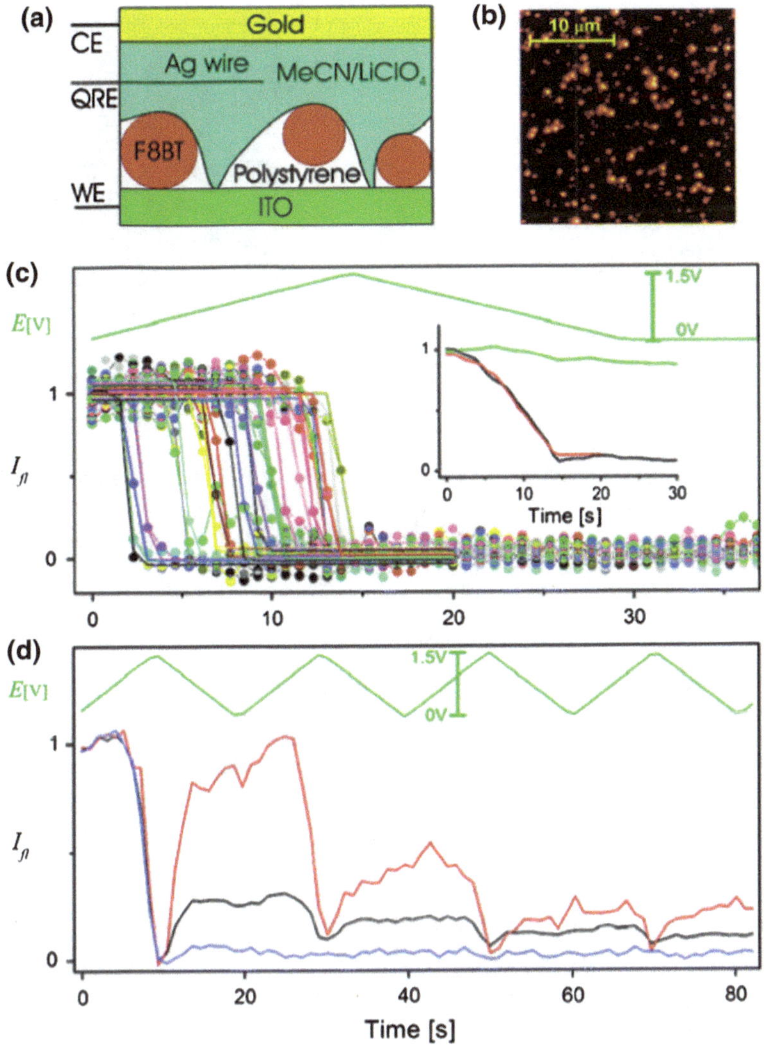

Fig. 2 **a** Cell configuration for the observation of single-molecule electrochemistry, SMS-EC. **b** Wide-field TIRF image for a typical SMS-EC cell. **c** Examples of normalized single-molecule fluorescence intensity trajectories (*data points*) obtained while applying the bias shown by the *top green curve*, and corresponding fits to a sigmoidal function. *Inset* Ensemble average of 200 trajectories (*black curve*) and of their corresponding fits (*red curve*). The *green curve* shows the ensemble average of 200 trajectories obtained at zero bias. **d** Ensemble average of normalized single-molecule fluorescence intensity trajectories (*black curve*) and sub-ensembles. Reproduced from Ref. [34]. Copyright 2006 American Chemical Society, reproduced with permission

Most examples of single-molecule spectroelectrochemistry have focused on studying surface bound molecules or individual catalytic sites. In contrast, single-molecule spectroelectrochemistry can also be performed with freely diffusing

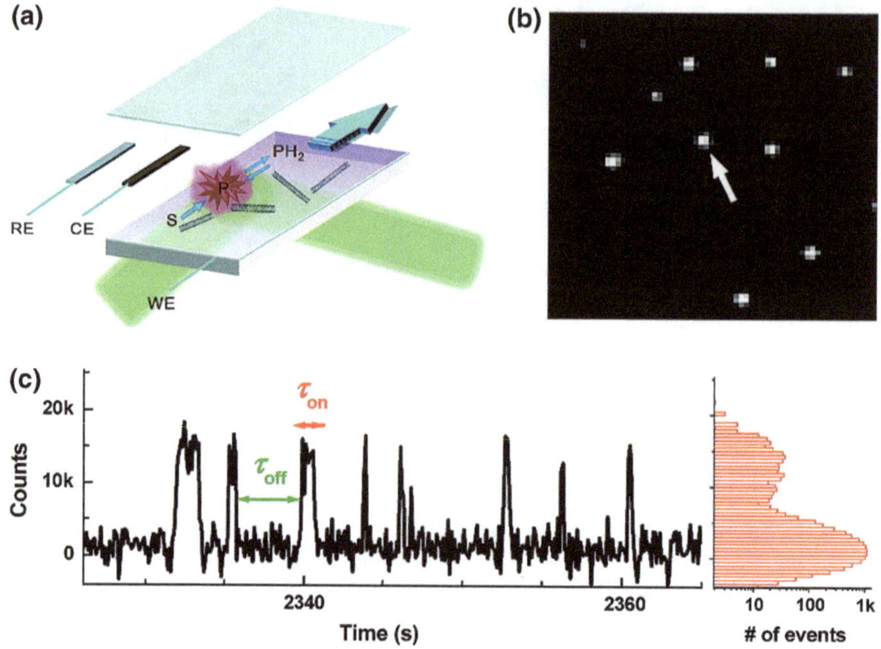

Fig. 3 Real-time single-molecule detection of SWNT electrocatalysis. **a** Schematic illustration of electrochemical TIRF microscopy flow cell made between an ITO-coated quartz slide and a coverslip. **b** Fluorescence image (17 × 19 μm^2) of SWNT electrocatalysis at −0.35 V with 0.05 μM resazurin in phosphate buffer. **c** *Left* segment of the fluorescence trajectory from the fluorescence spot marked by the *arrow* in (**b**). *Right* histogram of an 18 min long trajectory. Reproduced from Ref. [48]. Copyright 2009 American Chemical Society, reproduced with permission

molecules by targeting a confocal volume directly above an actively controlled ITO electrode [36, 52]. Ackerman and coworkers showed distinct changes in fluorescent bursts from cresyl violet as a function of applied potential as depicted in Fig. 4. When the electrode is held at reducing potentials, a few stochastic events are observed in stark contrast to the large number of single-molecule events observed at oxidizing potentials. These experiments illustrate the capacity to sweep the potential and observe the change in single-molecule spectroscopic behavior in real time, a manipulation that is difficult to realize with immobilized redox species.

Single-molecule microscopy coupled to electrochemistry is particularly useful in examining heterogeneous behavior that would not be observable in ensemble measurements. This strategy has been applied with great success to studying the electrocatalysis at nanoscale structures, with the resolution only being limited by the optical diffraction limit, and it has been extended to proteins, allowing the observation of thermodynamic and kinetic dispersion in the behavior of immobilized proteins by combining fluorescence and electrochemistry.

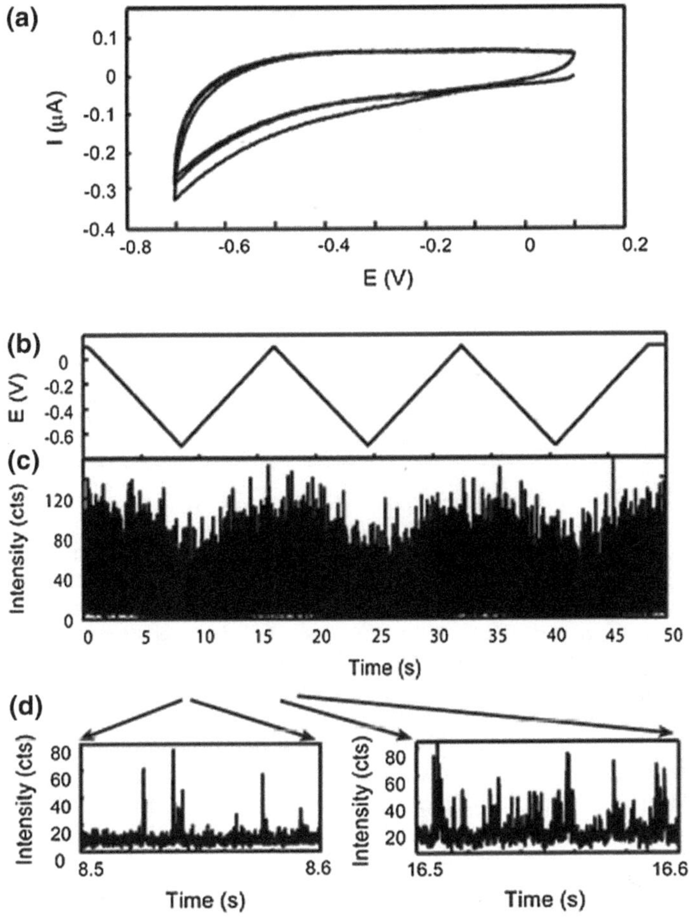

Fig. 4 Single-molecule fluorescence spectroelectrochemistry of cresyl violet: **a** cyclic voltammograms of 1.2 nM cresyl violet in pH 6.2, 20 mM sodium phosphate at the ITO electrode; **b** potential versus time plot of the CV scan; **c** single-molecule fluorescence intensity burst trajectory of cresyl violet versus time; **d** two 0.1 s segments of the fluorescence time trajectory. Reproduced from Ref. [36]. Copyright 2008 Royal Society of Chemistry, reproduced with permission

Spectroelectrochemical probing of proteins has been extended to the single-protein level by monitoring single functionalized proteins on electrode. By using fluorescence, Akkilic et al. determined the optical midpoint potential from single Cy5-labeled azurin molecules to be 45.7 ± 0.5 mV versus SCE by fitting the fluorescence versus potential response of 66 individual azurin molecules to a sigmoidal response function [53]. In this case, the heterogeneity at the single-molecule level was in agreement with the width of the response from an ensemble measurement of ~1200 molecules.

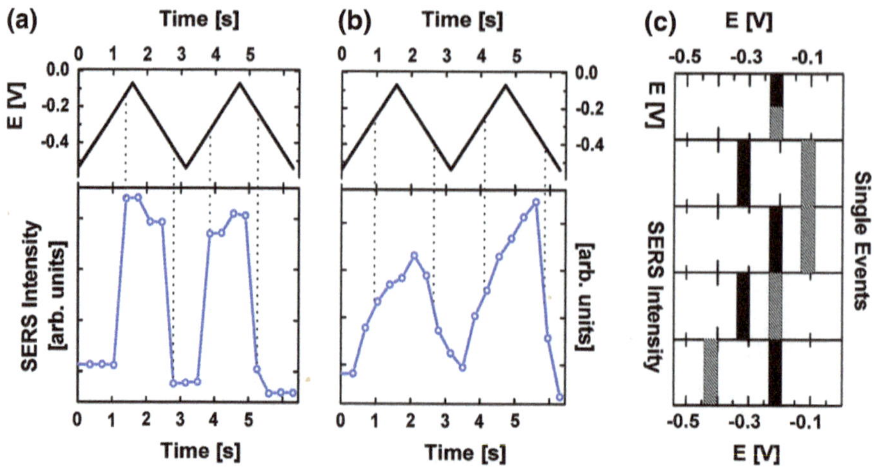

Fig. 5 **a**, **b** Variations in redox properties for two consecutive periods of cyclic potential sweep, as determined from the SERS intensity of the 590 cm^{-1} peak of *Niles Blue*: **a** single molecule; **b** many molecules. **c** Temporal variation of single molecules. Each row illustrates the temporal variation of the oxidation potential in two successive redox events for a single NB molecule (*black, first event*; *dashed, second event*). The top panel corresponds to a molecule whose oxidation potential did not change, while those of the other four did. Reproduced from Ref. [54]. Copyright 2014 American Chemical Society, reproduced with permission

While fluorescence remains the dominant technique for single-molecule electrochemical experiments, surface-enhanced Raman spectroscopy (SERS) has also been used to investigate single-molecule electrochemistry [38, 39]. This is advantageous, because it expands the field of single-molecule spectroelectrochemistry beyond those fluorophores that exhibit a change in fluorescence with redox state. Willets' group examined electrochemically modulated single-molecule SERS, SM-SERS, of Nile Blue A on Ag nanoparticles [54]. Figure 5 shows the difference between the voltage sweep SERS response for single molecules, Fig. 5a, versus many molecules, Fig. 5b. Single molecules exhibit rectangular SERS signal modulations with potential, indicating a well-defined threshold for SERS. In contrast, electrochemical SERS acquired over many molecules produces a gradual increase and decrease in SERS intensity with signal, presumably due to averaging over the distribution of single-molecule thresholds. Clearly, the number of molecular candidates, which fit the criteria for SM-SERS, is larger than the number for which single-molecule electrofluorochromism can be observed, which makes Raman-based single-molecule electrochemical techniques an exciting new development.

Single-molecule spectroelectrochemical measurements show great promise in determining the thermodynamic and kinetic heterogeneity of complex molecular systems at the nanoscale. In particular, information that is usually averaged out by ensemble measurements can be accessed by investigating single-electron transfer events with spectroscopic measurements, as shown for protein monolayers, catalytic surfaces, and other types of monolayers and films. Typically, single-molecule electrochemical microscopy is performed on immobilized redox-active species, but single-molecule spectroelectrochemistry has also been shown in the diffusive regime [55]. Ultimately, there are significant barriers to coupling single-molecule electrochemistry with fluorescence in many systems, because direct application requires a molecular system that exhibits bright fluorescence in only one redox state. However, recent developments in single-molecule Raman spectroscopy and its application to single-molecule electrochemistry enhance the number of possible candidates for single-molecule spectroelectrochemical investigations, suggesting that the field has a bright future.

3 Spectroelectrochemistry in Nanophotonic Structures

3.1 Enzymes in Nanophotonic Structures

The sensitivity limit of optical emission experiments is single-molecule fluorescence, which, as discussed in Sect. 2.2, has been used to monitor the stochastic turnover of individual oxidase enzyme molecules [56]. In correlation-based single-molecule experiments employing confocal optics, the femtoliter-scale probe volume limits the fluorophore concentration to ≤ 1 nM. In order to use single-molecule techniques to study higher concentrations relevant to enzyme catalysis, biofuel production, photosynthesis, and other biomolecular processes, the observation volume must be reduced significantly. Metallic nanostructures (channels/pores) can provide the required reduction in observation volume, due to the steep dropoff of the optical excitation intensity within the zeptoliter-volume observation region defined by the metal aperture of a zero-mode waveguide, ZMW. ZMWs, defined as metallic nanopores with a critical dimension less than a cutoff wavelength determined by the dimensions of the nanopore and the refractive index of the filling medium [57], create a $\sim$zeptoliter size region within an evanescent excitation region, in which binding/catalytic events can be observed at a single-molecule substrate concentrations, even for molecules with characteristic concentrations ≥ 1 μM.

In our laboratory, we have examined single molecules of the functioning enzyme subunit monomeric sarcosine oxidase, MSOX, by immobilizing single MSOX molecules in the SiO_2 region of an array of poly(ethylene glycol)-passivated Au ZMWs, Scheme 1. These structures enable the observation of the catalytic properties of the enzyme in the oxidation of sarcosine at the single-molecule level [58, 59]. The Au layer of the ZMW was over-etched into the substrate to lessen the position dependence of the magnitude of a single-molecule fluorescence. Although Al is the common choice for metallic overlayer in ZMW construction, Au was used here to access its unique surface-binding chemistry. Enzyme turnover was signaled by the on–off fluorescence transition, depending on whether the enzyme's flavin adenine dinucleotide (FAD) cofactor was oxidized (emissive) or reduced (dark).

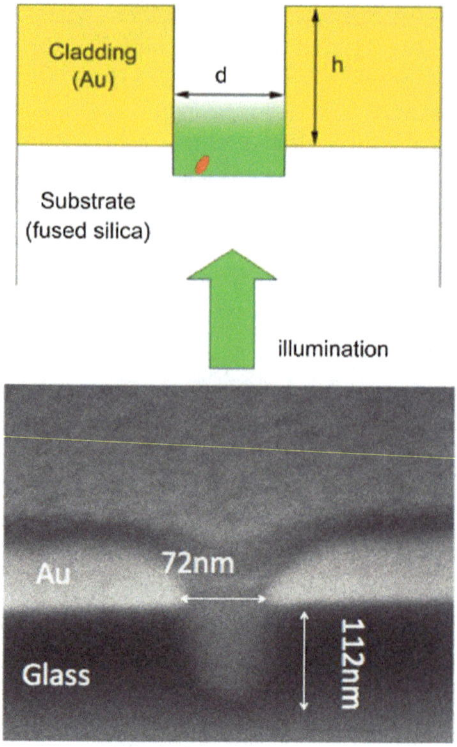

Scheme 1 Enzyme turnover was signaled by the on–off fluorescence transition, depending on whether the enzyme's flavin adenine dinucleotide (FAD) cofactor was oxidized (emissive) or reduced (*dark*)

For MSOX reactions involving both the nominal substrate (sarcosine, or N-methylglycine) and an analogous substrate (proline), statistical analysis of single-molecule temporal trajectories revealed static heterogeneity of single-enzyme reaction rates, but no dynamic disorder. In addition, the single-molecule data confirmed the independence of reductive and oxidative reactions. The observed dynamics of MSOX are consistent with single-molecule Michaelis–Menten kinetics under the steady-state limit, showing that sarcosine and L-proline, two distinct substrates of the enzyme, can be distinguished by distributions of waiting times for the reductive half-reaction, at single-molecule resolution. These structures open the way, for example, to systematic studies of the effect of molecular crowding on enzyme dynamics. ZMWs have proven to be excellent devices for single-molecule studies, especially for studying single-molecule behavior, such as static heterogeneity and dynamic disorder, not accessible with ensembles.

In these experiments, redox chemistry was strictly molecular, with the redox-active FAD cofactor being successively reduced (during oxidation of the enzyme substrate) and oxidized by adventitious molecular O_2 to return it to its native initial oxidation state and prepare the enzyme for the next turnover. As such, they did not take advantage of the control of reactivity possible by controlled potential in heterogeneous electron transfer reactions. Nevertheless, we hypothesized that the metal cladding of the zero-mode waveguide could function simultaneously as the working electrode of a nanoscale electrochemical cell. Thus, the opaque metal can be exploited to play a dual role—as an electrochemical element in heterogeneous electron transfer and as the optical cladding element in the ZMW, as described below.

3.2 Immobilized Luminescent Factors

The MSOX experiments illustrate the possibilities inherent in coupling nanophotonic structures, like ZMWs, with redox reactions. The next logical step involves a true electrochemical conversion in which the metal cladding layer plays a dual role as an optical screen and a working electrode. In these types of direct electrofluorochromism experiments, the molecule must exhibit a change in emission intensity, resulting from a combination of the excitation spectrum and the emission quantum efficiency, when it changes its redox state. FAD is a particularly interesting molecule which exhibits this behavior, and it is important for its role as a redox cofactor. The isoalloxazine chromophore in FAD undergoes a $2e^{-}/2H^{+}$ reduction from the luminescent oxidized state to the reduced state, $FADH_2$, which is much less strongly emissive, making it a suitable candidate for single-molecule spectroelectrochemistry. There are significant differences in the spectroscopic properties, especially the excited state dynamics, of FAD in proteins like MSOX described in Sect. 3.1, compared to its behavior in free solution. Most notably, free FAD exhibits two significant excited state lifetimes—a nanosecond lifetime similar to that observed in the protein, and a picosecond component that has been ascribed to intramolecular energy transfer resulting from a hinge motion that brings the isoalloxazine ring in close proximity to the adenine moiety [60]. In addition, given the proton-coupled electron transfer mechanism for FAD reduction, the equilibrium potential and voltammetry of free FAD in bulk exhibits a strong pH dependence [61]. Thus, FAD, with its rich set of dynamics and environmental sensitivity, makes an interesting subject for detailed studies of its single-molecule electron transfer behavior when confined in electrochemically active ZMW nanophotonic structures.

Unlike the single-enzyme experiments detailed in Sect. 3.1 that were carried out at open circuit conditions, using the ZMW substrate as a working electrode for active potential control experiments requires further design consideration. While Al is generally the most common metal used in ZMW fabrication due to its short skin depth [62], the ideally polarizable region of Al is not compatible with the types of voltammetric experiments of interest. A better choice is Au, which has sufficient skin depth to ensure adequate optical field attenuation and also has well-studied surface derivatization schemes, especially Au-thiol self-assembly chemistry [63], available for functionalizing molecules or enzymes.

In the work from our laboratory, Au-thiol chemistry was used to functionalize the Au surface with a terminal amine which was subsequently reacted with pyrroloquinoline quinone, PQQ, and *m*-aminophenylboronic acid. FAD was then bound to this monolayer by reacting the sugar group with the boronic ester as

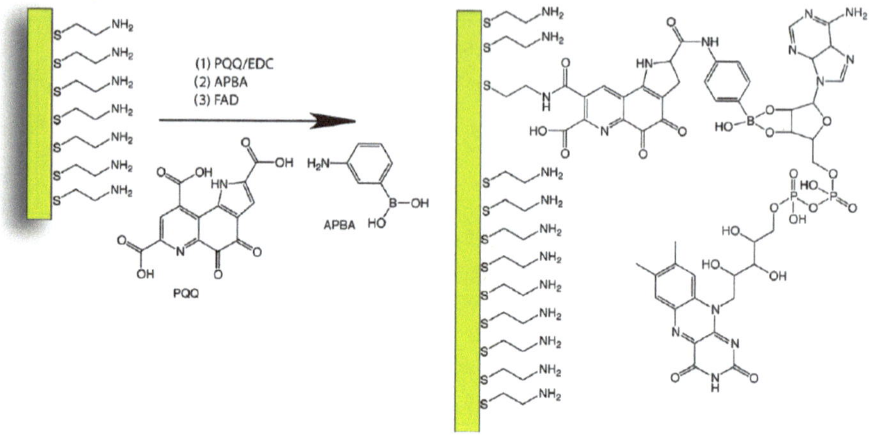

Scheme 2 FAD was then bound to this monolayer by reacting the sugar group with the boronic ester

shown in Scheme 2. The surface functionalization was optimized to produce an array of nanopores in which a significant number of pores contained only a single bound FAD molecule, by optimizing the Poisson distribution governing molecular occupancy. Starting from the macroscopic equilibrium potential at planar Au electrodes, $E_{eq} = -0.48$ V versus Ag/AgCl, fluorescence trajectories were acquired at −0.2, −0.48, and −0.8 V versus Ag/AgCl. Pores containing only a single molecule were chosen, based on their statistical behavior, for further analysis. As shown in Fig. 6, the fluorescence trajectories show distinct differences as a function of potential. At a reducing potential (−0.8 V vs. Ag/AgCl), the molecular fluorescence is generally at a low (background) level, while at an oxidizing potential (−0.2 V vs. Ag/AgCl), the bound FAD molecules spend most of the time in the highly emissive state. Then, when E_{appl} is poised near the equilibrium potential, the fluorescence frequently transitions between the low and high levels. Thus, the general trend is that as the potential is set at increasingly positive values, single immobilized FAD molecules are more likely to be found in the oxidized state.

We hypothesized that the total amount of time the molecule spends in the fluorescent (oxidized) or non-fluorescent (reduced) state would display complementary sigmoidal relationships with E_{appl} and that the sum of on → off and off → on transitions would reflect the joint probability distributions of the oxidized and reduced states, thus giving a peak near E_{eq}. Figure 7 illustrates these fundamental features of the single-molecule spectroelectrochemistry of ZMW-immobilized FAD. Comparison of the fluorescence trajectories of single FAD molecules at different potentials shown in Fig. 6 clearly indicates that the transition rate between FAD and $FADH_2$ is potential dependent. Consistent with this hypothesis, Fig. 7 shows that the total on-time duration increases with increasingly positive E_{appl}. In addition, the transition rate (sum of on → off and off → on transitions) reaches a maximum near E_{eq}.

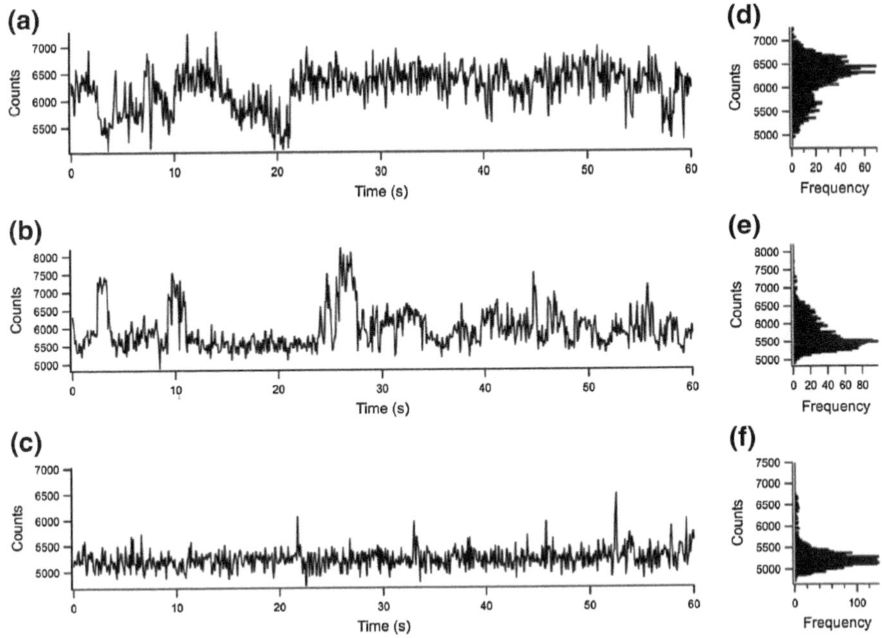

Fig. 6 *Left* 60 s long fluorescence trajectories at applied potentials of **a** −0.2 V; **b** −0.48 V; **c** −0.8 V versus Ag/AgCl. *Right* Intensity histograms for the full 150-s acquisition at applied potentials of **d** −0.2 V; **e** −0.48 V; **f** −0.8 V. Reproduced from Ref. [35]. Copyright 2013 Royal Society of Chemistry, reproduced with permission

Overall, these results demonstrate that coupling single-molecule spectroscopy and voltammetry for immobilized electrofluorochromic species in nanophotonic structures shows great promise, although challenges remain. Most pointedly, the average time for single-molecule photobleaching was found to be ~ 110 s under these experimental conditions, an event which naturally limits the observation time for each experiment. This limitation can largely be circumvented by examining freely diffusing redox species, as demonstrated in Fig. 4.

3.3 Freely Diffusing Molecules

One way to work around the limitations of surface-functionalized single-molecule spectroelectrochemistry, as discussed in Sect. 3.1, is to examine freely diffusing single molecules. Due to the zeptoliter excitation volumes characteristic of the ZMW, single-molecule occupancy can be achieved at modest (μM) concentrations. For example, at an excitation volume of 100 zL, a concentration of 17 μM produces single-molecular occupancy conditions, $\langle N \rangle \sim 1$.

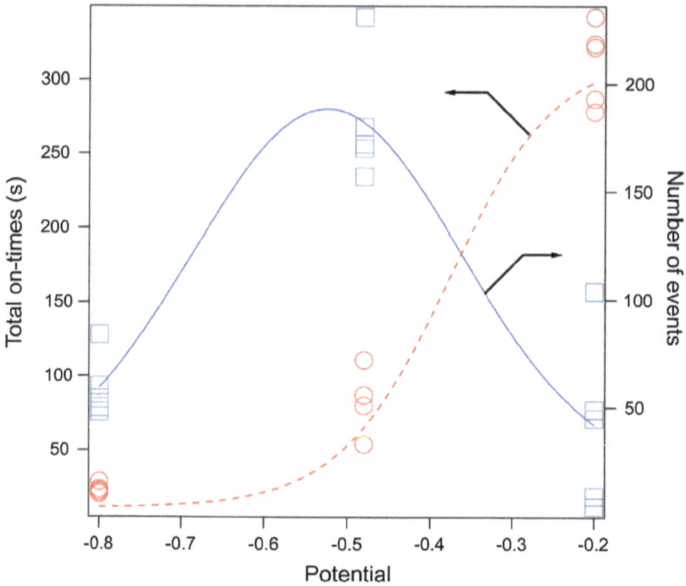

Fig. 7 Potential dependence of total on-time duration (*left*, *red*, *circles*) and the number of transition events (*right*, *blue*, *squares*) for 15 separate 6-min trajectories; 5 at each potential. The total on-time duration and number of transitions are fit to sigmoid (*red*, *dashed*) and Gaussian (*blue*, *solid*) functions, respectively. Reproduced from Ref. [35]. Copyright 2013 Royal Society of Chemistry, reproduced with permission

Counterbalancing the advantage of working at relatively high concentration is the relatively small residence time, $\langle\tau\rangle$ ~2–5 μs, in which any given single molecule resides in the ZMW excitation volume. Thus, it is important to consider temporal resolution when switching from a surface-functionalized single molecule to the study of freely diffusing molecules. Given this time scale, fluorescence measurements were acquired with avalanche photodiodes rather than an electron-multiplied CCD, in order to obtain the necessary temporal resolution to observe single molecules. The frequency of molecule–electrode interactions is another important consideration in this experiment. Given a nanopore of radius r ~20 nm and purely diffusive transport, the average number of wall collisions a single molecule exhibits is of the order ~10^6 s^{-1}. Thus, it is highly probable that each single molecule entering the pore under these conditions will experience one or more electrode collisions. Furthermore, it is possible to carry out electrochemistry in these structures without supporting electrolyte, thereby extending the Debye screening length to distances that exceed the pore dimensions, thus accessing ion migration as a transport mechanism which can further enhance electrochemical currents [18].

Initial experiments were carried out under static potential control. In this case, the ZMW electrode was held at constant potential for the entire experiment, and fluorescence of freely diffusing flavin mononucleotide, FMN, was measured after

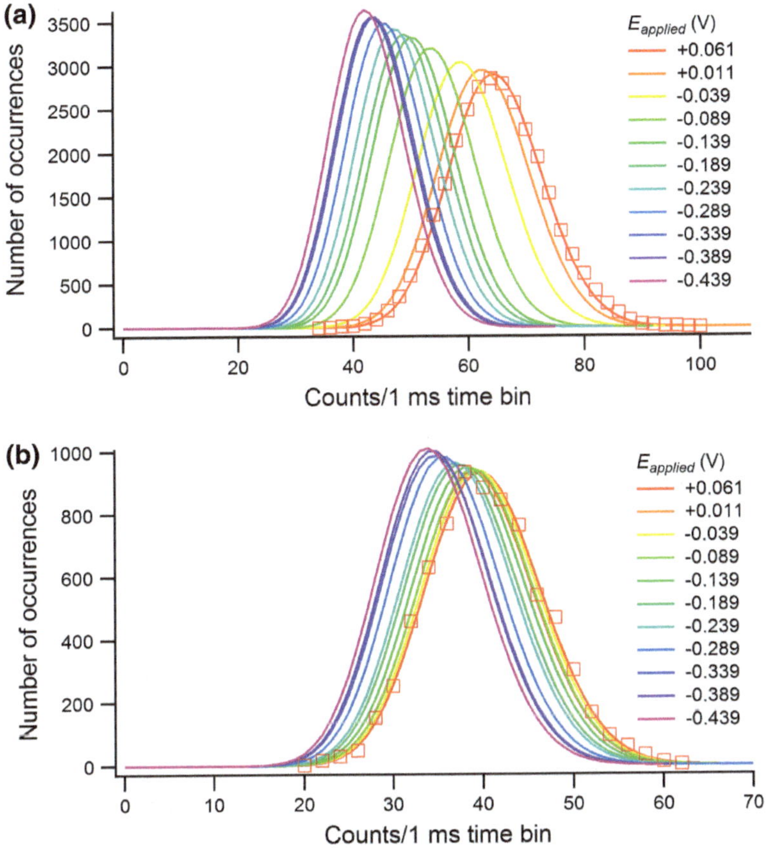

Fig. 8 Poisson distribution fits to histograms of fluorescence intensities at different values of E_{appl} for **a** 10 μM and **b** 1 μM FMN in 100 mM citrate buffer at pH 2.9. Experimental data are included for E_{appl} = +0.061 V to illustrate the quality of the fit. Reproduced from Ref. [55]. Copyright 2015 Royal Society of Chemistry, reproduced with permission

allowing the system to reach steady state. Figure 8 illustrates the distribution of intensities observed at occupancies $\langle N \rangle = 1.2$ and $\langle N \rangle = 0.12$, achieved at bulk concentrations of 10 and 1 μM, respectively, over a range of potentials. As the potential is increased toward oxidizing overpotential, the fluorescence intensity increases, consistent with the behavior of immobilized FAD described in Sect. 3.2, as oxidizing potentials favor the creation of highly emissive FMN, at the expense of reduced $FMNH_2$. These experiments were run with FMN, which lacks the adenine moiety, instead of FAD, to avoid problems with the intramolecular quenching behavior of FAD in free solution.

Further single-occupancy spectroelectrochemical investigations were carried out in ZMWs while under dynamic potential control. In this case, the fluorescence analog of cyclic voltammetry, cyclic potential sweep single-molecule fluorescence,

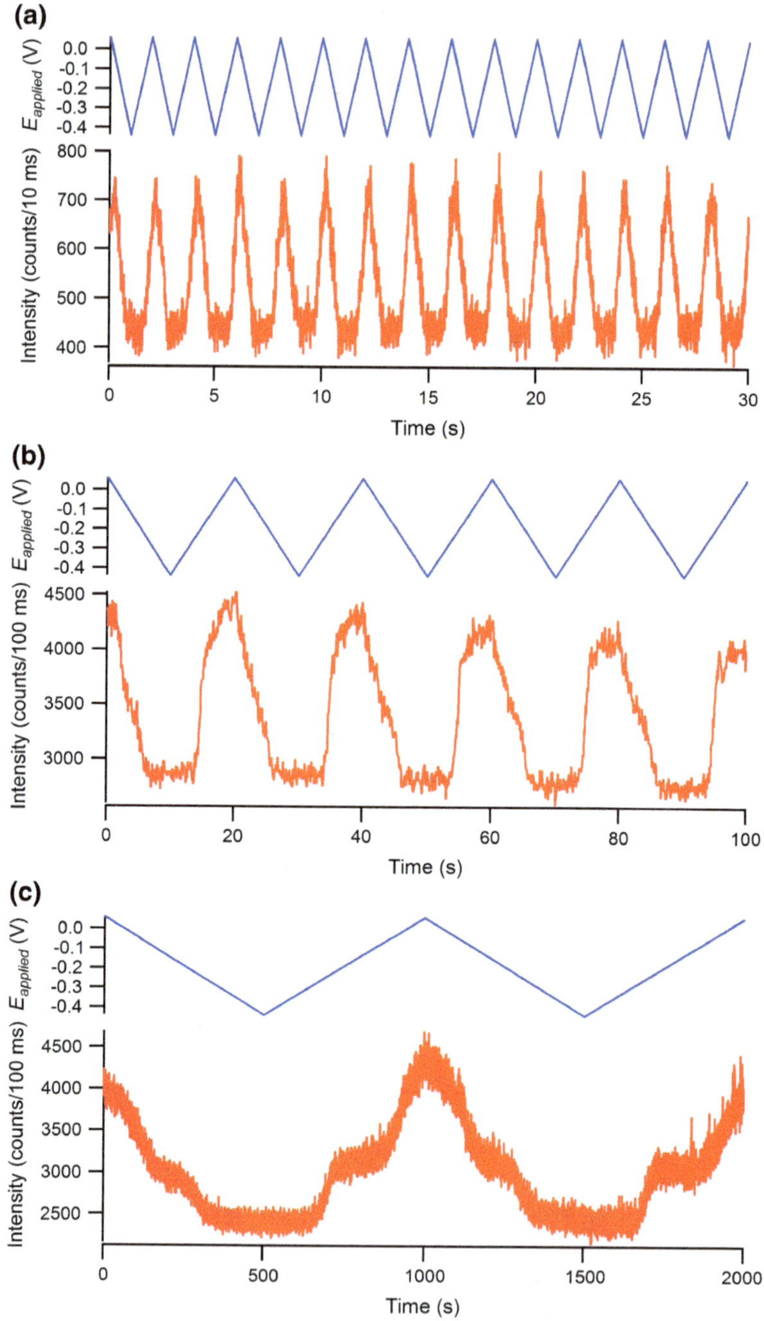
(a)
$E_{applied}$ (V)
0.0
-0.1
-0.2
-0.3
-0.4
Intensity (counts/10 ms)
800
700
600
500
400
0
5
10
15
20
25
30
Time (s)
(b)
$E_{applied}$ (V)
0.0
-0.1
-0.2
-0.3
-0.4
Intensity (counts/100 ms)
4500
4000
3500
3000
0
20
40
60
80
100
Time (s)
(c)
$E_{applied}$ (V)
0.0
-0.1
-0.2
-0.3
-0.4
Intensity (counts/100 ms)
4500
4000
3500
3000
2500
0
500
1000
1500
2000
Time (s)

◀ **Fig. 9** Cyclic potential sweep fluorescence measurements of 10 μM FMN in 100 mM citrate buffer at pH 2.9 at scan rates of **a** 0.5 V/s, **b** 0.05 V/s, and **c** 0.001 V/s. In each panel, the potential program is shown at the top, and the fluorescence response at the bottom. Reproduced from Ref. [55]. Copyright 2015 Royal Society of Chemistry, reproduced with permission

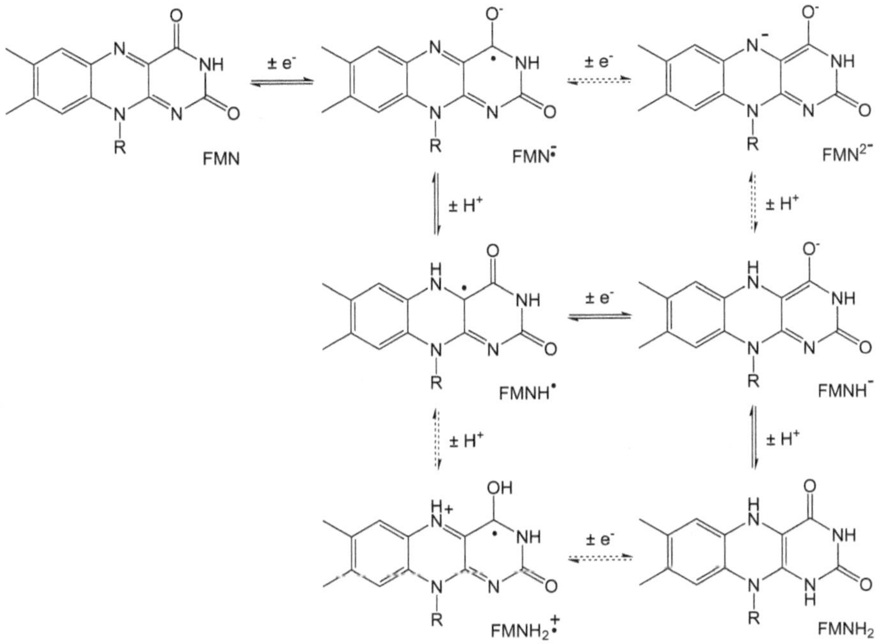

Scheme 3 FMN undergoes a net pH-dependent $2e^{-}/2H^{+}$ reduction to $FMNH_2$, at pH 2.9, there are three possible reaction pathways for FMN reduction and five possible intermediate forms

was performed and showed a strong correlation between E_{appl} and fluorescence intensity. Figure 9 shows the ZMW florescence response to a triangular waveform at different scan rates. At 0.5 V s^{-1}, fast and symmetric fluorescence response was observed, while at slower scan rates, 0.05 and 0.001 V s^{-1}, more asymmetric responses were obtained, which we believe reflect the multiplicity of detailed mechanistic paths for the $2e^{-}/2H^{+}$ FMN redox process. In particular, while FMN undergoes a net pH-dependent $2e^{-}/2H^{+}$ reduction to $FMNH_2$, at pH 2.9, there are three possible reaction pathways for FMN reduction and five possible intermediate forms as shown in Scheme 3. We hypothesize that the reduction proceeds primarily through an intermediate neutral semiquinone species. The shoulder observed in Fig. 9c, which is not observed in the bulk voltammetry, suggests that the semiquinone species may be stabilized by the confined nature of the nanopore environment.

These experiments studying the single-molecule redox behavior under confinement in artificially constructed nanophotonic structures are just getting started. Nevertheless, it is already clear that the extremely small volumes and fast response of the ZMWs to the applied potential make them an interesting substrate for studying single-molecule electrochemistry of both immobilized as well as freely diffusing redox species.

4 Conclusions and Prospects for the Future

4.1 Luminescence Versus Direct Current Measurement

Amperometry is the most straightforward technique used for analytical electrochemistry, since the faradic current produced is proportional to the number of charge transfer events per second at the electrode which, in turn, is proportional to the concentration of analyte species in the detection volume. Although single-molecule sensitivity has been achieved in well-designed electrode geometries [31], measurement of analytes at nM to μM target concentrations is challenging. As detailed in this chapter, when luminescence detection is coupled to electrochemistry, it is possible to take advantage of the inherently low background of luminescence [34, 35]. Furthermore, by coupling redox reactions of a target analyte with that of a fluorescence reporter [28, 64], it is not strictly necessary for the analyte to be electrofluorochromic. This strategy has been demonstrated using electrogenerated chemiluminescence (ECL) [65] and fluorescence-enabled electrochemical microscopy (FEEM) [66, 67] on electrode arrays, allowing simultaneous monitoring of multiple redox reactions using imaging. The coupling of amperometry to luminescence readout eliminates the need for electrical connection to the electrode array, as with current measurements [68], and because photons can be multiplied/amplified during detection, the Johnson noise, which is typically the noise floor of current-based measurements, is rendered irrelevant. Moreover, electrochemical imaging via fluorescence microscopy can enhance the spatial resolution for electrochemical processes, allowing the luminescence to report on multiple electrochemical reactions occurring in parallel [69]. With the spatial information provided by luminescence imaging, the distribution of target redox species, as well as the heterogeneity of their spatial distribution, can be monitored [70, 71]. Taken together, these factors provide strong motivation for incurring the additional experimental complexity and expense of luminescence readout.

4.2 Stochastic Nature of Single-Electron Transfer Events

The stochastic character of single-molecule behavior, as revealed through single-molecule fluorescence, is well recognized for its importance in understanding the transport, interaction, and reactivity of the molecules. The development of techniques sensitive to single-electron transfer events could bring similar insights about the stochastic character of charge transfer events on the electrode surface.

Unlike the continuous current-voltage/time response in ensemble measurements that average over a large number of charge transfer events, the response in stochastic electrochemistry is discrete, reflecting the random nature of individual charge transfer events [34, 72]. In this regard, it is important to distinguish between single-molecule electrochemistry, as demonstrated in a technical tour-de-force by Bard and coworkers [17, 73], and the measurement of single-electron transfer events. Recently, Lemay's group has delineated the properties associated with the stochasticity of electrochemical behavior in a beautiful series of experiments in a nanoscale thin-layer cell (typical electrode separation ~50 nm) [14, 32, 74]. Since electroactive molecules can enter and leave the nanofluidic TLC randomly, the observed redox cycling produces an inherently stochastic current response. Discrete features observed for this response were attributed to the stochastic motion of individual redox-active molecules.

Besides the Brownian character of the motion delivering molecules from one electrode to the other in these experiments, other dynamic events, such as adsorption to the channel wall or electrode surfaces, can also impact the current response [32, 74]. Adsorption decreases the apparent diffusion coefficient of the redox-active species, diminishing the current per molecule compared to that obtained under purely diffusive conditions [16]. In addition, the adsorption–desorption events add extra noise [74], which can also be exploited to understand the adsorption behavior. Electrochemical correlation of the response from the top and bottom electrodes can be used to differentiate the stochastic signal from the background noise, leading to enhanced sensitivity and electrochemical detection (stochastic sensing) at concentrations as low as ~100 pM [14]. Thus, stochastic electrochemistry offers an excellent complement to fluorescence detection for ultrasensitive electrochemical measurements.

4.3 Additional Opportunities for Nanophotonic Electrochemistry

The coupling of electron transfer reactions to zero-mode waveguides, described in this chapter, is clearly a first step toward exploiting nanophotonic and nanoelectronic structures in electrochemistry. This chapter has focused on coupling to nanophotonic structures due to its focus on luminescence and electrochemistry, but there are many novel manipulations that can be accomplished by coupling to nanoelectronic structures. For example, electrochemically deposited Au-Ag-Au bimetallic atomic scale junctions (ASJs) are the small-size limit of nanowires [75, 76]. They can be fabricated by overgrowing nanojunctions then using electromigration to back-thin the junction [77]. If these steps are carried out under electrochemical potential control, the relationship between junction stability and applied potential can provide a reliable method of regulating the size and, thus, properties of the nanojunction [78]. In general, more positive potentials decrease junction stability and increase the rate at which conductance decays. Conductance behavior under these labile conditions is principally determined by Ag oxidation potential, electrochemical potential-induced surface stress and the nature of any adsorbate at the Ag ASJ–solution interface. Junctions fabricated at more negative potentials experience only slight changes in conductance, likely due to surface atom diffusion and stress-induced structural rearrangement. Electrochemical potential also plays a significant role in determining adsorption–desorption kinetics, as our group showed by studying adsorption–desorption behavior of surface pyridine at steady-state at Au-Ag-Au ASJs by fluctuation spectroscopy [79].

The use of electrochemical potential to control the fabrication and behavior of ASJs holds significant promise. Coupling fluctuation spectroscopy to mesoscopic molecular assemblies supported on ASJs reveals the central role played by molecular adsorption–desorption, which carries both analytical (concentration) and dynamic (kinetics) information. Future studies can address other materials, such as Pt, and solvent systems, e.g., ionic liquids, capable of supporting ultrafast atomic rearrangements and, thus, high-frequency formation and breaking of atomic junctions [80]. In addition, there is a significant interest in using photoinduced charge injection, for example, from adsorbed redox-active dyes, to modulate the inherent conductivity of atomic junctions, thus establishing a connection between nanoelectronic structures and the nanophotonic structures discussed throughout the chapter.

In addition, there are a number of other exciting possibilities yet to be explored in the coupling of nanophotonic structures to electrochemical phenomena. The use of zero-mode waveguides is just the first example. Essentially, any nanophotonic structure which is comprised of electronic conductors and/or semiconductors has the potential to be used electrochemically. In the ZMW architectures, it is the metallic cladding layer that doubles as a working electrode that permits luminescence to be coupled to electrochemistry in mesoscopic assemblies. However, there are a large number of nanophotonic structures that could also be envisioned. Also, exciting are the possibilities of coupling nanoscale electrochemical phenomena to optical metamaterials—materials in which the 3D arrangement of elements, typically metallic, enables non-classical optical phenomena. Optical metamaterials have already been used to demonstrate such intriguing phenomena as optical black holes, beam shifters, field rotators, light concentrators, lossless waveguide bends, invisibility cloaks, and illusion optics [81]. Because these non-classical optical phenomena make it possible to control the 4D spatiotemporal electric and magnetic field distributions in nanostructures [82], it is only a matter of time before researchers start to exploit these capabilities in combination with electrochemistry. The combined future for luminescence and nanoscale electrochemistry looks bright indeed.

Acknowledgements Work described in this chapter from the authors' laboratory was supported by the US National Science Foundation and the US Department of Energy Office of Science under various grants.

References

1. Correia, A.N., Mascaro, L.H., Machado, S.A.S., Mazo, L.H., Avaca, L.A.: Ultramicroelectrodes. 1. Theoretical revision and outlook. Quim. Nova **18**, 475–480 (1995)
2. Mozos, J.L., Wan, C.C., Taraschi, G., Wang, J., Guo, H.: Transport through a single-atom junction. J. Phys. Cond. Matter **10**, 2663–2671 (1998)
3. Oliver, D.J., Maassen, J., El Ouali, M., Paul, W., Hagedorn, T., Miyahara, Y., Qi, Y., Guo, H., Grutter, P.: Conductivity of an atomically defined metallic interface. Proc. Natl. Acad. Sci. USA **109**, 19097–19102 (2012)
4. Kemery, P.J., Steehler, J.K., Bohn, P.W.: Electric field mediated transport in nanometer diameter channels. Langmuir **14**, 2884–2889 (1998)
5. Kuo, T.C., Cannon, D.M., Chen, Y.N., Tulock, J.J., Shannon, M.A., Sweedler, J.V., Bohn, P. W.: Gateable nanofluidic interconnects for multilayered microfluidic separation systems. Anal. Chem. **75**, 1861–1867 (2003)
6. Kuo, T.C., Cannon, D.M., Shannon, M.A., Bohn, P.W., Sweedler, J.V.: Hybrid three-dimensional nanofluidic/microfluidic devices using molecular gates. Sens. Actuators, A Phys. **102**, 223–233 (2003)
7. Xiao, X.Y., Xu, B.Q., Tao, N.J.: Measurement of single molecule conductance: benzenedithiol and benzenedimethanethiol. Nano Lett. **4**, 267–271 (2004)
8. Schafer, R., Hecker, K., Hegger, H., Langheinrich, W.: Experimental study of mesoscopic fluctuations in nonlinear conductance and magnetoconductance. Phys. Rev. B **53**, 15964–15970 (1996)
9. Hoeben, F.J.M., Meijer, F.S., Dekker, C., Albracht, S.P.J., Heering, H.A., Lemay, S.G.: Toward single-enzyme molecule electrochemistry: NiFe-hydrogenase protein film voltammetry at nanoelectrodes. ACS Nano **2**, 2497–2504 (2008)
10. Bard, A.J., Fan, F.R.F., Kwak, J., Lev, O.: Scanning electrochemical microscopy—introduction and principles. Anal. Chem. **61**, 132–138 (1989)
11. Amemiya, S., Bard, A.J., Fan, F.R.F., Mirkin, M.V., Unwin, P.R.: Scanning electrochemical microscopy. Ann. Rev. Anal. Chem. **1**, 95–131 (2008)
12. Bard, A.J., Faulkner, L.R.: Electrochemical methods. John Wiley & Sons, New York (2001)
13. Sun, P., Laforge, F.O., Mirkin, M.V.: Scanning electrochemical microscopy in the 21st century. PCCP **9**, 802–823 (2007)
14. Zevenbergen, M.A.G., Singh, P.S., Goluch, E.D., Wolfrum, B.L., Lemay, S.G.: Stochastic sensing of single molecules in a nanofluidic electrochemical device. Nano Lett. **11**, 2881–2886 (2011)
15. Sun, P., Mirkin, M.V.: Electrochemistry of individual molecules in zeptoliter volumes. J. Am. Chem. Soc. **130**, 8241–8250 (2008)
16. Zevenbergen, M.A.G., Singh, P.S., Goluch, E.D., Wolfrum, B.L., Lemay, S.G.: Electrochemical correlation spectroscopy in nanofluidic cavities. Anal. Chem. **81**, 8203–8212 (2009)
17. Fan, F.R.F., Bard, A.J.: Electrochemical detection of single molecules. Science **267**, 871–874 (1995)
18. Ma, C.X., Contento, N.M., Bohn, P.W.: Redox cycling on recessed ring-disk nanoelectrode arrays in the absence of supporting electrolyte. J. Am. Chem. Soc. **136**, 7225–7228 (2014)
19. Lemay, S.G., Kang, S., Mathwig, K., Singh, P.S.: Single-molecule electrochemistry: present status and outlook. Acc. Chem. Res. **46**, 369–377 (2013)
20. Mathwig, K., Mampallil, D., Kang, S., Lemay, S.G.: Electrical cross-correlation spectroscopy: measuring picoliter-per-minute flows in nanochannels. Phys. Rev. Lett. **109**, 118302 (2012)
21. Wolfrum, B., Zevenbergen, M., Lemay, S.: Nanofluidic redox cycling amplification for the selective detection of catechol. Anal. Chem. **80**, 972–977 (2008)
22. Bard, A.J., Crayston, J.A., Kittlesen, G.P., Shea, T.V., Wrighton, M.S.: Digital-simulation of the measured electrochemical response of reversible redox couples at microelectrode arrays—

consequences arising from closely spaced ultramicroelectrodes. Anal. Chem. **58**, 2321–2331 (1986)

23. Niwa, O., Morita, M., Tabei, H.: Electrochemical-behavior of reversible redox species at interdigitated array electrodes with different geometries—consideration of redox cycling and collection efficiency. Anal. Chem. **62**, 447–452 (1990)
24. Goluch, E.D., Wolfrum, B., Singh, P.S., Zevenbergen, M.A.G., Lemay, S.G.: Redox cycling in nanofluidic channels using interdigitated electrodes. Anal. Bioanal. Chem. **394**, 447–456 (2009)
25. Rahimi, M., Mikkelsen, S.R.: Cyclic biamperometry at micro-interdigitated electrodes. Anal. Chem. **83**, 7555–7559 (2011)
26. Ma, C., Contento, N.M., Gibson, L.R., Bohn, P.W.: Recessed ring-disk nanoelectrode arrays integrated in nanofluidic structures for selective electrochemical detection. Anal. Chem. **85**, 9882–9888 (2013)
27. Ma, C.X., Contento, N.M., Gibson, L.R., Bohn, P.W.: Redox cycling in nanoscale-recessed ring-disk electrode arrays for enhanced electrochemical sensitivity. ACS Nano **7**, 5483–5490 (2013)
28. Ma, C.X., Zaino, L.P., Bohn, P.W.: Self-induced redox cycling coupled luminescence on nanopore recessed disk-multiscale bipolar electrodes. Chem. Sci. **6**, 3173–3179 (2015)
29. Menshykau, D., Cortina-Puig, M., del Campo, F.J., Munoz, F.X., Compton, R.G.: Plane-recessed disk electrodes and their arrays in transient generator-collector mode: the measurement of the rate of the chemical reaction of electrochemically generated species. J. Electroanal. Chem. **648**, 28–35 (2010)
30. Zhu, F., Yan, J.W., Lu, M., Zhou, Y.L., Yang, Y., Mao, B.W.: A strategy for selective detection based on interferent depleting and redox cycling using the plane-recessed microdisk array electrodes. Electrochim. Acta **56**, 8101–8107 (2011)
31. Mathwig, K., Aartsma, T.J., Canters, G.W., Lemay, S.G.: Nanoscale methods for single-molecule electrochemistry. Ann. Rev. Anal. Chem. **7**(7), 383–404 (2014)
32. Singh, P.S., Chan, H.S.M., Kang, S., Lemay, S.G.: Stochastic amperometric fluctuations as a probe for dynamic adsorption in nanofluidic electrochemical systems. J. Am. Chem. Soc. **133**, 18289–18295 (2011)
33. Lu, H.P., Xun, L.Y., Xie, X.S.: Single-molecule enzymatic dynamics. Science **282**, 1877–1882 (1998)
34. Palacios, R.E., Fan, F.R.F., Bard, A.J., Barbara, P.F.: Single-molecule spectroelectrochemistry (SMS-EC). J. Am. Chem. Soc. **128**, 9028–9029 (2006)
35. Zhao, J., Zaino, L.P., Bohn, P.W.: Potential-dependent single molecule blinking dynamics for flavin adenine dinucleotide covalently immobilized in zero-mode waveguide array of working electrodes. Faraday Discuss. **164**, 57–69 (2013)
36. Lei, C.H., Hu, D.H., Ackerman, E.J.: Single-molecule fluorescence spectroelectrochemistry of cresyl violet. Chem. Commun. **43**, 5490–5492 (2008)
37. Liu, J., Hill, C.M., Pan, S.L., Liu, H.Y.: Interfacial charge transfer events of BODIPY molecules: single molecule spectroelectrochemistry and substrate effects. PCCP **16**, 23150–23156 (2014)
38. Cortes, E., Etchegoin, P.G., Le Ru, E.C., Fainstein, A., Vela, M.E., Salvarezza, R.C.: Monitoring the electrochemistry of single molecules by surface-enhanced Raman spectroscopy. J. Am. Chem. Soc. **132**, 18034–18037 (2010)
39. Cortes, E., Etchegoin, P.G., Le Ru, E.C., Fainstein, A., Vela, M.E., Salvarezza, R.C.: Strong correlation between molecular configurations and charge-transfer processes probed at the single-molecule level by surface-enhanced Raman scattering. J. Am. Chem. Soc. **135**, 2809–2815 (2013)
40. Chang, Y.L., Palacios, R.E., Fan, F.R.F., Bard, A.J., Barbara, P.F.: Electrogenerated chemiluminescence of single conjugated polymer nanoparticles. J. Am. Chem. Soc. **130**, 8906–8907 (2008)

41. Hill, C.M., Bennett, R., Zhou, C., Street, S., Zheng, J., Pan, S.L.: Single Ag nanoparticle spectroelectrochemistry via dark-field scattering and fluorescence microscopies. J. Phys. Chem. C **119**, 6760–6768 (2015)
42. Hill, C.M., Pan, S.L.: A dark-field scattering spectroelectrochemical technique for tracking the electrodeposition of single silver nanoparticles. J. Am. Chem. Soc. **135**, 17250–17253 (2013)
43. Engelkamp, H., Hatzakis, N.S., Hofkens, J., De Schryver, F.C., Nolte, R.J.M., Rowan, A.E.: Do enzymes sleep and work? Chem. Commun. **9**, 935–940 (2006)
44. Gupta, A., Aartsma, T.J., Canters, G.W.: One at a time: intramolecular electron-transfer kinetics in small laccase observed during turnover. J. Am. Chem. Soc. **136**, 2707–2710 (2014)
45. Goldsmith, R.H., Tabares, L.C., Kostrz, D., Dennison, C., Aartsma, T.J., Canters, G.W., Moerner, W.E.: Redox cycling and kinetic analysis of single molecules of solution-phase nitrite reductase. Proc. Natl. Acad. Sci. USA **108**, 17269–17274 (2011)
46. Xie, X.S., Lu, H.P.: Single-molecule enzymology. J. Biol. Chem. **274**, 15967–15970 (1999)
47. Shen, H., Xu, W.L., Chen, P.: Single-molecule nanoscale electrocatalysis. PCCP **12**, 6555–6563 (2010)
48. Xu, W., Shen, H., Kim, Y.J., Zhou, X., Liu, G., Park, J., Chen, P.: Single-molecule electrocatalysis by single-walled carbon nanotubes. Nano Lett. **9**, 3968–3973 (2009)
49. Wu, R.X., Chen, R.Y., Qin, C.B., Gao, Y., Qiao, Z.X., Zhang, G.F., Xiao, L.T., Jia, S.T.: An electric field induced reversible single-molecule fluorescence switch. Chem. Commun. **51**, 7368–7371 (2015)
50. Zhang, G.F., Xiao, L.T., Chen, R.Y., Gao, Y., Wang, X.B., Jia, S.T.: Single-molecule interfacial electron transfer dynamics manipulated by an external electric current. PCCP **13**, 13815–13820 (2011)
51. Chen, R.Y., Wu, R.X., Zhang, G.F., Gao, Y., Xiao, L.T., Jia, S.T.: Electron transfer-based single molecule fluorescence as a probe for nano-environment dynamics. Sensors **14**, 2449–2467 (2014)
52. Lei, C.H., Hu, D.H., Ackerman, E.: Clay nanoparticle-supported single-molecule fluorescence spectroelectrochemistry. Nano Lett. **9**, 655–658 (2009)
53. Akkilic, N., Kamran, M., Stan, R., Sanghamitra, N.J.M.: Voltage-controlled fluorescence switching of a single redox protein. Biosens. Bioelectron. **67**, 747–751 (2015)
54. Wilson, A.J., Willets, K.A.: Visualizing site-specific redox potentials on the surface of plasmonic nanoparticle aggregates with superlocalization SERS microscopy. Nano Lett. **14**, 939–945 (2014)
55. Zaino, L.P.I., Grismer, D.A., Han, D., Crouch, G.M., Bohn, P.W.: Single molecule spectroelectrochemistry of freely diffusing flavin mononucleotide in zero-dimensional nanophotonic structures. Faraday Disc. **184**, 101–115 (2015)
56. Zhao, J., Branagan, S.P., Bohn, P.W.: Single-molecule enzyme dynamics of monomeric sarcosine oxidase in a gold-based zero-mode waveguide. Appl. Spectrosc. **66**, 163–169 (2012)
57. Levene, M.J., Korlach, J., Turner, S.W., Foquet, M., Craighead, H.G., Webb, W.W.: Zero-mode waveguides for single-molecule analysis at high concentrations. Science **299**, 682–686 (2003)
58. Wagner, M., Trickey, P., Chen, Z., Mathews, F.S., Jorns, M.S.: Monomeric sarcosine oxidase: 1. Flavin reactivity and active site binding determinants. Biochemistry **39**, 8813–8824 (2000)
59. Wagner, M.A., Jorns, M.S.: Monomeric sarcosine oxidase: 2. Kinetic studies with sarcosine, alternate substrates, and a substrate analogue. Biochemistry **39**, 8825–8829 (2000)
60. Kao, Y.-T., Saxena, C., He, T.-F., Guo, L., Wang, L., Sancar, A., Zhong, D.: Ultrafast dynamics of flavins in five redox states. J. Am. Chem. Soc. **130**, 13132–13139 (2008)
61. Tan, S.L.J., Kan, J.M., Webster, R.D.: Differences in proton-coupled electron-transfer reactions of flavin mononucleotide (FMN) and flavin adenine dinucleotide (FAD) between buffered and unbuffered aqueous solutions. J. Phys. Chem. B **117**, 13755–13766 (2013)
62. Jin, E.X., Xu, X.F.: Finitte-difference time-domain studies on optical transmission through planar nano-apertures in a metal film. Jpn. J. Appl. Phys. Part **1**(43), 407–417 (2004)

63. Nuzzo, R.G., Allara, D.L.: Adsorption of bifunctional organic disulfides on gold surfaces. J. Am. Chem. Soc. **105**, 4481–4483 (1983)
64. Xu, W., Ma, C., Bohn, P.W.: Coupling of independent electrochemical reactions and fluorescence at closed bipolar interdigitated electrode arrays. ChemElectrochem **3**, 422–428 (2016)
65. Mavre, F., Chow, K.F., Sheridan, E., Chang, B.Y., Crooks, J.A., Crooks, R.M.: A theoretical and experimental framework for understanding electrogenerated chemiluminescence (ECL) emission at bipolar electrodes. Anal. Chem. **81**, 6218–6225 (2009)
66. Oja, S.M., Guerrette, J.P., David, M.R., Zhang, B.: Fluorescence-enabled electrochemical microscopy with dihydroresorufin as a fluorogenic indicator. Anal. Chem. **86**, 6040–6048 (2014)
67. Guerrette, J.P., Percival, S.J., Zhang, B.: Fluorescence coupling for direct imaging of electrocatalytic heterogeneity. J. Am. Chem. Soc. **135**, 855–861 (2013)
68. Chow, K.F., Mavre, F., Crooks, J.A., Chang, B.Y., Crooks, R.M.: A large-scale, wireless electrochemical bipolar electrode microarray. J. Am. Chem. Soc. **131**, 8364–8365 (2009)
69. Reddington, E., Sapienza, A., Gurau, B., Viswanathan, R., Sarangapani, S., Smotkin, E.S., Mallouk, T.E.: Combinatorial electrochemistry: a highly parallel, optical screening method for discovery of better electrocatalysts. Science **280**, 1735–1737 (1998)
70. Murphy, J.N., Cheng, A.K.H., Yu, H.Z., Bizzotto, D.: On the nature of DNA self-assembled monolayers on Au: measuring surface heterogeneity with electrochemical in situ fluorescence microscopy. J. Am. Chem. Soc. **131**, 4042–4050 (2009)
71. Salverda, J.M., Patil, A.V., Mizzon, G., Kuznetsova, S., Zauner, G., Akkilic, N., Canters, G. W., Davis, J.J., Heering, H.A., Aartsma, T.J.: Fluorescent cyclic voltammetry of immobilized Azurin: direct observation of thermodynamic and kinetic heterogeneity. Angew. Chem. Int. Ed. **49**, 5776–5779 (2010)
72. Bard, A.J.: Toward single enzyme molecule electrochemistry. ACS Nano **2**, 2437–2440 (2008)
73. Fan, F.R.F., Kwak, J., Bard, A.J.: Single molecule electrochemistry. J. Am. Chem. Soc. **118**, 9669–9675 (1996)
74. Singh, P.S., Katelhon, E., Mathwig, K., Wolfrum, B., Lemay, S.G.: Stochasticity in single-molecule nanoelectrochemistry: origins, consequences, and solutions. ACS Nano **6**, 9662–9671 (2012)
75. Castle, P.J., Bohn, P.W.: Interfacial scattering at electrochemically fabricated atom-scale junctions between thin gold film electrodes in a microfluidic channel. Anal. Chem. **77**, 243–249 (2005)
76. Shi, P., Zhang, J., Lin, H.-Y., Bohn, P.W.: The effect of molecular adsorption on electrical conductance of single Au nanowires fabricated by electron beam lithography and focused ion beam etching. Small **6**, 2598–2603 (2010)
77. Hwang, T.-W., Bohn, P.W.: Robust Au-Ag-Au bimetallic atom-scale junctions fabricated by self-limited Ag electrodeposition at Au nanogaps. ACS Nano **5**, 8434–8441 (2011)
78. Hwang, T.-W., Bohn, P.W.: Potential-dependent restructuring and chemical-noise at Au-Ag-Au atomic-scale junctions. ACS Nano **8**, 1718–1727 (2014)
79. Hwang, T.-W., Branagan, S.P., Bohn, P.W.: Chemical noise produced by equilibrium adsorption/desorption of surface pyridine at Au-Ag-Au bimetallic atom-scale junctions studied by fluctuation spectroscopy. J. Am. Chem. Soc. **135**, 4522–4529 (2013)
80. Jo, S.H., Chang, T., Ebong, I., Bhadviya, B.B., Mazumder, P., Lu, W.: Nanoscale memristor device as synapse in neuromorphic systems. Nano Lett. **10**, 1297–1301 (2010)
81. Liu, Y., Zhang, X.: Metamaterials: a new frontier of science and technology. Chem. Soc. Rev. **40**, 2494–2507 (2011)
82. Chen, H., Chan, C.T., Sheng, P.: Transformation optics and metamaterials. Nat. Mater. **9**, 387–396 (2010)

Electrochemically Monitored Photoluminescence of Conjugated Polymers

Francisco Montilla and Francisco Huerta

Abstract The coupling of different spectroscopic techniques such as UV–Vis, Raman or FTIR to electrochemical methods became a useful strategy to study structure, kinetics and dynamics at electrochemical interfaces during the past two decades. However, the application of fluorescence spectroscopy to electrochemistry emerged more recently. The so-called in situ electrochemical fluorescence can be successfully applied to study photophysical processes of conjugated polymers in the electrochemical environment. The present chapter reviews the major optical and electrochemical characteristics of these materials and focuses, particularly, on the relation between electronic structure and chemical characteristics by means of the oligomer approach. In situ electrochemical fluorescence is highly sensitive to the redox state of the explored polymers. Owing to this key feature, it is possible to obtain information from their photoluminescent processes even at very low doping levels. Apart from the study of electrochemical phenomena, it will be emphasized that in situ fluorescence constitutes an exceptional tool to examine relevant interactions between electric charges and photoexcited states.

1 Introduction

Synthetic polymers have constituted one of the most attractive materials for scientists and engineers since their discovery in the nineteenth century. These compounds are molecular materials whose properties can be modulated from their constituent monomers in order to provide new features or to enhance the existing

F. Montilla (✉)
Departamento de Química Física e Instituto Universitario de Materiales de Alicante, Universidad de Alicante, Apdo. de Correos 99, 03080 Alicante, Spain
e-mail: francisco.montilla@ua.es

F. Huerta
Departamento de Ingenieria Textil y Papelera, Universitat Politecnica de Valencia, Plaza Ferrandiz y Carbonell, 1, 03801 Alcoy, Spain

F. Miomandre and P. Audebert (eds.), *Luminescence in Electrochemistry*,
DOI 10.1007/978-3-319-49137-0_4

ones: chemical or mechanical stability, mechanical processability or self-organization capability, among others. Conventional polymers are often electric insulators, although the possibility of acquiring electrical abilities led to the development of conjugated conducting polymers during the last decades of the past century.

Polyacetylene was the object of the initial studies during the 1970s because, at least hypothetically, compounds with an infinite chain of sp^2 hybridized carbons (with one free electron per atom) could yield a zero-gap unidimensional material. It was believed that these materials could be suitable candidates to achieve high-temperature superconductivity.

Chemically obtained polyacetylene was, instead of the expected superconductor, a black powder with deficient properties: solid, insoluble and poor electrical conductor [1]. These characteristics appear because of the tendency of the system to stabilize electrons in pairs, i.e., associating two monomer units. The dimerization process, known as *Peierls instability* or *Jahn-Teller effect*, in either the physicist's or the chemist's terminology, leads to the formation of a bandgap in the electronic structure of the conjugated chain. This bandgap was a serious drawback for the first devised applications of conjugated polymers, for example synthetic materials based on light carbon structures as a replacement of metals for aerospace purposes [2, 3].

The electrical conductivity of native conjugated polymers could be strongly enhanced after oxidation by exposure to halogen vapors, a treatment that introduces charged defects in the conjugated backbone, as discovered in 1977 by the Nobel Prize winners Heeger, MacDiarmid and Shirakawa [4]. In particular, it was observed that chemically oxidized (doped) polyacetylenes showed a conductivity up to seven orders of magnitude higher than the native polymer and not far from that of conventional metals, $\sim 10^5$ S cm^{-1} [5, 6]. Unfortunately, polyacetylene presented low chemical stability in air, and, in the quest for more stable alternatives, doped polythiophenes, polyanilines or polypyrroles were intensively investigated during the 1980s and 1990s [1]. In particular, polyaniline derivatives attracted attention because their conductivity was regulated by both redox state and acid–base chemistry, even though the modest figures obtained (usually below 10^3 S cm^{-1}) often forced to discard them for practical purposes as synthetic metals [3].

The redox doping process, besides the dramatic change in conductivity, modifies some other properties of conjugated polymers. For example, it is known that the insertion of charges yields new intergap optical transitions or enhanced affinity toward ionic species and solvents, thus producing the swelling of the polymer. These special properties led to an intensive research in the fields of electrochromic devices or electrochemical actuators, among others [7].

Interestingly, conjugated polymers in their undoped state have also remarkable optical and electronic properties due to the semiconducting character, which was very promising for the optoelectronics industry. The most common semiconducting materials are either inorganic solids as silicon (for transistors or photovoltaics) or semiconducting crystalline salts as gallium nitride (for diodes). However, the bandgap of these materials is determined by the crystalline structure and is often difficult to modulate. A clear example of these limitations is the historical development of the light-emitting diodes (LEDs). H.J. Round discovered electroluminescence in 1907 using a silicon carbide crystal [8], but the first commercial red emitter could not be commercialized until 1968. Such device employed gallium arsenide phosphide as the active material, whose bandgap is lower than 2 eV. Red LEDs dominated the market for years because there was no easy way to obtain stable compounds for green and blue emitters. Almost a century after LED discovery, Shuji Nakamura presented in 1994 the first blue LED based on InGaN (bandgap 2.5–3.3 eV). Nakamura, Akasaki and Amano were awarded the 2014 Nobel Prize in physics for their work.

In contrast to the inorganic electronics, devices based on organic materials progressed much faster. In 1990, the group of R. Friend in Cambridge developed the first organic polymer LED (OLED) based on poly(phenylene vinylene), an orange-red emitter. The work opened the possibility of using conjugated polymers in novel electronic devices where the classical inorganic semiconductors were replaced by organic materials, which are more versatile. This was the starting point for a novel research field in materials science: *organic electronics* or *plastic electronics*. In contrast with the evolution of inorganic LEDs, only two years after Friend's pioneering work, Grem et al. [9] announced the discovery of the blue OLED. It was demonstrated that the modulation of the bandgap in conjugated polymers (and hence of the optoelectronic properties) is based on the adequate choice of the precursor monomers or, alternatively, on their chemical modification. The set of procedures involved in this task has been known as *bandgap engineering* from then [10–12]. The fruit of the intensive research done in bandgap engineering is the variety of luminescent conjugated polymers that are commercially available covering the entire visible electromagnetic spectrum. Some examples of these polymers are shown in Fig. 1.

Although great part of the early work on organic electronics was focused on electroluminescent diodes, the properties of the organic constituents make them attractive for other optoelectronic devices. Indeed, Heeger reported in 1992 the photoinduced charge transfer between polymers and fullerenes, which was the starting point for the development of polymer-based photovoltaic systems [13]. In several publications dated from 1995 to 1999, Friend also suggested other electronic applications of conjugated polymers as, for example, photodiodes, lasers or organic transistors [14–16]. Most of these applications fall within the field of solid-state physics and have been well developed during the present century [17].

The applications of conjugated polymers can be divided into two groups according to the nature of the chemical environment surrounding these materials: on the one hand, *wet* applications, which are often related to the ability of the polymers to be reversibly doped by chemical or electrochemical redox reactions in the presence of a solvent. Some examples of wet applications are polymer batteries, electrochromic devices and mechanical actuators.

On the other hand, we find *dry* applications for which polymers, in their native semiconducting state, are employed for solid-state electronic devices such as organic light-emitting diodes (OLED), organic photovoltaic cells (OPV), transistors

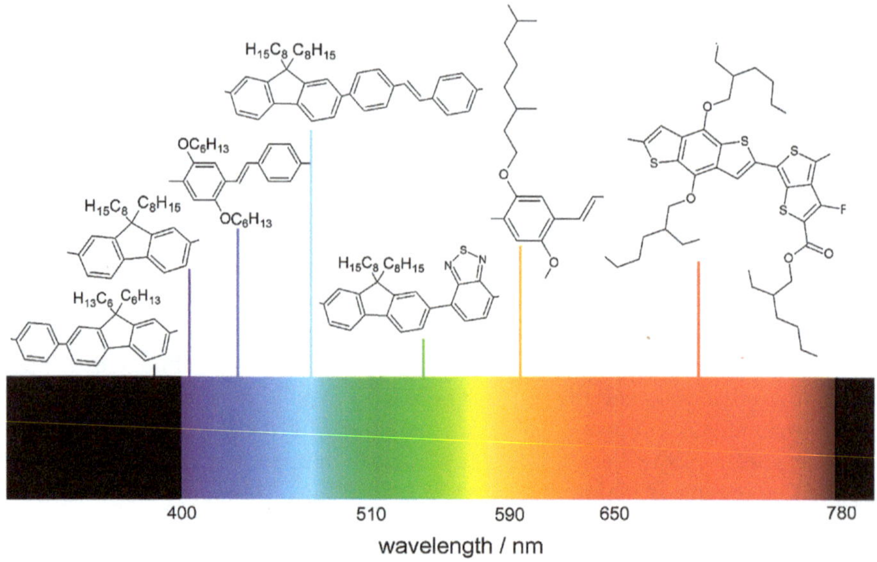

Fig. 1 Conjugated polymers with light-emitting properties ranging the complete visible spectrum. All these materials are commercially available

or lasers [18]. Since the organic material is not doped, no formal change in its oxidation state takes place during operation.

The two major dry applications of conjugated polymers are OLED and OPV. In typical OLED devices, the conjugated polymer is placed between two electrodes in a sandwich-like configuration. The anode is usually a transparent ITO film, and the cathode is made of a low work function metal such as Ca or Al. Charge carriers (holes and electrons) are injected in the polymer–electrode interface because of the potential difference imposed between the external electrodes. Both species then drift under the electric field and are recombined in the bulk polymer layer. The recombination process involves the formation of bounded hole–electron pairs (called excitons), which may decay radiatively. In most cases, the emission efficiency in OLEDs is limited because of several non-radiative decay paths of the excitons [19–23]. Organic photovoltaic cells present a similar architecture to OLEDs, but in this case, the process is opposite. The polymer absorbs photons, and the hole–electron pairs generated are separated in free charges, which will be collected at the external electrodes [18].

When organic conjugated polymers are employed as the active materials in both OLED and OPV devices, understanding the interaction of charged species with the photogenerated species is of paramount importance. For example, fast exciton migration is required in photovoltaic cells, because the exciton must move to a donor–acceptor interface for charge generation. However, the quenching of luminescence by exciton–charge interaction during this process may lead to poor performance in organic light-emitting diodes.

The electrochemical control of the polymer redox state provides a solid method for the quantitative and selective insertion of charged moieties in the polymer backbone. The coupling of electrochemical and spectroscopic techniques and, particularly, the in situ electrochemical fluorescence allows exploring the interaction between the photogenerated species and the electric charges. We will show in this chapter how electrochemistry may serve as a bridge connecting *dry* and *wet* applications. To achieve this, the major optical features and properties of conjugated polymers will be presented first. Here, the relation between electronic structure and chemical characteristics of polymer chains will be discussed.

Then, we will show that electrochemical methods constitute a major characterization tool for conjugated polymers and, particularly, that the main electrochemical features of the voltammetric profiles can be related to some optoelectronic properties.

A comprehensive review on the development of in situ electrochemical fluorescence spectroelectrochemistry and the major results on the application of this technique to the characterization of conjugated polymers will be presented next.

In the final sections, it will be shown the major features of the in situ electrochemical fluorescence spectroscopy of conjugated polymers and the way these experimental results can be modeled to study relevant interactions between charged species and photoexcitations.

2 Optical Properties of Conjugated Polymers

Upon photoexcitation, conjugated polymers suffer an electronic vertical transition (Franck-Condon) from HOMO to LUMO level ($S_0 \rightarrow S_1$). The excited state is an intrachain electron–hole pair bounded by Coulomb attraction, called exciton. These excitons get delocalization (and stabilization) along several monomer units within a conformational subunit, in a process known as *lattice relaxation*. The excited states are characterized by the transformation of benzenoid to quinoid domains in those conjugated segments (Scheme 1).

The conjugation length of the polymer segments has been determined from experimental and theoretical methods making use of the oligomer approach [24, 25]. This approach considers that key physical properties of conjugated polymers can be determined by studying homologue oligomers with increasing sizes until a convergence limit is reached, the so-called polymer limit. The optical bandgap (which is related to the conjugation extent) can be obtained from UV–Vis

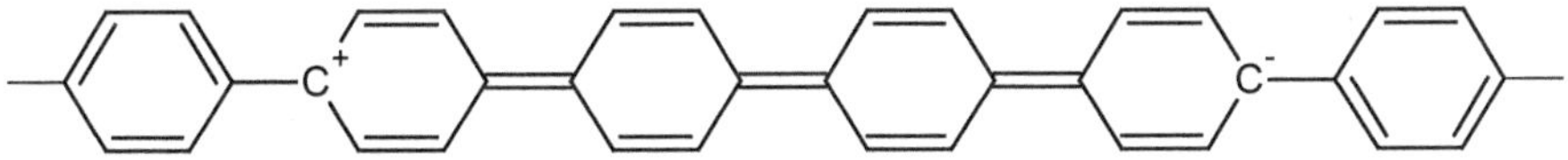

Scheme 1 Molecular structure of an exciton in a poly-*para*-phenylene segment

spectroscopy by determining the energy of the $\pi\pi^*$ transition in oligomers of increasing molecular weights, as shown in Fig. 2 for a set of oligothiophene derivatives.

In this case, the energy of the optical transition decreases as a function of $1/n$, where n is the number of monomers in the conjugated oligomer. Although most research groups used to perform linear extrapolation to obtain the limit of an ideal infinite conjugated polymer (for $1/n \rightarrow 0$), real optical bandgaps of polymers were always significantly larger than extrapolated values. With a number of suitable approximations, including solvent effects, Gierschner and coworkers reported accurate values for the effective conjugation length [27, 28]. Figure 3 shows the spectral positions of the electronic transition for a series of oligothiophenes.

In Gierschner's contributions, the evolution of the spectral positions was fitted to an expression derived by Kuhn from classical mechanics calculations. In this model, the double bonds of the polyene chain are considered as identical oscillators, each vibrating at a characteristic energy, E_0 [27]:

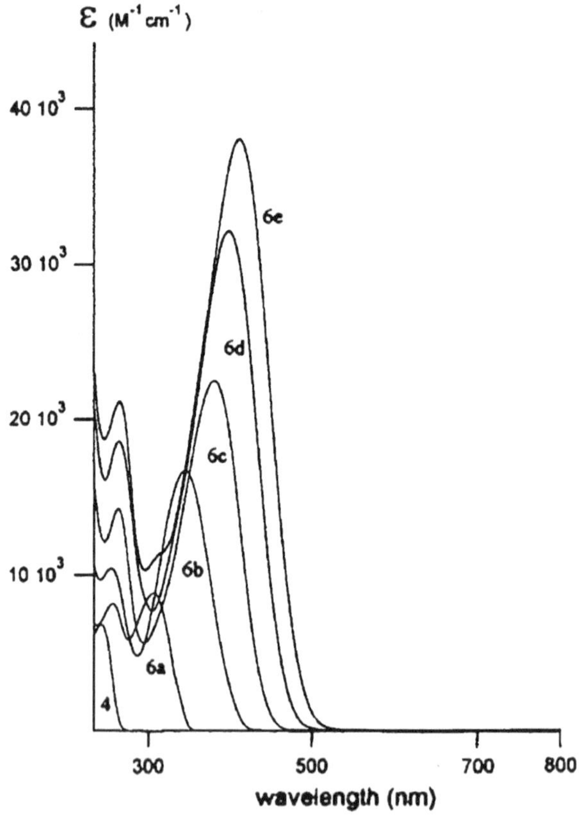

Fig. 2 UV–Vis spectra of chlorinated oligo-3-octylthiophenes dissolved in chloroform (*4* monomer; *6a* dimer; *6b* trimer; *6c* tetramer; *6d* pentamer and *6e* hexamer). Reprinted with permission from [26]. Copyright 1998, American Chemical Society

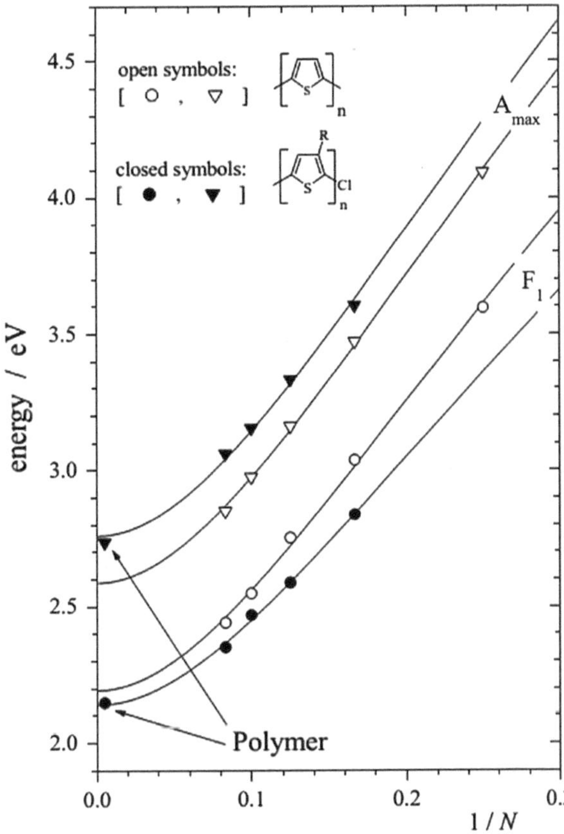

Fig. 3 Spectral position of the spectral maxima A_{max} and the first fluorescent emission maxima F_1 of unsubstituted oligothiophenes (*open symbols*) and regioregular oligo-3-octyl-thiophenes (*closed symbols*) dissolved in chloroform. Adapted from [27] with permission of Wiley

$$E = E_0\sqrt{1 + 2\frac{k}{k_0}\cos\frac{\pi}{N+1}} \quad (1)$$

where N is the number of adjacent double bonds of a polymer, with a coupled force constant k, while k_0 is the force constant of an isolated oscillator.

It was observed that, in most cases, the number of double bonds involved in the exciton of conjugated segments is around 18–22, regardless of the chemical nature of the conjugated polymer [27].

These photogenerated excitons may suffer several de-excitation processes such as internal conversion, non-radiative vibrational relaxation, intersystem crossover or luminescent decay (photoluminescence). Now, we will focus on the latest process: the fluorescent emission of conjugated polymers.

Figure 4 shows a typical fluorescence spectrum recorded for a conjugated MDMO-PPV polymer. The main $\pi\pi^*$ electronic transition is observed as a peak centered at 550 nm, while the vibronic fine structure appearing beyond 575 nm is caused by intramolecular phonon modes coupled to the exciton. In contrast, the

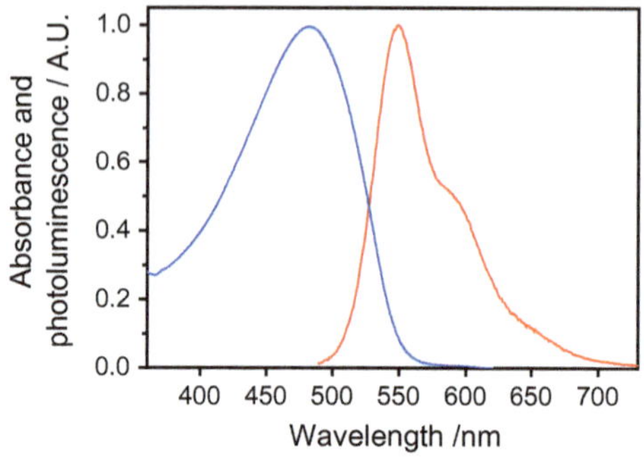

Fig. 4 Normalized UV–Vis absorption (*blue line*) and fluorescence spectra of MDMO-PPV in chloroform (*red line*)

absorption band is usually featureless due to the overlapping of absorptions coming from different polymer segments.

The spectra in Fig. 4 were obtained in dilute solution. Under such conditions, polymer chains remain isolated by the solvent, with no relevant interactions between them. Consequently, the emission is produced from excitations residing in isolated chains, which are known as *intrachain excitons*. However, in most potential applications of conjugated polymers, the materials must be deposited as films on suitable surfaces and more complex photophysical processes can be observed when polymers are in solid state. Since polymeric chains tend to twist and coil when they are deposited, the material should not be considered as composed of long isolated conjugated chains. Actually, polymers are a collection of several structural subunits with shorter conjugated lengths and separated by structural disorders that break conjugation. During the lifetime of an exciton (typically hundreds of ps for most conjugated polymers), several electronic energy transfers lead the exciton to the lowest energy chromophores, which are known as *red-sites*, see Scheme 2 [29–32].

Photoluminescence of conjugated polymers in solid state is strongly dominated by these low-energy exciton traps, despite the multichromophoric character of the light absorbed in regions with different energy. Mirzov and Scheblykin [33] studied the Stokes shifting of the emission in MEH-PPV and concluded that the emission maximum shifts to red when the intermolecular distance between polymer chains becomes lower than 100 nm, as shown in Fig. 5. Consequently, red-sites seem formed by conjugated chain aggregates, causing a decrease in the exciton energy due to π-stacking [34]. These aggregates appear when π-electrons are delocalized over multiple chains in the ground state. The emission from these sites usually preserves some vibronic structure, but it appears redshifted from the emission of isolated chains in solution [35].

In addition to the emission from intrachain excitons or from aggregates, other photoexcited species involving several chains of the conjugated polymer can be formed. Such species are known as *interchain excited-state species*.

The terminology used to describe the different interactions has been extensively revised by Schwartz [29]. Among the existing interchain species, the most studied are probably excimers and exciplexes. These species are photogenerated dimers

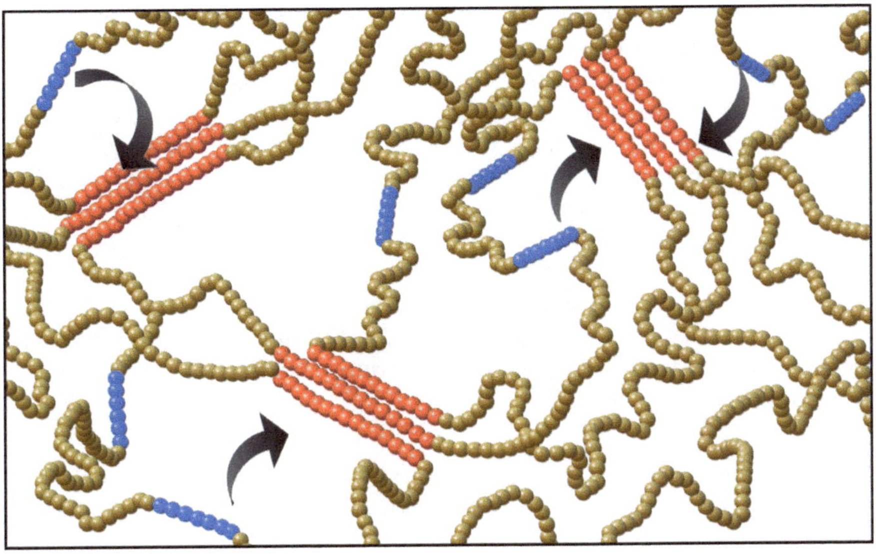

Scheme 2 Representation of the energy transfer between high-energy chromophores (*blue-sites*) and low-energy chromophores (*red-sites*), during the lifetime of an exciton on conjugated polymers

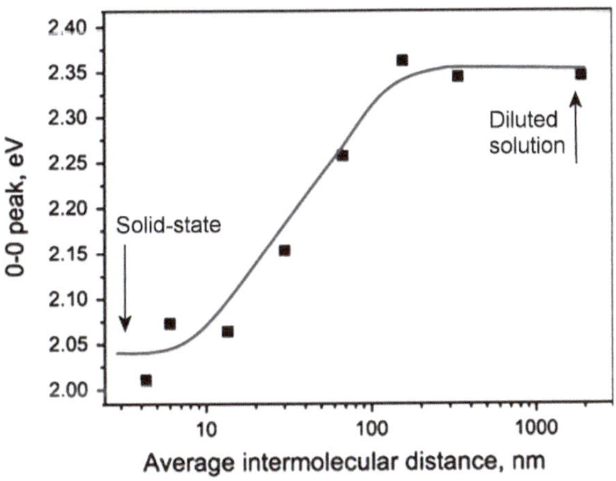

Fig. 5 Energy of the 0-0 transition of the fluorescence spectrum of MEH-PPV films as a function of intermolecular distance. Adapted from [33] with permission of The Royal Society of Chemistry

formed when the components are in the excited state, but not in ground state. Upon photoexcitation, charge transfer between polymer segments can occur and the process leaves a positive charge (or positive polaron) on one segment and an electron (or negative polaron) on the other. These interchain species are known as *polaron pairs*. Excimers appear when π-electrons are shared fairly by vicinal chains, whereas exciplexes are characterized by an unequal sharing of the π-electron density between the segments, i.e., by a partial charge transfer [36].

In most cases, the formation of interchain species leads to non-emitting de-excitation processes (mainly charge separation) [19], but some polymers show emission from excimers [19, 30, 37–39]. Excimer emission is characterized by a featureless band, which appears redshifted with respect to the characteristic excitonic emission [40]. A clear example of this kind of event is shown in the spectra in Fig. 6, which were obtained from a PPV polymer functionalized with nitrile groups (CN-PPV) [41, 42]. This material shows strong electron-accepting properties, and some calculations have demonstrated that the spacing between chains in solid state is smaller than in other PPVs facilitating the formation of bounded polaron pairs in vicinal chains [43]. While CN-PPV solutions present the usual vibronic features of common conjugated polymers (see Fig. 6a), CN-PPV films present a characteristic featureless and redshifted fluorescence spectrum, indicative of the presence of excimeric species (Fig. 6b) [41].

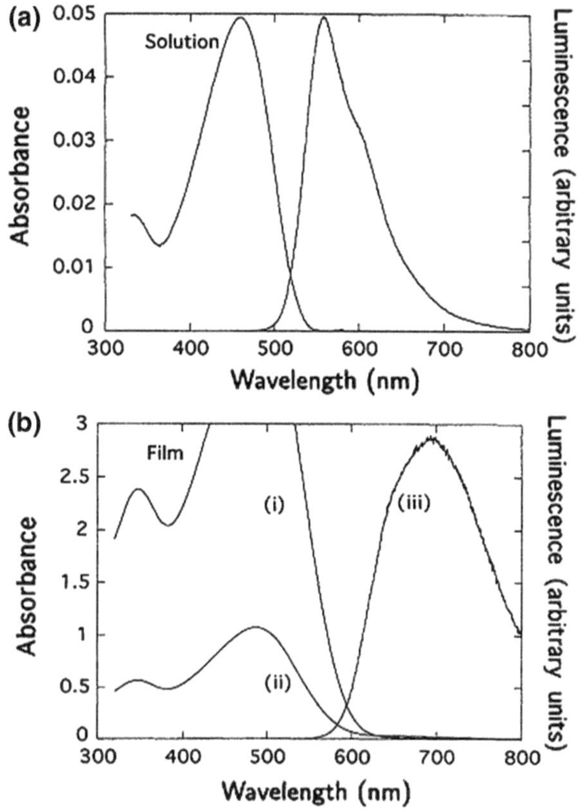

Fig. 6 **a** Absorption and emission spectra (excitation 488 nm) of CN-PPV in toluene solution, **b** absorption and emission spectra (excitation 488 nm) of CN-PPV films. *Curves* (*i*) and (*ii*) are the absorption spectrum for thick and thin films, respectively, and *curve* (*iii*) is the photoluminescence spectra. Adapted from [41] with permission of the American Physical Society

3 Electrochemical Properties of Conjugated Polymers

Electrochemical techniques have been widely used for the characterization of conjugated polymers. These techniques allow the fine control of the redox state of the polymer through the reversible injection of charges, a process that is also known as *electrochemical doping*. The most typical experiment for the characterization of a conjugated polymer consists in its deposition from a solution by dip or spin coating onto a planar electrode, such as ITO or glassy carbon and the subsequent application of a potential program. A characteristic voltammetric response for a fluorene-based conjugated polymer film is shown in Fig. 7.

Here, the stabilized curve on the right corresponds to the reversible oxidation processes of the conjugated chain, or *p*-doping, whereas the left curve shows the *n*-doping process for the same polymer. The onset potential for the *p*-doping is related to the polymer ionization potential or, in other words, with the extraction of one electron from the HOMO level. In the same way, the onset potential for the *n*-doping is linked to the polymer electron affinity and corresponds to the injection of electrons into the LUMO level. Therefore, voltammetric measurements have constituted a simple and suitable way to determine absolute energy levels of conjugated polymers [44–46].

Cyclic voltammetry responses of these systems often differ from ideal curves of absorbed films on solid electrodes. The expected electrochemical features of films are constituted by symmetrical anodic and cathodic waves, with matching peak potentials and peak currents. In contrast to this expected behavior, voltammograms of polymers in solid state exhibit a marked hysteresis between anodic and cathodic waves, which are followed by broad flat current plateaus at further potentials. When polymers are electrochemically doped, the transition from insulating to conducting states generates wide electroactive areas and the oligomer approach has been used to explain the observed electrochemical properties [47–49]. The following two rules can be applied from such approach:

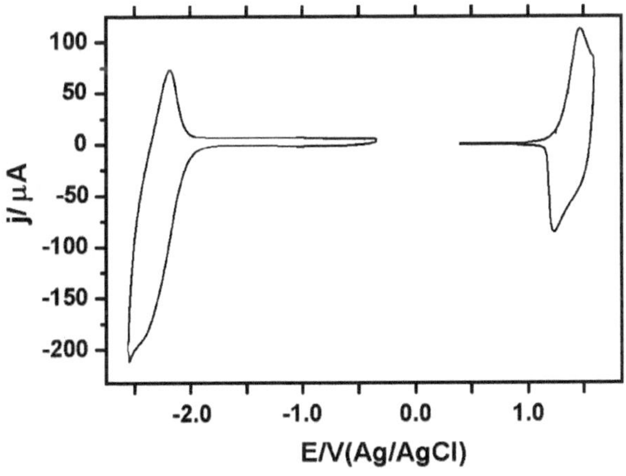

Fig. 7 Stabilized cyclic voltammograms of ITO/polyfluorene–phenylene film in a solution of 0.1M tetrafluoroborate tetrabutylammonium in acetonitrile

Rule 1: the potential required for the formation of a redox state in the oligomer (i.e., the transition from neutral to +1 state) decreases at increasing conjugated chain lengths. These potential values converge to a limit that is characteristic of each conjugated polymer.
Rule 2: The number of accessible redox states increases at increasing chain lengths.

A clear example of these rules is presented in Fig. 8 for a set of conjugated fluorene-based oligomers [50]. According to Rule 1, the first oxidation process appears at lower potentials for longer oligomers. The same tendency is followed by the second, third and successive redox steps. It can be also observed that, according to Rule 2, dimers allow oxidation states +1 and +2, but trimers are able to accept up to three positive charges and tetramers up to four positive charges. Oligomers having more than four units allow a maximum oxidation degree of +4 (Fig. 8a).

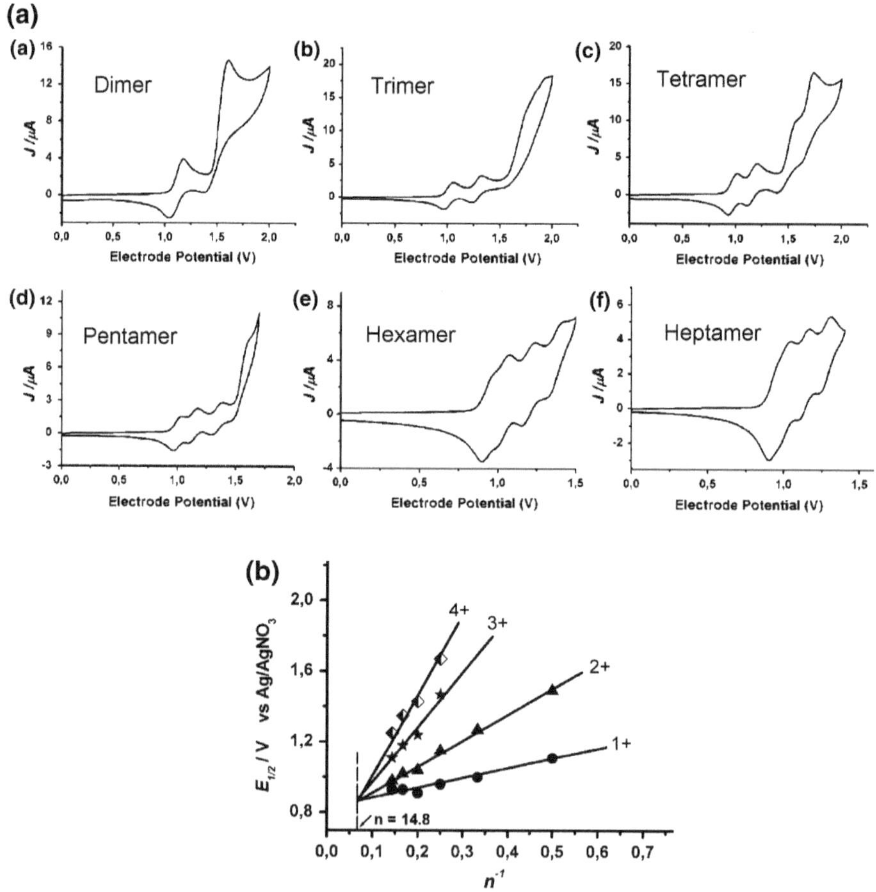

Fig. 8 **a** Cyclic voltammetry of oligofluorenes; **b** relation between the redox half-wave potentials and the reciprocal number of fluorene units for different redox levels. Adapted from [50] with permission of Wiley

The representation of the half-wave potentials for the different oxidation states versus the reciprocal of the number of monomer units shows a linear relationship between both parameters (Fig. 8b). In this case, the difference between the half-wave potentials for oxidized states converges for $n = 14$ monomer units. This value is known as the *saturation length* and means that the polymer is able to accommodate up to four charges in a segment constituted by 14 monomer units without electrostatic repulsion [50]. Furthermore, the shape of the cyclic voltammograms recorded for solid-state conducting polymers is characterized by the overlapping of redox processes coming from the different conjugated units present in the layer.

From the point of view of solid-state physics, the electrochemical injection of charge into conducting polymers leads to the separation of HOMO and LUMO and to the formation of new energy levels within the bandgap, which are known as polarons. Polarons can be treated as quasiparticles and are responsible for the flow of current within the material in solid state. In contrast to free carriers, polarons interact with a cloud of phonons that drags their movement and consequently reduces their mobility, an effect that is known as *electron–phonon coupling*. The electrochemical doping produces a conductivity rise in the conjugated polymer that uses to reach a maximum at around 0.5 charges per ring, with a mobility of charge carriers between 10^{-6} and 10^{-1} $cm^2\ V^{-1}\ s^{-1}$. Usually, this mobility remains below 10^{-5} $cm^2\ V^{-1}\ s^{-1}$ at small doping levels (those obtained for less than 0.01 charge per monomer unit) but sharply increases up to near 10^{-1} $cm^2\ V^{-1}\ s^{-1}$ at doping levels of 0.1 charge/monomer [51–54].

4 In Situ Electrochemical Fluorescence Spectroscopy: A Short Survey on Its Application to the Characterization of Conjugated Polymers

Photoluminescence spectroscopy is one of the less explored areas of spectroelectrochemistry. Indeed, the IUPAC Physical Chemistry Division published in 1998 a survey of spectroelectrochemical techniques, where photoluminescence spectroscopy coupled to electrochemical techniques was not included [55].

Some initial attempts to couple fluorescence and electrochemical systems appeared in reports published in the 1980s [56, 57]. There, it was studied the electrochemical modulation of the fluorescence coming from zinc tetraphenylporphyrin films or from passivating films obtained from the reduction in benzotriazolate on copper electrodes [58].

Fluorescence spectroscopy was also used for the characterization of electrochemical liquid–liquid interfaces (interface between two immiscible electrolyte solutions, ITIES) to investigate the transfer of fluorescent ions across a dichloroethane–water interface [59, 60]. Later, it was reported the electrochemical modulation of the fluorescence of a polyallylamine hydrogel containing attached flavins

[61]. There, a portion of electroactive flavins was reduced to the non-fluorescent leucoflavins. However, the emission of the hydrogel only decreased at around 25%, indicating that most fluorescent flavins stayed far from the electrode surface and did not suffered electrochemical reduction.

The electrochemical fluorescence switching of species like *o*-toluidine and a perylene derivative were reported during the 2000s [62, 63]. The authors presented in those papers a multipurpose thin-layer spectroelectrochemical cells valid for several in situ techniques such as UV–Vis, Raman and fluorescence spectroscopy.

The ability for the electrochemical switching of fluorescence of several triazines and tetrazines promoted the development of a new kind of system called electrofluorochromic device. This structure is equivalent to an electrochromic window and was reported in a series of papers appeared before 2010 [64–67]. Other electrofluorochromic devices based on polyfluorenes whose emission can be switched from yellow to non-fluorescent with high contrast ratios have been recently reported [68–70], and a review on this particular application appeared in 2013 [71].

In addition to electrofluorochromic windows, further applications of fluorescence coupled to electrochemical systems include spectroelectrochemical sensing, a field that has been developed mainly by Heineman and coworkers [72–75]. Fluorescent spectroelectrochemical sensors are based on selective Nafion® films that act as preconcentration agents for fluorescent model analytes, such as Ru and Fe complexes or pyrene derivatives. Interestingly, Ding et al. have shown that the ability of fluorescent benzothiadiazole–carbazole copolymers to undergo reversible electrochemical switching can be successfully applied to the detection of cyanide anions [76]. It is observed that the oxidation of this copolymer strongly quenches its fluorescence in conventional supporting electrolytes. However, the presence of cyanide ions weakens the quenching because of the interactions appearing between these species and the electron-deficient benzothiadiazole centers.

In situ electrochemical fluorescence can also be applied to gain insight into the photophysical properties of conjugated polymers. The interaction of charged excitations with photogenerated species is important to understand the performance of polymer-based optoelectronic devices. The operation of such systems is strongly affected by the generation, migration and annihilation of excitons [23]. For example, a fast exciton migration is required in photovoltaic cells, where the exciton must move to a donor–acceptor interface for the eventual generation of electrical charge. However, in light-emitting diodes the exciton migration is harmful because it can lead to luminescence quenching in defect sites or in the electrodes [77].

Some former papers devoted to the characterization of polythiophenes, poly(*N*-vinylcarbazole) and indole-based polymer films were published in the last years of the past century [78–81]. In those papers, a reversible decrease in the emission intensity of polymers is observed upon electrochemical doping.

Polyaniline (PANI) and its derivatives are traditionally the most studied conducting polymers, but in situ fluorescence has been employed as a proper characterization tool only in recent years. PANI in solution presents a slight fluorescent emission in its fully reduced form, leucoemeraldine. When the polymer is oxidized to the emeraldine form, the subsequent oxidation state of PANI, the fluorescence is quenched [82]. Antonel et al. [83, 84] observed that the fluorescence of leucoemeraldine films was also quenched when deposited onto metallic electrodes like platinum. Only when PANI films were thicker than 25 nm, a fluorescent emission could be detected, which increased in proportion with the polymer thickness [85].

Several studies on conducting polymers are focused on the so-called *first-cycle* or *memory effect*. When conducting polymers are aged in its semiconducting state, the *p*-doping features of the first voltammetric cycle appeared at more positive potentials than in the stabilized cycles [86–88]. Marmisolle and coworkers studied this effect in poly(*o*-methyl aniline) films and detected that the fluorescence signal seemed to increase during aging. They found a semilogarithmic relation of the emitted intensity with the time of the experiment [89].

The in situ electrochemical fluorescence technique is also an interesting tool for the characterization of conjugated polymers different from PANI. Particularly, light-emitting conjugated polymers mainly phenylene vinylene (PPV) and fluorene-based (PF) polymers have been the subject of recent research since they are widely employed in optoelectronic devices. Some of the polymer materials characterized by this technique are shown in Scheme 3.

PPVs are frequently used as emitters in the range from yellow to orange-red in OLED technology, but they have been also employed as chromophores and donors in bulk heterojunction organic photovoltaic cells [90–92].

Carter and coworkers studied for the first time the fluorescence of MEH-PPV during electrochemical *p*- and *n*-doping [93]. They reported that the quenching observed upon doping was due to the formation of metallic-like domains in the polymeric material.

BEH-PPV and MEH-PPV were studied by Palacios et al. by single-molecule fluorescence spectroelectrochemistry [94]. They observed that the electrochemical oxidation of isolated solid-state particles of conjugated polymers produced the quenching of their fluorescent emission over a narrow potential range. However, this effect was limited to those polymer areas in close contact with the working electrode surface, and the material that remained away from the metallic support suffered a slower quenching process. In that case, the quenching was dominated by the penetration of electrolyte anions into the polymer nanoparticles.

The in situ fluorescence of MDMO-PPV was monitored by Montilla and Mallavia during voltammetric scans [95]. The fluorescence of the polymer in its native state remains high, corresponding to its typical photoluminescence. As observed for other conjugated polymers, the *p*-doping induces the emission quenching.

Concomitant to the doping process, conjugated polymers suffer morphological modifications due to the transition from the semiconducting to the conducting state. These polymers present nonpolar, solvatophobic character, in their native state, but the injection of charges increases their affinity for polar solvents causing swelling processes. Structural and morphological changes that occur upon electrochemical doping have been well described by Otero and coworkers in its electrochemically stimulated conformational relaxation (ESCR) model [86, 96]. However, the

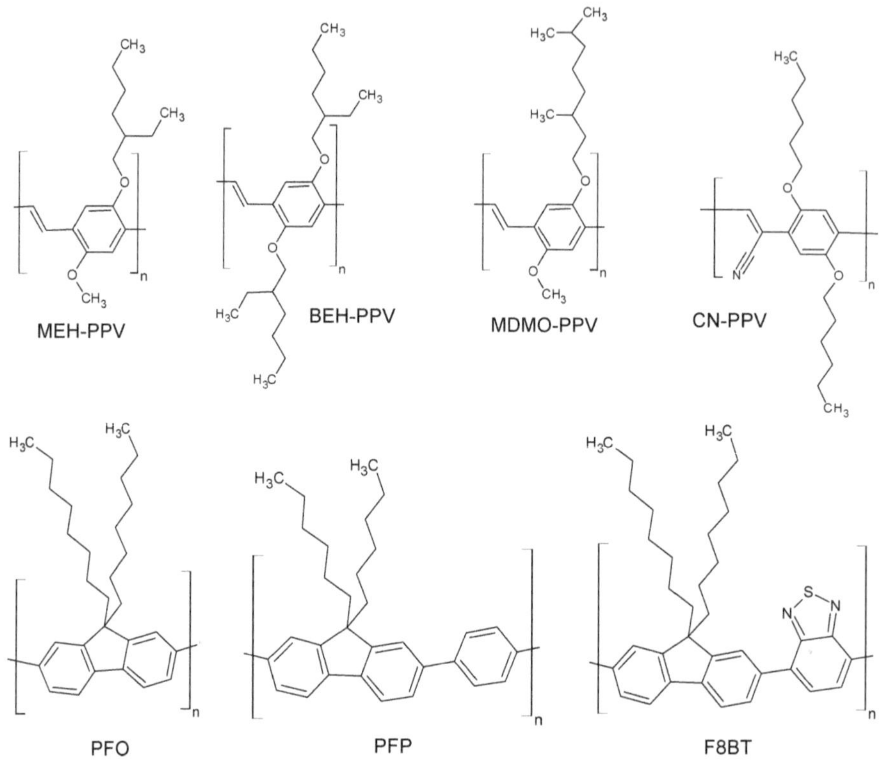

Scheme 3 Chemical structure of conjugated polymers characterised by in situ electrochemical fluorescence

fluorescence of MDMO-PPV cannot be completely restored after dedoping, which is indicative that the material retains part of the solvent after the first voltammetric cycle (commonly, the interaction of fluorescent molecules with polar solvents causes an additional deactivation of emissive states [97]). Interestingly, the emission returns after a resting time, without passing electrical current, in the semiconducting state. The fluorescence recovery follows a pseudo-first-order kinetics that can be modeled as a shrinking of the solvatophobic polymer [98].

The photoluminescence quenching that appears during the doping process of MEH-PPV and other conjugated polymers, such as poly(3-hexylthiophene), was quantitatively modeled by van Reenen et al. They employed an exciton diffusion model followed by a Förster resonance energy transfer or a charge transfer to polaronic sites [99].

Another broad group of light-emitting conjugated polymers is the one based on fluorene units (PFs), which presents a wide range of application in organic electronics [100–106]. The photophysics of these materials are of great importance, among other aspects, because they are strategically relevant for the development of polymer-based RGB OLEDs [107, 108]. The blue component of the organic diodes is usually made of polyfluorene, which presents low chemical stability under operation. It is known that PFs degrade fast and tend to produce undesirable low-energy emission bands (green or g-bands). As a result, the lifespan of the blue emitters is significantly shorter than that of red and green emitters [109].

The first paper devoted to the study of in situ fluorescence of PFs was published in 2006 [107]. There, the modulation of the photoluminescence of a blue-violet PFP was attained through the reversible electrochemical modification of the conjugated chain (either *p*- or *n*-doping). While the injection of holes quenched completely the emission, injected electrons quenched only up to one-third of the initial fluorescence of the polymer. Although the electrochemical *n*-doping of polyfluorenes initially induces the quenching of the blue emission, a parallel electrochemical reaction produces the growth of new green emission bands at around 500 nm. This green emission was assigned to cross-linked species favoring the formation of intramolecular excimers [108].

PFO, F8BT and other conjugated polymers have been extensively studied by Barbara and coworkers using single-molecule fluorescence spectroelectrochemistry [30, 35, 92, 94, 110–115]. It was reported that the oxidation of F8BT originates a fluorescence quenching over a narrow distribution of single-molecule half-wave potential values [110]. This observation reveals that the electrochemical potential was uniformly distributed throughout the electrode dimensions. In addition to this effect, it was observed that after the reversible hole injection, a small fraction of those holes remained deeply trapped inside the polymer. Those species, known as *deep traps*, were able to quench partially the fluorescence of the polymer particles [112].

In situ fluorescence studies on a fluorene-*alt*-phenylene polymer functionalized with perylenediimide pendant groups (PFP-PDI) were performed by Montilla et al. [116]. It was reported that the conjugated chain of this polymer acts as an electron donor and the attached perylene as an electron acceptor after photoexcitation. The result is significant because polymers with donor–acceptor features are promising materials for photovoltaic applications. Both PFP and PDI species are fluorescent when isolated, but the PFP-PDI polymer did not yield photoluminescence because of the photoinduced electron transfer from the conjugated chain to the perylenediimide group. When these PDI pendant groups were electrochemically reduced, the photoinduced electron transfer was inhibited and the fluorescence from the polymeric chain was recovered.

In addition to conventional spectroscopy, some advanced fluorescence techniques can be applied to spectroelectrochemical studies on conjugated polymers. Among such techniques, polarized fluorescence (or fluorescence anisotropy) has been applied successfully to the study of MDMO-PPV and PFP because it provides valuable information about exciton dynamics [117]. Usually, the emission from polymer films in their native (semiconducting) state is strongly depolarized due to resonance energy homotransfer, and consequently, the emission is fully isotropic. An anisotropy increase can be observed, upon electrochemical doping, which is accompanied by the partial quenching of emission. The intensification in the emission polarization has been attributed to the introduction of randomly distributed quenchers in the conjugated polymer chains. These quenchers cause the drop of the exciton lifetime but also reduce the mean free path of each exciton, and as a result, the anisotropic character of the emission is increased.

5 Major Features of In Situ Electrochemical Fluorescence Spectroscopy of Conjugated Polymers

Self-trapped excitations in conjugated polymers, appearing as either charged polaronic (i.e., electrons and holes) or neutral species (i.e., excitons), can interact by a combination of electronic and lattice energy contributions.

Many spectroscopic techniques can be used to explore the role and nature of polarons formed upon electrochemical doping [118]. Among them, common spectroelectrochemical techniques such as in situ UV–Vis, FTIR Raman or ESR have been applied to the understanding of chemical changes and electronic transitions occurring after the electrochemical injection of charges. The study of photogenerated excitations in conjugated polymers can be performed by complex time-resolved spectroscopic techniques such as pump–probe spectroscopy, ultrafast laser spectroscopy. However, few methods are able to provide data on the interaction between polarons and photogenerated excitations. In situ electrochemical fluorescence allows the study of the interactions taking place during the doping of conjugated polymers and provides useful information, even at very low doping levels, due to its high sensitivity. In addition, this technique is able to reveal polymer–solvent interactions or conformational relaxation effects occurring during electrochemical doping.

Apart from pure electrochemical phenomena, the study of relevant interactions between electric charges and photoexcited states opens a way to gain more insight into the most significant processes affecting the performance of organic LEDs and organic photovoltaic cells. The technique can be used to study polaron–exciton annihilation, to determine diffusion coefficients of excitons or effective annihilation radii, among other required parameters. In other words, the electrochemical monitoring of fluorescence can be applied to conjugated polymers at the two main levels: *wet* (electrochemical) and *dry* (organic electronics) environments.

Figure 9 constitutes a clear example of the former application. There, a cyclic voltammogram is presented for a MDMO-PPV deposited onto an ITO electrode and immersed in a fluorescence spectroelectrochemical cell. In addition, Fig. 9b shows in situ fluorescence spectra recorded at different potentials during the forward scan. The first spectrum taken at 0.38 V (a potential at which the polymer is in semiconducting state) is characterized by an emission maximum at 583 nm and a vibronic feature around 630 nm. The application of potentials higher than 0.40 V (the onset of *p*-doping) results in a progressive quenching of the emission. The complete quenching process is accomplished at early stages of polymer doping from 0.40 to 0.65 V. The polymer fluorescence raises in the reverse scan and

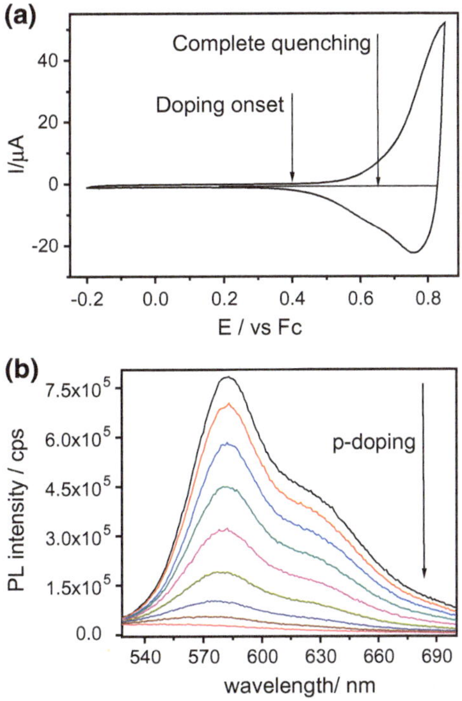

Fig. 9 **a** Cyclic voltammogram of a MDMO-PPV deposited on ITO electrode. **b** Fluorescence spectra recorded at different potentials in acetonitrile medium. The spectra were taken at 0.38 (initial spectrum), 0.41, 0.44, 0.47, 0.50, 0.53, 0.56, 0.59, 0.65 V. Excitation at 500 nm

reaches a maximum at 0.40 V, where the polymer recovers its semiconducting state.

Among the existing spectroelectrochemical techniques, fluorescence spectroscopy is one of the most sensitive to the redox state of the polymer, and Fig. 10 constitutes another clear example of this. There, the effect of the applied potential on the doping level of polyfluorene (PFO) on the detected fluorescence and on the UV–Vis spectral features is shown.

It can be observed that the quenching of the fluorescence emission starts at 0.60 V and reaches 20% level at 0.65 V, although no electrochemically induced UV–Vis bands can be detected in this potential range. Such bands appear at potentials above 0.68 V, where the differential UV–Vis spectrum presents two negative-going absorption features at 407 and 438 nm, both assigned to the bleaching of the neutral PFO absorption. The positive-going band at 643 nm and the shoulder at around 590 nm are a consequence of the absorption from isolated charge carriers. Fluorescence is completely quenched beyond 0.79 V, and new UV–Vis absorption bands raise at $\sim$580 nm, corresponding to the formation of polaronic species. In parallel to the MDMO-PPV polymer, the complete quenching of emission for PFO appears at the early stages of doping, and it is accomplished after one electron is transferred to a polymer segment with around 250 monomer units.

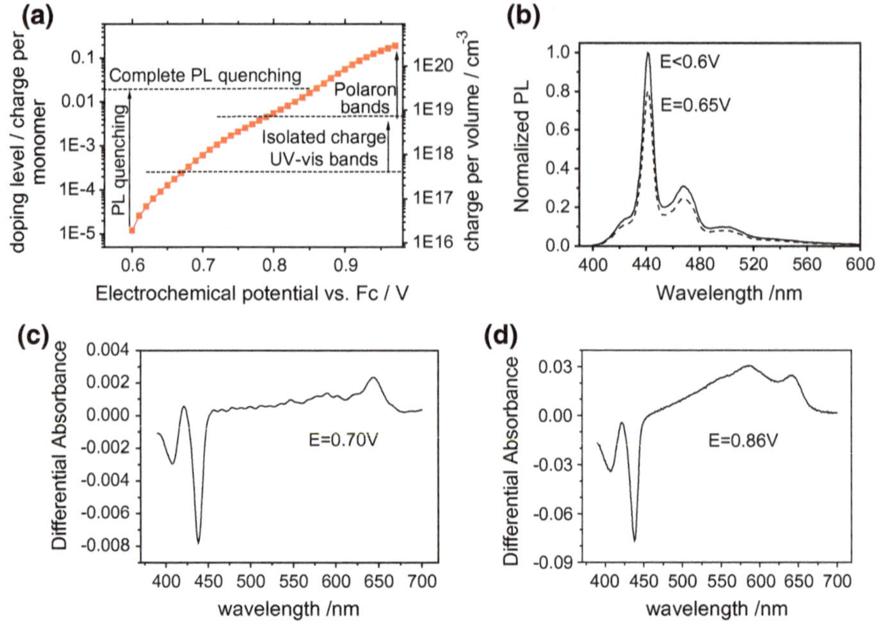

Fig. 10 **a** Doping level and charge density versus applied potential for a PFO-modified electrode; **b** fluorescence spectra of PFO deposited on ITO at different electrochemical potentials (excitation at 380 nm); **c**, **d** electrochemically induced UV–Vis band at different electrochemical potential, all spectra are referred to the single spectrum of PFO at 0.5 V. Adapted from [119], Creative Commons License (CC-BY)

The electrochemical quenching mechanism implies Förster resonance energy transfer from excitons (that act as energy donors) to polarons (the energy acceptors). The rate constant for this energy transfer depends on the distance between both species, R:

$$k_T = \frac{1}{\tau_0}\left[\frac{R_0}{R}\right]^6 \tag{2}$$

In this expression, τ_0 is the exciton lifetime in the absence of quencher and R_0 is the quenching critical radius (or Förster critical radius), that is the distance at which energy transfer and spontaneous decay of the excited state are equally probable [120].

The quenching critical radius can be obtained from experimental data by determining the spectral overlap between donors and acceptors. Figure 11 shows both the emission and the absorbance spectrum of doped PFO, where the spectral overlap extends from ca. 470 to 580 nm.

Equation (3) provides a simple way to obtain the value of such parameter for PFO:

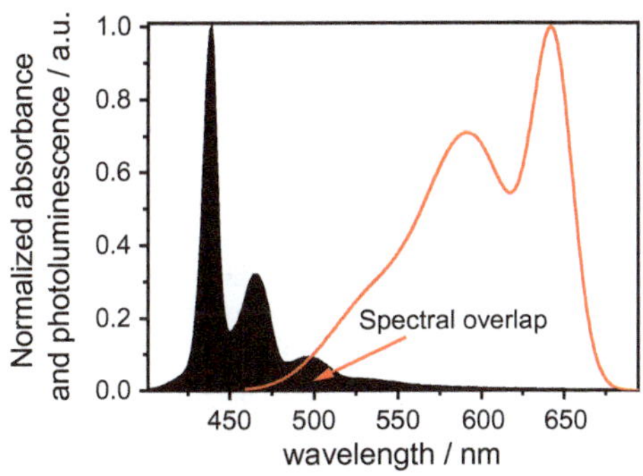

Fig. 11 Normalized fluorescence spectrum of pristine PFO (*black*) and absorbance of doped PFO polymer at 0.80 V (*red line*)

$$R_0(\text{Å}) = 0.211\left[Q_D\kappa^2 n^{-4} J\right]^{1/6} \tag{3}$$

Here, Q_D is the emission quantum yield, J is the spectral overlap integral in M^{-1} cm^{-1} nm^4, κ^2 is the dipole orientation factor (0.476 for randomly oriented chromophores in solid state), and n is the refractive index of the polymer [120]. The calculation yields R_0 = 30 Å, so the probability of quenching is more than 50% when excitons find polarons at lower distances.

Other characteristic features observed during polymer electrochemical doping include blueshifts of the emission maximum. Figure 12a shows the in situ fluorescent intensity recorded for MDMO-PPV at increasing potentials. In this particular case, the doping of the polymer stars at 0.38 V and the complete quenching occurs at 0.65 V.

The evolution of the normalized emission spectra at different potentials is shown in Fig. 12b. The emission maximum stays at around 583 nm for the early stages of quenching, but after 50% fluorescence is quenched (a level attained at 0.50 V), the peak shifts to lower wavelengths. This hypsochromic displacement reaches 8 nm at 0.58 V, a point at which the emission from the polymer is hardly detected.

The blueshifting is explained in terms of a decrease in the exciton lifetime upon electrochemical doping. Initially, with the polymer in its semiconducting state, the photogenerated excitons formed at different conjugated segments are able to migrate toward low-energy chromophores, i.e., red-sites. The electrochemical doping inserts charged species that quench the photogenerated excitations, thus decreasing the exciton lifetime. As mentioned previously, the quenching mechanism implies Förster energy transfer between excitons and polarons. Since the transfer efficiency depends strongly on the distance between these species, quenching is effective only if the exciton finds a polaron at a distance lower than 3 nm.

For MDMO-PPV, the exciton lifetime is still too high to reach red-sites at potentials within the range from 0.40 to 0.47 V. The blueshift observed at higher potentials reveals that the bulk density of electrochemical charges has reached a value high enough to produce direct quenching from red-sites or, in other words, that red-sites find polaronic moieties at a distance lower than 3 nm. Under those

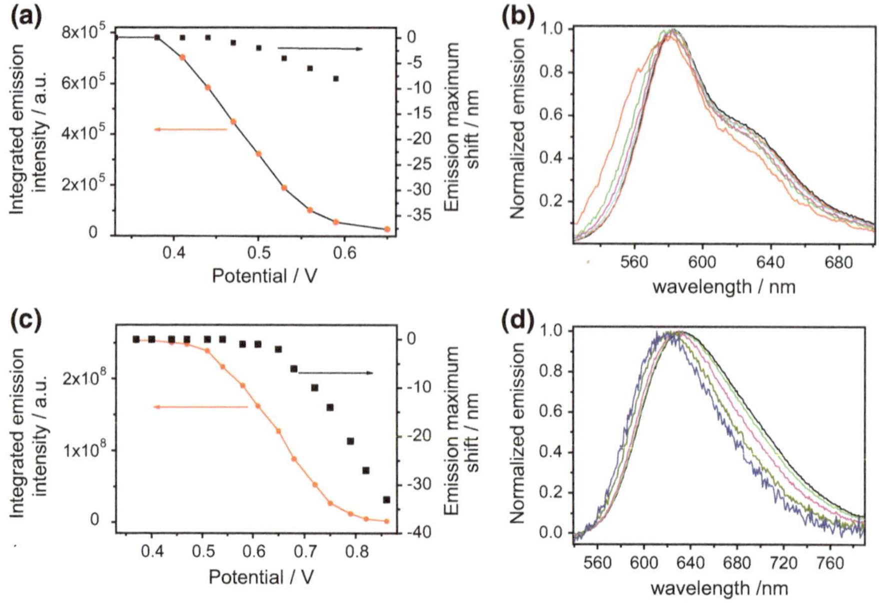

Fig. 12 **a** Evolution of the integrated emission intensity of MDMO-PPV (*red dots*) and emission maximum shift (*black solid squares*) as a function of the electrochemical potential. **b** Normalized fluorescence spectra recorded at different potentials. The spectra were taken at 0.38 (initial spectrum), 0.44, 0.47, 0.50, 0.53 V; excitation at 500 nm. **c** Evolution of the integrated emission intensity of CN-PPV (*red dots*) and emission maximum shift (*black solid squares*) as a function of the electrochemical potential. **d** Normalized fluorescence spectra recorded at different potentials. The spectra were taken at 0.40 (initial spectrum), 0.54, 0.65, 0.72 0.75 V; excitation at 500 nm

conditions, the emission takes place directly from non-aggregated polymer chains (blue-sites), while excitons that reach red-sites are quenched. The −0.03 eV shift means that emission from blue-sites arises from segments with around 50 monomers, as determined by Eq. (1).

The blueshift observed in the emission maximum is more evident during the electrochemical doping of CN-PPV, for which the fluorescence of the semiconducting state arises mainly from excimers. Figure 12c shows the evolution of the emission intensity at 632 nm during a forward potential scan. The emission is quenched by the electrochemical charges injected from 0.40 V, and the complete quenching is attained above 0.85 V. During the initial stages of quenching, the emission peak is centered on 632 nm, but after 50% deactivation (occurring at ca. 0.65 V), the emission shifts to the blue and reaches a value of −33 nm at 0.86 V. Figure 12d shows that, along with the shift of the maximum, a spectral narrowing occurs upon electrochemical doping. These results reveal that, at first, electrochemically injected polarons quench the low-energy excimers and, from 0.65 V, the emission is obtained directly from intrachain excitons, until the full quenching at 0.9 V.

6 Modeling and Quantification of Fluorescence Electrochemical Quenching

The electrochemical quenching of fluorescence can be modeled using parallel approaches to those employed for solid-state fluorophores. For conjugated polymers in their native state, the emitted fluorescence is proportional to the density of singlet excitons generated during a photoexcitation pulse $[S_1]$. The decrease in such exciton density over time obeys the following differential equation (monomolecular decay):

$$\frac{\mathrm{d}[S_1]}{\mathrm{d}t} = -\frac{[S_1]}{\tau_0} \tag{4}$$

where τ_0 is the characteristic lifetime for the photogenerated excited state. The solution of this differential equation leads to:

$$[S_1] = [S_1]_0 \exp\left(-\frac{1}{\tau_0}t\right) \tag{5}$$

where the initial condition is $[S_1] = [S_1]_0$ at $t = 0$.

The fluorescence intensity emitted after the excitation pulse, I, should be proportional to the density of excitons, and therefore:

$$I(t) = I(0)\exp\left(-\frac{1}{\tau_0}t\right) \tag{6}$$

Emitted intensity must follow an exponential trend, and, in fact, this is the most common case of fluorescence emission decay of conjugated polymers in semiconducting state [23, 121].

The introduction of quenching centers, such as electrochemically injected polarons, into the polymer promotes non-radiative ways for de-excitation. When a photogenerated singlet exciton (S_1) finds a quencher (p), an annihilation process takes place:

$$S_1 + p \xrightarrow{k_q} S_0 + p + \text{phonons} \tag{7}$$

The annihilation process follows a bimolecular mechanism, which is dependent on the diffusion rate of the quasiparticles in the organic solid. We will consider first that the annihilation rate constant, k_q, is time-independent. In such a case, the emission intensity would follow an exponential time decay:

$$I(t) = I(0)\exp\left[-\left(\frac{1}{\tau_0} + k_q[p]\right)t\right] \tag{8}$$

being $[p]$ the quencher density, i.e., the charge density required for electrochemical quenching.

In steady-state fluorescence experiments, the excitation source may be considered as an infinite sum of differential light pulses (a continuous light source), and then, the fluorescent emission intensity can be determined after integration of Eq. (8):

$$I = \int_0^{\infty} I(0)\exp\left[-\left(\frac{1}{\tau_0} + k_q[p]\right)t\right]\mathrm{d}t \tag{9}$$

$$I = I(0)\frac{1}{1 + \tau_0 k_q[p]} \tag{10}$$

here $I(0)$ and I represent the fluorescent intensity in the absence and in the presence of the quencher, respectively. Equation (10) can be then rewritten as:

$$\frac{I(0)}{I} = 1 + k_q \tau_0 [p] \tag{11}$$

an expression that is known as the *Stern–Volmer equation.*

The value of the annihilation constant, k_q, is dependent on the diffusion rate of the species implied in the annihilation process, excitons and quenchers, and can be expressed in the form originally proposed by Smoluchowski [122]:

$$k_q = 4\pi DR \tag{12}$$

In such relation, D is the mutual diffusion coefficient ($D_{\text{exciton}} + D_{\text{quencher}}$) and R is the distance at which the particles can interact, i.e., the exciton–polaron annihilation radius.

The annihilation rate constant is often considered time-independent. However, in most cases, the exciton–polaron annihilation implies a distance-dependent resonant energy transfer via dipole–dipole interaction between donor–acceptor couples. Based on such annihilation mechanism, Gösele et al. obtained an approximate solution for k_q and proposed an expression that includes a time-dependent factor, resembling the Smoluchowski equation [123]:

$$k_q(t) = 4\pi DR\left[1 + \frac{R}{(\pi Dt)^{1/2}}\right] \tag{13}$$

In this case, fluorescence decay in the presence of quenchers does not follow a simple exponential trend and the intensity emitted after a photoexcitation short pulse would follow the general time-dependent form:

$$I(t) = I(0)\exp\left[-\alpha t - 2\beta\sqrt{t}\right] \tag{14}$$

where α comes from the time-independent component of the quenching:

$$\alpha = \frac{1}{\tau_0} + 4\pi DR[p] \tag{15}$$

and β comes from the time-dependent component:

$$\beta = 4R^2\sqrt{\pi D}[p] \tag{16}$$

In this scenario, transient effects are detected at short times after the excitation pulse. Usually, these effects are irrelevant for low or moderate concentrations of quenchers, but they are noticeable at larger concentrations. The integration of Eq. (14) yields the ratio of the intensities of steady-state emission in the absence and in the presence of quencher and leads to an expression akin to the Stern–Volmer equation:

$$\frac{I_0}{I} = \frac{1 + 4\pi DR\tau_0[p]}{Y} \tag{17}$$

being Y:

$$Y = 1 - \frac{\beta\sqrt{\pi}}{\sqrt{\alpha}}\exp\left(\frac{\beta^2}{\alpha}\right)\mathrm{erfc}\left(\frac{\beta}{\sqrt{\alpha}}\right) \tag{18}$$

with erfc the complementary error function:

$$\mathrm{erfc}(x) = 1 - \mathrm{erf}(x) = 1 - \left(\frac{2}{\sqrt{\pi}}\int_0^x \exp(-u^2)\,\mathrm{d}u\right) \tag{19}$$

Equation (17), where the Y factor symbolizes the transient term, can be now compared with the Stern–Volmer Eq. (11). The Y expression is not easy to handle, but there exist some useful approximations. In the common case that $\beta/\sqrt{\alpha}$ gets close to zero, Y can be approximated to an exponential function [124]:

$$Y \approx \frac{1}{\exp(4R^2\pi\sqrt{\tau_0 D}[p])} \tag{20}$$

After combining (17) and (20), we obtain:

$$\frac{I_0}{I} = (1 + 4\pi DR\tau_0[p])\exp\left(4R^2\pi\sqrt{\tau_0 D}[p]\right) \tag{21}$$

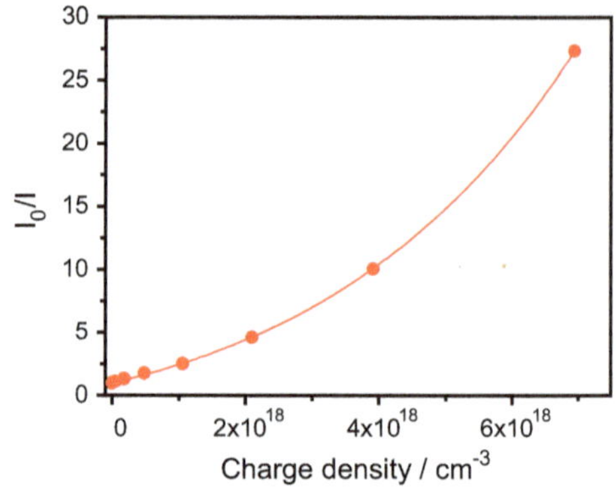

Fig. 13 Electrochemical Stern–Volmer plot obtained during the *p*-doping process of MDMO-PPV

The electrochemical quenching of steady-state fluorescence of conjugated polymers in solid state can be represented as Stern–Volmer plots by tracing *I*(0)/*I* against the injected charge. Typically, these plots exhibit fair linear regions at low doping levels, thus showing little transient effects [95, 107]. However, the represented lines often deviate upwards at higher values of injected charge. Figure 13 shows the electrochemical Stern–Volmer plot for MDMO-PPV obtained upon *p*-doping.

Data in Fig. 13 were fitted to Eq. (21) after fixing a value for the exciton lifetime of $\tau_0 = 360$ ps [36]. A mutual diffusion coefficient, *D*, of 0.0018 $cm^2\ s^{-1}$ and an exciton–polaron annihilation radius of 1.3 nm are then derived.

Individual diffusion coefficients of excitons can be calculated from these experiments, previous estimation of the diffusion coefficients of polarons from the values of charge carrier mobility in conjugated polymers.

Polaron mobilities ranging from 10^{-11} to 10^{-7} $cm^2\ V^{-1}\ s^{-1}$ have been reported for MDMO-PPV [125, 126]. It is worth noting that such values have been obtained usually in transistor-like devices and the nature of the charge carriers may differ from the electrochemically generated ones [53, 127]. Typical values of mobility for electrochemically generated carriers in conjugated polymers are in the range 10^{-6} to 10^{-4} $cm^2\ V^{-1}\ s^{-1}$ at low doping levels [51, 52].

It is assumed for the case of doped conjugated polymers that the electric field inside the polymer film remains low since, in the conditions of electrochemical doping, charge compensation inside the polymer is achieved by the presence of counterions [99]. In these circumstances, the values of diffusion coefficient of charged moieties can be calculated by making use of the Einstein relation:

$$D = \frac{\mu k T}{e} \tag{22}$$

where μ is the electrical mobility of the polaron, *k* the Boltzmann constant, *T* the absolute temperature, and *e* the charge of the carrier. Diffusion coefficients are in the order of 10^{-8} to 10^{-6} $cm^2\ s^{-1}$ for this polymer, and therefore, we may neglect their contribution to the mutual diffusion coefficient. In this way, the exciton diffusion coefficient obtained from the fitting curve in Fig. 13 matches the mutual diffusion coefficient (0.0018 $cm^2\ s^{-1}$). Such a value is comparable to those found for other PPVs using more complex methodologies, as fluorescence volume quenching [128–130].

7 Conclusion

In summary, in this chapter we have shown that in situ electrochemical fluorescence spectroscopy is useful to gain more insight into the photophysical processes associated with the doping of conjugated polymers. The characterization can be performed even at very low doping levels because of the high sensitivity of the technique. Interactions between polymers and solvents are also exposed since fluorescence recognizes changes in the polarity of the molecular environment surrounding the photoactive polymers. A suitable analysis of such interactions may be used to explore conformational relaxation effects appearing during the swelling of the polymers after electrochemical doping. Apart from electrochemical phenomena, in situ fluorescence spectroscopy allows the study of relevant interactions between electric charges and photoexcited states and the interpretation of significant processes influencing the performance of organic devices (particularly LEDs and photovoltaic cells). Exciton diffusion coefficients or effective annihilation radii, among others, are key parameters required to improve the operation of such devices.

References

1. Saxman, A.M., Liepins, R., Aldissi, M.: Polyacetylene: its synthesis. Doping and structure. Prog. Polym. Sci. **11**(1–2), 57–89 (1985)
2. Salaneck: Science and applications of conducting polymers. Papers from the Sixth European Industrial Workshop. CRC Press, Boca Raton (1991)
3. MacDiarmid, A.G., Epstein, A.J.: "Synthetic Metals": a novel role for organic polymers. Makromol. Chemie. Macromol. Symp. **51**(1), 11–28 (1991)
4. Heeger, A.J.: Semiconducting and metallic polymers: the fourth generation of polymeric materials. J. Phys. Chem. B **105**(36), 8475–8491 (2001)
5. Basescu, N., Liu, Z.-X., Moses, D., Heeger, A.J., Naarmann, H., Theophilou, N.: High electrical conductivity in doped polyacetylene. Nature **327**(6121), 403–405 (1987)
6. Tsukamoto, J.: Recent advances in highly conductive polyacetylene. Adv. Phys. **41**(6), 509–546 (1992)
7. Skotheim, T.A., Reynolds, J.R. (eds.): Conjugated Polymers: Processing and Applications, 3rd edn. CRC Press, Boca Raton (2006)
8. Schubert, E.F.: Light-Emitting Diodes. Cambridge University Press, Cambridge (2003)
9. Grem, G., Leditzky, G., Ullrich, B., Leising, G.: Blue electroluminescent device based on a conjugated polymer. Synth. Met. **51**(1–3), 383–389 (1992)

10. Colladet, K., Fourier, S., Cleij, T.J., Lutsen, L., Gelan, J., Vanderzande, D., Huong Nguyen, L., Neugebauer, H., Sariciftci, S., Aguirre, A., et al.: Low band gap donor—acceptor conjugated polymers toward organic solar cells applications. Macromolecules **40**(1), 65–72 (2007)
11. Beaujuge, P.M., Amb, C.M., Reynolds, J.R.: Spectral engineering in π-conjugated polymers with intramolecular donor-acceptor interactions. Acc. Chem. Res. **43**(11), 1396–1407 (2010)
12. van Mullekom, H.A.M., Vekemans, J.A.J.M., Meijer, E.W.: Band-gap engineering of donor-acceptor-substituted π-conjugated polymers. Chem. A Eur. J. **4**(7), 1235–1243 (1998)
13. Sariciftci, N.S., Smilowitz, L., Heeger, A.J., Wudl, F.: Photoinduced electron transfer from a conducting polymer to buckminsterfullerene. Science **258**(5087), 1474–1476 (1992)
14. Sirringhaus, H., Tessler, N., Friend, R.H.: Integrated optoelectronic devices based on conjugated polymers. Science **280**(5370), 1741–1744 (1998)
15. Tessler, N., Denton, G.J., Friend, R.H.: Lasing from conjugated-polymer microcavities. Nature **382**(6593), 695–697 (1996)
16. Sirringhaus, H., Brown, P.J., Friend, R.H., Nielsen, M.M., Bechgaard, K., Langeveld-Voss, B.M.W., Spiering, A.J.H., Janssen, R.A.J., Meijer, E.W., Herwig, P., et al.: Two-dimensional charge transport in self-organized, high-mobility conjugated polymers. Nature **401**(6754), 685–688 (1999)
17. Singh, T.B., Sariciftci, N.S.: Progress in plastic electronic devices. Annu. Rev. Mater. Res. **36**(1), 199–230 (2006)
18. Facchetti, A.: π-conjugated polymers for organic electronics and photovoltaic cell applications. Chem. Mater. **23**(3), 733–758 (2011)
19. Nguyen, T.-Q., Martini, I.B., Liu, J., Schwartz, B.J.: Controlling on exciton–exciton annihilation and aggregation in interchain interactions in conjugated polymers: the effects of chain morphology MEH–PPV films. J. Phys. Chem. B **104**(2), 237–255 (2000)
20. Hodgkiss, J.M., Albert-Seifried, S., Rao, A., Barker, A.J., Campbell, A.R., Marsh, R.A., Friend, R.H.: Exciton-charge annihilation in organic semiconductor films. Adv. Funct. Mater. **22**(8), 1567–1577 (2012)
21. Howard, I.A., Hodgkiss, J.M., Zhang, X.P., Kirov, K.R., Bronstein, H.A., Williams, C.K., Friend, R.H., Westenhoff, S., Greenham, N.C.: Charge recombination and exciton annihilation reactions in conjugated polymer blends. J. Am. Chem. Soc. **132**(1), 328–335 (2010)
22. List, E.J.W., Kim, C.H., Naik, A.K., Scherf, U., Leising, G., Graupner, W., Shinar, J.: Interaction of singlet excitons with polarons in wide band-gap organic semiconductors: a quantitative study. Phys. Rev. B **64**(15) (2001)
23. Shaw, P.E., Ruseckas, A., Peet, J., Bazan, G.C., Samuel, I.D.W.: Exciton-exciton annihilation in mixed-phase polyfluorene films. Adv. Funct. Mater. **20**(1), 155–161 (2010)
24. Müllen, K., Wegner, G.: Electronic materials: the oligomer approach. Wiley, Weinheim (1998)
25. Zade, S.S., Bendikov, M.: From oligomers to polymer: convergence in the HOMO-LUMO gaps of conjugated oligomers. Org. Lett. **8**(23), 5243–5246 (2006)
26. Bidan, G., De Nicola, A., Enée, V., Guillerez, S.: Synthesis and UV-visible properties of soluble regioregular oligo(3-octylthiophenes), monomer to hexamer. Chem. Mater. **10**(4), 1052–1058 (1998)
27. Gierschner, J., Cornil, J., Egelhaaf, H.J.: Optical bandgaps of π-conjugated organic materials at the polymer limit: experiment and theory. Adv. Mater. **19**(2), 173–191 (2007)
28. Gierschner, J., Mack, H.G., Lüer, L., Oelkrug, D.: Fluorescence and absorption spectra of oligophenylenevinylenes: vibronic coupling, band shapes, and solvatochromism. J. Chem. Phys. **116**(19), 8596–8609 (2002)
29. Schwartz, B.J.: Conjugated polymers as molecular materials: how chain conformation and film morphology influence energy transfer and interchain interactions. Annu. Rev. Phys. Chem. **54**, 141–172 (2003)

30. Kim, D.Y., Grey, J.K., Barbara, P.F.: A detailed single molecule spectroscopy study of the vibronic states and energy transfer pathways of the conjugated polymer MEH-PPV. Synth. Met. **156**(2–4), 336–345 (2006)
31. Fakis, M., Polyzos, I., Tsigaridas, G., Giannetas, V., Persephonis, P.: Excited state dynamics of a partially conjugated polymer studied by femtosecond fluorescence upconversion spectroscopy. Chem. Phys. Lett. **394**(4–6), 372–376 (2004)
32. Schindler, F., Lupton, J.M., Feldmann, J., Scherf, U.: A universal picture of chromophores in Pi-conjugated polymers derived from single-molecule spectroscopy. Proc. Natl. Acad. Sci. U.S.A. **101**(41), 14695–14700 (2004)
33. Mirzov, O., Scheblykin, I.G.: Photoluminescence spectra of a conjugated polymer: from films and solutions to single molecules. Phys. Chem. Chem. Phys. **8**(47), 5569–5576 (2006)
34. Dykstra, T.E., Kovalevskij, V., Yang, X., Scholes, G.D.: Excited state dynamics of a conformationally disordered conjugated polymer: a comparison of solutions and film. Chem. Phys. **318**(1–2), 21–32 (2005)
35. Barbara, P.F., Gesquiere, A.J., Park, S.-J., Lee, Y.J.: Single-molecule spectroscopy of conjugated polymers. Acc. Chem. Res. **38**(7), 602–610 (2005)
36. Offermans, T., van Hal, P.A., Meskers, S.C.J., Koetse, M.M., Janssen, R.A.J.: Exciplex dynamics in a blend of π-conjugated polymers with electron donating and accepting properties: MDMO-PPV and PCNEPV. Phys. Rev. B **72**(4), 045213 (2005)
37. Hsu, J.-H., Fann, W., Meng, H.-F., Chen, E.-S., Chang, E.-C., Chen, S.-A., To, K.-W.: Decay dynamics of interchain excited states in luminescent conjugated polymer CN-PPV. Chem. Phys. **269**(1–3), 367–379 (2001)
38. Aguiar, M., Fugihara, M.C., Hümmelgen, I.A., Péres, L.O., Garcia, J.R., Gruber, J., Akcelrud, L.: Interchain luminescence in poly(acetoxy-P-phenylene vinylene). J. Lumin. **96** (2–4), 219–225 (2002)
39. Vacha, M., Ha, J., Ito, Y., Shimada, T., Mo Son, J., Sato, H.: Excimer emission in electroluminescence and photoluminescence of polyfluorene derivatives: effects of local heating and exciton density. J. Appl. Phys. **97**(2), 023514 (2005)
40. Peteanu, L.A., Sherwood, G.A., Wildeman, J., Werner, J.H., Goodwin, P.M., Shreve, A.P.: Effects of aggregation on the emission spectra and dynamics of electrolumine scent materials. Adv. Mater. **1**(c), 1–2 (2010)
41. Samuel, I.D.W., Rumbles, G., Collison, C.J.: Efficient interchain photoluminescence in a high-electron-affinity conjugated polymer. Phys. Rev. B **52**(16), R11573–R11576 (1995)
42. Samuel, I.D.W., Rumbles, G., Collison, C.J., Crystall, B., Moratti, S.C., Holmes, A.B.: Luminescence efficiency and time dependence in a high electron affinity conjugated polymer. Synth. Met. **76**(1–3), 15–18 (1996)
43. Conwell, E.M.: Excimers in poly(phenylene vinylene) and its derivatives. Synth. Met. **85**(1–3), 995–999 (1997)
44. Micaroni, L., Nart, F.C., Hummelgen, I.A.: Considerations about the electrochemical estimation of the ionization potential of conducting polymers. J. Solid State Electrochem. **7** (1), 55–59 (2002)
45. Johansson, T., Mammo, W., Svensson, M., Andersson, M.R., Inganas, O.: Electrochemical bandgaps of substituted polythiophenes. J. Mater. Chem. **13**(6), 1316–1323 (2003)
46. Cardona, C.M., Li, W., Kaifer, A.E., Stockdale, D., Bazan, G.C.: Electrochemical considerations for determining absolute frontier orbital energy levels of conjugated polymers for solar cell applications. Adv. Mater. **23**(20), 2367–2371 (2011)
47. Elandaloussi, E.H., Frère, P., Richomme, P., Orduna, J., Garin, J., Roncali, J.: Effect of chain extension on the electrochemical and electronic properties of π-conjugated soluble thienylenevinylene oligomers. J. Am. Chem. Soc. **119**(44), 10774–10784 (1997)
48. Heinze, J., Tschuncky, P., Smie, A.: The oligomeric approach—the electrochemistry of conducting polymers in the light of recent research. J. Solid State Electrochem. **2**(2), 102–109 (1998)
49. Heinze, J., Frontana-Uribe, B.A., Ludwigs, S.: Electrochemistry of conducting polymers-persistent models and new concepts. Chem. Rev. **110**(8), 4724–4771 (2010)

50. Chi, C.Y., Wegner, G.: Chain-length dependence of the electrochemical properties of conjugated oligofluorenes. Macromol. Rapid Commun. **26**(19), 1532–1537 (2005)
51. Jiang, X., Harima, Y., Zhu, L., Kunugi, Y., Yamashita, K., Sakamoto, M., Sato, M.: Mobilities of charge carriers hopping between π-conjugated polymer chains. J. Mater. Chem. **11**(12), 3043–3048 (2001)
52. Shimotani, H., Diguet, G., Iwasa, Y.: Direct comparison of field-effect and electrochemical doping in regioregular poly(3-hexylthiophene). Appl. Phys. Lett. **86**(2), 022104 (2005)
53. Laquai, F., Wegner, G., Bässler, H.: What determines the mobility of charge carriers in conjugated polymers? Philos. Trans. A. Math. Phys. Eng. Sci. **365**(1855), 1473–1487 (2007)
54. Vanmaekelbergh, D., Houtepen, A.J., Kelly, J.J.: Electrochemical gating: a method to tune and monitor the (opto)electronic properties of functional materials. Electrochim. Acta **53**(3), 1140–1149 (2007)
55. Plieth, W., Wilson, G.S., de la Fe, C.G.: Spectroelectrochemistry: a survey of in situ spectroscopic techniques (technical report). Pure Appl. Chem. **70**(7), 1395–1414 (1998)
56. Pflug, J.S., Faulkner, L.R.: Simultaneous electrochemical and fluorometric monitoring of zinc tetraphenylporphyrin deposited on indium oxide and pyrolytic graphite electrodes. J. Am. Chem. Soc. **102**(19), 6143–6144 (1980)
57. Pflug, J.S., Faulkner, L.R., Seitz, W.R.: Electrochemical modulation of fluorescence from zinc tetraphenylporphyrin on electrode surfaces. J. Am. Chem. Soc. **105**(15), 4890–4896 (1983)
58. Rubim, J.C., Gutz, I.G.R., Sala, O.: Cyclic-fluorovoltammetry as a technical tool in the study of passivating films generated on electrode surfaces. J. Electroanal. Chem. Interfacial Electrochem. **190**(1–2), 55–63 (1985)
59. Kakiuchi, T., Takasu, Y., Senda, M.: Voltage-scan fluorometry of rose bengal ion at the 1,2-dichloroethane-water interface. Anal. Chem. **64**(24), 3096–3100 (1992)
60. Kakiuchi, T., Takasu, Y.: Differential cyclic voltfluorometry and chronofluorometry of the transfer of fluorescent ions across the 1,2-dichloroethane-water interface. Anal. Chem. **66** (11), 1853–1859 (1994)
61. Bonazzola, C., Brust, M., Calvo, E.J.: Electrochemical fluorescence modulation at a redox hydrogel modified electrode. J. Electroanal. Chem. **407**(1–2), 203–207 (1996)
62. Yu, J.S., Yang, C., Fang, H.Q.: Variable thickness thin-layer cell for electrochemistry and in situ UV-VIS absorption, luminescence and surface-enhanced raman spectroelectrochemistry. Anal. Chim. Acta **420**(1), 45–55 (2000)
63. Dias, M., Hudhomme, P., Levillain, E., Perrin, L., Sahin, Y., Sauvage, F.X., Wartelle, C.: Electrochemistry coupled to fluorescence spectroscopy: a new versatile approach. Electrochem. Commun. **6**(3), 325–330 (2004)
64. Kim, Y., Kim, E., Clavier, G., Audebert, P.: New tetrazine-based fluoroelectrochromic window; modulation of the fluorescence through applied potential. Chem. Commun. (Camb.) (34), 3612–3614 (2006)
65. Kim, Y., Do, J., Kim, E., Clavier, G., Galmiche, L., Audebert, P.: Tetrazine-based electrofluorochromic windows: modulation of the fluorescence through applied potential. J. Electroanal. Chem. **632**(1–2), 201–205 (2009)
66. Yoo, J., Kwon, T., Sarwade, B.D., Kim, Y., Kim, E.: Multistate fluorescence switching of S-triazine-bridged p-phenylene vinylene polymers. Appl. Phys. Lett. **91**(24), 1–4 (2007)
67. Kwon, T., Sarwade, B.D., Kim, Y., Yoo, J., Kim, E.: A P-phenylene vinylene polymer bridged by a triazine for electro-fluorescence switching. Mol. Cryst. Liq. Cryst. **486**(1), 101/[1143]–109/[1151] (2008)
68. Kuo, C.-P., Lin, Y.-S., Leung, M.: Electrochemical fluorescence switching properties of conjugated polymers composed of triphenylamine, fluorene, and cyclic urea moieties. J. Polym. Sci. Part A-Polymer Chem. **50**(24), 5068–5078 (2012)
69. Kuo, C.-P., Chuang, C.-N., Chang, C.-L., Leung, M., Lian, H.-Y., Wu, K.C.-W.: White-light electrofluorescence switching from electrochemically convertible yellow and blue fluorescent conjugated polymers. J. Mater. Chem. C **1**(11), 2121–2130 (2013)

70. Kuo, C.-P., Chang, C.-L., Hu, C.-W., Chuang, C.-N., Ho, K.-C., Leung, M.: Tunable electrofluorochromic device from electrochemically controlled complementary fluorescent conjugated polymer films. ACS Appl. Mater. Interfaces **6**(20), 17402–17409 (2014)
71. Audebert, P., Miomandre, F.: Electrofluorochromism: from molecular systems to set-up and display. Chem. Sci. **4**(2), 575–584 (2013)
72. Kaval, N., Seliskar, C.J., Heineman, W.R.: Spectroelectrochemical sensing based on multimode selectivity simultaneously achievable in a single device. 16. Sensing by fluorescence. Anal. Chem. **75**(22), 6334–6340 (2003)
73. Andria, S.E., Seliskar, C.J., Heineman, W.R.: Spectroelectrochemical sensing based on multimode selectivity simultaneously achievable in a single device. 21. Selective chemical sensing using sulfonated polystyrene-blockpoly(ethylene-ran-butylene)block-polystyrene thin films. Anal. Chem. **81**(23), 9599–9606 (2009)
74. Pinyayev, T.S., Seliskar, C.J., Heineman, W.R.: Fluorescence spectroelectrochemical sensor for 1-hydroxypyrene. Anal. Chem. **82**(23), 9743–9748 (2010)
75. Wilson, R.A., Seliskar, C.J., Talaska, G., Heineman, W.R.: Spectroelectrochemical sensing of pyrene metabolites 1-hydroxypyrene and 1-hydroxypyrene-glucuronide. Anal. Chem. **83** (10), 3725–3729 (2011)
76. Ding, G., Zhou, H., Xu, J., Lu, X.: Electrofluorochromic detection of cyanide anions using a benzothiadiazole-containing conjugated copolymer. Chem. Commun. (Camb.) **50**(6), 655–657 (2014)
77. Reineke, S., Thomschke, M., Lüssem, B., Leo, K.: White organic light-emitting diodes: status and perspective. Rev. Mod. Phys. **85**(3), 1245–1293 (2013)
78. Hayashi, S., Kaneto, K., Yoshino, K.: Quenching of photoluminescence in poly(thiophene) films by electrochemical doping. Solid State Commun. **61**(4), 249–251 (1987)
79. Kaneto, K., Hayashi, S., Yoshino, K.: Kinetics of photoluminescent excitons in polythiophene films during electrochemical doping. J. Phys. Soc. Japan **57**(3), 1119–1126 (1988)
80. Balaei, S., Aaron, J.J., Desbenemonvernay, A., Lacaze, P.C.: Exsitu and insitu fluorescence studies of electrosynthesized poly(N-vinylcarbazole) in the presence of various electrolytes. Synth. Met. **53**(1), 95–107 (1992)
81. Jennings, P., Jones, A.C., Mount, A.R.: In situ spectroelectrochemical studies of the fluorescence of 5-substituted indole trimer films. Phys. Chem. Chem. Phys. **2**(6), 1241–1248 (2000)
82. Antonel, P.S., Molina, F.V., Andrade, E.M.: Fluorescence of polyaniline films on platinum surfaces. Influence of redox state and conductive domains. J. Electroanal. Chem. **599**(1), 52–58 (2007)
83. Antonel, P.S.S., Andrade, E.M.M., Molina, F.V.V.: Potential and film thickness dependence of the photoluminescence of aryl amine polymers. Electrochim. Acta **49**(22–23), 3687–3692 (2004)
84. Antonel, P.S., Völker, E., Molina, F.V.: Photophysics of polyaniline: sequence-length distribution dependence of photoluminescence quenching as studied by fluorescence measurements and Monte Carlo simulations. Polymer (U.K.) **53**(13), 2619–2627 (2012)
85. Soledad Antonel, P., Andrade, E.M., Molina, F.V.: Fluorescence of polyaniline films on electrode surfaces: thickness dependence and surface influence. J. Electroanal. Chem. **632** (1–2), 72–79 (2009)
86. Suarez, I.J., Otero, T.F., Marquez, M.: Diffusion coefficients in swelling polypyrrole: ESCR and Cottrell models. J. Phys. Chem. B **109**(5), 1723–1729 (2005)
87. Mazeikiene, R., Malinauskas, A.: A study into electrochemical slow relaxation of polyaniline. Synth. Met. **129**(1), 61–66 (2002)
88. Malinauskas, A., Holze, R.: Suppression of the “first cycle effect” in self-doped polyaniline. Electrochim. Acta **43**(5–6), 515–520 (1998)
89. Marmisolle, W.A., Posadas, D., Florit, M.I.: An experimental study of the intrinsic fluorescence emission and electrochemically induced ageing in poly-O-methylaniline films. Electrochim. Acta **109**, 894–900 (2013)

90. Gesquiere, A.J., Park, S.J., Barbara, P.F.: Photochemistry and kinetics of single organic nanoparticles in the presence of charge carriers. Eur. Polym. J. **40**(5), 1013–1018 (2004)
91. Borchardt, J.K.: Developments in organic displays. Mater. Today **7**(9), 42–46 (2004)
92. Yu, Z., Barbara, P.F.: Low-temperature single-molecule spectroscopy of MEH-PPV conjugated polymer molecules. J. Phys. Chem. B **108**(31), 11321–11326 (2004)
93. Holt, A.L., Leger, J.M., Carter, S.A.: Electrochemical and optical characterization of P- and N-doped poly[2-methoxy-5-(2-ethylhexyloxy)-1,4-phenylenevinylene]. J. Chem. Phys. **123** (4) (2005)
94. Palacios, R.E., Chang, W.S., Grey, J.K., Chang, Y.L., Miller, W.L., Lu, C.Y., Henkelman, G., Zepeda, D., Ferraris, J., Barbara, P.F.: Detailed single-molecule spectroelectrochemical studies of the oxidation of conjugated polymers. J. Phys. Chem. B **113**(44), 14619–14628 (2009)
95. Montilla, F., Mallavia, R.: In situ electrochemical fluorescence studies of PPV. J. Phys. Chem. B **110**(51), 25791–25796 (2006)
96. Otero, T.F., Boyano, I.: Comparative study of conducting polymers by the ESCR model. J. Phys. Chem. B **107**(28), 6730–6738 (2003)
97. Esquembre, R., Pastor, I., Mallavia, R., Mateo, C.R.: Fluorometric detection of nitric oxide using 2,3-diaminonaphthalene incorporated in beta-cyclodextrin. J. Photochem. Photobiol. A-Chem. **173**(3), 384–389 (2005)
98. Wu, C., Yan, C.-Y.: Studies of the swelling and drying kinetics of thin gelatin gel films by in situ interferometry. Macromolecules **27**(16), 4516–4520 (1994)
99. van Reenen, S., Vitorino, M.V., Meskers, S.C.J., Janssen, R.A.J., Kemerink, M.: Photoluminescence quenching in films of conjugated polymers by electrochemical doping. Phys. Rev. B **89**(20) (2014)
100. Kervella, Y., Armand, M., Stephan, O.: Organic light-emitting electrochemical cells based on polyfluorene—investigation of the failure modes. J. Electrochem. Soc. **148**(11), H155–H160 (2001)
101. Pei, Q.B., Yang, Y.: Efficient photoluminescence and electroluminescence from a soluble polyfluorene. J. Am. Chem. Soc. **118**(31), 7416–7417 (1996)
102. Craig, M.R., de Kok, M.M., Hofstraat, J.W., Schenning, A.P.H.J., Meijer, E.W.: Improving color purity and stability in a blue emitting polyfluorene by monomer purification. J. Mater. Chem. **13**(12), 2861–2862 (2003)
103. Kim, J.S., Friend, R.H., Cacialli, F.: Improved operational stability of polyfluorene-based organic light-emitting diodes with plasma-treated indium-tin-oxide anodes. Appl. Phys. Lett. **74**(21), 3084–3086 (1999)
104. Wallace, J.U., Chen, S.H.: Fluorene-based conjugated oligomers for organic photonics and electronics. Polyfluorenes **212**, 145–186 (2008)
105. Kulkarni, A.P., Jenekhe, S.A.: Blue light-emitting diodes with good spectral stability based on blends of poly(9,9-dioctylfluorene): interplay between morphology, photophysics, and device performance. Macromolecules **36**(14), 5285–5296 (2003)
106. Akcelrud, L.: Electroluminescent polymers. Prog. Polym. Sci. **28**(6), 875–962 (2003)
107. Montilla, F., Pastor, I., Mateo, C.R., Morallon, E., Mallavia, R., Morallón, E., Mallavia, R.: Charge transport in luminescent polymers studied by in situ fluorescence spectroscopy. J. Phys. Chem. B **110**(12), 5914–5919 (2006)
108. Montilla, F., Mallavia, R.: On the origin of green emission bands in fluorene-based conjugated polymers. Adv. Funct. Mater. **17**(1), 71–78 (2007)
109. Technical papers|Cambridge Display Technology. https://www.cdtltd.co.uk/technology/technical-papers/. Accessed 26 Aug 2015
110. Palacios, R.E., Fan, F.R.F., Bard, A.J., Barbara, P.F.: Single-molecule spectroelectrochemistry (SMS-EC). J. Am. Chem. Soc. **128**(28), 9028–9029 (2006)
111. Gesquiere, A.J., Park, S.J., Barbara, P.F.: Single molecule charge transfer probed by cyclic voltammetry-single molecule spectroscopy. Abstr. Pap. Am. Chem. Soc. **227**, U260-U260 (2004)

112. Palacios, R.E., Fan, F.R.F., Grey, J.K., Suk, J., Bard, A.J., Barbara, P.F.: Charging and discharging of single conjugated-polymer nanoparticles. Nat. Mater. **6**(9), 680–685 (2007)
113. Palacios, R.E., Chang, W.-S., Grey, J.K., Chang, Y.-L., Miller, W.L., Lu, C.-Y., Henkelman, G., Zepeda, D., Ferraris, J., Barbara, P.F.: Detailed single-molecule spectroelectrochemical studies of the oxidation of conjugated polymers. J. Phys. Chem. B **113**(44), 14619–14628 (2009)
114. Gesquiere, A.J., Uwada, T., Asahi, T., Masuhara, H., Barbara, P.F.: Single molecule spectroscopy of organic dye nanoparticles. Nano Lett. **5**(7), 1321–1325 (2005)
115. Chang, Y.L., Palacios, R.E., Chen, J.T., Stevenson, K.J., Guo, S., Lackowski, W.M., Barbara, P.F.: Electrogenerated chemiluminescence of soliton waves in conjugated polymers. J. Am. Chem. Soc. **131**(40), 14166–+ (2009)
116. Montilla, F., Esquembre, R., Gomez, R., Blanco, R., Segura, J.L.: Spectroelectrochemical study of electron and energy transfer in poly(fluorene-alt-phenylene) with perylenediimide pendant groups. J. Phys. Chem. C **112**(42), 16668–16674 (2008)
117. Montilla, F., Frutos, L.M., Mateo, C.R., Mallavia, R.: Fluorescence emission anisotropy coupled to an electrochemical system: study of exciton dynamics in conjugated polymers. J. Phys. Chem. C **111**(49), 18405–18410 (2007)
118. Furukawa, Y.: Electronic absorption and vibrational spectroscopies of conjugated conducting polymers. J. Phys. Chem. **100**(39), 15644–15653 (1996)
119. Montilla, F., Ruseckas, A., Samuel, I.D.W.: Absorption cross-sections of hole polarons in glassy and beta-phase polyfluorene. Chem. Phys. Lett. **585**, 133–137 (2013)
120. Valeur, B.: Molecular Fluorescence: Principles and Applications, 1st edn. Wiley-VCH Verlag GmbH (2001)
121. Ward, A.J., Ruseckas, A., Samuel, I.D.W.: A shift from diffusion assisted to energy transfer controlled fluorescence quenching in polymer-fullerene photovoltaic blends. J. Phys. Chem. C **116**(45), 23931–23937 (2012)
122. Lakowicz, J.R.: Principles of Fluorescence Spectroscopy. Plenum Press, New York (2013)
123. Gosele, U., Hauser, M., Klein, U.K.A., Frey, R.: Diffusion and long-range energy-transfer. Chem. Phys. Lett. **34**(3), 519–522 (1975)
124. Castanho, M.A.R., Prieto, M.J.: Fluorescence quenching data interpretation in biological systems. Biochim. Biophys. Acta Biomembr. **1373**(1), 1–16 (1998)
125. Tuladhar, S.M., Poplavskyy, D., Choulis, S.A., Durrant, J.R., Bradley, D.D.C., Nelson, J.: Ambipolar charge transport in films of methanofullerene and poly(phenylenevinylene)/methanofullerene blends. Adv. Funct. Mater. **15**(7), 1171–1182 (2005)
126. Blom, P.W.M., de Jong, M.J.M., van Munster, M.G.: Electric-field and temperature dependence of the hole mobility in poly(p-phenylene vinylene). Phys. Rev. B **55**(2), R656–R659 (1997)
127. Arkhipov, V.I., Emelianova, E.V., Heremans, P., Bässler, H.: Analytic model of carrier mobility in doped disordered organic semiconductors. Phys. Rev. B **72**(23), 235202 (2005)
128. Mikhnenko, O.V., Blom, P.W.M., Nguyen, T.-Q.: Exciton diffusion in organic semiconductors. Energy Environ. Sci. **8**(7), 1867–1888 (2015)
129. Scully, S.R., McGehee, M.D.: Effects of optical interference and energy transfer on exciton diffusion length measurements in organic semiconductors. J. Appl. Phys. **100**(3), 034907 (2006)
130. Lewis, A.J., Ruseckas, A., Gaudin, O.P.M., Webster, G.R., Burn, P.L., Samuel, I.D.W.: Singlet exciton diffusion in MEH-PPV films studied by exciton-exciton annihilation. Org. Electron. **7**(6), 452–456 (2006)

Electrofluorochromic Devices with Organic Dyes and Conjugated Polymers

Seogjae Seo and Eunkyoung Kim

Abstract Electrofluorochromic materials exhibit reversible switch of fluorescence emission by the applied potentials. This unique property is based on the electrochemistry of organic materials. Especially, electrofluorochromic organic dyes and conjugated polymers can be categorized into three types, depending on the distance between the electroactive moiety and the fluorescent core. The electrofluorochromic properties such as working potential, emission color, ON/OFF ratio, stability, and switching efficiency are summarized for each type of material. Also, the working principle of each type is introduced with suitable examples. Finally, some applications of electrofluorochromic materials are presented in the last part of this chapter. Although the reports on applications are still scarce until now, we believe that challenges in display or sensing applications will follow these pioneering researches.

1 Introduction

Electrofluorochromic devices (EFDs) show an electrochemically driven reversible fluorescence changes accompanied with electrochemical redox conversion [1]. In this EFD, electrofluorochromic materials control the fluorescence emission wavelength or intensity by internal electron transfer or energy transfer between a redox-active unit and a fluorescent core [2] or by the intrinsic fluorescence change of the material [3]. When such electrofluorochromic materials are applied into electrochemical sensor or displays, the device can exhibit dual-mode signal by detecting fluorescence intensity. For example, the electrofluorochromic sensor provided the dual detection of electrochemical and fluorescence signals in biosensors [4], and electrofluorochromic displays provide a dual-mode image by incorporating with electrochromic display [5]. This chapter reviews materials and

S. Seo (✉) · E. Kim (✉)
Department of Chemical and Biomolecular Engineering, Yonsei University, 50 Yonsei-ro, Seodaemun-gu, Seoul 120-749, Korea
e-mail: eunkim@yonsei.ac.kr

F. Miomandre and P. Audebert (eds.), *Luminescence in Electrochemistry*,
DOI 10.1007/978-3-319-49137-0_5

working principles of electrofluorochromic devices with organic dyes and conjugated polymers.

The materials for electrofluorochromic devices can be classified into three categories, depending on the distance between the electroactive moiety and fluorescent core. When the fluorescent core itself is electroactive and changes fluorescence intensity according to the applied potentials, the electrofluorochromic devices can be illustrated as Type I of Fig. 1. Tetrazine derivatives are the most famous materials in this type [3, 6]. The yellow emission from the neutral-state tetrazine is quenched upon application of a reduction potential by converting the tetrazines to its anion radicals. Such tetrazines electrofluorochromic materials are easily dissolved into a solution of electrolyte due to their low molecular weights, and thus, the EFD for these materials is simply fabricated by applying the mixture of tetrazines and electrolyte solution in a gap between two transparent electrodes. For example, yellow-emissive electrofluorochromic 3-chloro-6-methoxy-1,2,4,5-tetrazine was used after dissolving it in a solution of polymer electrolyte [3]. The yellow emission from the tetrazine is reversibly controlled with the redox reaction of tetrazine within the electrolyte solution of EFD. The electrofluorochromic (EF) switching was quite reversible due to the high electrochemical stability of tetrazines [6]. Such an EFD based on redox electrochemistry of fluorescent dyes, however, was limited to the electroactive fluorescent molecules.

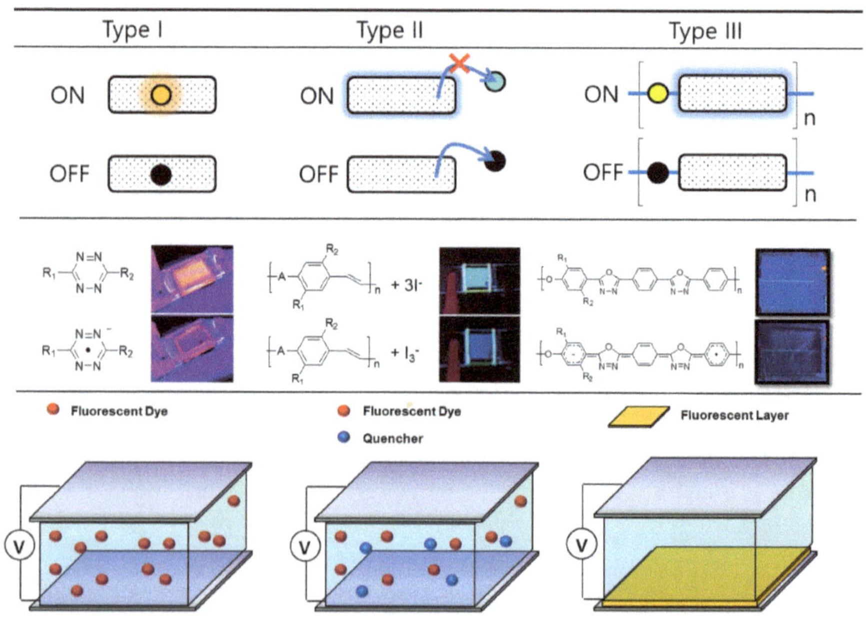

Fig. 1 Three different types of electrofluorochromic materials and examples of each case [2, 3, 7]

To achieve EF switching from a wide collection of fluorescent dyes including non-electroactive fluorescent dyes, the electrochemically controlled fluorescence energy transfer may be introduced in an EFD. Such electrofluorochromism by external electroactive quenchers represents another class and is classified as Type II. Since the fluorescent switch is based on the energy transfer between the fluorescent and external quencher, EF contrast in Type II is different from the EFD of Type I because the quenching efficiency must be strongly dependent on the energy transfer efficiency between the dyes and quenchers. For example, the degree of the fluorescence quenching of poly[(4,6-bis (formylphenoxy)-2-diphenylamino-s-triazine)co(5-(3′,7′-dimethyloctyloxy)-2-methoxy-1,4phenylene-vinylene)] (DTOPV) by the iodine as external electroactive quenchers is highly dependent on the redox state of iodine, so the emission intensity from the DTOPV showed on and off switching as a function of applied potential [2]. Such external quenching electrofluorochromic materials can provide a feasible fluorescence switch, but cannot be applied in fluorescence imaging or fluorescence displays, because the electrofluorochromic materials are not coated on the electrodes. For those applications, polymeric electrofluorochromic materials were coated as a thin film on an electrode. In this type, therefore, EFD is consisting of EF and electrolyte layers along with two electrodes. In this structure, the EF layer can be an electroactive and fluorescent layer of polymer, or a dye-anchored inorganic oxide layer, and can be classified as electrofluorochromic film type (Type III). The Type III materials showed fluorescence intensity changes or color changes upon electrochemical redox reaction [7]. For example, poly(1,3,4-oxadiazole) (POD) substituted with long alkyl chains showed reversible fluorescence switching between dark (n-doping) and bright (neutral) states upon application of redox potential (Fig. 1). Because the polymers can be coated on the electrodes, patterning of fluorescent layer is easy and thus fluorescence image patterns can be easily obtained and switched by electrofluorochromism. Another advantage of the electrofluorochromic polymer is the high fluorescence contrast because the electrogenerated radical species can be efficiently quenched through the energy transfer in film state [1].

Electrofluorochromism can be achieved from the 3 different types of molecular systems, and characteristics of the each type are different from each other. Thus, in this chapter, we introduce the required features for electrofluorochromic materials with the different types, and examples of organic dyes, organometallic dyes, and conjugated polymers with their characteristics. Also, some applications of electrofluorochromic materials will be following.

1.1 Key Parameters in Electrofluorochromism

Such electrofluorochromism from the EFD can be visualized as a fluorescence signal or displays from electrofluorochromic devices. As shown in Fig. 2, an electrofluorochromic window consists in a transparent working electrode, a counter electrode, and electrolyte. Actually, to display the fluorescence signal efficiently,

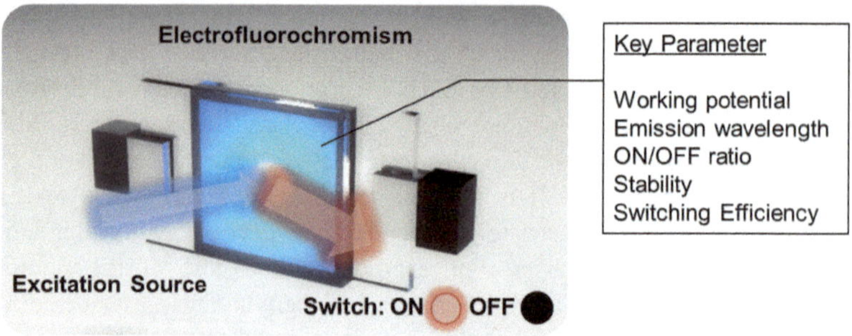

Fig. 2 Structure of an electrofluorochromic device which consists in two ITO electrodes and electrolyte packed between the ITO electrodes. The characteristics of a given electrofluorochromic device can be summarized by the working potential, emission wavelength, ON/OFF ratio, stability, and switching efficiency

the electrofluorochromic device mimics the structure of classical electrochromic windows [8]. The design of an electrofluorochromic device was initially established by Kim and Audebert et al. and then widely adopted in electrofluorochromism due to the simple structure and vivid fluorescence changes. A few works appeared in the late 1990s and early 2000s that presented the electrochemical monitoring of fluorescent molecules in solution. Lehn et al. reported fluorescence ON/OFF switch of a ruthenium–bipyridine complex using a quinone redox state [9], and Levillain et al. [10] reported an electrochemical monitoring of a perylene dipentylimide using fluorescence spectroscopy. However, in the early stage of researches, the investigation of electrofluorochromic properties was not visualized because they were examined at the molecular scale in solution not in electrofluorochromic devices. Thus, in this chapter, we introduce the electrofluorochromic properties in an EFD, such as ON/OFF ratio, cyclability, and efficiency. Because the quantitative investigations were carried out using the electrofluorochromic devices, we mainly focused on the relatively recent works.

To evaluate the electrofluorochromic property, the characteristics of electrofluorochromic materials can be determined from the working potential, emission wavelength, maximum ON/OFF ratio, long-term stability, and switching efficiency. The working potential (V) is the applied potential to operate the electrofluorochromic devices. Typically, the working potentials are obtained from CV data and empirically determined from the fluorescence changes of the switching device in response to the different applied potentials. Because the electrofluorochromism is based on the redox electrochemistry of an electroactive moiety, the working potential is typically explored from 0 V to the peak redox potential. Especially, the working potential can be dramatically reduced by using a three-electrode electrofluorochromic device, which contains a reference electrode [11]. Emission wavelength (λ_{emi}) is determined from the comparison with the fluorescence spectroscopy of a fluorescent dye. It is dependent on the emissive core of the dye. The maximum ON/OFF ratio is the maximum fluorescence contrast, determined by dividing the emission intensity at ON state (I_{ON}) with that of the OFF state (I_{OFF}) (Eq. 1).

$$\mathrm{ON/OFF\,ratio} = \frac{I_{ON}}{I_{OFF}} \tag{1}$$

Long-term stability is determined by monitoring the fluorescence contrast with repeated switching, which means the guaranteed switching cycle without a loss of fluorescence contrast. Electrofluorochromic efficiency (E_{EF}) can be derived from the quantitative analysis of the electrofluorochromism. Similarly to the coloration efficiency in electrochromism, the electrofluorochromism efficiency E_{EF} was determined by the following equation (Eq. 2), where Q represents the injected/ejected charge per unit area.

$$E_{EF} = \frac{I_{ON}}{I_{OFF}} \times \frac{1}{Q} \tag{2}$$

Because the total amount of electrofluorochromic materials in the electrolyte medium is known, the percentage of the electro-converted dye can be determined using the injected/ejected charge (Eq. 3), where Q_{Total} is total amount of charge that is required to fully convert the electrofluorochromic materials in the switching cell.

$$R(\%) = \frac{Q}{Q_{Total}} \times 100(\%) \tag{3}$$

2 Electrofluorochromism by Redox Chemistry of Fluorescent Dyes (Type I)

Figure 3 shows the mechanism of the electrofluorochromic switching for Type I. In these dyes, the redox electrochemistry of fluorescent dyes is accompanied with the large fluorescence quantum yield change. In general, the neutral state of fluorescent dyes exhibits higher emission than their anion radical or cation radical state, because the emissive conjugated system disappears at the anion radical or cation radical state. Also, the electrogenerated ion-radical species can quench the surrounding fluorescence emission through energy transfer. Using this unique property of Type I dyes, electrofluorochromic switching can be realized by the redox electrochemistry.

The device structures for Type I EF materials consist of the two transparent ITO electrodes and electrolytes. The electrolyte solution contains fluorescent dyes with supporting electrolyte salts. The prepared electrolyte is carefully injected between the ITO electrodes. To carefully control the applied potential, a reference electrode can be inserted between the two ITO electrodes. The Ag wire is used often as a pseudo-reference electrode in EFDs. The switching potentials can be dramatically reduced compared with two-electrode devices, and the precise control of redox reaction is possible in three-electrode EFDs [11]. The device structures are presented in Fig. 4.

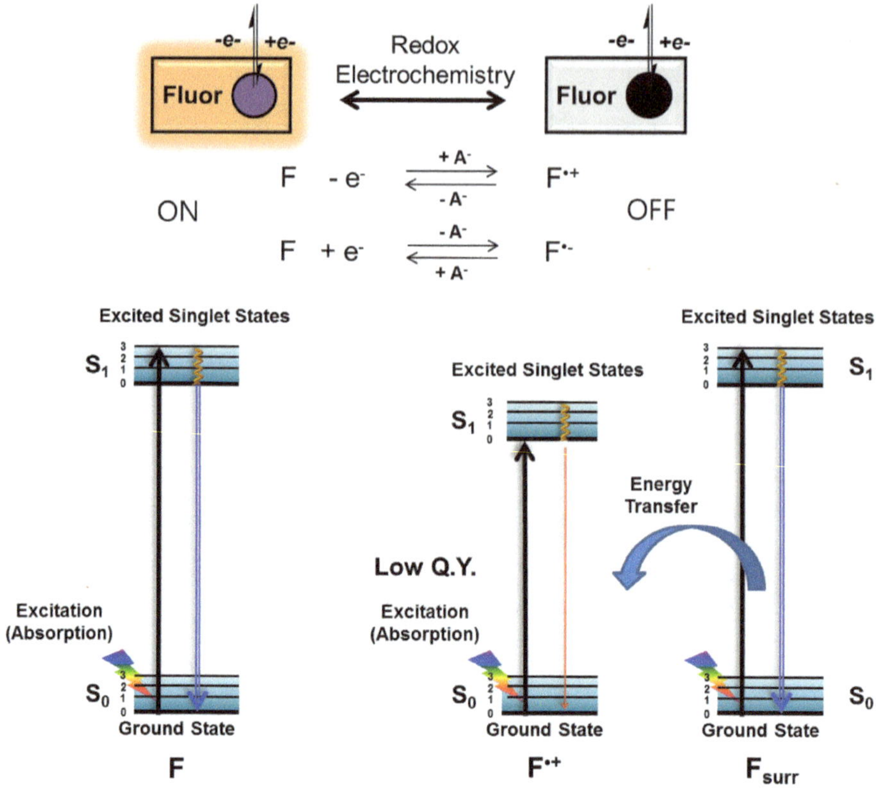

Fig. 3 Mechanism of electrofluorochromism of Type I dyes using the redox chemistry of fluorescent dyes (**F**)

2.1 Electrofluorochromism in Solution and Gel-like Medium

Because Type I electrofluorochromic materials use the intrinsic electrochemical conversion of a fluorescent core, small molecules such as tetrazines and viologens are included in Type I. Especially, the tetrazine derivatives are good candidates to realize electrofluorochromism. The electrochemical conversion between neutral state and anion radical state of tetrazine derivatives results in high fluorescence quantum yield changes. Using this property, Kim and Audebert et al. [3] provided the first electrofluorochromic window. As shown in Fig. 5a, the electrofluorochromic material, 3-chloro-6-methoxy-1,2,4,5-tetrazine, was mixed with polymer electrolyte, and then injected between the two ITO electrodes. The prepared device was operated by potential switching from +2 to −2 V. Under the excitation source ($\lambda_{exc} = 356$ nm), the yellow emission from the tetrazine was switched ON and OFF reversibly (Fig. 5b). This switching property can also be measured by the fluorescence spectra recorded

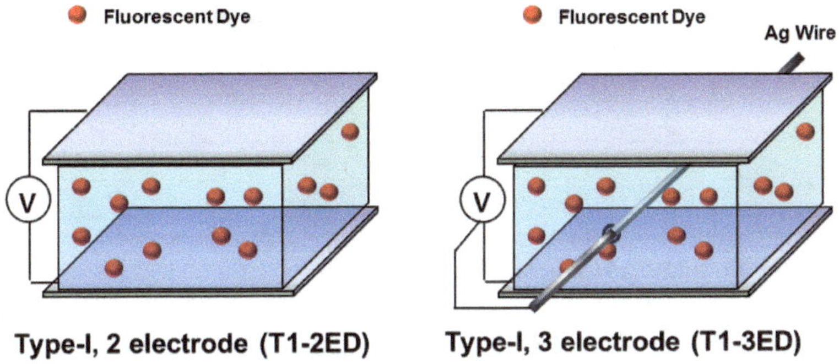

Fig. 4 Device structures used for Type I electrofluorochromic materials

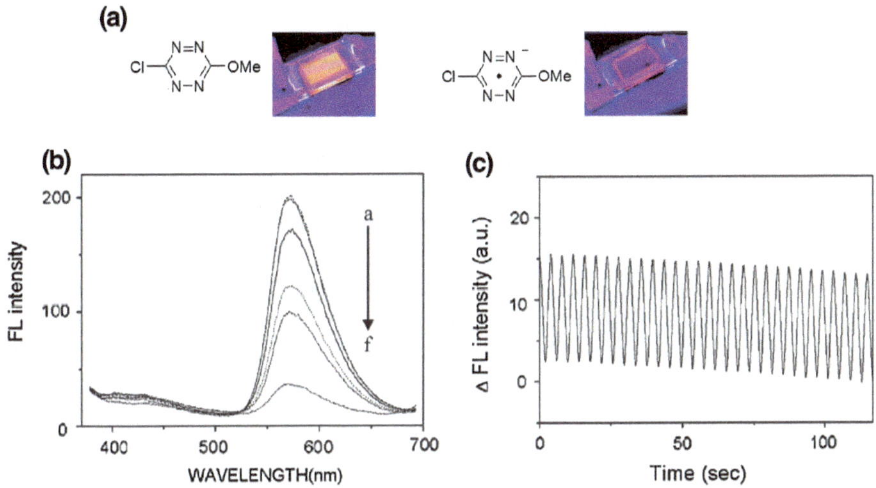

Fig. 5 **a** Operation of the window with different potentials (0/−3 V). **b** Electrofluorochromic contrast was measured using fluorescence spectra at different applied potentials (a: 1.25 V, b: 2 V, c: start, d: −1.25 V, e: −2 V, f: −3 V). **c** Fluorescence modulation of the window by potential switching from +2 to −2 V, 2s for each step [3]

at various potentials (Fig. 5c). It is clear that the fluorescence intensity is dependent upon the applied potential and almost reversibly extinct upon potential scanning toward negative values. Such a fluorescence intensity change is found to be reversible upon repetitive cycling between +2 and −2 V, as shown in Fig. 5d. Similarly with the 3-chloro-6-methoxy-1,2,4,5-tetrazine, tetrazine derivatives also exhibited electrofluorochromic switching in solution and gel-like medium [3, 6, 11–18]. The chemical structure and the electrofluorochromic properties are shown in Table 1.

Table 1 Electrofluorochromic properties of the representative tetrazine derivatives

Compounds	Electrofluorochromic properties			
	ON/OFF ratio	Cyclability (cycle)	λ_{emi} (nm)	Working potential (ON/OFF)
1[a]	5.6	>120	568	2 V/−3 V
2[a]	3.1	>120	–	2 V/−2 V
3[b]	14	>100	558	1.4 V/−1.4 V

[a]EFD structure: T1-2ED. [b]EFD structure: T1-3ED

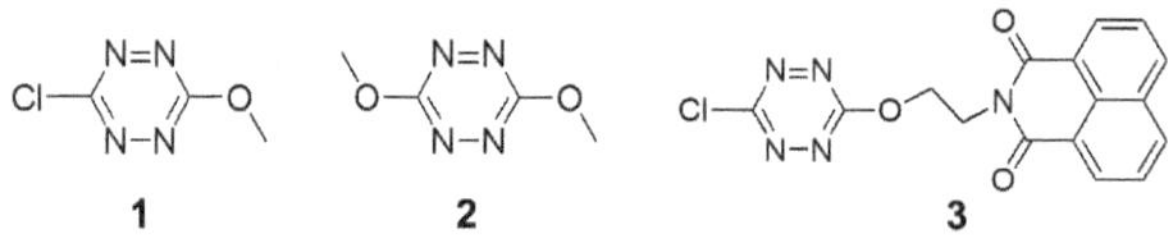

2.1.1 Quantitative Analysis During Electrofluorochromic Switching

Electrofluorochromic switching in the solution or gel-like electrolyte medium is accompanied with the electrochemical conversion. The electrofluorochromic device containing 1wt% **3** in polymer electrolyte was quantitatively examined by monitoring the injected/ejected charge and fluorescence changes. Q_{total} was calculated by multiplying the concentration of **3** by the volume of the switching cell, using the equation: $Q_{Total} = F \times C \times V$, where F is the Faraday constant (96485 C/mol), C is the concentration of **3**, and V is the volume of the switching cell (20 mm 16 mm × 0.52 mm). Upon the applied step potential from 1.4 to −0.8 V, the electrofluorochromic device was fully switched to exhibit maximum ON/OFF ratio. However, the percentage of reduced fluorophore (R) for the switching cell was calculated as 5.7%. This result shows that the electrofluorochromic switching is mainly related to the electrochemical conversion close to the surface of the working electrode. The vivid fluorescence switching was achieved, because the sensitivity was high and the switching was driven by the formation of anion radical within the electrode diffusion layer [11].

2.1.2 Electrofluorochromism with Ionic Liquid Crystals

Fluorescent liquid crystals with redox-active centers can modulate the photoluminescence through redox processes. π-conjugated ionic liquid crystals based on thienoviologens showed high fluorescence in the bulk state. **4** and **5** showed over 60% of Q.Y [19]. Their direct electrochemical reduction leads to fast and reversible bulk electrofluorochromic response. These materials can show electrofluorochromic switching in the gel-like medium. The fluorescence band was maximized at 530 nm. The emission properties of the thienoviologens depend on the fluorophore percentage: In the 2–6% (w/w) range, the spectrum is dominated by the emission band centered at 530 nm. Above this concentration range, a new band adds at 630 nm arising from the formation of thienoviologen aggregates. In the concentration range 2–6%, **4** showed high-contrast electrofluorochromic switching with ON/OFF ratio 11.3 when the applied potential was −1.5/0 V. The reversibility of the device is excellent up to about 500 cycles after which the contrast ratio decreases from 6.1 in the first cycles to 5.46, corresponding to a stability of more than 90%. A good stability (65%) is still retained up to 1000 cycles with a decrease in the contrast ratio to 4.0. The mechanism of electrofluorochromism is based on the reversible conversion between the emissive cation radical state and non-emissive dication state. This fluorescence change was not due to an increase in the absorbance of the reduced species, because the absorbance of the emissive cation radical state was higher than the non-emissive state at the emission wavelength.

Interestingly, the ionic material **4** and **5** can be used in EFD in bulk state. The room temperature thin-film absorption and emission spectra of both materials show broader and slightly red-shifted (~40 nm) absorption bands and hardly red-shifted emission bands (~100 nm) with respect to those in solution, as shown in Table 2. Because **4** and **5** have liquid crystal properties in the bulk phase, their photophysical properties are changed by their molecular structure and mesophase behavior. (Iso = isotropic liquid state; Col_{ro} = rectangular ordered columnar phase; SmA = Smectic A phase)

Especially, Iso/mesophase field-induced transition was shown with compound **5**. When a reductive voltage is applied on **5** in the isotropic liquid state (Iso), the characteristic emission band at 540 nm disappears and a strong band at about 610 nm takes place. This change of the fluorescence spectrum is similar to the aggregation-caused shifting effect, and this result suggests a field-induced molecular ordering. Notably, this field-induced transition process is not observed for compound **4** (Iso/columnar), suggesting that it is strongly dependent on the nanosegregation of the LC structures (Fig. 6).

Although the neutral species of **4** and **5** are non-emissive, the radical cations show higher emission than the dication state. Under the precise control of applied potential, the fluorescent intensity of **4** and **5** is increased (~10-fold increase) in

Table 2 Photophysical and electrofluorochromic properties of compounds **4** and **5**

Compounds	State	λ_{abs} (nm)	λ_{emi} (nm)	Q.Y	ON/OFF ratio	Stability
4[a]	Solution	430	530	0.61	11.3[b]	>1000
	Col_{1ro}	450	630	0.68	6.7	–
	Col_{2ro}	450	630	0.51	–	–
5[a]	Solution	430	530	0.60	–	–
	Iso				3.3[c]	–
	SmA	450	670	0.62	4	–
	Glassy	450	670	0.48	–	–

[a]EFD structure: T1-3ED. [b]The electrofluorochromic polymer gels were prepared by dissolving **5** (10wt%) in NMP with 0.3%–1.7 wt% of ferrocene and 35 wt% of polyvinyl formal (PVF). [c]When a reductive voltage is applied on **2** in the isotropic liquid state (Iso), Iso/mesophase field-induced transition was occurred

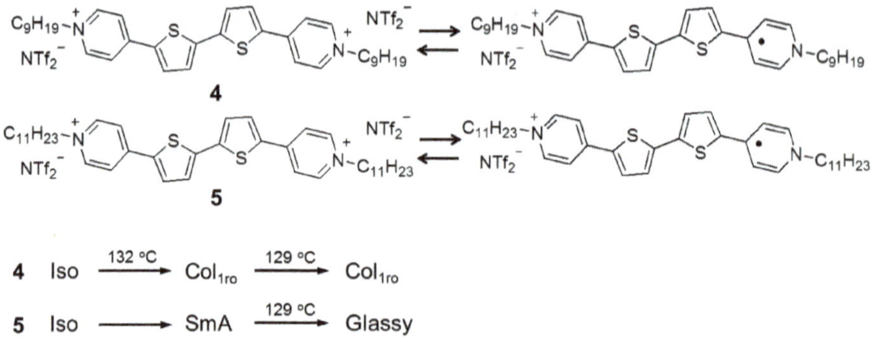

Fig. 6 Molecular structures and mesophase behavior on cooling from the isotropic liquid state. (Iso = isotropic liquid state; Co_{lro} = rectangular ordered columnar phase; SmA = Smectic A phase) [19]

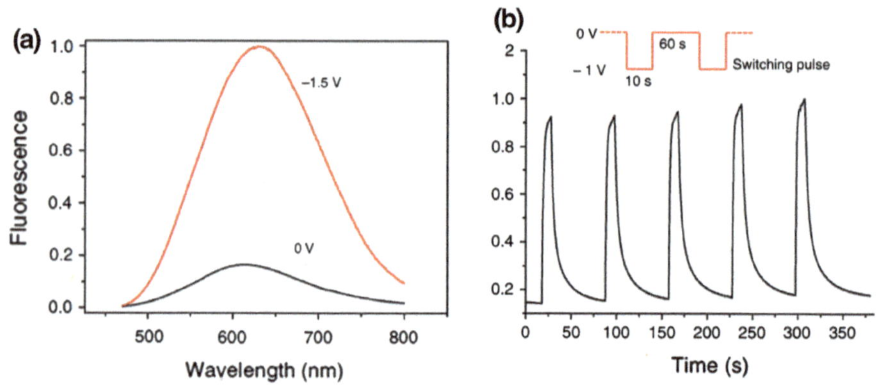

Fig. 7 **a** Fluorescence spectra at 0 V (*black line*) and −1.5 V (*red line*) of thin films (10 mm thick) of **4** in the columnar phase. **b** Repetitive switching cycles of **4** in the Col_{1ro} phase between −1.0 V for 10 s and 0 V for 60 s [19]

both mesophases because of the formation of the radical cation. This electrofluorochromic switching can be repeated several times by consecutive on/off switching cycles under very different pulse conditions, as shown in Fig. 7.

2.1.3 Multi-Color Electrofluorochromic Switching

Although the electrofluorochromic materials showed the ability to reversibly switch the fluorescence intensity, to apply in several sensing or displays, multi-color electrofluorochromic switching would be required. To obtain the multi-color switching, more than two states of reversible electrochemical conversion are required, and the redox electrochemical conversion should be demonstrated in the

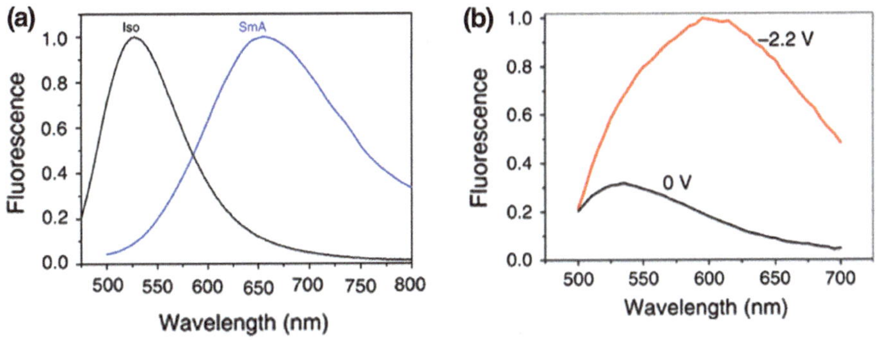

Fig. 8 **a** Aggregation-caused fluorescence shift **5** in the Iso state (*black line*) and in the SmA phase (*blue line*). **b** Fluorescence color tuning by electric field-induced transition effect. Fluorescence spectrum of a thin film of **5** in the isotropic liquid state (*black line*). Upon the application of a −2.2 V potential, the spectrum is hardly red-shifted (*red line*) due to the electric field-induced transition from the Iso state to the SmA phase [19]

same device. This should be in principle possible by using dyad-type chromophores and/or mixtures of fluorophores in a same solution or polymeric electrolyte (Fig. 8). The blend of **3** and **6** in the same electrolyte medium exhibited the mixed emission color without mutual quenching (this being due to the dilution of each fluorophore, which made their average distance longer than the Förster energy exchange radius). The optical properties of these blends were also the same as the individual components. No side product formation was noticed, indicating that the extra **6** is untouched by the energy transfer between **3** and **6**. Using the individual redox electrochemistry of **3** and **6**, multi-color switching is possible as shown in Fig. 9. Both fluorescence bands for **3** and **6** are observed at 558 and 385 nm, respectively, which correspond to yellow and blue color emission. With the applied potential to switch the tetrazine unit (1.4/−0.8 V), only the yellow fluorescence was switched ON and OFF. Similarly, with the applied potential (0/−1.4 V) the blue emission from the naphthalimide unit can be switched individually afterward. The maximum ON/OFF ratio of **3** and **6** is 14 and 6.3, respectively. The long-term stability of **3** is guaranteed over 100 cycles, but the stability of **6** was less than 50 cycles due to its lower electrochemical stability.

2.1.4 NIR Electrofluorochromic Switching

Expending the electrofluorochromic switching region to the near-infrared (NIR) region is possible with the NIR-emissive Type I dyes, such as polymethines and BODIPYs [20, 21]. Based on the same electrofluorochromic mechanism, the reversible redox electrochemistry of polymethines and BODIPY was used to

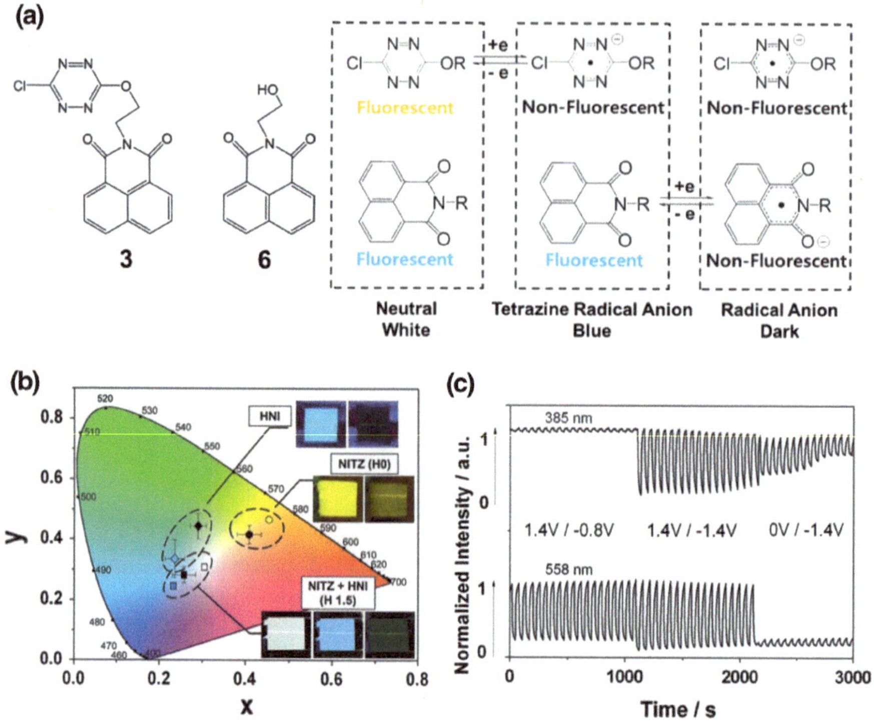

Fig. 9 **a** Chemical structures of the naphthalimide-tetrazine dyad (3) and (2-hydroxyethyl)-N-naphthalimide (6). **b** The reversible emission color change of the cell with **3** and **6** in chromaticity diagram as compared to the cell of **3** and **6**, measured at different potentials. **c** Fluorescence switching of the blended device between multi-color states [11]

achieve reversible electrochemical fluorescence switching in the NIR region. Optical properties in NIR have attractive advantages in bio-imaging, bio-analysis [22], and night vision devices [23, 24]. Especially, in the field of biomedical imaging, the use of NIR emission is a promising approach because it can provide noninvasive and background signal-free images [25]. A NIR-emissive polymethine dye, 3H-indolium, 2-[2-[2-chloro-3-[2-[1,3-dihydro-3,3-dimethyl-1-(phenylmethyl)-2H-indol-2-ylidene]ethylidene]-5-(1,1-dimethylethyl)-1-cyclohexen-1-yl]ethenyl]-3,3-dimethyl-1-(phenylme thyl)-, bromide (**7**), displayed high absorption and emission in the NIR region. Using the reversible redox reaction of polymethine dye (**7**), the NIR fluorescence ($\lambda_{em} = 820$ nm) was switched with an ON/OFF ratio of ~1.5 and a cyclability of ~100 cycles with a switching potential of 1.1/−0.5 V. The electrofluorochromic efficiency was also calculated as 557 cm^2/C (Fig. 10).

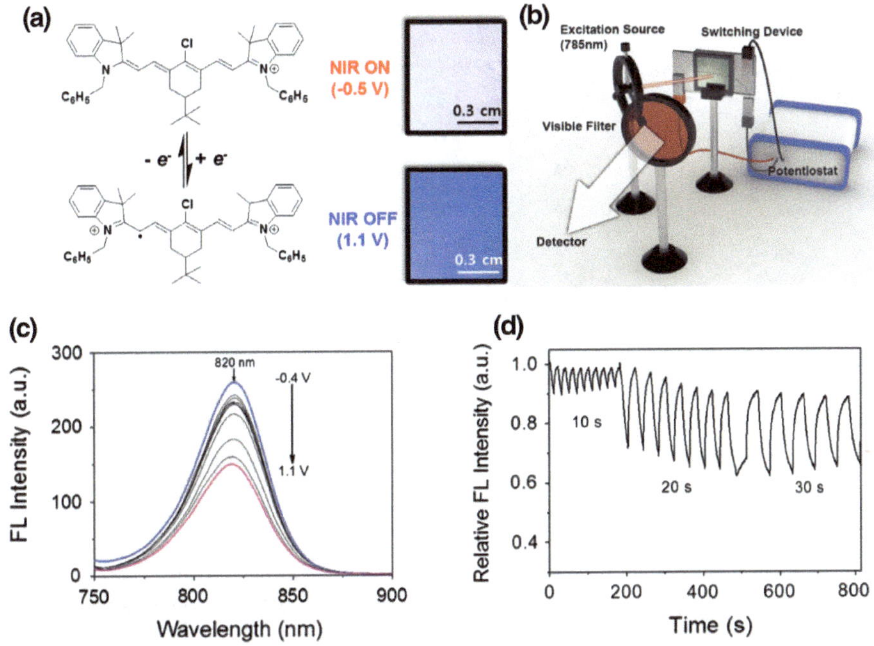

Fig. 10 **a** Chemical structure of a NIR electrofluorochromic polymethine dye (**7**). **b** The schematic diagram of the electrochemical fluorescence switching system, which consists in the excitation source, the switching device, the visible cutoff filter, and the detector. **c** Fluorescence changes in the switching device containing a polymethine dye. **d** The fluorescence switching responses of the NIR electrofluorochromic device [21]

In order to achieve high electrofluorochromic property in the NIR region, highly stable and NIR-emissive electrofluorochromic materials are required. The aza-boron-dipyrromethene (aza-BODIPY) dyes incorporating nitrofluorene and alkoxy decorations (**8–12**) showed a highly reversible one-electron reduction process and exhibited electrofluorochromic properties in the NIR range, depending on the substituents. The nitrofluorene ethynyl-substituted BODIPYs (**8**, **9**, **10**) showed smaller electrofluorochromic contrast than the alkoxy-containing BODIPYs (**11**, **12**) because of the difference in their intrinsic fluorescence contrast between the neutral and reduced states (radical anion). In addition, **11** and **12** showed a larger diffusion coefficient for ion transport, which enhanced the EF contrast and the response time for the fluorescence change at a given step potential. With optimized conditions, the NIR EF ON/OFF ratio of **11** reached a value of 6.9 and a long cyclability over 1000 cycles between −0.4 and +0.4 V switching potentials (Figs. 11 and 12) (Table 3).

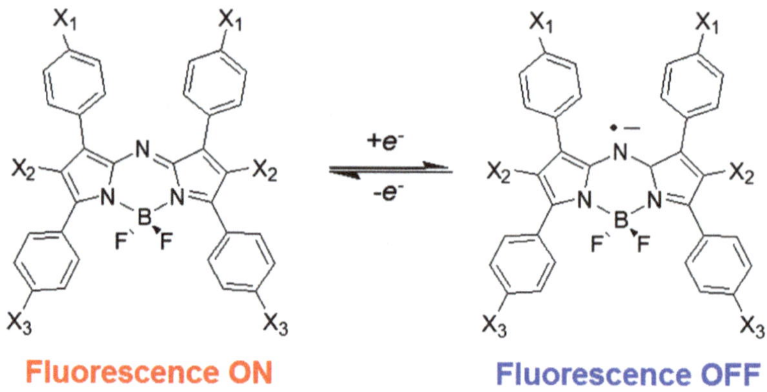

Fig. 11 Electrofluorochromism of the aza-boron-dipyrromethene (Aza-BODIPY) dyes [20]

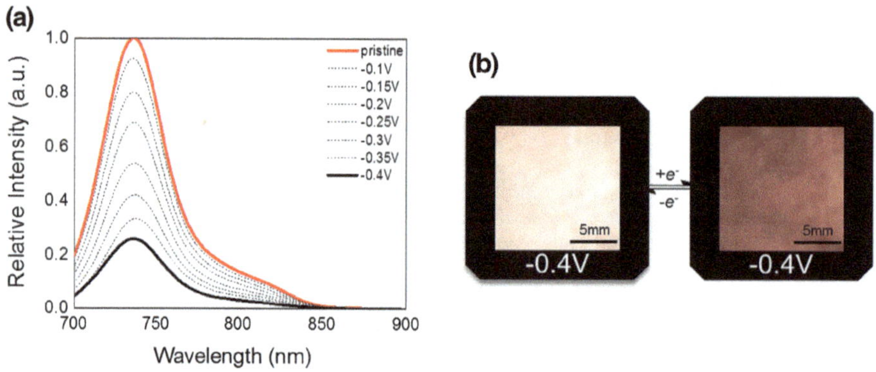

Fig. 12 **a** Fluorescence changes of the dyes **11** from neutral to reduced states, respectively. Each spectrum was obtained after applying a target potential for 50 s. **b** NIR images of the electrofluorochromic switching devices applied at their switching potential to show ON/OFF NIR emission switching [20]

Table 3 Electrofluorochromic properties of the NIR-emissive dyes

Compounds	Electrofluorochromic Properties				
	λ_{emi} (nm)	ON/OFF ratio	Cyclability (Cycle)	Working potential (V) (ON/OFF)	E_{EF} (cm^2/C)
7[a]	820	1.5	~100	−0.5/1.1 V	557
8[a]	711	1.1	>20	0.35/−0.35 V	1.1×10^4
9[a]	741	1.1	>20	0.36/−0.36 V	1.3×10^4
10[a]	803	1.5	>20	0.4/−0.4 V	6.3×10^3
11[a]	721	6.9	>1000	0.4/−0.4 V	1.4×10^3
12[a]	748	1.3	>1000	0.39/−0.39 V	7.2×10^3

[a]EFD structure: T1-3ED

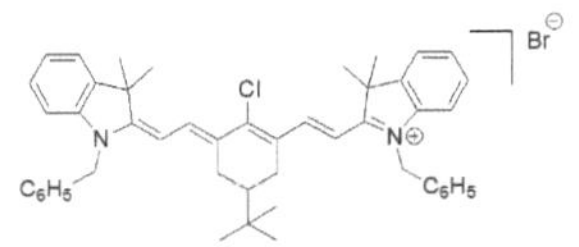

7

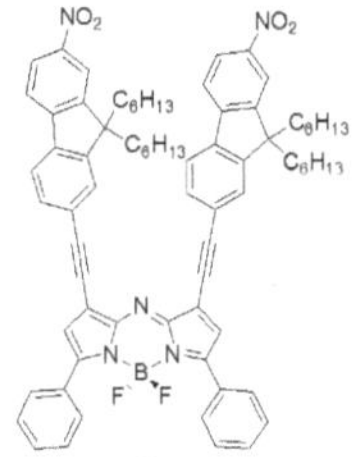

8

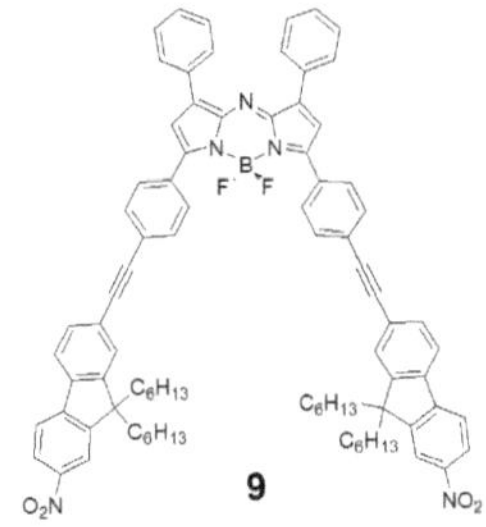

9

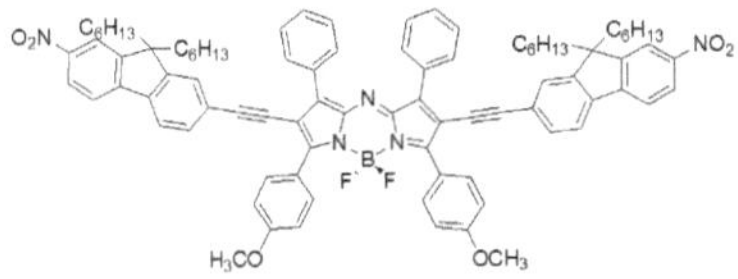

10

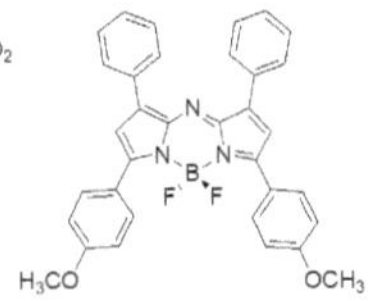

11

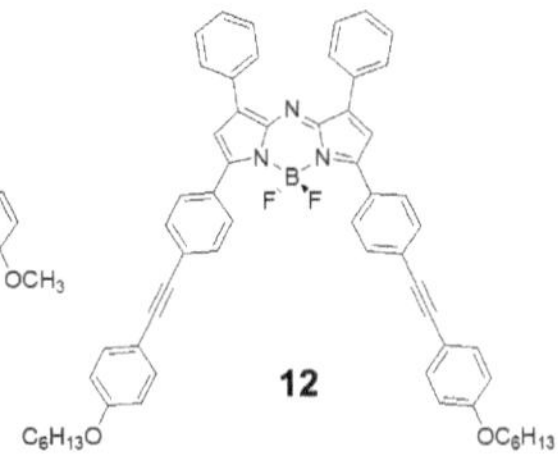

12

3 EF Switching of Fluorescent Dyes by External Quencher (Type II)

Electrofluorochromic switching can be driven not only by the redox electrochemistry of the dye, but also by external quenchers. As shown in Fig. 13, the redox reaction of external quenchers can be used in electrofluorochromic switching with the appropriate fluorescent dye. By the redox state of the quencher, the energy or electron transfer from the donor dye to the quencher is switched. Therefore, the redox electrochemistry of the quencher can result in electrofluorochromic switching. Because the kinetics of the quenching process follows the Stern–Volmer relationship, the different quenching efficiencies can be estimated by the following equation (Eq. 4), where I_0 is the intensity of fluorescence without quencher, I represents the fluorescence intensity with a quencher, k_q is the Stern–Volmer quenching constant, and $[Q]$ is the concentration of the quencher.

$$\frac{I_0}{I} = 1 + k_q [Q] \qquad (4)$$

Because fluorescence intensity should be only controlled by quenchers, the paired fluorescent materials with the quencher are non-electroactive stable materials that are not affected by the applied potential. For example, PPV fluorescent polymers and lanthanide (III) complex were used with the external quenchers such as iodine/iodide redox couple and viologen derivatives [2, 26].

The device structures for Type II EF materials are similarly made of the two transparent ITO electrodes and electrolytes. The electrolyte solution contains fluorescent dyes and redox-active quenchers with supporting electrolyte salts. The

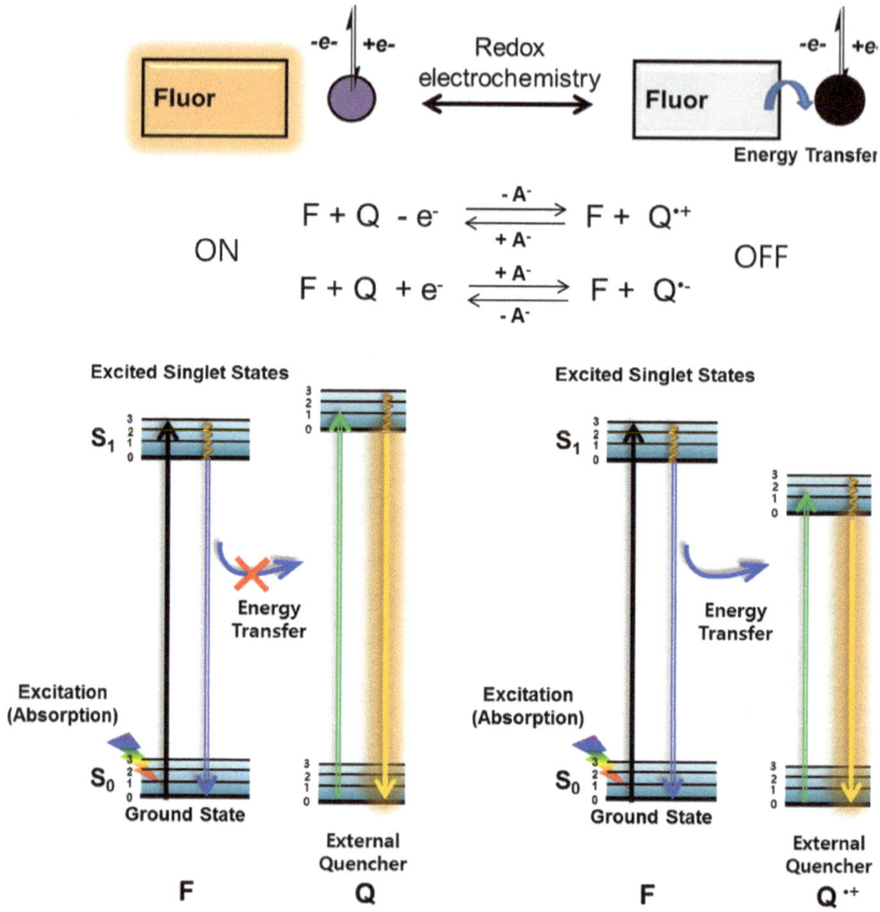

Fig. 13 Mechanism of electrofluorochromism of Type II dyes using the redox chemistry of fluorescent dyes

prepared electrolyte is carefully injected between the ITO electrodes. Three-electrode EFD can also be prepared using an Ag wire reference electrode, but it is not favorable with the corrosive electrochemical materials such as iodine/iodide redox couple. The device structures are presented in Fig. 14.

3.1 *Iodine/Iodide Redox Couple*

The redox reaction of iodine/iodide couple was used as an effective electrofluorochromic quencher for a methylene-bridged anthracene polymer (PMAn) and s-triazine-bridged p-phenylene vinylene polymers (DTOPV) [2, 27–29]. Based

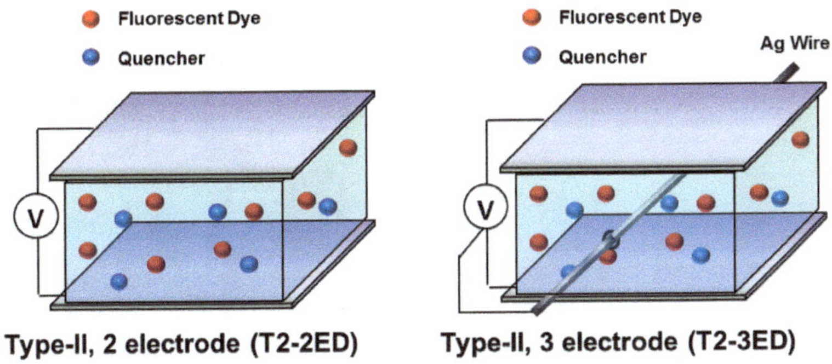

Fig. 14 Devices structures used for Type II electrofluorochromic materials

on Stern–Volmer plot for the fluorescence quenching of DTOPV by I^-, I_2, and I_3^- in chloroform, the Stern–Volmer quenching constants were determined as 45, 11500, and 13840 M^{-1} for I^-, I_2, and I_3^-, respectively. Thus, the degree of the fluorescence quenching of DTOPV by the iodine derivatives is highly dependent on the redox state of iodine. The ON/OFF ratio of iodine/iodide redox with DTOPV and PTOPV as a fluorescent pair was ~4 and ~8, respectively. The working voltage was −2.0/2.0 V. The device could maintain 20% of the starting ON/OFF ratio after 1500 cycles. This redox couple can be universally applied in fluorescent polymers; PPV polymers and PMAn showed stable electrofluorochromic switching, as shown in Table 4 (Fig. 15).

Table 4 Electrofluorochromic properties of the fluorescent materials with iodine/iodide redox couple

Compounds	Electrofluorochromic properties			
	λ_{emi} (nm)	ON/OFF ratio	Cyclability (Cycle)	Working potential (V) (ON/OFF)
13[a]	532	~4	>1500	−2.0/2.0 V
14[a]	520	~8	>1500	−2.0/2.0 V
15[a]	460	2.4	–	−1.5/1.5 V
16[a]	431	2.6	–	−2.0/2.0 V
17[a]	415	4.8	>50	−2.0/2.0 V

[a]EFD structure: T2-2ED

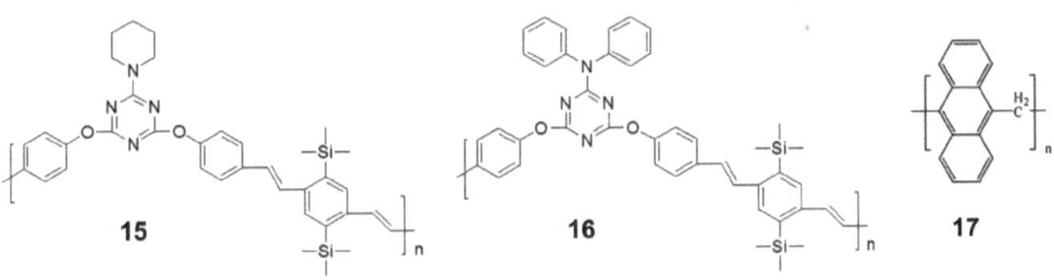

15 **16** **17**

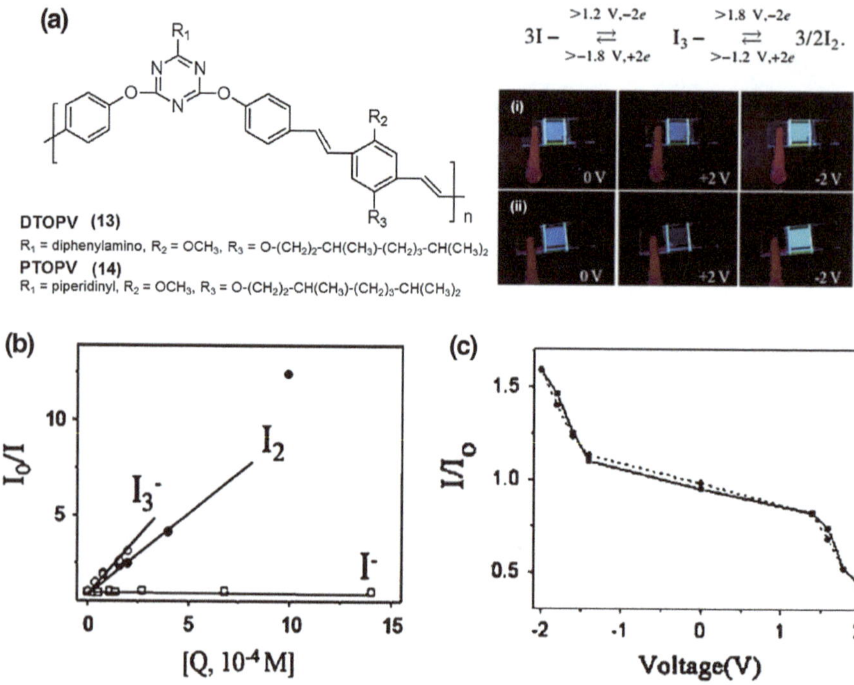

Fig. 15 **a** Chemical structures of DTOPV and PTOPV, and photographic images of the emission from the device containing (*i*) DTOPV and (*ii*) PTOPV, recorded at 0, +2, and −2 V under UV light. **b** Stern–Volmer plot for the fluorescence quenching of DTOPV in chloroform DTOPV = 6.68 × 10^{-5} M by I^-, I_2, and I_3^- (λ_{exc} = 440 nm, probed at 532 nm). **c** Fluorescence intensity of DTOPV at an emission wavelength of 532 nm. I_0 = emission intensity without potential application and I = emission intensity under potential application [2]

3.2 Viologens

In this type of electrofluorochromic switching, viologen derivatives can be used as external quenchers. Viologens are known as an efficient electrochromic material, which changes their color with the redox states. The electrochromic material, 1,1′-diheptyl-4,4′-bipyridinium dibromide (HV^{2+}), exhibits transparent state at oxidized state (HV^{2+}) and is converted to blue colored state with the reduction process to (HV^+ or HV). This bandgap change of the material can be used as a switch for the electrofluorochromism. Luminescent Eu(III) complex has unique luminescent properties such as line-like red emissions, long luminescence lifetimes, large Stokes shifts, and stable fluorescence emission with applied potentials. When Eu(III) complex was mixed with 1,1′-diheptyl-4,4′-bipyridinium dibromide (HV^{2+}), the device can show electrochromic properties and also electrofluorochromic switching by the HV^{2+} electrochromism [26, 30–32] (Table 5).

Table 5 Electrofluorochromic properties using redox electrochemistry of Viologens

Compounds	Electrofluorochromic properties			
	λ_{emi} (nm)	ON/OFF ratio	Cyclability (Cycle)	Working potential (V) (ON/OFF)
18[a]	615	380	–	0/−2.2 V
19[a]	615	52	–	0/−2.0 V
20[a]	450, 570	1.6	–	2.0/−2.0 V

[a]EFD structure: T2-2ED

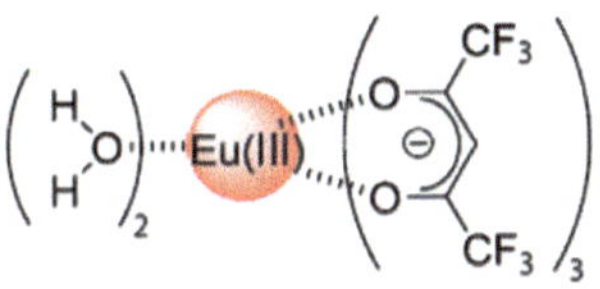

18

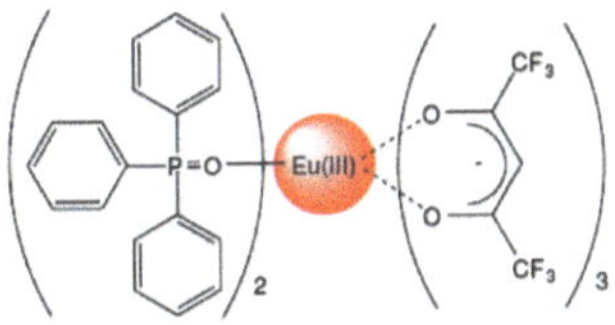

19

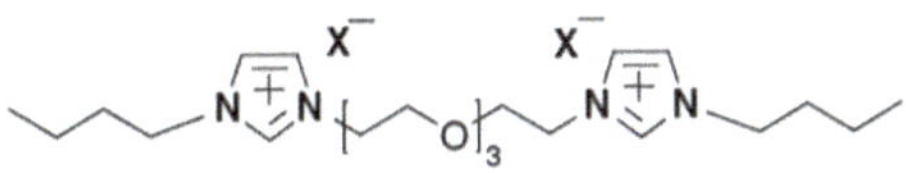

20

4 Electrofluorochromic Switching of Fluorescent Films (Type III)

4.1 *Electrofluorochromic Switching of a Dye-Anchored Film*

The solution and gel-like medium can provide feasible electrofluorochromic switching with Type I dyes, but mechanically improved solid-state devices are still required to achieve high durability [12, 33, 34]. Inorganic nanostructured hosts are good candidates for achieving such mechanical properties. Specially, for the electrofluorochromic switching, conductive TiO_2 nanoparticles (NPs) are good candidates because of an ultrahigh surface area and adsorption site. The chemical structure of **21** was designed to anchor the electrofluorochromic tetrazine to the metal oxide. The triethoxysilane group provided feasible method for surface functionalization. Figure 5 shows the schematic images for the surface treatment by a sol-gel process. To anchor a sufficient amount of **21** on the conductive electrode surface, anatase TiO_2 NPs were deposited on ITO glass. The anchoring process was carried out by dipping the TiO_2 electrode in a 5 mM solution of **21** and 15 mM of acetic acid for 30 min [12] (Fig. 16).

The tetrazine-modified electrode with **21** showed a bright fluorescence emission compared with the non-treated electrode. The fluorescence band was observed at 566 nm, which originated from the tetrazine unit. Furthermore, the electrochemical behavior of the tetrazine-anchored electrode was identical to the typical tetrazine

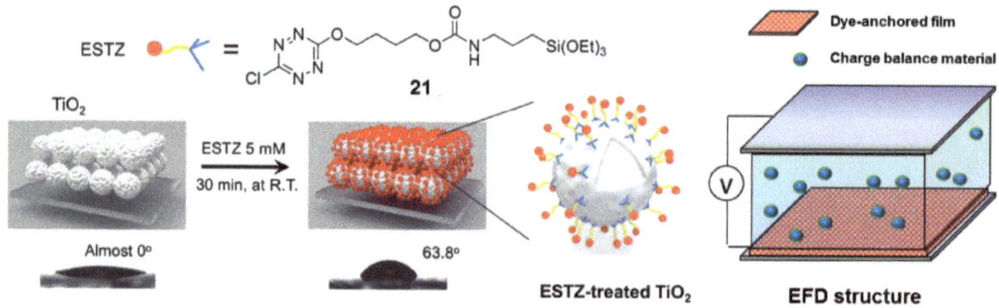

Fig. 16 Preparation of the tetrazine modified solid-state electrode, and the structure of EFD [12]

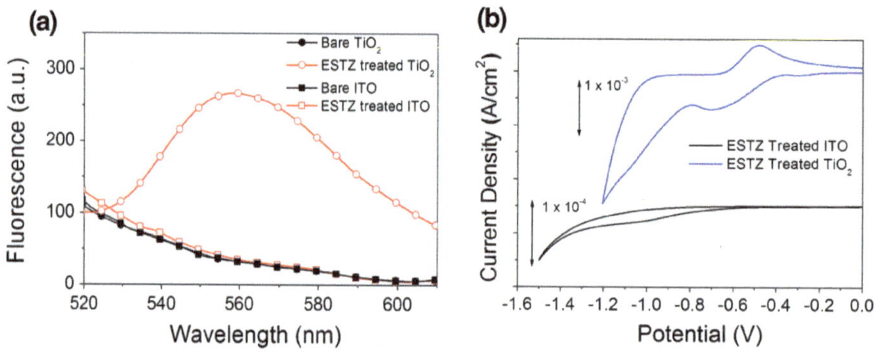

Fig. 17 **a** Fluorescence spectra of the tetrazine-anchored TiO_2 electrode and tetrazine-anchored ITO electrode (λ_{exc} = 480 nm). **b** Cyclic voltammograms of the tetrazine treated electrodes in acetonitrile with 0.1 M $TBAPF_6$ recorded at a scan rate of 50 mV/s [12]

derivatives (Fig. 17). It is worth noting that the tetrazine-anchored ITO electrode did not show the characteristics of the tetrazine unit, even after the treatment. This means that the amount of anchored dye in bare ITO electrode was much lower than in TiO_2 electrode.

The prepared electrofluorochromic material-anchored electrode was used in EFD with the electrolyte and counter electrode. It showed ON/OFF ratio of 4.0 and cyclability greater than 100 cycles with a working potential of 1.2/−1.2 V. This electrofluorochromic property is compatible with the electrofluorochromic system in solution or gel-like medium. It is noteworthy that the additional charge balancing was found to be essential to achieve a reversible switching. As a charge balancing redox mediator, TEMPO was added in the electrolyte to enable charge transport (Fig. 18). Because TiO_2 is not a highly conductive material, a redox mediator should be added to generate a completely reversible electrochemical conversion in TiO_2-based devices.

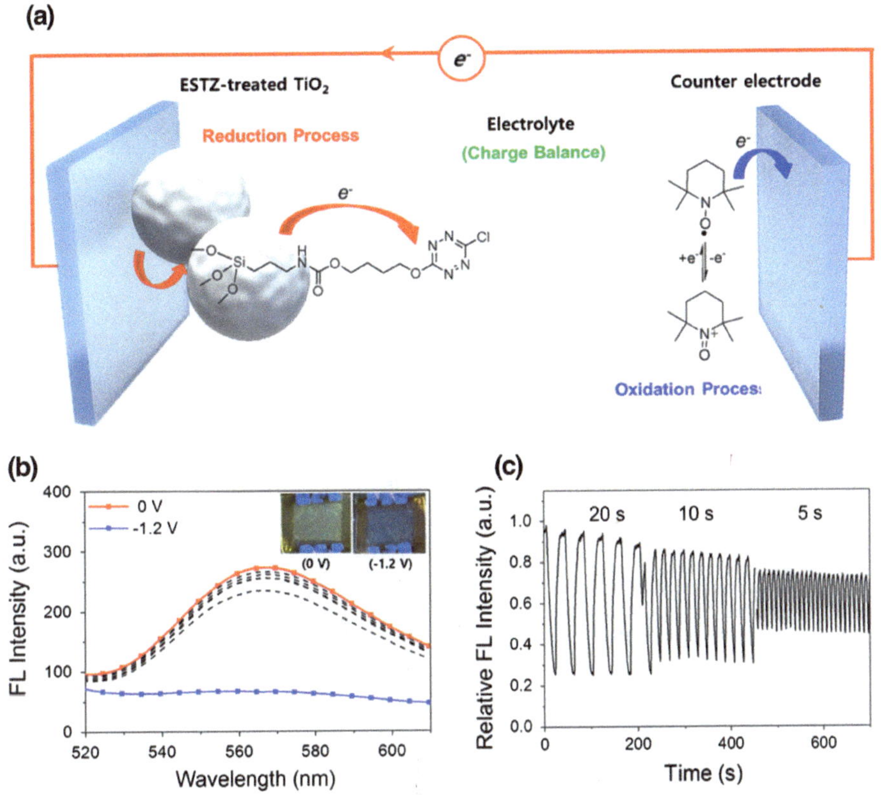

Fig. 18 **a** Schematic diagram of the mediation and charge balance process in an electrofluorochromic switching device. **b** Fluorescence changes in the switching device at different applied potentials. The *inset* shows the image of the reversible fluorescence switching cells, at given potential, under 365 nm UV. **c** The fluorescence switching responses of the device at potentials between -1.2 and 1.2 V [12]

4.2 *The Electrofluorochromic Switching in Fluorescent Polymers*

The electrofluorochromic switching in fluorescent polymers is based on a decrease in the quantum yield and energy transfer from the fluorescent group to the acceptor group. To achieve intrinsically switchable fluorescent polymers, the polymer can be composed of a conjugated fluorescent moiety, a spacer, and an electroactive moiety. Poly(1,3,4-oxadiazole) [7], propylenedioxythiophene phenylene copolymers [35], and triphenylamine (TPA)-modified fluorenes have been reported so far [7, 35–41]. The device structures for Type III EF materials consist of the fluorescent layer-coated working electrode and the counter electrode. The electrolyte solution contains supporting salts, and redox-active charge balancing materials, while additional redox-active

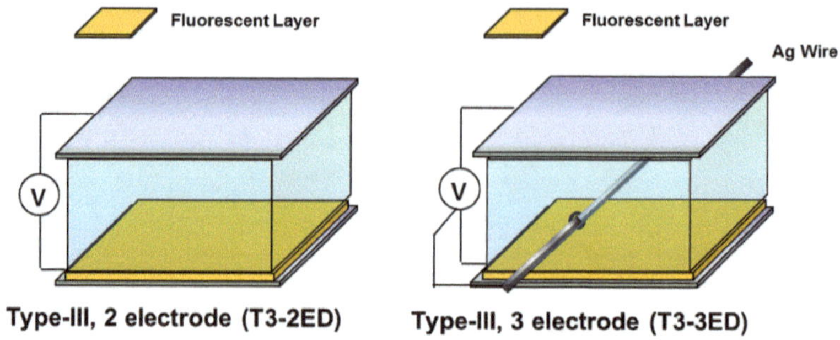

Fig. 19 Devices structures used for Type III polymeric electrofluorochromic materials

quenchers can be included. The prepared electrolyte is carefully injected between the ITO electrodes. Three-electrode EFD can also be prepared using an Ag wire reference electrode. The device structures are presented in Fig. 19.

4.2.1 Electrofluorochromism of Poly(1,3,4-Oxadiazole) Thin Film

Poly(1,3,4-oxadiazole) (POD) thin films can be prepared by spin coating the soluble PODs that are substituted with long alkyl chains, as shown in compound **22**. The high solubility of the polymers enabled a simple fabrication of an electrochemical cell, which showed reversible fluorescence switching between dark (n-doping) and bright (neutral) states with a maximum ON/OFF ratio of 2.5 and a cyclability longer than 1000 cycles, with the EFD structure, T3-3ED. The switching mechanism of electrofluorochromism is presented in Fig. 20. Based on the reversible electrochemical conversion, the electrochemical fluorescence switching properties of the POD films were examined by monitoring their photoluminescence at different applied external potentials (Fig. 21) with the T3-3ED EFD and solid polymer electrolyte. When the potential was applied from +1.8 to −1.8 V, the bright fluorescence from the **22** film was extinguished gradually and was completely extinguished when the potential decreased beyond −1.8 V. As shown in Fig. 21,

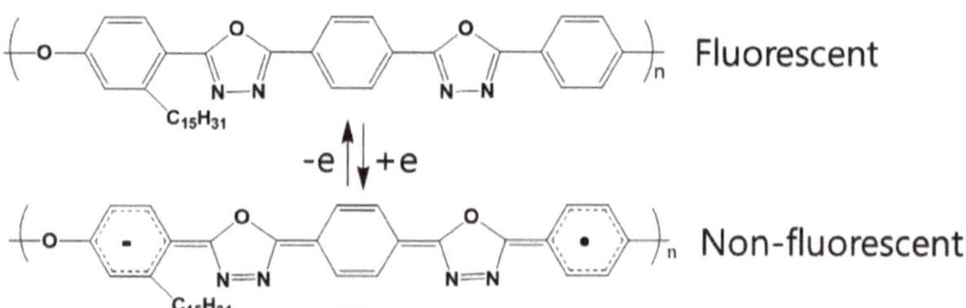

22

Fig. 20 Electrofluorochromic mechanism of POD [7]

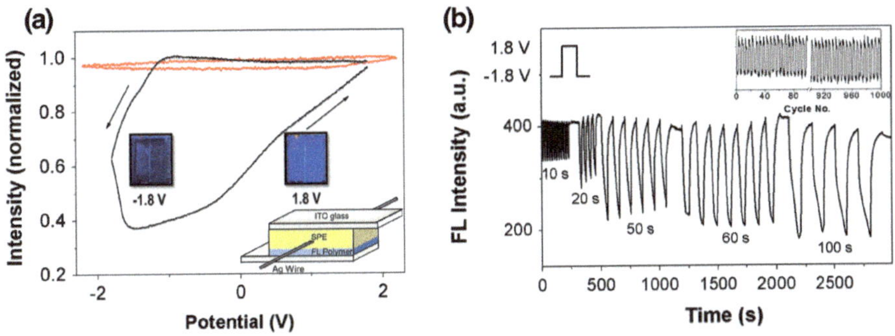

Fig. 21 **a** Cyclic electrofluorescence spectra of **22** (*black line*). **b** *Inset* (*bottom*) shows photographic images of EFD containing **22** film between +1.8 and −1.8 V, under UV (365 nm). **b** Fluorescence switching responses of cell containing **22** under potentials between 1.8 and −1.8 V applied for 10, 20, 50, 60, and 100 s at each voltage. *Inset*: the cyclability of the **22** polymer [7]

the completely quenched fluorescence was recovered upon oxidation process. The cell with the **22** film recovered its original emission when the applied potential was 1.8 V.

4.2.2 Electrochromism and Electrofluorochromism of a Chiral Propylenedioxythiophene-Phenylene Copolymer

Several soluble π-conjugated polymers can also be used in electrofluorochromic switching. The propylenedioxythiophene-phenylene (ProDOT-Ph) with alkyl substituents on the propylene bridge showed electrofluorochromism in a thin-film state. (S)-2-Methyl-1-butoxy group substituted ProDOT-Ph (**23**) showed electrofluorochromic property with an ON/OFF ratio of 3.3 at 525 nm emission wavelength, when the switching potential range was 0/1.0 V. Especially, the electrochemical conversion in **23** was accompanied by electrochromism altogether. The optical bandgap of the π-conjugated polymers are changed with their redox states, which results into dramatic color changes and fluorescence changes. The electrochemical conversion is presented in Fig. 22.

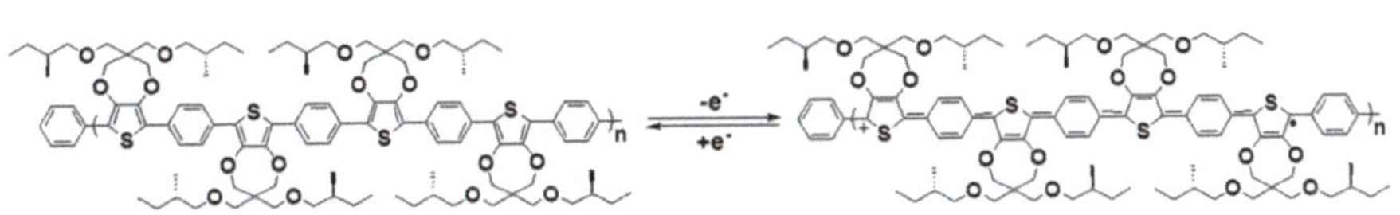

23

Fig. 22 Electrochromism and electrofluorochromism arise from the reversible redox reactions in the electrochemical device [35]

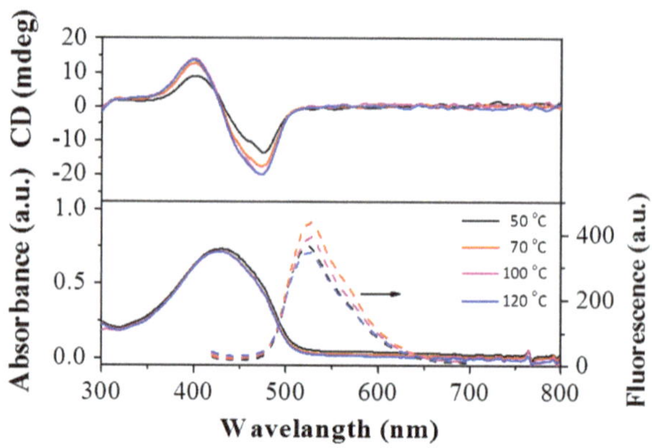

Fig. 23 CD, UV/vis absorbance, and fluorescence spectra of the 23 after annealing 2 h at different temperatures and cooling slowly [35]

In the polymeric materials, packing density also affects the fluorescence intensity. The fluorescence intensity of **23** increased, as the intermolecular packing becomes loose, by increasing the annealing temperature. When the annealing temperature was increased higher than the glass transition Tg (100, 120 °C) and the film cooled slowly afterward, the circular dichroism (CD) intensity was maximized, and the fluorescence intensity decreased significantly because of the fluorescence quenching by the intermolecular assembly in the conjugated polymer films (Fig. 23). This unique property stems from chirality and is originated from the chiral group substituted on the propylene bridge. Using redox electrochemistry, the chiroptical switching is possible due to the reversible structural disorder in the doped state. As shown in Fig. 24, the $\pi - \pi$ stacking on the (100) plane is displaced by the addition of counteranions, showing a distorted structure. This structural disorder is reversibly recovered to the original state by electrochemical reaction. The three different optical properties, chirality (CD, at 460 nm), color (absorbance, at 425 nm), and fluorescence (at 525 nm), were switched ON/OFF in a single device (Fig. 24).

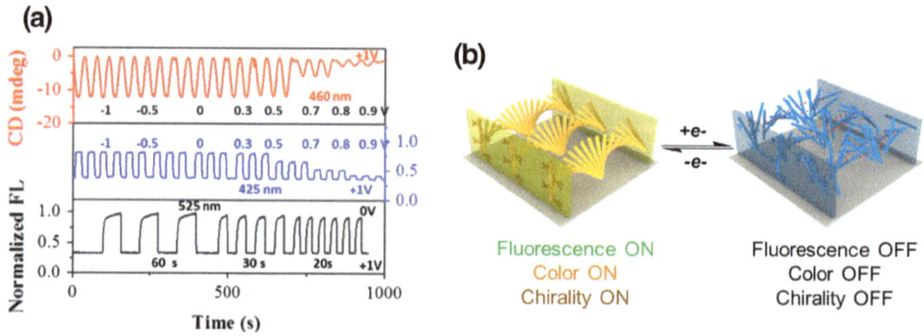

Fig. 24 **a** Multiswitching of **23**. The chirality (CD, at 460 nm, *red line*) and color (absorbance, at 425 nm, *blue line*) switching was recorded with applied potential steps from 1 V to the different potentials (−1 V ~ +0.9 V) with 20 s switching time. The fluorescence (at 525 nm, *black line*) switching was recorded with applied potential steps from 1 to 0 V with different switching time (60, 30, and 20 s). (EFD: T3-3ED). **b** Schematic images of **23** at undoped and doped states [35]

4.2.3 TPA-Modified Poly(Fluorene)s

To achieve intrinsically switchable fluorescent polymers, the polymer can be composed of a conjugated fluorescent moiety, a spacer, and an electroactive moiety. Triphenylamine (TPA) can be a suitable electroactive moiety in electrofluorochromic polymers [36–41]. The radical cation generated on TPA can efficiently quench the surrounding fluorescence to a complete disappearance, because of exciton migration or energy transfer. The quenched emission is simultaneously recovered upon the reduction process. The electrofluorochromic polymers can be prepared as thin-film layers and switched by the applied potentials. The advantages of the thin-layer electrofluorochromic materials are the easy fabrication of patterned switching devices, and simple color blending. The color blending can be achieved by mixing the electrofluorochromic polymers with an appropriate ratio. The green-emissive **24** and blue-emissive **26** were blended to realize white light-emitting electrofluorochromic layers. Because the electrofluorochromic polymers **24** and **25** have a benzothiadiazole (BT) unit, through the energy transfer from

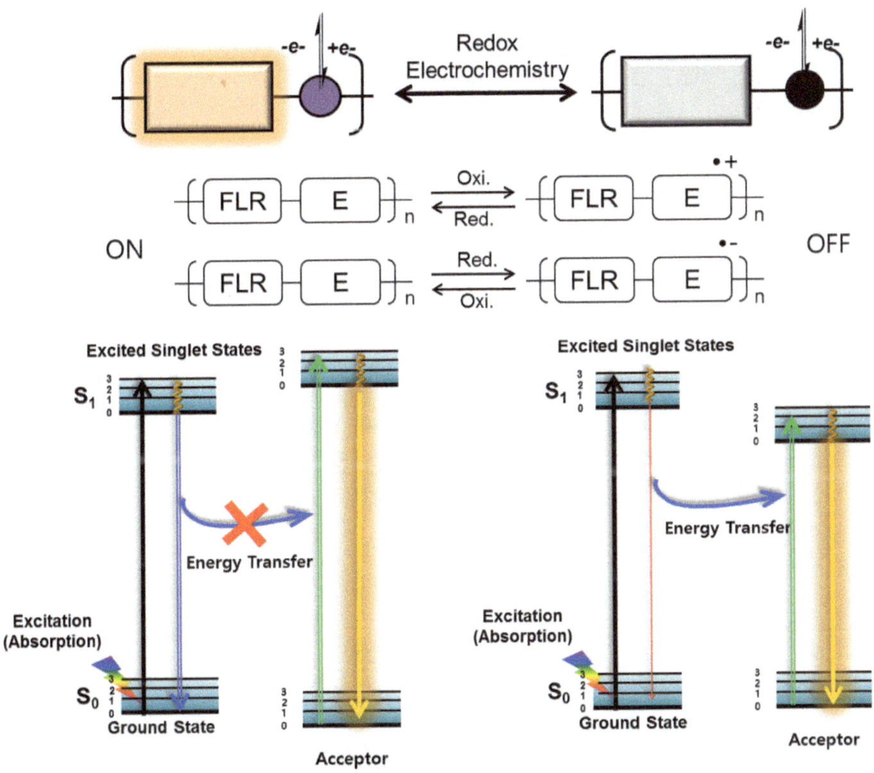

Fig. 25 Mechanism of electrofluorochromism of Type III fluorescent polymers which are consisting of a fluorescent moiety, a spacer, and an electroactive moiety

the fluorene segments to the BT unit the polymers exhibit green or yellow emission. The **24–26** polymers are similarly constituted of fluorene and triphenylamine groups, so their blend can emit the mixed color without mutual quenching (Fig. 25). The chemical structures are presented in Fig. 26, [39]. Also, the fluorescence emission from the polymer blend can be switched by the reversible electrochemistry of the TPA group. However, in this electrofluorochromic system, multi-color switching is not observed possibly because both electrofluorescent polymers were switched at almost the same potential (Fig. 27a). In the case of thin-film electrofluorochromic materials, the coated polymers can be easily patterned on the electrode. Each patterned material can be switched ON/OFF, so it can be used in fluorescent displays (Fig. 27b).

ON/OFF ratio of the thin-film electrofluorochromic materials are usually much higher than the gel-like electrofluorochromic device, because complete quenching is possible without the background fluorescence. The TPA-based polymers **27** showed ON/OFF ratio of about 150. The electrofluorochromic properties of the fluorescent polymers are presented in Table 6 [7, 35, 37–43] (Fig. 28).

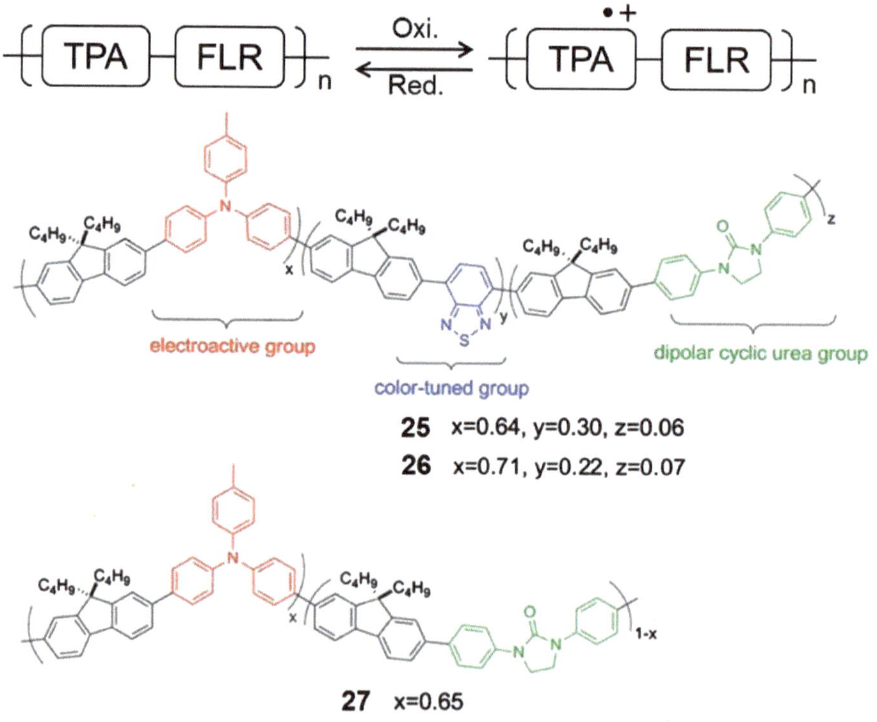

Fig. 26 Mechanism of electrofluorochromism by the TPA group, and the TPA-based polymers [39]

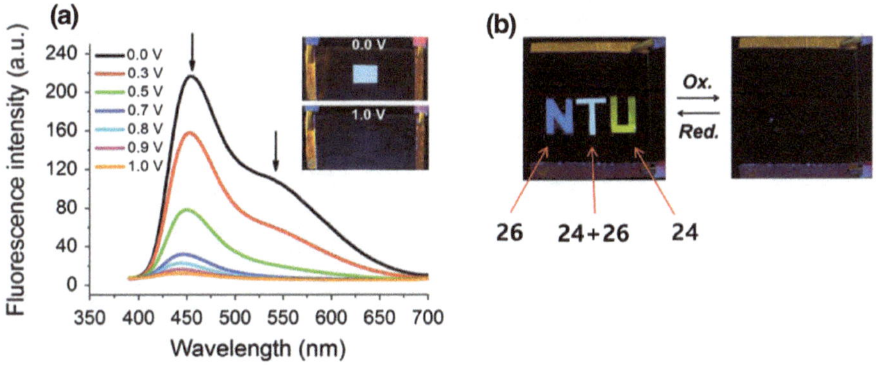

Fig. 27 **a** Fluorescence intensity changes of the electrofluorochromic switching of the polymer blend (**24** + **26**) at different applied potentials. Photographs are the images of the fluorescence changes of the device at indicated potentials under UV excitation (365 nm). **b** Photographs of a single-layer ITO-coated glass electrofluorochromic device, using **26**, **24** + **26**, and **24** for N, T, and U letters, respectively [39]

Table 6 Electrofluorochromic properties of the electrofluorochromic polymers

Compounds	Electrofluorochromic properties			
	λ_{emi} (nm)	ON/OFF ratio	Cyclability (Cycle)	Working potential (V) (ON/OFF)
24[a]	565	21.4	>50	−1.0/1.0 V
25[a]	556	–	–	–
26[a]	443	16.3	>50	−1.0/1.0 V
24 + **26**[a,b]	443, 565	14.6	>50	−1.0/1.0 V
27[a]	470	242	>10	−1.5/1.9 V
28[a]	435	192	>10	−1.7/2.2 V
29[a)]	467	51.3	10	0/1.3 V

[a]EFD: T3-2ED. [b]The polymer blended to present white color emission

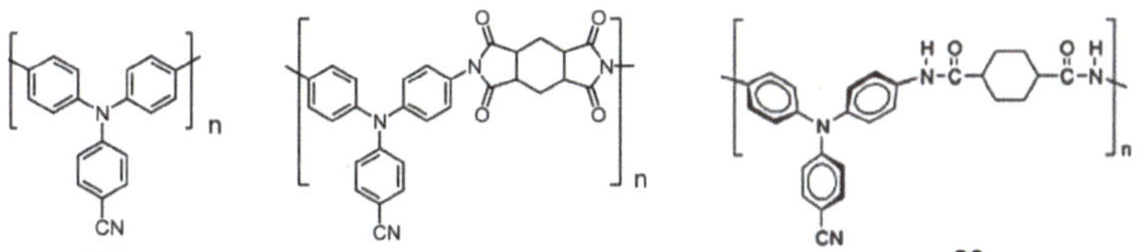

27 **28** **29**

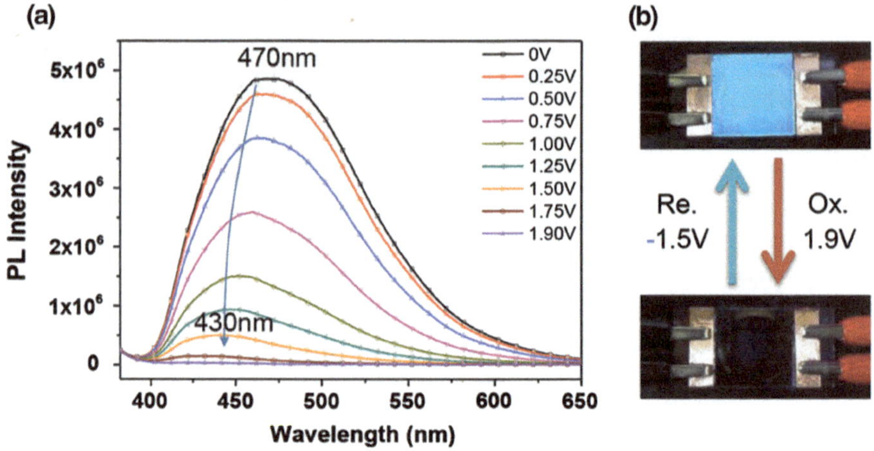

Fig. 28 **a** FL spectrums of **27** with different applied potentials. **b** Electrofluorochromic behavior of the device (**27**, polymer film 200 ± 10 nm in thickness) [37]

4.3 Generation of Electrofluorochromic Patterns by Photopatterning of Fluorescent Polymers

Photoreactive fluorescent polymers afford a convenient method of image formation and patterning with a simple photopatterning process. In electrofluorochromic switching, patterning appears to be a very promising strategy to record information, as it can open up opportunities for new displays with controllable light output, fluorescent displays, sensors, and spectroscopy. Also, it is very promising to secure document applications [7, 42]. Although an imprinting method can be used with most polymers [44, 45], photopatterning is preferred for high-contrast fluorescence patterns because the photo-irradiated region can be completely quenched. When the photo-irradiated region is selectively washed by the solvent, the pattern will result in not only a high-contrast fluorescence pattern but also a lithographic pattern with a thickness difference.

Poly(1,3,4-oxadiazole)s (PODs) substituted with long alkyl chains possess electrofluorochromic properties, as well as patternability through photochemical cleavage. The photochemical cleavage of the oxadiazole unit in POD allowed photopatterning of the POD film upon exposure to UV source. The patterned POD films can display the patterned image reversibly upon the applied potentials (Fig. 29).

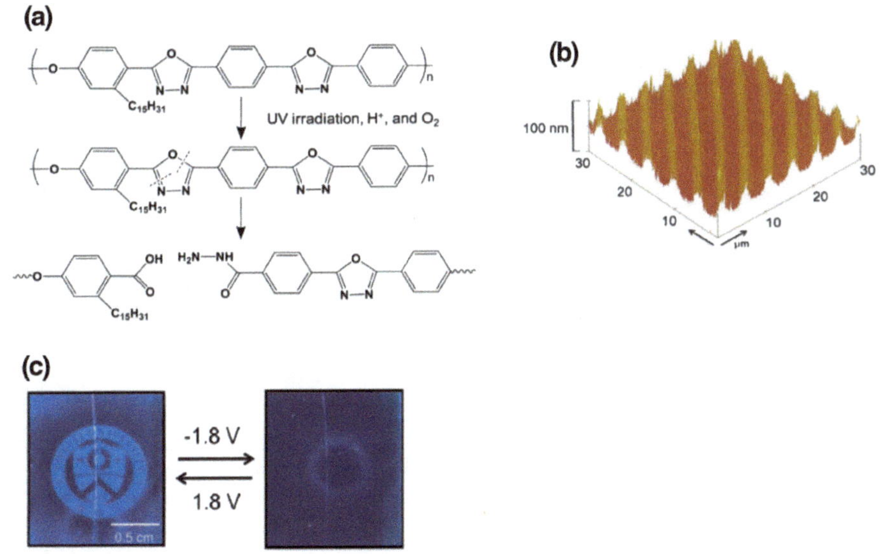

Fig. 29 **a** Possible mechanism of POD photopatterning. **b** AFM image of the 5-μm pattern of 22. **c** Reversible fluorescence switching image of the patterned POD at +1.8 and −1.8 V [7]

5 Applications

5.1 Dual-Mode Displays

Dual-mode display (DMD) devices are designed to compensate the main disadvantage of electrochromic devices (ECDs). Reflective displays such as ECDs and printed paper rely on external light sources and can generally be viewed only when there is sufficient ambient light available. To compensate this weakness, DMD which can be operated in both emissive and reflective modes are promising as next-generation displays, digital signaling, and sensors [5, 31]. In order to fabricate DMD devices with simple configurations, electrochromic and electrofluorochromic materials can be combined in a single device. Because both types are based on the redox electrochemistry, the devices can be simply combined in a single cell. For example, poly(3,4-ethylenedioxythiophene) (PEDOT) having blue to transparent electrochromism (EC) and 9-methylanthracene polymer (PMAn; **17**) having electrofluorochromism in blue light emission under 390 nm light excitation, were combined in a single device to show DMD. In Fig. 30, the DMD structure is presented along with the cyclic voltammogram. To examine the electrochemical behavior of the electroactive layers, cyclic voltammetry was used to identify the redox potential ranges and to estimate the electrochemical behavior of the electroactive layers. PMAn-coated ITO glass was connected to working electrode and PEDOT-coated ITO glass electrode was used as counter electrode. The current was increased dramatically in the presence of PMAn layer with TBAI and I_2. This is

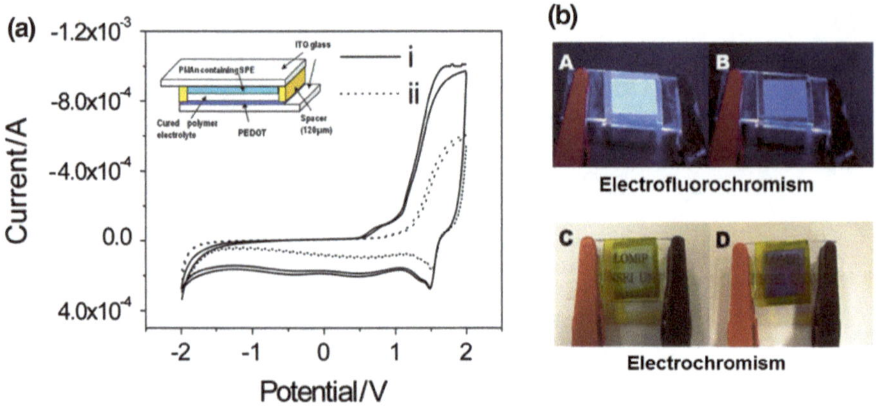

Fig. 30 **a** CV of two-electrode cells consisting of PMAn, PEDOT, and an electrolyte layer containing electrolyte (i) without additives and (ii) with TBAI and I_2 (*inset*; schematic of dual-mode switching device) **b** Photographic images of alternating electrochemical switching of color and fluorescence in one cell. A, C: −2 V, fluorescence ON (under UV light) and color bleached (*under room light*), respectively. B, D: +2 V, fluorescence OFF (under UV light) and colored (*under room light*), respectively [5]

probably due to the electroactive iodine derivatives that work as PMAn quenchers and dopants for PEDOT [5]. The photographs of the DMDs show that the device can be operated in reflective mode and emissive mode.

5.2 Electrofluorochromic Sensing

Electrofluorochromism can be applied in detection of anions using the interaction between a nucleophilic anion and a fluorescent polymer [4, 46–48]. The interaction between the CN^- and electropositive aromatic (benzothiadiazole) is a reversible process. Based on the interaction, the selective detection of CN^- was designed with an electrofluorochromic polymer **30**. The possible mechanism of the detection is presented in Fig. 31. The detection experimental setup is composed of a three-electrode spectroelectrochemical cell filled with electrolyte. The polymer **30** was coated as a thin film on an ITO electrode and immersed into the electrolyte. Fluorescence intensities were measured with the applied potential. The detection procedure was as follows: At zero potential, the stabilized PL intensity was recorded as I_0 when the system reached equilibrium. At a certain potential, the ratio of the corresponding stabilized PL intensity (I) to I_0 was defined as the normalized intensity. The values of I/I_0 obtained at various potentials were plotted against the potential. Then, the detection electrode was inserted into another cell filled with the corresponding electrolyte containing a certain concentration of TBACN, and then, the stabilized PL intensities were recorded using the same series of potentials,

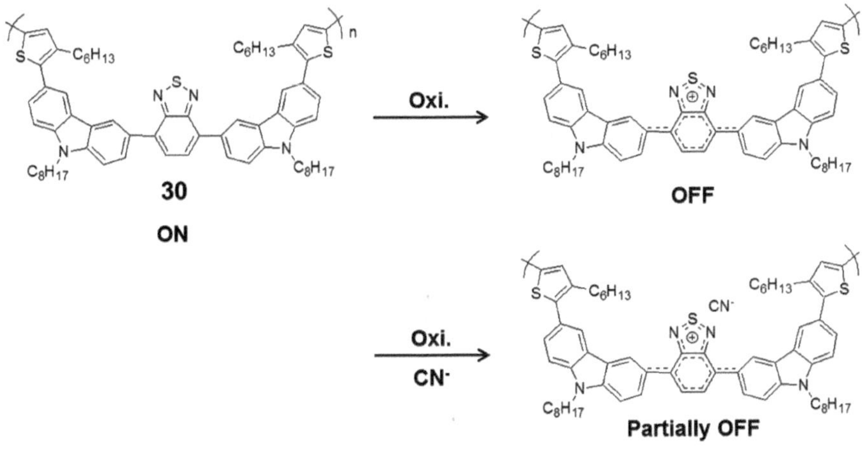

Fig. 31 Possible mechanism of electrofluorochromic sensing of CN^- [4]

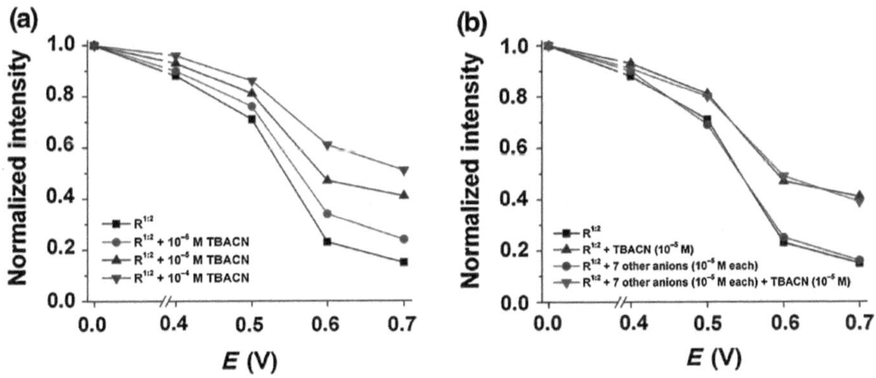

Fig. 32 Normalized fluorescence intensities (l_{exc} = 450 nm, l_{emi} = 580 nm) of **30** as a function of the potential, E. **a** The curves were obtained using electrolytes and electrolytes + TBACN. The TBACN concentrations are 10^{-6}, 10^{-5}, and 10^{-4} M. **b** The curves were obtained using electrolyte and electrolyte +TBACN (10^{-5} M), and electrolyte + seven types of other TBA-based anions (F-,Cl-, Br-,I-,AcO-,NO_3^-, and HSO_4^-; 10^{-5} M each), with and without TBACN (10^{-5} M) [4]

giving another I/I_0 versus E curve (Fig. 32). As a result, the fluorescence intensity reduction caused by oxidative quenching can be weakened by adding CN^- into the electrolyte. This phenomenon can be attributed to the interaction between nucleophilic CN^- and electron-deficient benzothiadiazole. The sensitive and selective CN^- detection is thus demonstrated using the electrofluorochromic switching.

6 Concluding Remarks

In this chapter, we focused on electrofluorochromic materials that can be categorized into three types, depending on the distance between the electroactive moiety and the fluorescent core. To compare the materials, the electrofluorochromic properties, such as working potential, emission color, ON/OFF ratio, stability, and switching efficiency, are defined by several equations. Electrofluorochromism by redox chemistry of fluorescent dyes (Type I) includes small molecules such as tetrazine derivatives and BODIPYs. The ON/OFF ratio varies from ~ 2 to ~ 10, and the highest value was 14 with tetrazine derivative. The EF efficiency was 1.3×10^4 with aza-BODIPY. The cyclability of the materials depends on the reversibility. Over 1000 cycles are guaranteed in most materials. EF switching of fluorescence by external quencher (Type II) uses redox electrochemistry of quenchers. Iodine/iodide redox couple and viologens can quench the emission through energy transfer. In this type of electrofluorochromism, the existence of very different quenching efficiencies in relation with the different redox states is essential. The kinetics of the quenching process follows the Stern–Volmer relationship; the different quenching efficiencies can be estimated through equations accordingly. The ON/OFF ratio for Type II varies from ~ 2 to ~ 300, and the highest value was 380 with Eu (III) complex and viologen system. The third type is made of electrofluorochromic polymer films (Type III). The EF layers can be prepared as thin-film layers, thus these materials are easily patterned. The photopatternable fluorescent polymers can provide high-resolution patterns with electrofluorochromism. Based on the electrofluorochromism, some applications are reported. The ON/OFF ratio for Type III devices varies from ~ 2 to ~ 250, and the highest value was 242 with a TPA polymer. These EFDs could be potentially applied in displays and sensing. Dual-mode displays with high readability and electrofluorochromic anion sensing were realized using the unique characteristics of electrofluorochromism. The reports on the applications are still scarce until now, but we believe that novel challenges in display or sensing applications will follow these pioneering researches.

References

1. Audebert, P., Miomandre, F.: Electrofluorochromism: from molecular systems to set-up and display. Chem. Sci. **4**, 575–584 (2013)
2. Yoo, J., Kwon, T., Sarwade, B.D., Kim, Y., Kim, E.: Multistate fluorescence switching of s-triazine-bridged p-phenylene vinylene polymers. Appl. Phys. Lett. **91**, 241107 (2007)
3. Kim, Y., Kim, E., Clavier, G., Audebert, P.: New tetrazine-based fluoroelectrochromic window; modulation of the fluorescence through applied potential. Chem. Commun. **34**, 3612–3614 (2006)
4. Ding, G., Lin, T., Zhou, R., Dong, Y., Xu, J., Lu, X.: Electrofluorochromic detection of Cyanide Anions using a Nanoporous Polymer Electrode and the detection mechanism. Chem. —A Eur. J. **20**, 13226–13233 (2014)

5. Kim, Y., Kim, J., You, J., Kim, E.: Electrochemical modulation of color and Fluorescence in one cell using conducting Polymers. Mol. Cryst. Liq. Cryst. **538**, 39–44 (2011)
6. Clavier, G., Audebert, P.: s-Tetrazines as building blocks for new functional molecules and molecular materials. Chem. Rev. **110**, 3299–3314 (2010)
7. Seo, S., Kim, Y., You, J., Sarwade, B.D., Wadgaonkar, P.P., Menon, S.K., More, A.S., Kim, E.: Electrochemical Fluorescence switching from a patternable poly(1,3,4-oxadiazole) thin film. Macromol. Rapid Commun. **32**, 637–643 (2011)
8. Quinton, C., Alain-Rizzo, V., Dumas-Verdes, C., Miomandre, F., Clavier, G., Audebert, P.: Redox-controlled fluorescence modulation (electrofluorochromism) in triphenylamine derivatives. RSC Adv. **4**, 34332–34342 (2014)
9. Goulle, V., Harriman, A., Lehn, J.M.: An electro-photoswitch: redox switching of the luminescence of a bipyridine metal complex. J. Chem. Soc. Chem. Commun. **12**, 1034–1036 (1993)
10. Dias, M., Hudhomme, P., Levillain, E., Perrin, L., Sahin, Y., Sauvage, F.-X., Wartelle, C.: Electrochemistry coupled to fluorescence spectroscopy: a new versatile approach. Electrochem. Commun. **6**, 325–330 (2004)
11. Seo, S., Kim, Y., Zhou, Q., Clavier, G., Audebert, P., Kim, E.: White Electrofluorescence switching from Electrochemically convertible yellow Fluorescent dyad. Adv. Funct. Mater. **22**, 3556–3561 (2012)
12. Seo, S., Allain, C., Na, J., Kim, S., Yang, X., Park, C., Malinge, J., Audebert, P., Kim, E.: Electrofluorescence switching of tetrazine-modified TiO_2 nanoparticles. Nanoscale **5**, 7321–7327 (2013)
13. Kim, Y., Do, J., Kim, E., Clavier, G., Galmiche, L., Audebert, P.: Tetrazine-based electrofluorochromic windows: modulation of the fluorescence through applied potential. J. Electroanal. Chem. **632**, 201–205 (2009)
14. Janowska, I., Miomandre, F., Clavier, G., Audebert, P., Zakrzewski, J., Thi, K.H., Ledoux-Rak, I.: Donor—acceptor—donor Tetrazines containing a Ferrocene unit: synthesis, electrochemical and Spectroscopic properties. J. Phys. Chem. A **110**, 12971–12975 (2006)
15. Gong, Y.-H., Audebert, P., Tang, J., Miomandre, F., Clavier, G., Badré, S., Méallet-Renault, R., Marrot, J.: New tetrazines substituted by heteroatoms including the first tetrazine based cyclophane: synthesis and electrochemical properties. J. Electroanal. Chem. **592**, 147–152 (2006)
16. Audebert, P., Miomandre, F., Clavier, G.: First example of an electroactive polymer issued from an oligothiophene substituted tetrazine. Electrochem. Commun. **6**(2), 144–147 (2004)
17. Dumas-Verdes, C., Miomandre, F., Lépicier, E., Galangau, O., Vu, T.T., Clavier, G., Méallet-Renault, R., Audebert, P.: BODIPY-Tetrazine multichromophoric derivatives. Eur. J. Org. Chem. **2010**, 2525–2535 (2010)
18. Quinton, C., Alain-Rizzo, V., Dumas-Verdes, C., Miomandre, F., Audebert, P.: Tetrazine–triphenylamine dyads: influence of the nature of the linker on their properties. Electrochim. Acta **110**, 693–701 (2013)
19. Beneduci, A., Cospito, S., La Deda, M., Veltri, L., Chidichimo, G.: Electrofluorochromism in π-conjugated ionic liquid crystals. Nat. Commun. **5** (2014)
20. Lim, H., Seo, S., Pascal, S., Bellier, Q., Rigaut, S., Park, C., Kim, E. NIR Electrofluorochromic properties of Aza-Boron-dipyrromethene Dyes. Sci. Rep. **6** (2016)
21. Seo, S., Pascal, S., Park, C., Shin, K., Yang, X., Maury, O., Sarwade, B.D., Andraud, C., Kim, E.: NIR electrochemical fluorescence switching from polymethine dyes. Chem. Sci. **5**, 1538–1544 (2014)
22. Yang, J., Choi, J., Bang, D., Kim, E., Lim, E.-K., Park, H., Suh, J.-S., Lee, K., Yoo, K.-H., Kim, E.-K., Huh, Y.-M., Haam, S.: Convertible organic nanoparticles for near-infrared Photothermal ablation of cancer cells. Angew. Chem. Int. Ed. **50**, 441–444 (2011)
23. Hertel, D., Marechal, H., Tefera, D. A., Fan, W., Hicks, R.: A low-cost VIS-NIR true color night vision video system based on a wide dynamic range CMOS imager. In: Intelligent Vehicles Symposium, 2009 IEEE (pp. 273–278). (2009). doi:10.1109/ivs.2009.5164290

24. Lin, K.-T., Tseng, S.-C., Chen, H.-L., Lai, Y.-S., Chen, S.-H., Tseng, Y.-C., Chu, T.-W., Lin, M.-Y., Lu, Y.-P.: Ultrahigh-sensitivity CdS photoconductors with instant response and ultralow power consumption for detection in low-light environments. J. Mater. Chem. C. **1**, 4244–4251 (2013)
25. Kim, S., Lim, Y.T., Soltesz, E.G., De Grand, A.M., Lee, J., Nakayama, A., Parker, J.A., Mihaljevic, T., Laurence, R.G., Dor, D.M.: Near-infrared fluorescent type II quantum dots for sentinel lymph node mapping. Nat. Biotechnol. **22**, 93–97 (2003)
26. Nakamura, K., Kanazawa, K., Kobayashi, N.: Electrochemically-switchable emission and absorption by using luminescent Lanthanide(III) complex and electrochromic molecule toward novel display device with dual emissive and reflective mode. Displays **34**, 389–395 (2013)
27. Yoo, J., Jadhav, P., Kim, E.: Fluorescence Switching of a Copolymer of s-Triazine and p-Phenylene Vinylene by Redox Reaction of Iodide Ions. Mol. Cryst. Liq. Cryst. **491**, 114–121 (2008)
28. Kwon, T., Sarwade, B., Kim, Y., Yoo, J., Kim, E.: A p-Phenylene Vinylene Polymer Bridged by a Triazine for Electro-Fluorescence Switching. Mol. Cryst. Liq. Cryst. **486**, 101–109 (2008)
29. You, J., Kim, Y., Kim, E.: Electrochemical fluorescence switching from Anthracene Polymer Films. Mol. Cryst. Liq. Cryst. **520**, 128–135 (2010)
30. Nakamura, K., Kanazawa, K., Kobayashi, N.: Electrochemically controllable emission and coloration by using europium (III) complex and viologen derivatives. Chem. Commun. **47** (36), 10064–10066 (2011)
31. Kanazawa, K., Nakamura, K., Kobayashi, N.: Dual emissive-reflective display materials with large emission switching using highly luminescent lanthanide (III) complex and electrochromic material. Japan. J. of Appl. Phy. **52**, 05DA14 (2013)
32. Lee, W.R., Kim, Y., Kim, J.Y., Kim, T.H., Ahn, K.D., Kim, E.: Electro-fluorescence switching of Bis-imidazolium onic liquids. J. Nanosci. Nanotechnol. **8**, 4630–4634 (2008)
33. Malinge, J., Allain, C., Brosseau, A., Audebert, P.: White fluorescence from Core-Shell Silica Nanoparticles. Angew. Chem. **124**, 8662–8665 (2012)
34. Malinge, J., Allain, C., Galmiche, L., Miomandre, F., Audebert, P.: Preparation, Photophysical, Electrochemical, and Sensing Properties of Luminescent Tetrazine-Doped Silica Nanoparticles. Chem. Mater. **23**, 4599–4605 (2011)
35. Yang, X., Seo, S., Park, C., Kim, E.: Electrical Chiral Assembly Switching of Soluble Conjugated Polymers from Propylenedioxythiophene-Phenylene Copolymers. Macromolecules **47**, 7043–7051 (2014)
36. Wu, J.-H., Liou, G.-S.: Fluorescence: High-Performance Electrofluorochromic Devices Based on Electrochromism and Photoluminescence-Active Novel Poly(4-Cyanotriphenylamine) (Adv. Funct. Mater. 41/2014). Adv. Funct. Mater. **24**, 6406 (2014)
37. Wu, J.-H., Liou, G.-S.: High-Performance Electrofluorochromic Devices Based on Electrochromism and Photoluminescence-Active Novel Poly(4-Cyanotriphenylamine). Adv. Funct. Mater. **24**, 6422–6429 (2014)
38. Yen, H.-J., Liou, G.-S.: Flexible electrofluorochromic devices with the highest contrast ratio based on aggregation-enhanced emission (AEE)-active cyanotriphenylamine-based polymers. Chem. Commun. **49**, 9797–9799 (2013)
39. Kuo, C.-P., Chuang, C.-N., Chang, C.-L., Leung, M.-K., Lian, H.-Y., Chia-Wen, WuK: White-light electrofluorescence switching from electrochemically convertible yellow and blue fluorescent conjugated polymers. J. Mater. Chem. C. **1**, 2121–2130 (2013)
40. Kuo, C.-P., Leung, M.-K.: Electrofluorescence switching from a multilayer thin film by spin-assisted layer-by-layer assembly of an anionic fluorescent conjugated polyelectrolyte with poly(diallyldimethylammonium chloride). Phys. Chem. Chem. Phys. **16**, 79–87 (2014)
41. Kuo, C.P., Lin, Y.S., Leung, M.K.: Electrochemical fluorescence switching properties of conjugated polymers composed of triphenylamine, fluorene, and cyclic urea moieties. J. Polym. Sci. Pol. Chem. **50**, 5068–5078 (2012)

42. Seo, S., Shin, H., Park, C., Lim, H., Kim, E.: Electrofluorescence switching of fluorescent polymer film. Macromol. Res. **21**, 284–289 (2013)
43. Kawabata, K., Goto, H.: Dynamically Controllable Emission of Polymer Nanofibers: Electrofluorescence Chromism and Polarized Emission of Polycarbazole Derivatives. Chem. —A Eur. J. **18**, 15065–15072 (2012)
44. Bhuvana, T., Kim, B., Yang, X., Shin, H., Kim, E.: Electroactive subwavelength gratings (ESWGs) from conjugated polymers for color and intensity modulation. Nanoscale **4**, 3679–3686 (2012)
45. Bhuvana, T., Kim, B., Yang, X., Shin, H., Kim, E.: Reversible Full-Color Generation with Patterned Yellow Electrochromic Polymers. Angew. Chem. Int. Ed. **52**, 1180–1184 (2013)
46. Ding, G., Zhou, H., Xu, J., Lu, X.: Electrofluorochromic detection of cyanide anions using a benzothiadiazole-containing conjugated copolymer. Chem. Commun. **50**, 655–657 (2014)
47. Thomas, S.W., Joly, G.D., Swager, T.M.: Chemical sensors based on amplifying Fluorescent Conjugated Polymers. Chem. Rev. **107**, 1339–1386 (2007)
48. Wang, S., Xu, H., Yang, Q., Song, Y., Li, Y.: A triphenylamine-based colorimetric and turn-on fluorescent probe for detection of cyanide anions in live cells. RSC Advances **5**, 47990–47996 (2015)

Control of Emission and Coloration in Electrochemical Systems and Its Applications

Kenji Kanazawa, Kazuki Nakamura and Norihisa Kobayashi

Abstract This chapter describes the mechanisms of both emission and coloration control in electrochemical systems as well as their photophysical properties (e.g., response time, durability) and their applications. The first section provides a short overview on emission and coloration control upon various kinds of external stimuli. Then, we review the control of both emission and coloration processes based on electrochemical systems in our work.

1 Introduction

Multifunctional materials with stimuli-responsive optical properties have attracted considerable research interest, since they are prominent candidates for chemical sensors [1, 2], biochemical sensors [3], molecular logic gates [4–6], molecular memories [7, 8], and display devices [9–13]. These materials can change their photophysical properties, such as absorbance, absorption wavelength, fluorescence intensity, fluorescence wavelength, in response to various changes in conditions (e.g., pH [14–16] and solvent polarity [17, 18]) or external triggers, such as chemical [14, 19, 20], mechanical [21–23], vapor [24–26], photo [27–30], thermal [31–36], and electrical stimuli [37–42]. Therefore, condition changes and external triggers can be used to achieve the reversible switching of the optical properties of these materials. In particular, the mechanisms of luminescence change in these systems consist mainly of electron [2, 43–46] and/or energy transfers [44, 46–48] induced by external stimuli.

Among the various external stimuli that can be used, the authors have opted to focus on the electrochemically triggered chromic (i.e., electrochromic [37–42]) and luminescent (i.e., electrofluorochromic [47]) effects because such stimuli can be applied rapidly and reversibly. These particular advantages of the electrochemical

K. Kanazawa · K. Nakamura · N. Kobayashi (✉)
Graduate School of Advanced Integration Science,
Chiba University, Inage-Ku, Chiba 263-8522, Japan
e-mail: koban@faculty.chiba-u.jp

F. Miomandre and P. Audebert (eds.), *Luminescence in Electrochemistry*,
DOI 10.1007/978-3-319-49137-0_6

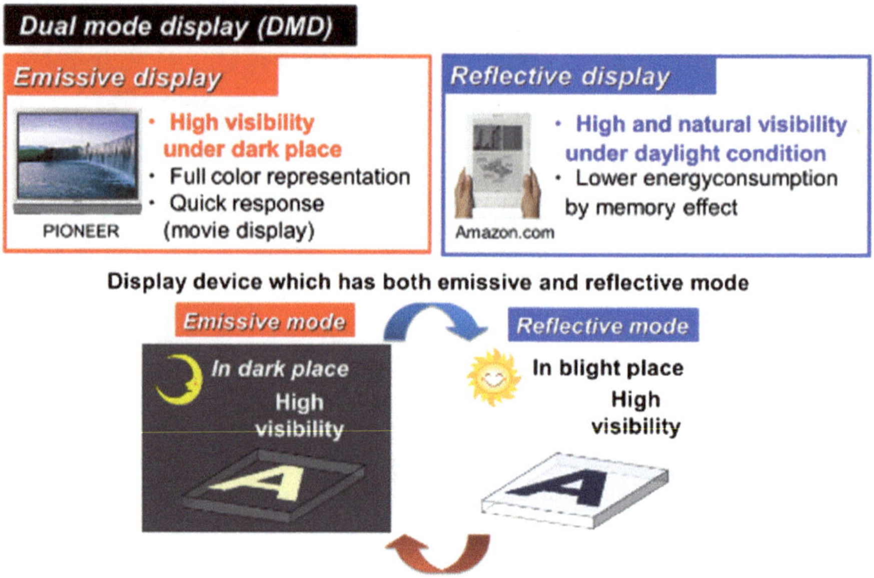

Fig. 1 Schematic illustration of the DMD concept

stimuli are very suitable for novel dual-mode display (DMD) devices, allowing the user to benefit from emissive display in dark place and reflective display in bright place (like reading outside on a sunny day) as shown in Fig. 1 [10–13, 49–54].

1.1 Electron and Energy Transfer

As described above, the mechanism of luminescence control in response to external stimuli is mainly based on electron and/or energy transfers. Therefore, the electron- and energy-transfer mechanisms are introduced in this section.

Electron transfer, particularly photoinduced electron transfer (PET) [43–47], is a process where an electron is transferred from the excited state of a donor to an acceptor. Photoluminescence is observed when an excited electron, for instance, one in the lowest unoccupied molecular orbital (LUMO), transitions to the highest occupied molecular orbital (HOMO), releasing its excess energy as light. As shown in scheme of the PET process (Fig. 2), this can also happen with the participation of an external orbital from another part of the molecule or from a different molecule, if it is located on an energy level between those of the HOMO and LUMO of the fluorophore. For example, a PET can take place from such an external orbital to the half-filled HOMO of the excited fluorophore, as shown in the process (i) in Fig. 2a. A further electron transfer from the LUMO of the fluorophore to the external orbital then recreates the stable ground state, as shown in the process (ii) in Fig. 2a.

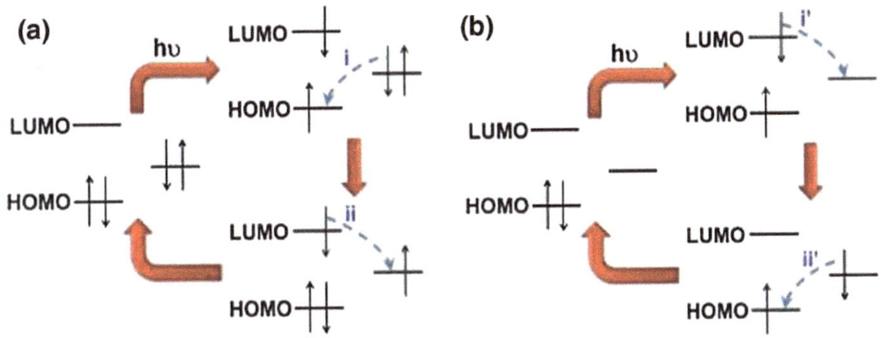

Fig. 2 PET processes with the participation of the HOMO and the LUMO of the fluorophore and an **a** occupied or **b** empty external molecular orbital

Consequently, in this example, luminescence quenching occurs because of the non-radiative transition from the excited to the ground state. A similar process can also take place involving an empty external orbital located between the HOMO and the LUMO of the fluorophore. In this case, a PET from the excited LUMO to the empty external orbital can occur (process (i′) in Fig. 2b), followed by a further electron transfer from this external orbital to the HOMO of the fluorophore (process (ii′) in Fig. 2b). In this case, fluorescence quenching occurs by non-radiative de-excitation.

On the other hand, energy transfer, particularly fluorescence resonant energy transfer (FRET) [44, 46, 47], is a process in which energy is transferred from a donor fluorophore (D) to an excited electronic state of a lower-energy chromophore, or acceptor (A), via an induced long-range dipole–dipole interaction (left in Fig. 3). The rate of this energy transfer depends on three criteria: (i) the overlap between the donor's emission and acceptor's absorption spectra (right in Fig. 3); (ii) the distance between the donor and the acceptor (below 10 nm); and (iii) the transition

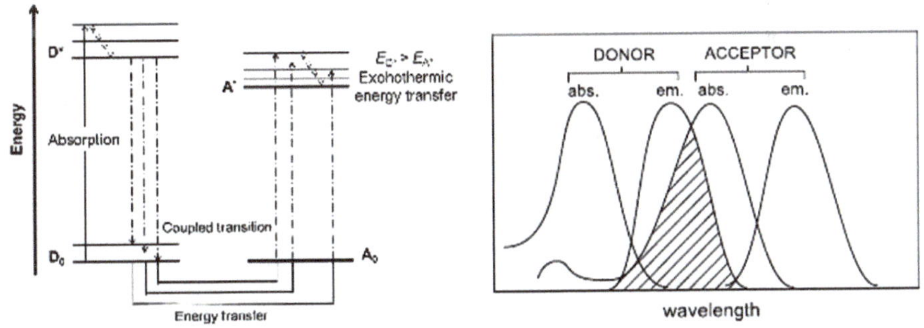

Fig. 3 Energetically favorable conditions for the energy-transfer process; $D^* + A_0 \rightarrow D_0 + A^*$. Illustration of the overlap integral between the emission spectrum of the donor and the absorption spectrum of the acceptor [44]

dipole orientations of the donor and acceptor (which must be roughly parallel). The FRET efficiency, η_{ENT}, is given by:

$$\eta_{\mathrm{ENT}} = \frac{R_0^6}{R_0^6 + r^6} \text{ or } 1 - \frac{\tau_{\mathrm{DA}}}{\tau_{\mathrm{D}}}$$

$$R_0^6 = 8.79 \times 10^{23} \left[\kappa^2 n^{-4} \Phi_{\mathrm{D}} J(\lambda)\right]$$

$$J(\lambda) = \int F_{\mathrm{D}}(\lambda)\varepsilon_{\mathrm{A}}(\lambda)\lambda^4 \, d\lambda$$

where R_0, r, τ_{DA}, τ_{D}, κ, n, Φ_{D}, $J(\lambda)$, $F_{\mathrm{D}}(\lambda)$, and $\varepsilon_{\mathrm{A}}(\lambda)$ are the distance between the donor and acceptor (the R_0 value corresponds to a 50% energy-transfer efficiency), the fluorescence lifetime of the donor in the presence of the acceptor, the fluorescence lifetime of the donor alone, the dipole orientation factor (κ^2 = 2/3), the reflective index of the medium, the emission quantum yields of the donor alone, the spectral overlap integral that represents the degree of overlap between the donor's fluorescence spectrum and the accepter's absorption spectrum, and the molecular absorbance coefficient of the acceptor, respectively.

1.2 *Photoresponsive Systems*

Photoresponsive multifunctional systems that show both tunable emission and coloration have been reported previously; these systems consist of a photochromic diarylethene derivative associated with a luminescent moiety [27, 28, 55–58]. As shown in the two examples of fluorophore–diarylethene systems in Figs. 4 [56] and 5 [27], turn-on/turn-off fluorescence can be triggered along with the change of coloration upon irradiation with light of appropriate wavelengths; the initial fluorescence of the open-ring isomer is switched off by the photocyclization of the diarylethene molecule. The resulting closed-ring isomer displays a very weak fluorescence due to the intramolecular electron (Fig. 4) or energy (Fig. 5) transfer from the fluorophore to the closed-ring diarylethene. The fluorescence can be readily restored by irradiating the system with visible light.

1.3 *Thermoresponsive Systems*

Various thermosensitive materials that change their colors on exposure to thermal stimuli (i.e., thermochromic materials) have been developed by making use of phase transitions of liquid crystals, conformational changes in organic and inorganic compounds, and fluoran dye-developer system. Among them, the aggregation structure of a thermochromic system containing a fluoran dye and its developer

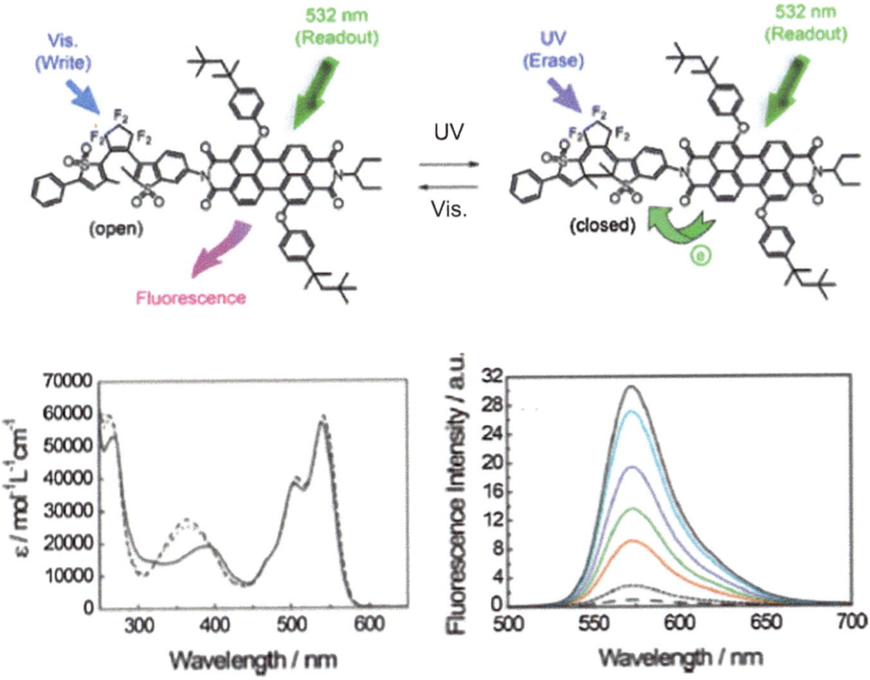

Fig. 4 Photochemical changes of a diarylethene-perylenebisimide dyad. Absorption and fluorescence spectral changes of the represented molecule in a binary solution (1,4-dioxane-methanol 50/50) upon irradiation with 445- and 365-nm light: open-ring isomer (*solid line*), closed-ring isomer (*dashed line*), and photostationary state under irradiation with 445-nm light (*dotted line*) [56]

undergoes changes in response to thermal stimuli that are accompanied by vivid and highly reversible color changes. The author's group reported on a thermoresponsive multifunctional material that allowed for the dual control of its emission and coloration using a polymeric composite material containing a fluoran dye, its developer molecule, and a luminescent europium(III) [Eu(III)] complex (Fig. 6) [36]. The emission and coloration of this material could be reversibly changed and maintained by the thermochromic reaction taking place within the fluoran dye-developer system. The coloration and decoloration of a film of this composite material were caused by the temperature-related control of the nanoscale aggregation of the fluoran dye and its developer, which had a long alkyl chain. The control of the photoluminescence could also be achieved by thermochromism via intermolecular FRET from the excited Eu(III) ions to the colored fluoran dye.

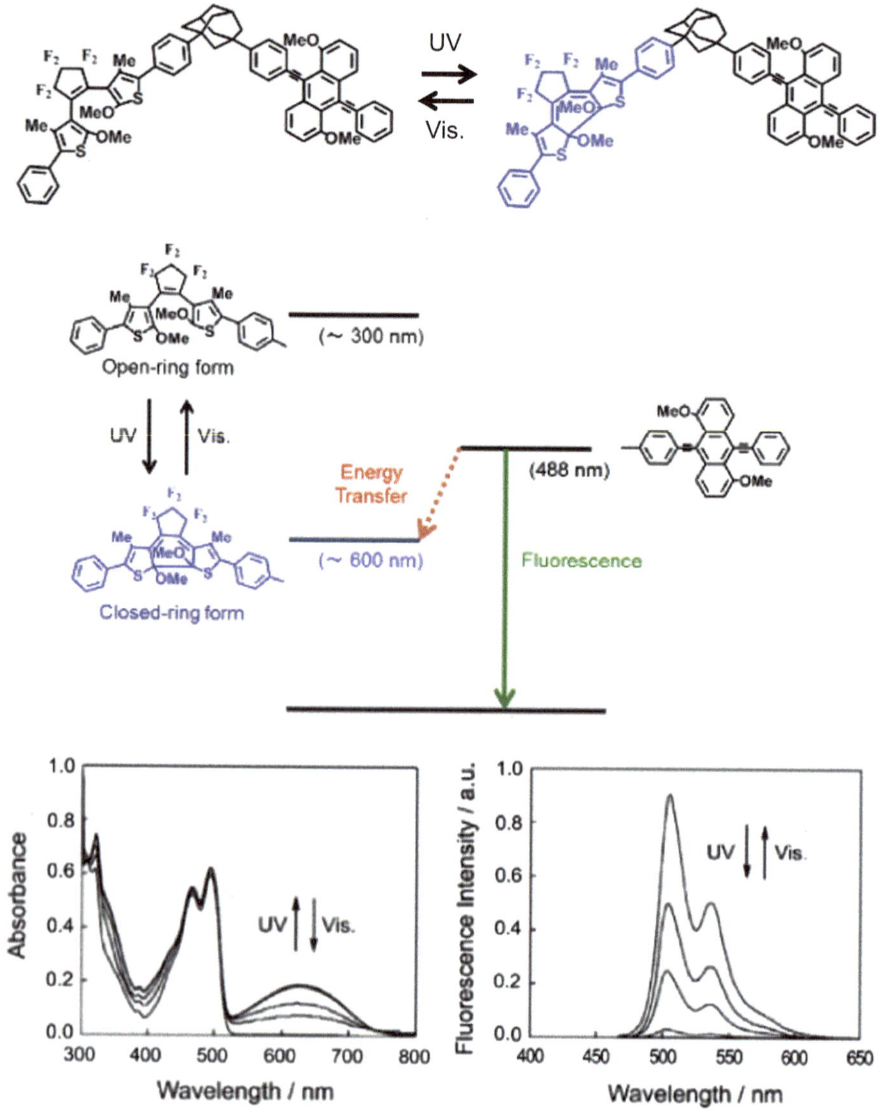

Fig. 5 Photochemical changes of diarylethene-bis(phenylethynyl)-anthracene. Energy diagrams of the molecules. Absorption and fluorescence spectral changes in toluene upon irradiation with 313-nm light. The concentration of the molecule in toluene was 1.2×10^{-5} M [27]

1.4 Electroresponsive Systems

In electrochemical system, molecules which display luminescence control can be broadly classified into intrinsically switchable electroactive fluorophores (Type 1) and switchable molecular dyads (Type 2); the mechanisms of the electroswitching of both

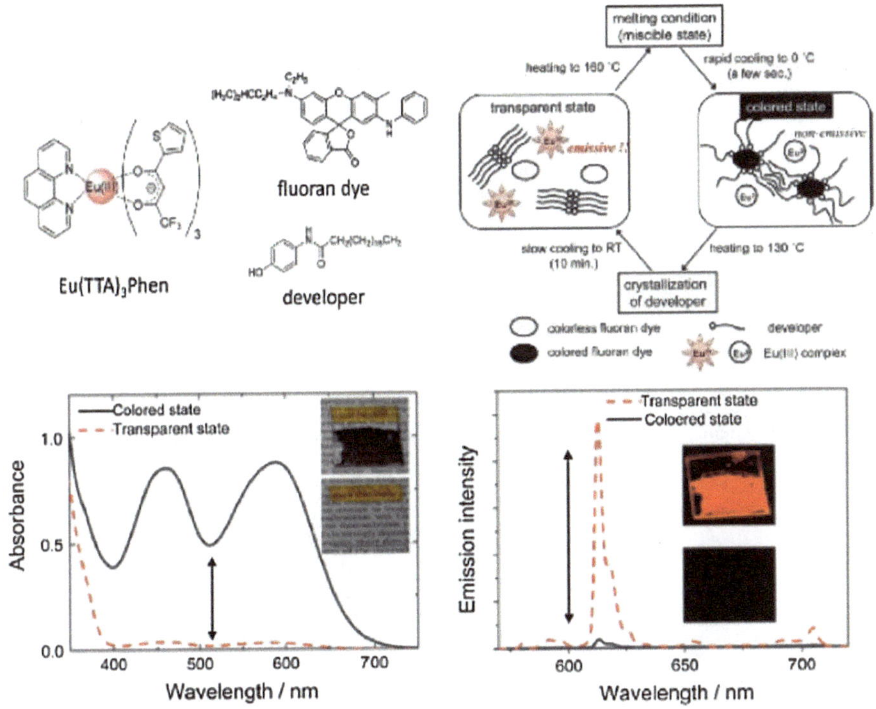

Fig. 6 Schematic illustrations of the coloring and decoloring (bleaching) processes. Absorption and emission spectra of a composite film in the transparent and colored states. The wavelength of the excitation light was 350 nm [36]

types are shown in Fig. 7. For Type 1 molecule, such as tetrazines [47, 59–62], triphenylamines [47, 63, 64], and conjugated molecules [47, 65–67] the redox-active fluorophore switches from emitting state to dark state via an intrinsic electron transfer during the redox process. These states in the molecules can be controlled directly by the formation of ions radicals or neutral radicals through electrochemical conversion. On the other hand, the Type 2 fluorophore is linked to a redox-active moiety that acts as a fluorescence switch through a photoinduced electron or energy transfer between the excited fluorophore and the redox-active moiety [47, 68–71].

A triad of tetrathiafulvalene (TTF) and anthracene units is an example of a switchable molecule based on an electron-transfer mechanism, the luminescence of the anthracene being modulated by the electrochemical reaction of a TTF unit (Fig. 8) [68]. In particular, the mechanism of luminescence modulation in this molecule (a D–A system) can be explained by intramolecular electron transfer between anthracene as a phosphor unit (acceptor (A) unit) and TTF as an electron donor (D) unit. The mechanism is proposed as follows: In the non-fluorescent state, a photoinduced electron transfer occurs from the TTF unit to the excited state of the anthracene unit (Fig. 8a); these D–A molecules exhibit weak fluorescence before oxidation (blue solid line in Fig. 8b). When the TTF is oxidized, since the

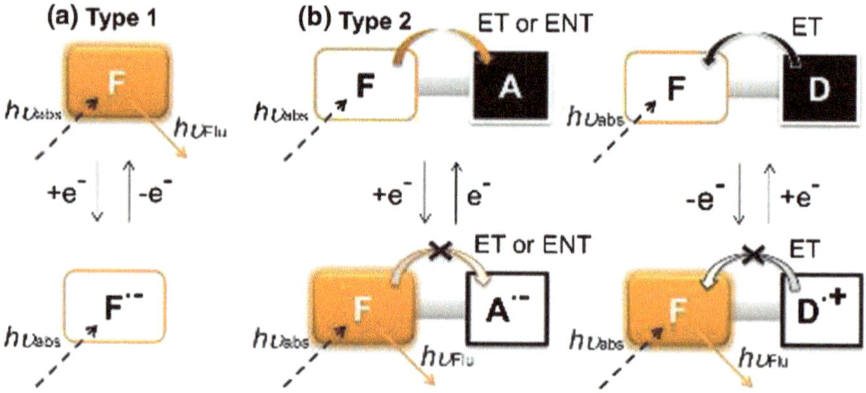

Fig. 7 Electrochemical photoswitching systems (*ET* electron transfer, *ENT* energy transfer)

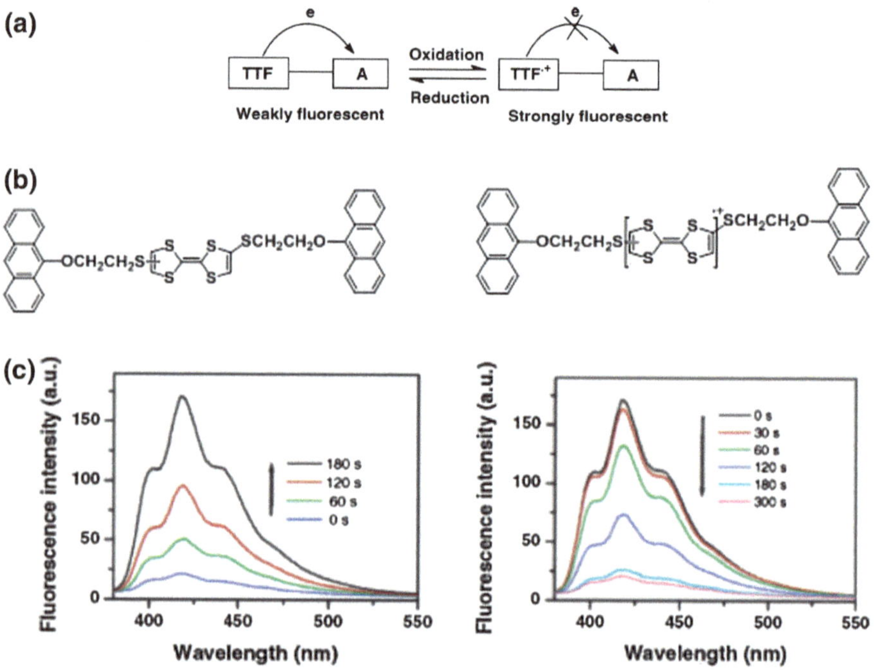

Fig. 8 **a** Electron transfer system of a tetrathiafulvalene-anthracene dye, **b** fluorescence spectra of the solution of a triad tetrathiafulvalene-anthracene dye in THF (4.94×10^{-5} M) containing *n*-Bu_4NPF_6 (27.8 mM) after applying an oxidation potential of 0.7 V (vs. Ag wire), and **c** fluorescence spectra of a solution of the triadic tetrathiafulvalene-anthracene dye that had been oxidized electrochemically for 3.0 min after applying a potential of 0.2 V (vs. a Ag wire) [68]

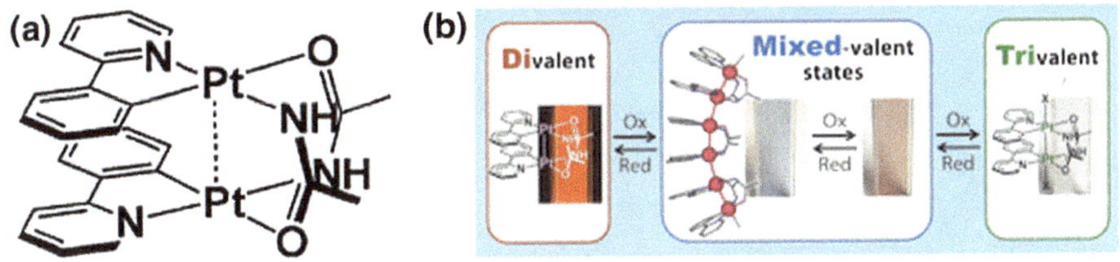

Fig. 9 **a** Structure of dinuclear Pt motif complex and **b** changes in optical properties based on electrochemical reaction [54]

photoinduced electron transfer cannot occur in the cation radical species of TTF ($TTF^{+\cdot}$), a fluorescence enhancement is observed (Fig. 8a and black solid line in Fig. 8b). Further, the luminescence intensity can be reversibly changed by the electrochemical reaction as shown in Fig. 8c.

As an instance of molecular structure changes switchable through an electron transfer, Kato et al. reported the first example of a redox-multistable platinum (Pt) complex that is switchable between luminescent and multicolored states by electrochemical redox reactions (Fig. 9) [54]. The resulting cyclometalated dinuclear motif linked by acetamidato (aam) bridging ligands, $[Pt_2(\mu\text{-aam})_2(ppy)_2]$ (ppy = 2-phenylpyridinate) (Fig. 9) was found to exhibit multistep redox states that were readily interconverted by electrochemical and chemical oxidation and reduction. In particular, the color changes from transparent ["trivalent state," Pt(+3)] to pink ["trimer-of dimer-structure," Pt(+2.67)], blue ["trimer-of dimer-structure," Pt(+2.33)], yellow ["trimer-of dimer-structure," Pt(+2.5)], and orange ["divalent state," Pt(+2)] with changes in valences and structures of Pt complex were observed by applying various potentials as shown in Fig. 9. Their outstanding luminescence, which arises from electronic Pt–Pt interactions, can be reversibly controlled; the Pt^{II}–Pt^{II} state is luminescent, and the Pt^{III}–Pt^{III} state is non-luminescent.

In the following section, the electrochemical control of the emission and coloration in two kinds of the author's system are described in details. In particular, the authors firstly investigated the control of the emission and coloration along with the intramolecular change of the electric structure in Sect. 2, and next, FRET-induced optical changes are described in Sects. 3 and 4. Finally, emission control induced by electron transfer between a fluorophore and titanium oxide (TiO_2) is mentioned in Sect. 5 as a recent example of our approach of luminescence control.

2 Electrochrochemical Modulation of Luminescence and Coloration Using a Dual Function Fluoran Dye

In this section, electrochemically induced switching of the emission and coloration of a fluoran dye (Yellow-1, Fig. 10 for molecular structure of Yellow-1) associated with changes in the intramolecular structure is described [72, 73]. The emission and

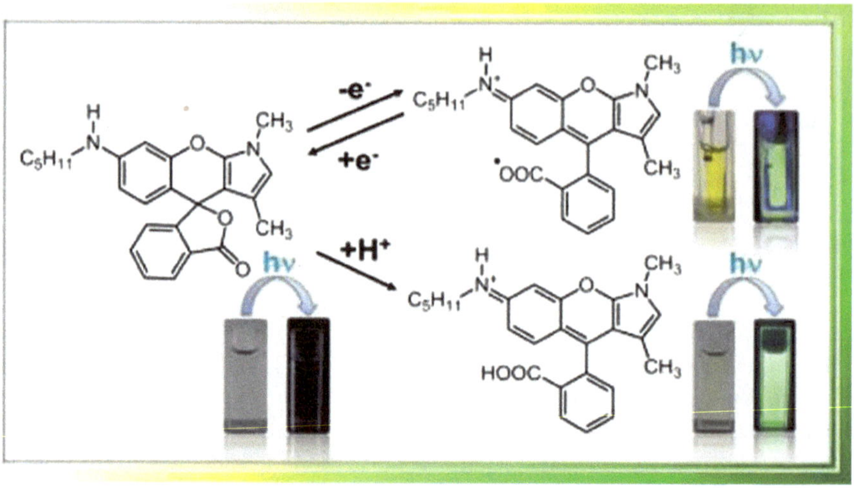

Fig. 10 Schematic illustration of electrochemical luminescence and coloration switching of Yellow-1 [72]

coloration are controlled simultaneously through the electrochemical- or protonation-induced closing and opening of the lactone ring in the Yellow-1 as shown in Fig. 10.

Kohn–Sham density functional theory (DFT) [74–77] and time-dependent DFT (TD-DFT) [78–80] were combined to study the switching of the emission and coloration mechanism. The molecular geometry was optimized based on DFT calculations with Gaussian 09 program (Revision A.02) [81] at the Becke's three-parameter hybrid functional RB3LYP [82–85]/6-31G(d) [86] level for the neutral state of Yellow-1 and at the RB3LYP/6-31+G(d) level for the protonation-induced lactone-ring opened structure. These results could be also considered to be the demonstration that the electrochemically opened lactone ring is the protonated state of Yellow-1 because the corresponding absorption and emission spectra (optical properties) were consistently similar. All the solvent effects were simulated using a polarized continuum model with dimethyl sulfoxide (DMSO) as the organic solvent [87, 88].

Figure 11 shows the frontier molecular orbital (FMO) of a neutral Yellow-1, whose HOMO−1 and HOMO were mainly located on the benzopyranopyrrole ring and nitrogen atom in the molecule, indicating that the pyrrole or nitrogen unit/element in the molecule had been oxidized, while the LUMO and LUMO+1 were predominantly located on the 1(3H)-isobenzofuranone ring in the molecule. In addition, the LUMO+2 was located on the benzopyran ring in the molecule. From the spread of these FMOs, quenching of Yellow-1 was possibly attributed to the lack of overlap between the HOMOs and LUMOs. The HOMO–LUMO gap for the neutral Yellow-1 was 4.05 eV (Fig. 11), which was consistent with the experimentally obtained value (E_g = 4.28 eV) from the absorption spectrum, as given in

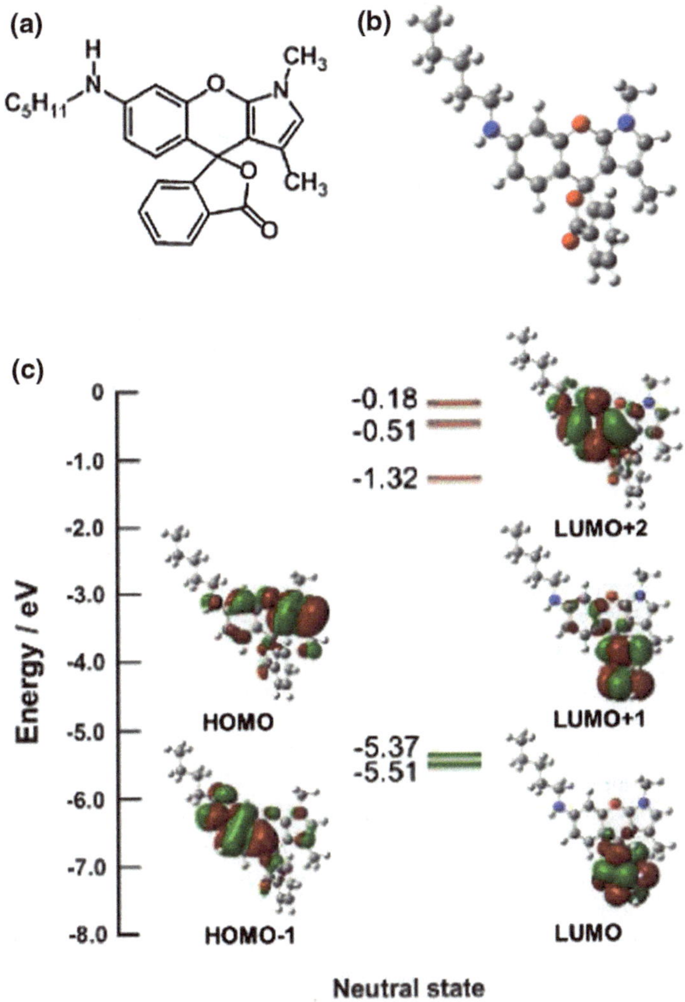

Fig. 11 **a** Molecular structure, **b** optimized molecular structure, **c** FMOs of neutral Yellow-1 calculated by RB3LYP/6-31G, and **d** DFT methods shown with corresponding relative energies. *Key* C, *gray*; H, *white*; O, *red*; N, *blue* [72]. With permission from the ACS

Table 1. This result suggested that the absorption bands were not in the visible range of the spectrum. TD-DFT and RB3LYP/6-31G(d) were used to simulate vertical transitions in order to validate the mechanism of emission property for the neutral Yellow-1. The calculated excitation wavelength for the $S_0 \rightarrow S_1$ transition was 360 nm and the corresponding oscillator strength (*f*) was 0.0013 (absorption in DMSO), which was close to the experimental result (absorption in propylene carbonate: PC). This small *f* values and the lack of overlap between either the HOMO −1 or HOMO and the LUMO (Fig. 11) indicated that the transition from the ground

Table 1 Photophysical properties of Yellow-1 [73]

Compd.	State	E_{ox} (V)[a]	E_{red} (V)[a]	HOMO (eV)		LUMO (eV)		E_g (eV)	
				exp.	calc.	exp.	calc.	exp.	calc.
Yellow-1	Neutral	0.61	–	−5.35[b]	−5.37	−1.07[c]	−1.32	4.28[d]	4.05
	Oxidized	–	−0.75	−6.48[c]	–	−3.99[b]	–	2.49[e]	–

[a]Oxidation and reduction onset potentials determined from Yellow-1 versus Ag/Ag^+. E_{ox} and E_{red} of BQ were determined from each peak from CV measurements (vs. Ag/Ag^+)
[b]Calculated from the equation: HOMO (neutral state) or LUMO (oxidized state) = $[-(E_{ox}$ or $E_{red} - 0.06) - 4.8]$ eV in which 0.06 V is the half-wave potential of Fc/Fc^+ versus Ag/Ag^+ in PC and 4.8 eV is the energy level of Fc under vacuum. The LUMO of BQ was calculated from the equation $[-(E_{1/2} - 0.06) - 4.8]$ eV, where $E_{1/2}$ is the half-wave potential of BQ
[c]LUMO (neutral state) = HOMO + E_g, HOMO (oxidized state) = LUMO − E_g
[d]The optical E_g in the neutral state was estimated using the onset of the absorption spectra
[e]E_g in the oxidized state was estimated from the intersection point between the absorption and emission spectra. HOMO, LUMO, and E_g in the theoretical calculation were estimated by DFT calculations

state to the lowest lying excited one was strongly forbidden, i.e., the S_1 state was not directly accessible through photoexcitation. Instead of that, the S_1 state could only be populated through the internal conversion (IC) of higher allowed excited states such as S_5 ($f = 0.0153$). According to Kasha's rule, IC quickly occurred, so S_5 would relax to S_1 before possibly relaxing to S_0 [44, 46, 89–92]. Therefore, the lowest lying excited state, S_1 was non-emissive and would not radiatively decay to the ground state, S_0. (In this case, a non-radiative deactivation channel would act as a drainpipe for the excited-state energy.) This means that neutral Yellow-1 is probably non-fluorescent.

On the other hand, the LUMO was localized in the benzopyranopyrrole ring of the molecule in the case of protonation-induced ring-opening, as shown in Fig. 12. Furthermore, HOMO−1, HOMO, and LUMO were widely spread over the benzopyranopyrrole ring in the Yellow-1 and showed significant overlap, indicating that the experimentally observed emission from the oxidized Yellow-1 reflected the locally excited (LE) state. Moreover, the HOMO–LUMO, E_g, for the oxidized Yellow-1 was found to be 3.27 eV (379 nm), as shown in Fig. 12. Thus, the experimental and theoretical E_g obtained for the oxidized Yellow-1 were smaller than those obtained for the neutral one as given in Table 1 and Fig. 11, leading to the appearance of absorption bands in the visible range of the spectrum. As in the case of the neutral Yellow-1, TD-DFT and RB3LP/6-31+G(d) were used to simulate vertical transitions in order to validate the mechanism of emission control for the protonated Yellow-1. The calculated excitation wavelength for the $S_0 \rightarrow S_1$ transition was 409.5 nm (absorption in DMSO), which was close to the experimental result of 465 nm (absorption in PC). This $S_0 \rightarrow S_1$ transition was also fully allowed, as indicated by the overlap between the HOMO−1 or HOMO and LUMO and because $f = 0.2660$, meaning that the reverse transition, that is radiation transition of the $S_1 \rightarrow S_0$ transition, was also fully allowed. Thus, the protonated Yellow-1 was potentially fluorescent.

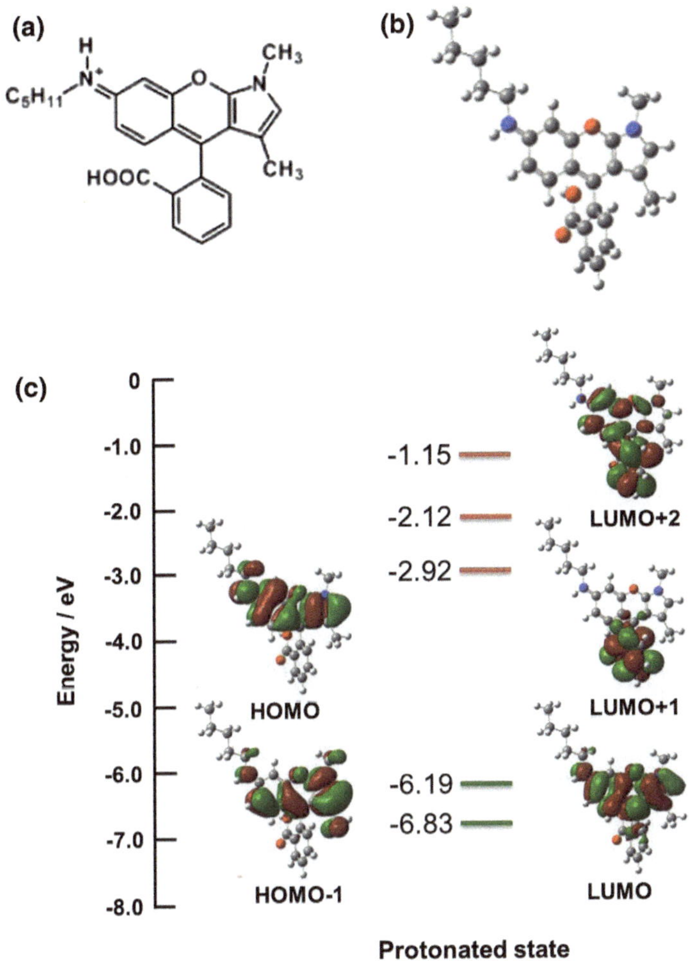

Fig. 12 **a** Molecular structure, **b** optimized molecular structure, **c** FMOs of oxidized Yellow-1 calculated by RB3LYP/6-31G, and DFT methods shown with corresponding relative energies. *Key* C, *gray*; H, *white*; O, *red*; N, *blue* [72]

3 Electrochemical Modulation of Luminescence and Coloration Based on Energy-Transfer Mechanism

3.1 Electroswitchable Emission and Coloration Using Luminescent Lanthanide(III) Complexes and Electrochromic Viologen Derivatives

Luminescent lanthanide(III) [Ln(III)] complexes consist of antenna ligands with a high molar absorption coefficient ($\varepsilon > 10{,}000\ M^{-1}\ cm^{-1}$) and luminescent Ln(III)

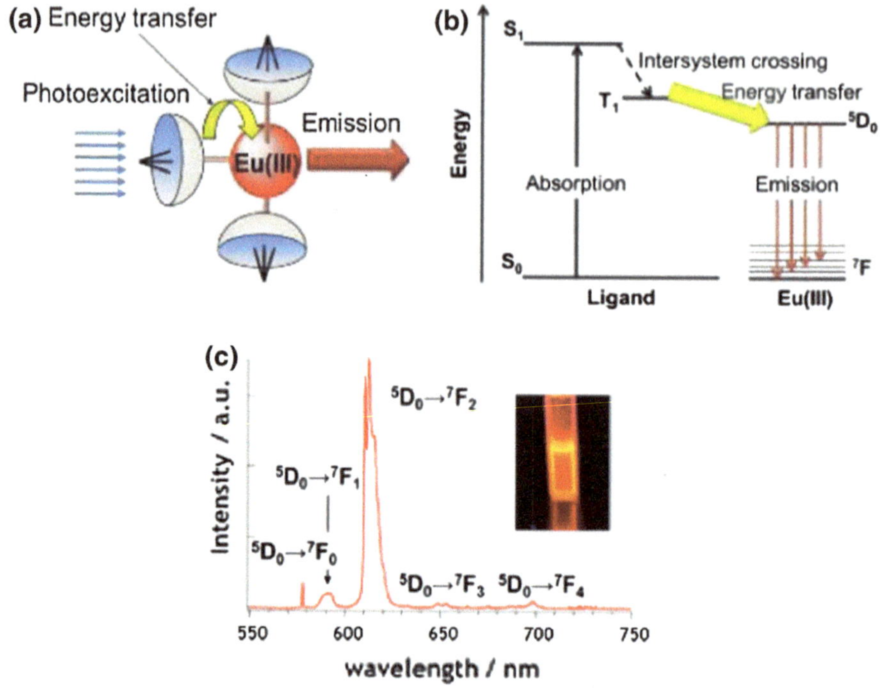

Fig. 13 **a** A schematic representation of the antenna effect occurring in Eu(III) complexes possessing aromatic light-harvesting chromophores and **b** Jablonski diagram of a Eu(III) complex. S = singlet state, T = triplet state. **c** Emission spectrum of the Eu(III) complex

ion (Fig. 13a). The luminescence of Ln(III) complexes is observed through the efficient intramolecular energy transfer from the antenna ligand to the Ln(III) ion (Fig. 13), resulting into a bright emission with a high emission quantum yield [93–102]. In addition to this, since the Ln(III) complexes possess unique optical properties such as narrow emission bands, long luminescence lifetimes, and high transparencies in the visible region (large Stokes shift), luminescent Ln(III) complexes have been regarded as attractive luminescent molecules for widespread applications, particularly for phosphors, bioassays, and sensors development. From these advantages attributed to Ln(III) complex, the authors elected to work on a luminescent molecule of this type, namely a europium(III) [Eu(III)] complex.

In this chapter, we demonstrate electrochemical control of both emission and coloration by combining a luminescent Eu(III) complex and an electrochromic molecule (such as viologen) as shown in Fig. 14 [51, 52].

In order to investigate the electrochromic properties of the composite material, cyclic voltammograms (CVs) and changes in the absorbance at 600 nm during cyclic potential sweeps of tris(hexafluoroacetylacetonato)europium(III) bis(triphenylphospine oxide) [Eu(hfa)$_3$(TPPO)$_2$], heptyl viologen (HV^{2+}), and Eu(hfa)$_3$(TPPO)$_2$/HV^{2+} were measured in a 200 mmol L^{-1} tetra-*n*-butylammonium

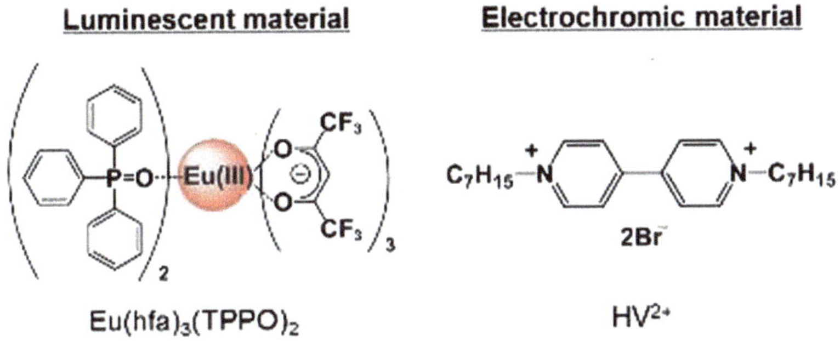

Fig. 14 Chemical structures of luminescent and electrochromic compounds

perchlorate (TBAP)/PC electrolyte (Fig. 15a). The $Eu(hfa)_3(TPPO)_2$ solution did not show obvious reductive and oxidative reactions in the measured potential region (1.0 to −0.65 V), and hence, there were no absorbance changes during the potential sweep. The HV^{2+} solution showed one reductive peak at −0.5 V and one corresponding oxidative peak at −0.25 V. As the reductive current increased from −0.3 V, the absorbance at 600 nm of the HV^{2+} solution increased. On the basis of typical electrochromic behaviors of HV^{2+}, this absorption band is attributed to the absorbance of $HV^{+\cdot}$ (Fig. 15b) [103]. On the other hand, a reductive peak and a corresponding oxidative peak of the $Eu(hfa)_3(TPPO)_2/HV^{2+}$ solution were observed at −0.53 and −0.24 V, respectively. As the reductive current increased from about −0.3 V, the absorbance change at 600 nm was almost identical to the case of the HV^{2+} solution. In absorption spectrum of colored $Eu(hfa)_3(TPPO)_2/HV^{2+}$ solution, new absorption bands at around 600 nm were observed, which were in agreement with those of the $HV^{+\cdot}$ (Fig. 15c, d). The coloration of HV^{2+} in the presence of the Eu(III) complex showed the typical electrochromic behavior of HV^{2+}. From these results, electrochromism of HV^{2+} was found to occur in the presence of Eu(III) complex.

In order to demonstrate the emission and coloration control based on the EC reaction of HV^{2+}, we fabricated a two-electrode cell by sandwiching a PC solution of $Eu(hfa)_3(TPPO)_2/HV^{2+}$ and TBAP between two indium tin oxide (ITO) glasses. From the CV measurement of the two-electrode cell, electrochromic reaction of the HV^{2+} occurred under applied voltage of −2.0 V. Figure 16a shows the absorption spectra of the $Eu(hfa)_3(TPPO)_2/HV^{2+}$ cell. No absorption bands were present in the visible region when the circuit was open, thereby indicating that the cell was colorless (photograph in Fig. 16a); this can be considered as a "reflection-off" representation. When a bias voltage of −2.0 V was applied for 10 s, new absorption bands appeared at around 400 and 600 nm; they could be assigned to the absorption bands of the reduced species of $HV^{+\cdot}$. The cell color changed from clear colorless to cyan as the number of new absorption bands increased; it was considered as a "reflection-on" state.

On the other hand, strong red emission bands of the Eu(III) complex were observed at around 580, 590, 615, and 650 nm by the excitation of hfa ligands

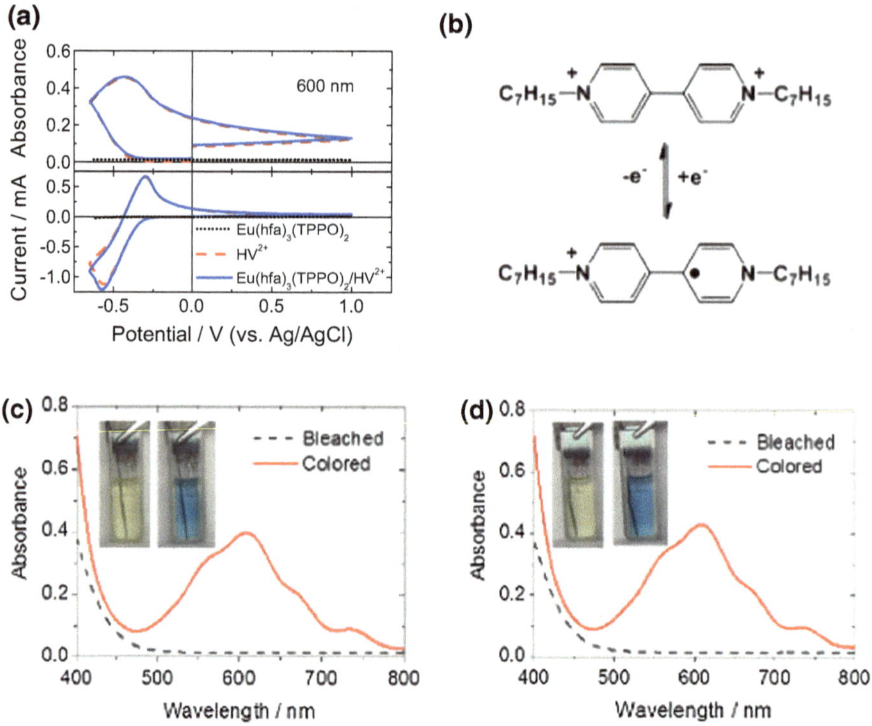

Fig. 15 **a** Change in absorbance at 600 nm (*top*) and cyclic voltammograms (*bottom*) of Eu (hfa)$_3$(TPPO)$_2$, HV^{2+}, and Eu(hfa)$_3$(TPPO)$_2$/HV^{2+} in a 200 mmol L^{-1} TBAP/PC electrolyte solution [52], **b** electrochemical reaction formula of viologen, **c-d** absorption spectra of **c** HV^{2+} and **d** Eu(hfa)$_3$(TPPO)$_2$/HV^{2+} in a 200 mmol L^{-1} TBAP/PC electrolyte solution with a three-electrode cell by application potential at −0.57 V (vs. Ag/AgCl) for 6 s

(337 nm) when the cell was colorless (open-circuit condition, Fig. 16b); this was considered the "emission-on" state (photograph in Fig. 16b). These emission bands were attributed to the *f*–*f* transition of $^5D_0 \rightarrow {}^7F_J$ (J = 0, 1, 2, and 3). On the other hand, when the cell was in the colored state (−2.0 V), the strong red emission of Eu (III) complex was completely quenched, resulting in an "emission-off" state (photograph in Fig. 16b). As the emission quantum yields of the transparent and colored states are summarized in Table 2, the emission intensity of the cell under −2.0 V application decreased to less than 1/40 in the cell under open-circuit conditions, resulting in low emission quantum yields (from 0.52 to less than 0.01). This resulted in efficient quenching of Eu(III) complex. On the other hand, the emission intensity, quantum yield, and emission lifetime of the Eu(III) complex without HV^{2+} in the two-electrode cell did not change by the bias voltage as shown in Fig. 16c and Table 2. Therefore, the quenching of the Eu(III) emission was actually due to the colored $HV^{+\cdot}$. As shown in Fig. 17, the absorption band at around 600 nm attributed to the colored $HV^{+\cdot}$ overlapped well with the emission bands of

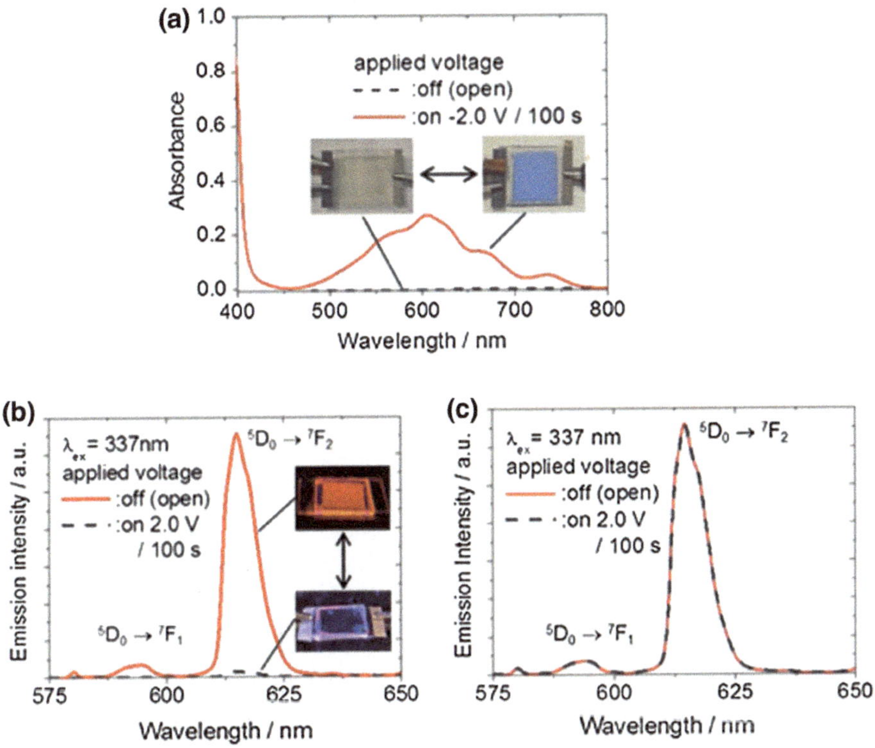

Fig. 16 **a** Absorption spectra of HV^{2+}/Eu(hfa)$_3$(TPPO)$_2$ and emission spectra of **b** Eu(hfa)$_3$(TPPO)$_2$/HV^{2+} and **c** Eu(hfa)$_3$(TPPO)$_2$ solution-based on the two-electrode cell [*red solid line* open-circuit condition, *black dashed line* under bias voltage of −2.0 V (100 s)]. *Insets* were figures of coloration and emission changes, respectively [52]

Table 2 Emission quantum yields (ϕ) and emission lifetimes (τ) of the samples [52]

Compound	Voltage (V)	ϕ	τ (μs)
Eu(hfa)$_3$(TPPO)$_2$	Open	0.52[a]	790
Eu(hfa)$_3$(TPPO)$_2$	−2.0	0.52[b]	790
Eu(hfa)$_3$(TPPO)$_2$/HV^{2+}	Open	0.52[a]	790
Eu(hfa)$_3$(TPPO)$_2$/HV^{2+}	−2.0 (50 s)	0.012[b]	0.55

[a]Emission quantum yields were estimated by using a quartz cell whose path length was 1.0 mm
[b]The values were relative emission intensities of 613 nm

the Eu(III) complex. Such an overlapping of the absorption and emission bands facilitated efficient fluorescence resonance energy transfers (FRETs) from the excited state of the Eu(III) ions to the ground state of $HV^{+\cdot}$.

As shown in the Jablonski diagrams of emissive and non-emissive states in Fig. 18, emission of Eu(hfa)$_3$(TPPO)$_2$ was observed when HV^{2+} species were in the

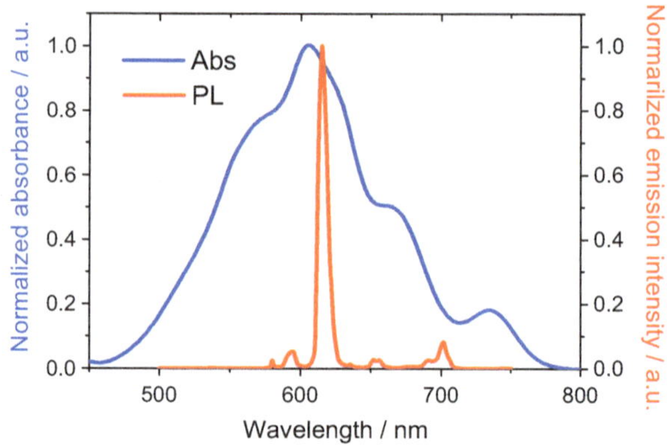

Fig. 17 Normalized absorption spectrum of monocation radical electrochromic molecules and emission spectrum of $Eu(hfa)_3(TPPO)_2/HV^{+\cdot}$

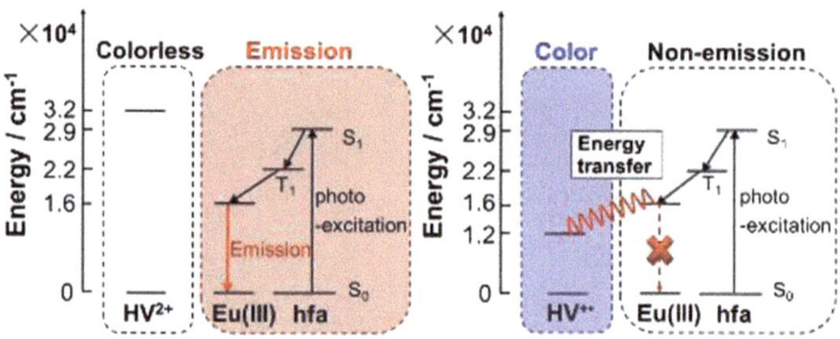

Fig. 18 Jablonski diagrams of HV^{2+}, $Eu(hfa)_3(TPPO)_2$, and $HV^{+\cdot}$, respectively

colorless state because the energy gap between HOMO and the LUMO of the HV^{2+} state was much larger than the energy gap between excited and ground states of the Eu(III) ion. In this situation, excitation energy of the Eu(III) ion could not be transferred to HV^{2+} molecule. On the other hand, when there was colored $HV^{+\cdot}$, the red emission of the Eu(III) complex could not be observed because the HOMO–LUMO gap of $HV^{+\cdot}$ was smaller than the energy gap between the excited and ground states of Eu(III) ion and excitation energy of Eu(III) ion could be transferred from colored $HV^{+\cdot}$.

In order to investigate the detailed mechanism of the emission control, time dependence of the absorbance at 600 nm and emission intensity at 615 nm were measured when a bias voltage at −2.0 V was applied to the cell (Fig. 19a). As shown in this figure, emission intensity decreased with increase of the absorbance and the time of applied voltage. The emission lifetimes of the Eu(III) complex were also measured before and after voltage application (Fig. 19b). The emission lifetimes immediately decreased after the bias voltage was applied (1 and 50 s). The emission lifetimes at 0, 1, and 50 s after voltage application were 790, 364, and

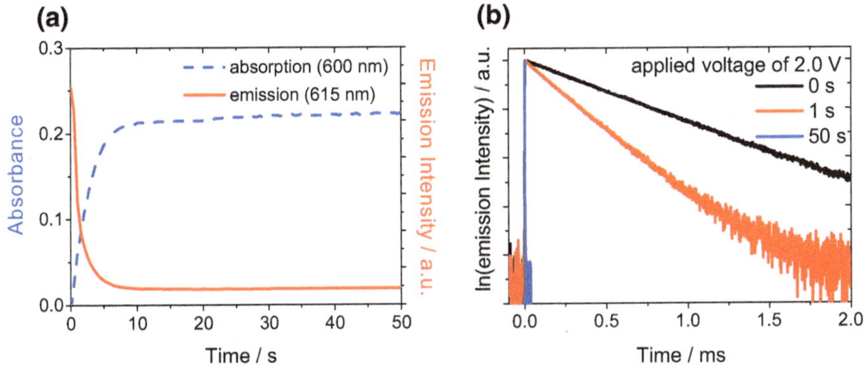

Fig. 19 **a** Changes of emission intensity at 615 nm (*red solid line*) and absorbance at 600 nm (*blue dashed line*) during the application voltage of −2.0 V and **b** emission decay profiles of the two–electrode cell before application voltage (0 s), 1, and 50 s after application voltage at −2.0 V shown on a logarithmic scale [52]

0.55 μs, respectively (the values are shown in Table 2). From consideration of decrease of emission lifetime and large overlapping of Eu(III) emission with absorption spectrum of $HV^{+\cdot}$, the quenching of the red emission from the Eu(III) complex would be caused by intermolecular energy transfer as shown in Fig. 18. Furthermore, the quenching efficiency (η_q) and energy-transfer efficiency (η_{ENT}) of our system in the two-electrode cell were calculated by using the following equations:

$$\eta_q = 1 - \frac{\phi}{\phi_0}$$

$$\eta_{ENT} = 1 - \frac{\tau}{\tau_0}$$

where ϕ_0, ϕ, τ_0, and τ are the emission quantum yield in the emissive state, emission quantum yield in the quenched state, emission lifetime in the emissive state, and emission lifetime in the quenched state, respectively. The estimated quenching efficiency was over 99.9%, resulting in high emission quenching in the cell. On the other hand, the energy-transfer efficiency was found to be 99%. This excellent agreement of the η_q and η_{ENT} clearly indicated that the dominant cause of the emission switching was not the absorption of Eu(III) emission by colored $HV^{+\cdot}$. The large decrease of the emission lifetimes also strongly suggested the energy-transfer mechanism, because the emission lifetime usually does not change through the absorption mechanism. As described above, the absorption bands of colored $HV^{+\cdot}$ species appeared at around 600 nm, and this overlapped well with the emission bands of Eu(III) complex. Such an overlapping of the absorption and emission bands enable efficient energy transfers from the excited state of Eu(III) complex to the ground state of the $HV^{+\cdot}$.

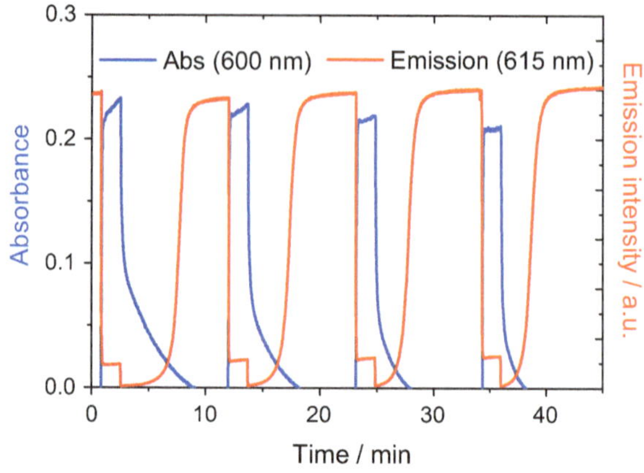

Fig. 20 Changes in absorbance at 600 nm and emission intensity at 615 nm of the two-electrode cell [−2.0 V (100 s)/open circuit (600 s)]. The excitation wavelength was 337 nm

Further, in order to investigate the repetition stability of the electrochemical switching of the emission and coloration, we measured the changes in emission intensity at 615 nm and absorbance at 600 nm of the two-electrode cell when the −2.0 V (coloring process) was repeatedly applied in an open circuit (Fig. 20). In the first cases, as the absorbance at 600 nm gradually increased, the emission of Eu(III) complex rapidly quenched within 5 s. After the circuit was open, absorbance of the cell gradually decreased with the increase of HV^{2+} in 15 min. When the absorbance of $HV^{+\cdot}$ decreased, the emission of Eu(III) complex was recovered to its initial intensity.

From the repetition test for the emission and absorbance changes, the authors found that these were repeatable. However, the stability and response time still need to be improved by introducing counter electrode reaction materials, immobilizing luminescent and/or electrochromic materials on the same electrode, etc. In Sect. 4, the authors employed the counter electrochromic material to improve the repetition stability and switching response time of both emission and coloration changing.

3.2 Application of Electrochemical Control of Luminescence and Coloration Using a Luminescence Eu(III) Complex and Electrochromic Viologen

In order to bring further on the demonstration that the dual-mode representation of numerical characters is possible, we fabricated a prototype DMD cell by using a 7-segment cell [50]. The cell was fabricated by sandwiching the PC solution of tris (hexafluoroacetylacetonato)europium(III) $[Eu(hfa)_3(H_2O)_2]/HV^{2+}$ and TBAP between a 7-segment patterned ITO electrode and another planar ITO electrode (Fig. 21). When a negative voltage of −2.2 V was applied to certain segments in

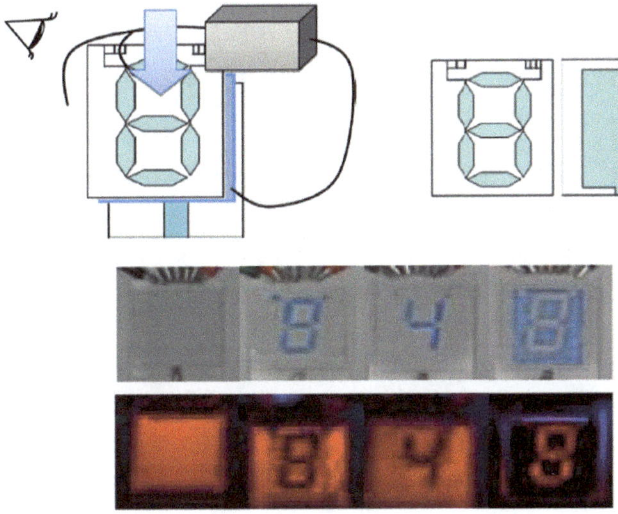

Fig. 21 Schematic illustrations of the 7-segment patterned electrode and device configuration for the demonstration of the electrochemical modulation of emission and coloration representation. Photographs show the demonstration of the electrochemical modulation of emission and coloration. The excitation light wavelength was 365 nm [53]

order to reduce the HV^{2+} molecule, the voltage-applied regions were colored from clear transparency to cyan by the HV^{2+} electrochromism; this resulted in numerical representations with a reflective mode (Fig. 21). Under UV light (365 nm) irradiation, a strong red emission was observed from the cell surface except in the colored area, which led to emissive representations of the numerical character.

3.3 *Multicolor Representation of Emission and Coloration*

In order to bring further on the development of fundamental knowledge and its application to electrochemical control of emission and coloration, we demonstrated the possibility of fabricating a prototype device for multicolor DMD representations by introducing other emission and coloration materials [53]. DMD cells showing red emission/magenta coloration, blue emission/cyan coloration, and green emission/cyan coloration were fabricated by using electrolyte solution of Eu$(hfa)_3(H_2O)_2$/dimethyl terephthalate (DMT), 9,10-diphenylanthracene (DPA)/HV^{2+}, and tris(acetylacetonate)dihydrated terbium(III) $[Tb(acac)_3(H_2O)_2]/HV^{2+}$, respectively (Fig. 22). The DMT molecule is known to be a superior electrochromic

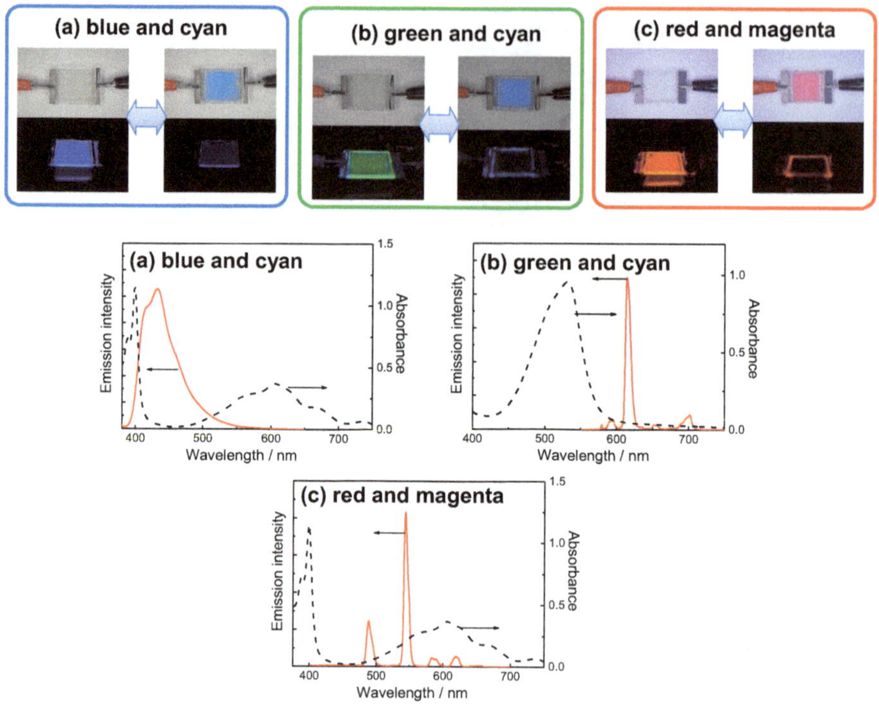

Fig. 22 Photographs of multicolor DMD cells showing red emission/magenta coloration, and green emission/cyan coloration. **a** Emission spectrum of DPA and absorption spectrum of reduced $HV^{+\cdot}$ species, **b** emission spectrum of $Tb(acac)_3(H_2O)_2$ and absorption spectrum of reduced $HV^{+\cdot}$ species, and **c** emission spectrum of $Eu(hfa)_3(H_2O)_2$ and absorption spectrum of reduced DMT [53]

molecule for magenta coloration. The devices showed obvious coloration by application of reduction voltage (−4.0 V for magenta coloration, −2.0 V for cyan coloration). On the other hand, the RGB photoluminescence of the devices was also well switched by the electrochromic reaction. In these systems, the emission spectra of the DPA and Tb(III) complex overlapped with the absorption spectra of the colored $HV^{+\cdot}$ species (Fig. 22); therefore, the FRET mechanism would contribute to their emission switching. On the other hand, the overlapping of the Eu(III) emission with the absorption band of the colored DMT was much less than for other systems (Fig. 22), although the Eu(III) emission was well quenched by the DMT coloration. Photoinduced electron transfer would be the cause of the emission switching in this case.

4 Introducing Electrochromic Material on Counter Electrode to Realize Short Response and Improvement of Reversibility

In Sect. 3, the authors mainly discussed about the mechanism of emission and coloration as well as the response times, reversibility, applications, and various emission colors and colorations of optical changes based on electrochemical reactions of EC materials. Recently, it was shown that electroswitching of emission and coloration is also possible when a photoluminescent material and an electrochemically active material are combined. However, upon the application of long-term electrical stimuli, most of these systems, including our previous system, lack a quick response and a high reversibility. Since electrochemical switching of fluorescence and absorption is time-consuming (over 100 s), it is important to improve the electroswitching rate performance.

Here, in Sect. 4, the authors aim to achieve an electrochemically modulated optical cell with a quick response and a high reversibility. Problems pertaining to responsivity and reversibility in the case of the $Eu(hfa)_3(TPPO)_2$ complex and HV^{2+} system (as discussed in Sect. 3) were caused by the absence of efficient counter electrode materials that could accelerate redox reactions of the electrochromic materials. In general, electrochemical-reaction-based devices such as electrochromic devices require both reducible and oxidizable materials. When an elcctrochemical reduction occurs at the cathode, an electrochemical oxidation should simultaneously occur at the anode and both oxidation and reduction should involve the same amount of charge and have comparable speeds.

Herein, the authors introduced Prussian blue (PB) film [41, 104] as a counter reaction material against HV^{2+} electrochromism into a cell containing a luminescent $Eu(hfa)_3(TPPO)_2$ complex and an electrochromic material HV^{2+} (Fig. 23) [51]. The electrochromism of PB is well known, and the color of PB is cyan in the oxidized state and transparent in the reduced state. Thus, when HV^{2+} turns cyan upon reduction, the PB-modified counter electrode also turns cyan upon oxidation, thereby improving the color contrast between the colored and transparent states of the cell. Thus, both emission and coloration control are achieved solely through the electrochromism of HV^{2+} and PB. Photoluminescence control was also achieved as a result of HV^{2+} and/or PB electrochromism, via fluorescence resonance energy transfer (FRET) from the excited states of Eu(III) ions to colored $HV^{+\cdot}$ and/or PB.

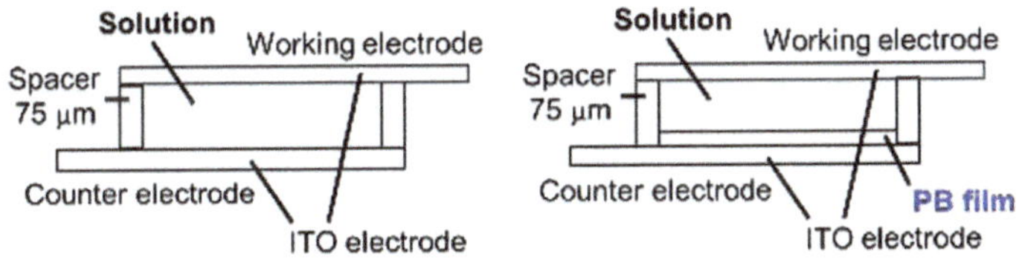

Fig. 23 Construction of ITO/ITO (*left*) and ITO/PB (*right*) cells

In addition, the PB-modified electrode must act as a counter electrode for the HV^{2+} redox reaction in order to facilitate a quick response and high switching stability of the cell. Therefore, the response time and reversibility are expected to be improved by the redox reaction involving HV^{2+}.

In order to investigate the response time of the coloring and bleaching processes of the ITO/ITO and ITO/PB cells, the absorption changes at 600 nm during the processes were measured (Fig. 24). In the case of the former cell without appropriate redox material at the counter electrode, the coloration process from the transparent state required the application of −2.0 V for 100 s to reach an absorbance of 0.4. The bleaching process required an even longer time, 1000 s, even under short circuit (Fig. 24a). In contrast, coloration of the ITO/PB cell was achieved by application of only −0.8 V for 10 s, and bleaching was complete within only 10 s under short circuit (Fig. 24b). The response times of both coloration and bleaching of the ITO/PB cell were dramatically improved by the introduction of the PB-modified electrode, which acts as a counter electrode reaction material. The stable redox properties of the PB-modified electrode were responsible for the quantitatively balanced HV^{2+} redox reaction. Consequently, the coloration and bleaching rates of the ITO/PB cell improved because the balanced redox system accelerated the electrochemical reaction of the cell.

The stabilities of the switching between the emissive and reflective modes were then investigated. The changes in the absorption (600 nm) and emission (615 nm) of the ITO/PB cell under the sequential application of voltages of −0.8 V (10 s, coloring process) and 1.4 V (10 s, bleaching process) were monitored (Fig. 25). Under application of a −0.8-V potential, as the absorption band increased because of the electrochromic reaction of HV^{2+} and PB, the red emission of the Eu(III) complex was rapidly quenched within 5 s. After the voltage of 1.4 V was applied, the cell was completely bleached within 10 s. When the absorbance of $HV^{+\cdot}$ and colored PB decreased, the Eu(III) complex emission was recovered to its initial

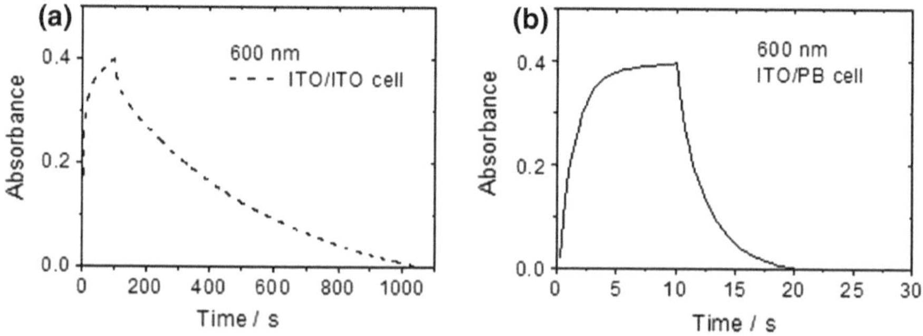

Fig. 24 Change in absorbance of the **a** ITO/ITO and **b** ITO/PB cells during the coloring and bleaching processes [51]

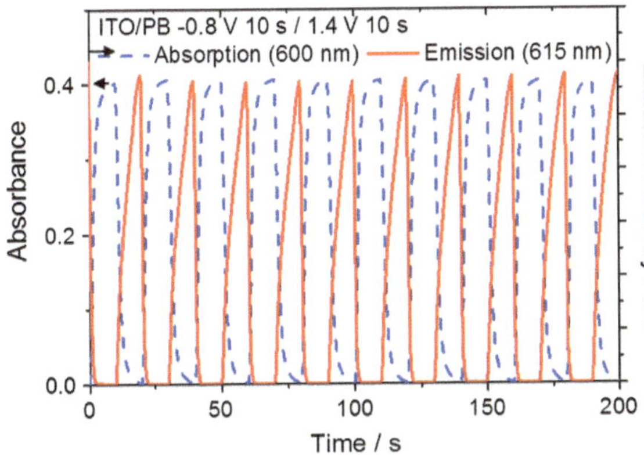

Fig. 25 Changes in absorbance at 600 nm (*dashed line*) and emission intensity at 615 nm (*solid line*) of the ITO/PB cell upon applying the bias voltages of −0.8 V (10 s) and 1.4 V (10 s), respectively [51]

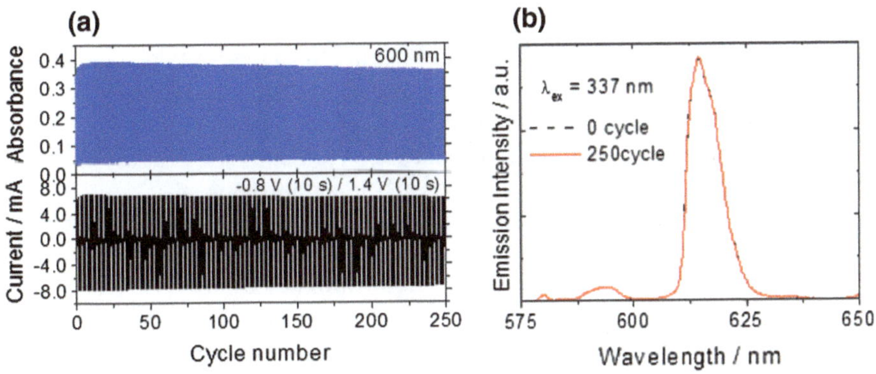

Fig. 26 **a** Changes in absorbance at 600 nm (*top*) and current (*bottom*) of ITO/PB cell upon application of successive voltage steps at −0.8 V (10 s) and 1.4 V (10 s) and **b** emission spectra at #0 (*black dashed line*) and #250 cycle (*red solid line*) under irradiation at 337 nm [51]

intensity, indicating a high reversibility of the emission–coloration switching. The absorbance of both the transparent and colored states of the cell was almost maintained after 250 cycles; the change of the emission intensity was also maintained within the cycles (Fig. 26). These results indicated a high reversibility for the ITO/PB cell in comparison with the ITO/ITO cell, which results from the quantitatively balanced redox reaction of the PB that prevented the solution from undergoing an irreversible electrochemical reaction, while the PB demonstrated its high switching stability.

5 Electrochemical Luminescence Control in a Eu(III) Complex-Modified Electrode

In Sect. 4, we reported the demonstration and improvement of the electrochemical switching of emission and coloration. However, the switching time and repetition stability of electrochemical emission and coloration control were still low for practical use. Since electrochemical switching of fluorescence and absorption is time-consuming (times of both processes take about 10 s), it is important to improve the electrochemical switching performance. The response time of the fluorescence modulation is dependent on the FRET and/or electron transfer between luminescent and electrochemically active molecules that are involved in the electrochemical redox reaction.

In order to achieve a rapid fluorescence modulation response, it is necessary to immobilize both luminescent and electrochemically active molecules on the same electrode; in particular, immobilizing these molecules on an electrode is a powerful method to decrease the time for the electrochemical reaction [62, 65, 66]. Various methods for achieving this immobilization have been reported, but most have typically been based on FRET from a host material such as nanocrystalline titania microspheres [105–108], clay [109–111], and zeolite [112–115] to luminescent species such as Eu(III) ions. However, the low energy-transfer efficiency of these materials makes it difficult to achieve large on–off emission contrast.

In order to increase the response rate and the on–off emission contrast of Eu(III) luminescence, the authors herein focus on the immobilization of a novel Eu(III) complex on a titanium dioxide (TiO_2) nanoparticle-modified electrode. Titanium dioxide nanoparticles have already widely been used in applications such as dye-sensitized solar cells [116–121] and electrochromic displays [117, 122–126]. A TiO_2 nanoparticle-coated film was chosen since it offers a large surface area for the adsorption and the reaction between adsorbed molecules on TiO_2. The immobilization of an Eu(III) complex on a TiO_2 electrode is therefore expected to increase the fluorescence intensity, as the typically high molecular extinction coefficient of antenna ligands in the Eu(III) complex allows them to efficiently transfer energy to Eu(III) ions, leading to a high contrast emission modulation.

As shown in Fig. 27, a modified Eu(III) complex on a TiO_2 electrode was prepared by immersing it in *N*,*N*-dimethylformamide of 2,2′-bipyridine-4,4′-dicarboxylic acid (0.5 mmol L^{-1}) at room temperature for 24 h and subsequent immersing in boiling methanol containing diaquatris(thienoyltrifluoroacetonate)-europium(III) complex (1 mmol L^{-1}) for 24 h.

In order to investigate the electrochemical properties of the TiO_2 electrode and $Eu(tta)_3dcbpy$-modified TiO_2 electrode, their cyclic voltammograms (CVs) and changes in absorbance at 600 nm were measured (Fig. 28). These data revealed that with both electrodes there was an increase in reduction currents from −0.3 V (vs. Ag/AgCl), with a slight increase in absorbance observed at 600 nm during potential scanning in the negative direction. Furthermore, when the potential was scanned from −0.8 to 0 V, an oxidation current was found at a potential of −0.61 and

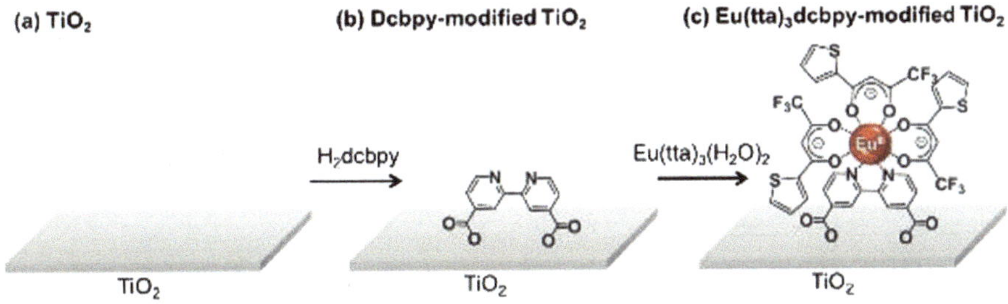

Fig. 27 Formation of a Eu(tta)$_3$dcbpy-modified TiO_2 electrode [127]

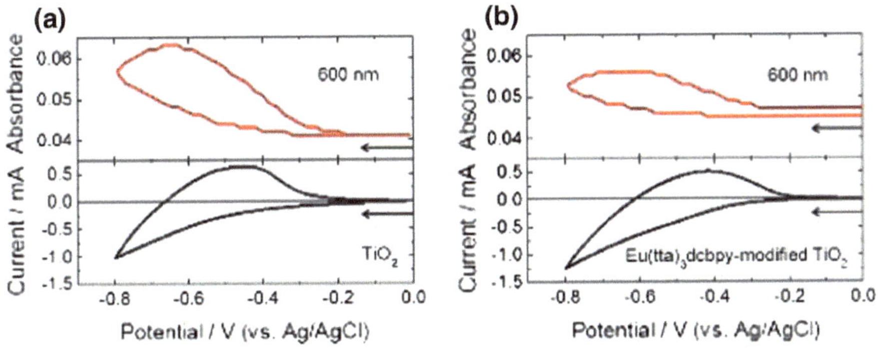

Fig. 28 Changes in absorbance at 600 nm (*top*) and CVs (*bottom*) of **a** a TiO_2-modified electrode and **b** a Eu(tta)$_3$ dcbpy-modified TiO_2 electrode in a PC solution containing 0.3 mol L^{-1} of $LiClO_4$. Scan rate was 50 mV/s [127]

−0.67 V in the TiO_2 and Eu(tta)$_3$dcbpy-modified TiO_2 electrodes, respectively. As this oxidation current increased, the absorbance at 600 nm decreased and recovered to its initial state for both electrodes. The absorption spectra of the TiO_2 and Eu (tta)$_3$dcbpy-modified TiO_2 electrodes also exhibited a slight increase from 400 to 700 nm upon application of a potential of −0.8 V for 5 s. Furthermore, when a potential of 0 V was applied to these cells for 10 s, the absorbance of each electrode was restored to its respective initial value. Based on this change in absorbance and the corresponding CVs, this electrochemical and optical behavior was considered to be linked to the electrochemical reduction of the TiO_2 film.

The excitation and emission spectra of the Eu(tta)$_3$dcbpy-modified TiO_2 electrode along with its change in emission intensity under an applied potential sweep, were performed in order to determine how the electrochemical reaction of the modified electrode affects emission from the Eu(III) complex (the measurement setup was as shown in Fig. 29a). As can be seen in the excitation spectrum in Fig. 29b, the peak observed at around 377 nm corresponded to the absorption of the tta ligands in the Eu(III) complex. When these were excited by a 377-nm light source without potential application, strong red emission was observed from the Eu

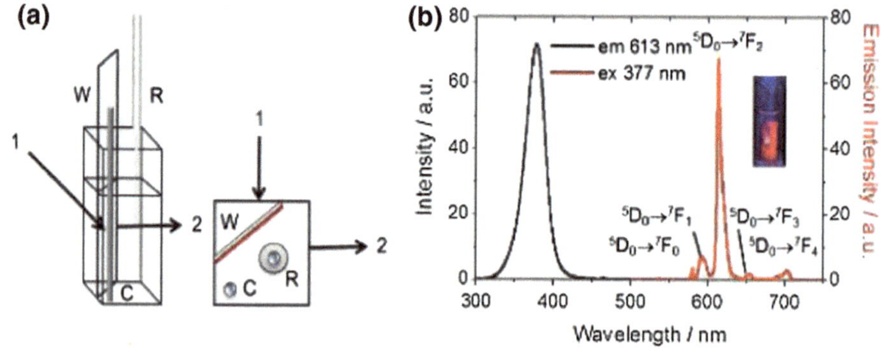

Fig. 29 **a** Schematic representation of the measurement configuration of the fluorescence spectroelectrochemical cell: connection for (W) working, (R) reference, and (C) counter electrodes: Eu(tta)$_3$dcbpy complex-modified electrode, Ag/AgCl, and Pt wire, respectively. *1* Excitation source and *2* emission detector. **b** Excitation (*black solid line*) and emission spectra (*red solid line*) of a Eu(tta)$_3$dcbpy-modified TiO_2 electrode. The photograph shows the change in emission under UV light (365 nm) [127]

(III) complex, as evidenced by the red solid line and picture in Fig. 29b. The sharp emission bands that can be seen around 613 nm were all attributable to the intra-configurational *f–f* transitions of the Eu(III) complex.

The change in emission intensity (613 nm) of the modified electrode was also monitored under a sweeping of the applied potential (Fig. 30a). Interestingly, when the reductive current increased, starting from −0.24 V (vs. Ag/AgCl), the red emission intensity from Eu(III) complex decreased. Furthermore, upon continuous sweeping of the potential from −0.8 to 0 V, the red emission was recovered gradually from −0.55 V in relation with the oxidative current flowed and returned to its initial value at 0 V. From these results, luminescence control would be most likely due to the electrochemical reaction of TiO_2. Additionally, as shown by the change in emission intensity with the application of different potentials for 5 s (Fig. 30b), emission from the Eu(III) complex was completely quenched at −0.8 V (Fig. 30c, and photograph in Fig. 30c). The on–off emission contrast was calculated to be 74:1 (Fig. 30c). This high on–off rate was attributed to the immobilization of the Eu(III) complex on the TiO_2 electrode. The emission quenching efficiency (η_q) was also calculated using the following equation:

$$\eta_q = 1 - \frac{\phi_5}{\phi_0}$$

where ϕ_0 and ϕ_5 are the emission quantum yields before and after the application of −0.8 V for 5 s (Table 3), respectively. The estimated quenching efficiency which was over 98%, suggested that emission switching was successfully achieved in the modified electrode.

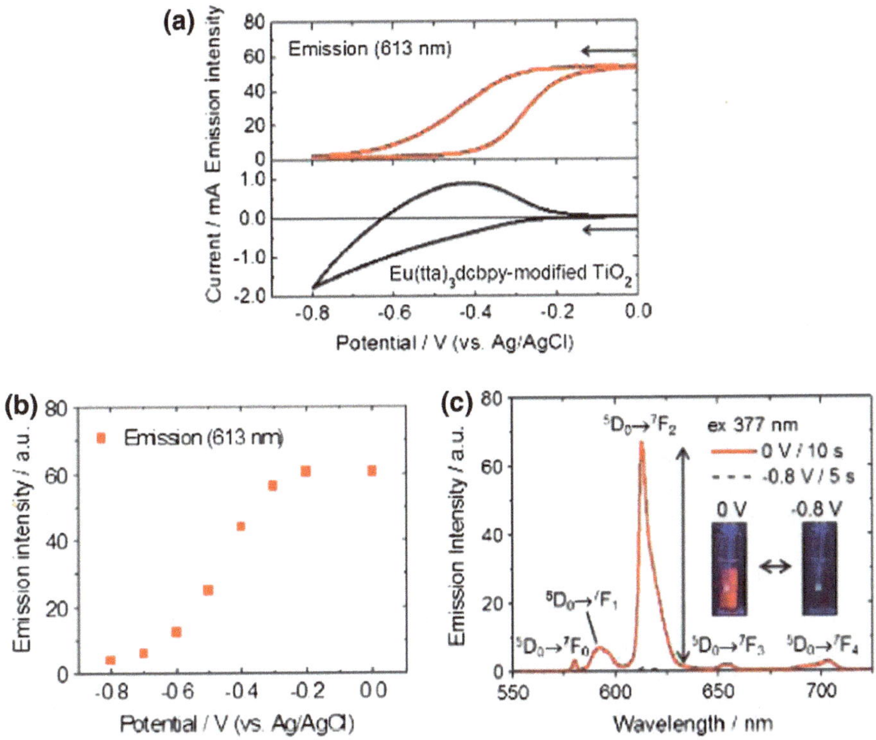

Fig. 30 **a** Change in emission intensity at 613 nm (*top*) and CV (*bottom*) at a scan rate of 50 mV/s of a Eu(tta)$_3$dcbpy-modified TiO_2 electrode in PC solution with 0.2 mol L^{-1} $LiClO_4$, **b** emission intensity at 613 nm with the application of different potentials for 5 s, and **c** emission spectra (*red solid line* 0 V for 10 s, *black dashed line* −0.8 V for 5 s) of a Eu(tta)$_3$dcbpy-modified TiO_2 electrode. The excitation wavelength was 377 nm. Photographs showed the change in emission with the electrochemical reaction under UV light (365 nm) [127]

Table 3 Emission quantum yields (ϕ) and emission lifetimes (τ) of the samples [127]

Compound	State	Potential (V)	ϕ (%)	τ_1 (μs)	τ_2 (μs)
Eu(tta)$_3$(H$_2$O)$_2$	Solution	Open	26[a]	553	–
Eu(hfa)$_3$dcbpy-TiO$_2$	w/o electrolyte	Open	3.7[b]	65 ± 1.66	225 ± 1.89
Eu(hfa)$_3$dcbpy-TiO$_2$	In electrolyte	Open	1.8[a]	46 ± 1.91	227 ± 4.63
Eu(hfa)$_3$dcbpy-TiO$_2$	In electrolyte	−0.8	<0.03[b]	1.1 ± 0.07	26 ± 0.60

[a]Modified on ITO electrode. The emission decay curves were analyzed using a biexponential curve fitting of $[I(t) = A_1 \exp(-t/\tau_1) + A_2 \exp(-t/\tau_2)]$
[b]The value in parentheses represents $A_1/(A_1 + A_2)$

In order to study the change in emission intensity at 613 nm with the application of a negative potential, and the response time of this electrochemically induced emission control, the changes in the emission intensity and chronoamperometric

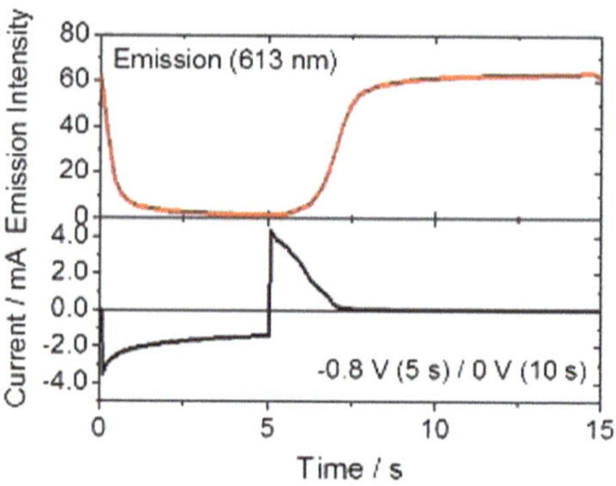

Fig. 31 Change in emission intensity at 613 nm (top) and chronoamperometric curve (bottom) of a Eu(tta)$_3$ dcbpy-modified TiO_2 electrode in a PC solution with 0.2 mol L^{-1} $LiClO_4$ at an applied potential of –0.8 V (vs. Ag/AgCl) for 5 s, followed by 0 V for 10 s [127]

response of the Eu(tta)$_3$dcbpy-modified TiO_2 electrode were measured under sequential application of −0.8 V (vs. Ag/AgCl) for 5 s and 0 V for 10 s. As shown in Fig. 30, the application of −0.8 V for 1 s was sufficient to reduce the initial emission intensity up to 90%. This rapid quenching response was again caused by the immobilization of the Eu(III) complex on the TiO_2 electrode. With subsequent removal of this applied potential, the initial emission intensity was recovered within 10 s, but up to 90% was recovered within the first 3 s. Thus, this modified electrode exhibited luminescence modulation with a quick response (Fig. 31).

By taking the energy levels of the Eu(III) complex and TiO_2 into consideration (Table 4 and Fig. 32), we assumed the mechanism of fluorescence switching as follows. When the reduction potential of the TiO_2 was applied to the modified electrode, electrons were injected into the CB of TiO_2 and were excited to a higher energy level than the HOMO of the tta ligands. By photoexcitation of tta ligands, in this situation, an electron in the CB could be transferred to a half-filled HOMO of photoexcited tta ligands, forming the reduced state of the Eu(tta)$_3$dcbpy. An electron in the LUMO of the tta ligands, which was situated above the CB of TiO_2, was subsequently transferred to the CB because the reduced state of the Eu(tta)$_3$(H_2O)$_2$ complex was not so stable in comparison with TiO_2 and the tta molecule itself. As a

Table 4 Optical and electrochemical properties of the samples [127]

Compound	$\lambda_{onset,abs}$ (nm)	$E_{red,onset}$ (V)[a]	HOMO (eV)[b]	LUMO (eV)[c]	E_g (eV)[d]
TiO_2 electrode	388[e]	−0.30	−7.30	−4.10	3.20[e]
Tta	412	−0.55	−6.86	−3.85	3.01
Eu(tta)$_3$(H_2O)$_2$	404	−0.79	−6.68	−3.61	3.07

[a]Potentials determined from the CV of the samples versus Ag/AgCl in a $LiClO_4$/PC solution at a scan rate of 50 mV s^{-1}
[b]HOMO = LUMO − E_g
[c]Calculated using LUMO = [−($E_{red,onset}$ − 0.40) − 4.8] eV, where 0.40 V is the half-wave potential of Fc/Fc^+ in 0.2 M $LiClO_4$/PC solution; 4.8 eV has been taken as the energy level of the Fc/Fc^+ versus vacuum
[d]HOMO–LUMO gap (E_g), as estimated from $\lambda_{onset,abs}$: $E_g = 1240/\lambda_{onset,abs}$
[e]Ref. [128]

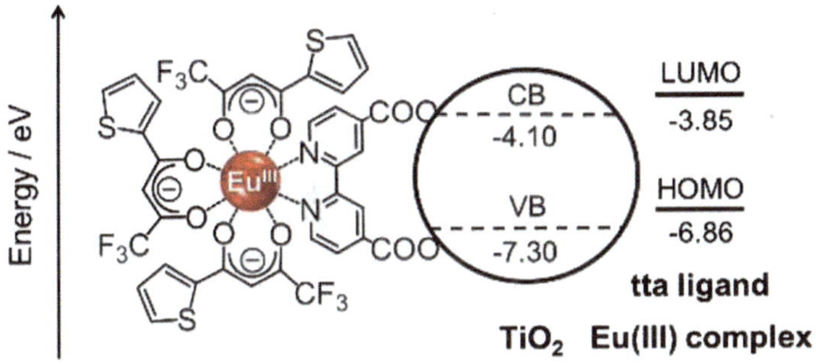

Fig. 32 Energy diagram for a $Eu(tta)_3dcbpy$-modified TiO_2 electrode [127]

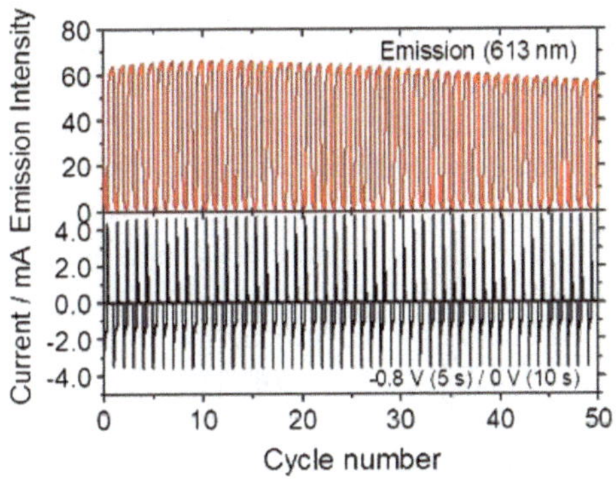

Fig. 33 Changes in emission intensity at 613 nm (*top*) and current (*bottom*) of a $Eu(tta)_3dcbpy$-modified TiO_2 electrode in a PC solution with 0.2 mol L^{-1} $LiClO_4$ during a repeated double-potential step sequence between −0.8 and 0 V for 5 and 10 s [127]

result, the photoexcited Eu(III) complex would return to the ground state without radiation of light [129]. This electron-transfer process was considered to be the most likely reason for the observed quenching of the red emission.

In contrast, red luminescence from the Eu(III) complex was clearly observed when the reduction potential of TiO_2 was not applied. The direct photoinduced electron transfer from the half-filled LUMO of the photoexcited tta ligand to the CB of TiO_2, which led to the formation of the oxidation state of the Eu(III) complex, would be an unfavorable process. This was because the electrochemical oxidation of the Eu(III) complex was not observed on the CV measurements. Therefore, the red luminescence of the Eu(III) complex could be obtained stably without electron transfer between tta ligands and TiO_2.

Finally, to verify the stability of the electrochemical emission modulation, the change in the emission intensity of the $Eu(tta)_3dcbpy$-modified TiO_2 electrode was monitored at 613 nm. As shown in Fig. 33, this change in emission intensity was found to be fairly constant over 50 cycles, indicating that the $Eu(tta)_3dcbpy$-modified electrode is a high reversible emission modulation system, generated by electrochemical stimuli.

6 Conclusion and Future Scopes

This chapter mainly presented the mechanism of emission and coloration monitoring and control, in the author's investigated electrochemical systems.

Single-molecular fluoran dye (Yellow-1) exhibited both green emission and yellow coloration with molecular structure changes induced by electrochemical reaction. The photophysical properties corresponding to the different molecular structures in the neutral and oxidized states were verified by simulating FMOs and applying the density functional theory. From the FMOs, it was clear that non-fluorescent features corresponding to the closed state of the lactone ring are due to the lack of overlap between the HOMO wave function and the LUMO wave function, while fluorescence characteristics corresponding to the open state of the lactone ring could be attributed to the significant overlap between the HOMO and LUMO wave functions. Thus, the emission of the Yellow-1 could be switched by controlling the overlap between the HOMO and the LUMO, and the corresponding change in the IC process after photoexcitation.

On the other hand, composite systems which had a luminescent Eu(III) complex and an heptyl viologen (HV^{2+}) have been investigated as electrochromic materials. The emission and coloration in this system were controlled by the electrochromism of HV^{2+}. In particular, the mechanism of emission switching was found to be the fluorescence resonance energy transfer (FRET) from the excited states of Eu(III) ions to colored species of HV^{2+} (resp. $HV^{+\cdot}$). This stems from the observation of large overlapping of the emission spectrum of the Eu(III) complex and the absorption spectra of the colored species of electrochromic materials, and the decrease of the emission lifetime with the increase of time for application of coloration voltage. Further, values of quenching efficiency and FRET efficiency also indicated that the mechanism of luminescence modulation is FRET. In particular, $Eu(hfa)_3(TPPO)_2/HV^{2+}$ system was effectively quenched (almost 1.00) by HV^{2+} electrochromism. These results could be attributed to the large overlapping between the Eu(III) complex emission and the absorption spectrum of $HV^{+\cdot}$.

Furthermore, the response time and the repetition stability were also investigated in this composite system of Eu(III) complex and HV^{2+}. In particular, the response time and the repetition stability of emission and coloration controls were low. It was because the counter electrode materials that could enable redox reactions of electrochromic materials were absent. Therefore, Prussian blue (PB) electrode as a counter electrode for the HV^{2+} redox reaction was introduced to improve their performance. Consequently, the response time and repetition stability were improved as follows:

- The time of coloring/quenching processes went as follows: 100 s → 10 s
- The time of bleaching/emission processes went as follows: 450 s → 10 s
- The repetition stability was the following: During 5 cycles, the absorbance decreased gradually and then became stable for 250 cycles (maintained at initial value).

These improvements were likely to be consequences of the well charge-balanced and reversible electrochemical reaction of the ITO/PB cell, as well as the quantitatively balanced redox reaction of HV^{2+} and the PB film.

To further shorten the response time of emission and coloration controls, the author tried to immobilize luminescent Eu(III) complex on a TiO_2 electrode. It was because the speed of the electrochemical reaction is generally dependent on the diffusion of the materials, and this was an attempt to suppress it. The Eu $(tta)_3$dcbpy-modified electrode emission was successfully modulated by the electrochemical reaction. The mechanism of luminescence modulation would be electron transfer between the Eu(III) complex and TiO_2 electrode because HOMO of the tta ligands in Eu(III) complex locates between conduction band (CB) and the valence band (VB) of TiO_2, and the decrease of the emission lifetime was induced by electron injection into TiO_2. About the response times of luminescence modulation, the modified electrode required 3.0 s for recovering its initial emission intensity up to 90%. Thus, short response times of luminescence modulation resulted from the immobilization of the Eu(III) complex.

In this work, the authors have presented novel multifunctional materials showing electrochemically modulated emission and coloration and a novel dual-mode display system. The authors hope this article will be of some help in awakening interests for further investigation and development of applications to sensors and display devices such as digital signage, e-paper.

References

1. Tyer McQuade, D., Pullen, A.E., Swager, T.M.: Conjugated polymer-based chemical sensors. Chem. Rev. **100**, 2537–2574 (2000)
2. Martinez-Mañez, R., Sancenon, F.: Fluorogenic and chromogenic chemosensors and reagents for anions. Chem. Rev. **103**, 4419–4476 (2003)
3. Rizzo, M.A., Springer, G.H., Granada, B., Piston, D.W.: An improved cyan fluorescent protein variant useful for FRET. Nat. Biotechnol. **22**, 445–449 (2004)
4. De Silva, A.P., Gunaratne, H.Q.N., McCoy, C.P.: A molecular photonic and gate based on fluorescent signaling. Nature **364**, 42–44 (1993)
5. Credi, A., Balzani, V., Langford, S.J., Stoddart, J.F.: Logic operations at the molecular level. An xor gate based on a molecular machine. J. Am. Chem. Soc. **119**, 2679–2681 (1997)
6. De Silva AP, McClenaghan ND (2004) Molecular-scale logic gates. Chem –Eur J 10: 574–586
7. Irie, M.: Diarylethenes for memories and switches. Chem. Rev. **100**, 1685–1716 (2000)
8. Irie, M., Fukaminato, T., Sasaki, T., Tamai, N., Kawai, T.: Organic chemistry: a digital fluorescent molecular photoswitch. Nature **420**, 759–760 (2002)
9. Bechinger, C., Ferrere, S., Zaban, A., Sprague, J., Gregg, B.A.: Photoelectrochromic windows and displays. Nature **383**, 608–610 (1996)
10. Wang, X.J., Lau, W.M., Wong, K.Y.: Display device with dual emissive and reflective modes. Appl. Phys. Lett. **87**, 113502 (2005)
11. Watanabe, Y., Nakamura, K., Kobayashi, N.: Fabrication of novel reflective-emissive dual-mode display cell based on electrochemical reaction. Chem. Lett. **39**, 1309–1311 (2010)

12. Koyuncu, S., Usluer, O., Can, M., Demic, S., Icli, S., Serdar Sariciftci, N.: Electrochromic and electroluminescent devices based on a novel branched quasi-dendric fluorene-carbazole-2,5-bis(2-thienyl)-1H-pyrrole system. J. Mater. Chem. **21**, 2684–2693 (2011)
13. Puodziukynaite, E., Oberest, J.L., Dyer, A.L., Reynolds, J.R.: Establishing dual electrogenerated chemiluminescence and multicolor electrochromism in functional ionic transition-metal complexes. J. Am. Chem. Soc. **134**, 968–978 (2012)
14. Miesenböck, G., De Angelis, D.A., Rothman, J.E.: Visualizing secretion and synaptic transmission with pH-sensitive green fluorescent proteins. Nature **394**, 192–195 (1998)
15. Lee, K., Asher, S.A.: Photonic crystal chemical sensors: pH and ionic strength. J. Am. Chem. Soc. **122**, 9534–9537 (2000)
16. Zhang, X., Rehm, S., Safont-Sempere, M.M., Würthner, F.: Photonic crystal chemical sensors: pH and ionic strength. Nat. Chem. **1**, 623–629 (2009)
17. Reichardt, C.: Solvatochromic dyes as solvent polarity indicators. Chem. Rev. **94**, 2319–2358 (1994)
18. Yamaguchi, S., Shirasaka, T., Akiyama, S., Tamao, K.: Dibenzoborole-containing π-electron systems: remarkable fluorescence change based on the "on/off" control of the p_π–π^* conjugation. J. Am. Chem. Soc. **124**, 8816–8817 (2002)
19. Han, J., Burgess, K.: Fluorescent indicators for intracellular pH. Chem. Rev. **110**, 2709–2728 (2010)
20. Nakai, H., Kitagawa, K., Nakamori, H., Tokunaga, T., Matsumoto, T., Nozaki, K., Ogo, S.: Reversible switching of the luminescence of a photoresponsive gadolinium(III) complex. Angew. Chem. **52**, 8722–8725 (2013)
21. Carpick, R.W., Sasaki, D.Y., Bums, A.R.: First observation of mechanochromism at the nanometer scale. Langmuir **16**, 1270–1278 (2000)
22. Kaupp, G.: Mechanochemistry: the varied applications of mechanical bond-breaking. CrystEngComm **11**, 388–403 (2009)
23. Zhang, X., Chi, Z., Zhang, Y., Liu, S., Xu, J.: Reversible switching emissions of tetraphenylethene derivatives among multiple colors with solvent vapor, mechanical, and thermal stimuli. J. Mater. Chem. C **21**, 8338–8346 (2011)
24. Wenger, O.S.: Vapochromism in organometallic and coordination complexes: chemical sensors for volatile organic compounds. Chem. Rev. **113**, 3686–3733 (2013)
25. Xu, J., Jia, L., Jin, N., Ma, Y., Liu, X., Wu, W., Liu, W., Tang, Y., Zhou, F.: Fixed-component lanthanide-hybrid-fabricated full-color photoluminescent films as vapoluminescent sensors. Chem. –Eur. J. **19**, 4556–4562 (2013)
26. Kobayashi, A., Kato, M.: Vapochromic platinum(II) complexes: crystal engineering toward intelligent sensing devices. Eur. J. Inorg. Chem. **2014**, 4469–4483 (2014)
27. Fukaminato, T., Sasaki, T., Kawai, Y., Tamai, N., Irie, M.: Digital photoswitching of fluorescence based on the photochromism of diarylethene derivatives at a single-molecule level. J. Am. Chem. Soc. **126**, 14843–14849 (2004)
28. Matsuda, K., Irie, M.: Diarylethene as a photoswitching unit. J. Photochem. Photobiol. C: Photochem. Rev. **5**, 169–182 (2004)
29. Amimoto, K., Kawato, T.: Photochromism of organic compounds in the crystal state. J. Photochem. Photobiol. C: Photochem. Rev. **6**, 207–226 (2005)
30. Kishimoto, Y., Abe, J.: A fast photochromic molecule that colors only under UV light. J. Am. Chem. Soc. **131**, 4227–4229 (2009)
31. Hirata, S., Watanabe, T.: Reversible thermoresponsive recording of fluorescent images (TRF). Adv. Mater. **18**, 2725–2729 (2006)
32. Yamamoto, S., Furuya, H., Tsutsui, K., Ueno, S., Sato, K.: In situ observation of thermochromic behavior of binary mixtures of phenolic long-chain molecules and fluoran dye for rewritable paper application. Cryst. Growth Des. **8**, 2256–2263 (2008)
33. Hirata, S., Lee, K.-S., Watanabe, T.: Reversible fluorescent on–off recording in a highly transparent polymeric material utilizing fluorescent resonance energy transfer (FRET) induced by heat treatment. Adv. Funct. Mater. **18**, 2869–2879 (2008)

34. Zhao, Y., Gao, H., Fan, Y., Zhou, T., Su, Z., Liu, Y., Wang, Y.: Thermally induced reversible phase transformations accompanied by emission switching between different colors of two aromatic-amine compounds. Adv. Mater. **21**, 3165–3169 (2009)
35. Azizian, F., Field, A.J., Heron, B.M., Kilner, C.: Intrinsically thermochromic fluorans. Chem. Commun. **48**, 750–752 (2012)
36. Nakamura, K., Kobayashi, Y., Kanazawa, K., Kobayashi, N.: Thermoswitchable emission and coloration of a composite material containing a europium(III) complex and a fluoran dye. J. Mater. Chem. C **1**, 617–620 (2013)
37. Mortimer, R.J.: Organic electrochromic materials. Electrochim. Acta **44**, 2971–2981 (1999)
38. Rauch, R.D.: Enhanced redox stability and electrochromic properties of aromatic polyamides based on N,N-bis(4-carboxyphenyl)-N′,N′-bis(4-tert-butylphenyl)-1,4-phenylenediamine. Electrochim. Acta **44**, 3165–3176 (1999)
39. Granqvist, C.G.: Electrochromic tungsten oxide films: review of progress 1993–1998. Sol. Energy Mater. Sol. Cells **60**, 201–262 (2000)
40. Argun, A.A., Aubert, P.-H., Thompson, B.C., Schwendeman, I., Gaupp, C.L., Hwang, J., Pinto, N.J., Tanner, D.B., MacDiarmid, A.G., Reynolds, J.R.: Multicolored electrochromism in polymers: structures and devices. Chem. Mater. **16**, 4401–4412 (2004)
41. Monk, R.M.S., Mortimer, R.J., Rosseinsky, D.R.: Electrochromism and Electrochromic devices. Cambridge University Press, Cambridge (2007)
42. Kobayashi, N., Miura, S., Nishimura, M., Urano, H.: Organic electrochromism for a new color electronic paper. Sol. Energy Mater. Sol. Cells **92**, 136–139 (2008)
43. Imanori, H., Sakata, Y.: Donor-linked fullerenes: photoinduced electron transfer and its potential application. Adv. Mater. **9**, 537–546 (1997)
44. Valeur, B.: Molecular Fluorescence: Principles and Applications. Wiley-VCH Verlag GmbH, New York (2001)
45. De Silva, A.P., Moody, T.S., Wright, G.D.: Fluorescent pet (photoinduced electron transfer) sensors as potent analytical tools. Analyst **134**, 2385–2393 (2009)
46. Turro, N.J., Ramamurthy, V., Scaiano, J.C.: Principles of Molecular Phorochemistry An Introduction. University Science Books, Mill-Valley, California (2009)
47. Audebert, P., Miomandre, F.: Electrofluorochromism: from molecular systems to set-up and display. Chem. Sci. **4**, 575–584 (2013)
48. Jares-Erijman, E.A., Jovin, T.M.: FRET imaging. Nat. Biotechnol. **21**, 1387–1395 (2003)
49. Watanabe, Y., Nakamura, K., Kobayashi, N.: Improvement in reflective-emissive dual-mode properties of electrochemical displays by electrode modification. Phys. Chem. Chem. Phys. **13**, 19420–19425 (2011)
50. Nakamura, K., Kanazawa, K., Kobayashi, N.: Electrochemically controllable emission and coloration by using europium(III) complex and viologen derivatives. Chem. Commun. **47**, 10064–10066 (2011)
51. Kanazawa, K., Nakamura, K., Kobayashi, N.: Electroswitching of emission and coloration with quick response and high reversibility in an electrochemical cell. Chem. -Asian J. **7**, 2551–2554 (2012)
52. Kanazawa, K., Nakamura, K., Kobayashi, N.: Dual emissive-reflective display materials with large emission switching using highly luminescent lanthanide(III) complex and electrochromic material. Jpn. J. Appl. Phys. **52**, 05DA14 (2013)
53. Nakamura, K., Kanazawa, K., Kobayashi, N.: Electrochemically-switchable emission and absorption by using luminescent lanthanide(III) complex and electrochromic molecule toward novel display device with dual emissive and reflective mode. Displays **34**, 389–395 (2013)
54. Yoshida, M., Yashiro, N., Shitama, H., Kobayashi, A., Kato, M.: Redox-active dinuclear platinum complex exhibiting multicolored electrochromism and luminescence. Chem. –Eur. J. **22**, 491–495 (2016)
55. Hasegawa, Y., Nakagawa, T., Kawai, T.: Recent progress of luminescent metal complexes with photochromic units. Coord. Chem. Rev. **254**, 2643–2651 (2010)

56. Fukaminato, T., Doi, T., Tamaoki, N., Okuno, K., Ishibashi, Y., Miyasaka, H., Irie, M.: Single-molecule fluorescence photoswitching of a diarylethene-perylenebisimide dyad: non-destructive fluorescence readout. J. Am. Chem. Soc. **133**, 4984–4990 (2011)
57. Ouhenia-Ouadahi, K., Yasukuni, R., Yu, P., Laurent, G., Pavageau, C., Grand, J., Guérin, J., Léaustic, A., Félidj, N., Aubard, J., Nakatani, K.: Photochromic-fluorescent-plasmonic nanomaterials: towards integrated three-component photoactive hybrid nanosystems. Chem. Commun. **50**, 7299–7302 (2014)
58. Fukaminato, T., Hirose, T., Doi, T., Hazama, M., Matsuda, K., Irie, M.: Molecular design strategy toward diarylethenes that photoswitch with visible light. J. Am. Chem. Soc. **136**, 17145–17154 (2014)
59. Miomandre, F., Méallet-Renault, R., Vachon, J.J., Pansu, R.B., Audebert, P.: Fluorescence microscopy coupled to electrochemistry: a powerful tool for the controlled electrochemical switch of fluorescent molecules. Chem. Commun., 1913–1915 (2008)
60. Miomandre, F., Lépicier, E., Munteanu, S., Galangau, O., Audibert, J.F., Méallet-Renault, R., Audebert, P., Pansu, R.B.: Electrochemical monitoring of the fluorescence emission of tetrazine and bodipy dyes using total internal reflection fluorescence microscopy coupled to electrochemistry. ACS Appl. Mater. Interfaces **3**, 690–696 (2011)
61. Miomandre, F., Audibert, J.F., Zhou, Q., Audebert, P., Martin, P., Lacroix, J.C.: Tunable electrofluorochromic device from electrochemically controlled complementary fluorescent conjugated polymer films. Electrochim. Acta Interfaces **110**, 56–62 (2013)
62. Seo, S., Allain, H., Na, J., Kim, S., Yang, X., Park, C., Malinge, J., Audebert, P., Kim, E.: Electrofluorescence switching of tetrazine-modified TiO_2 nanoparticles. Nanoscale **5**, 72321–72327 (2013)
63. Quinton, C., Alain-Rizzo, V., Dumas-Verdes, C., Miomandre, F., Clavier, G., Audebert, P.: Redox-controlled fluorescence modulation (electrofluorochromism) in triphenylamine derivatives. RSC Adv. **4**, 34332–34342 (2014)
64. Quinton, C., Alain-Rizzo, V., Dumas-Verdes, C., Miomandre, F., Clavier, G., Audebert, P.: Redox- and protonation-induced fluorescence switch in a new triphenylamine with six stable active or non-active forms. Chem. –Eur. J. **21**, 2230–2240 (2015)
65. Seo, S., Kim, Y., You, J., Sarwade, B.D., Wadgaonkar, P.P., Menon, S.K., More, A.S., Kim, E.: Electrochemical fluorescence switching from a patternable poly(1,3,4-oxadiazole) thin film. Macromol. Rapid Commun. **32**, 637–643 (2011)
66. Yang, X., Seo, S., Park, C., Kim, E.: Electrical chiral assembly switching of soluble conjugated polymers from propylenedioxythiophene-phenylene copolymers. Macromolecule **47**, 7043–7051 (2014)
67. Mai, S., Wu, J., Liu, J., Xu, Z., Wu, X., Luo, G., Zheng, J., Xu, C.: AIEE-active and electrochromic bifunctional polymer and a device composed thereof synchronously achieve electrochemical fluorescence switching and electrochromic switching. ACS Appl. Mater. Interfaces **7**, 27511–27517 (2015)
68. Zhang, G., Zhang, D., Guo, X., Zhu, D.: A new redox-fluorescence switch based on a triad with tetrathiafulvalene and anthracene units. Org. Lett. **6**, 1209–1212 (2004)
69. Zapata, F., Caballero, A., Espinosa, A., Tarraga, A., Molina, P.: A redox-fluorescent molecular switch based on a heterobimetallic Ir(III) complex with a ferrocenyl azaheterocycle as ancillary ligand. Dalton Trans., 3900–3902 (2009)
70. Tropiano, M., Kilah, N.L., Morten, M., Rahman, H., Davis, J.J., Beer, P.D., Faulkner, S.: Reversible luminescence switching of a redox-active ferrocene–europium dyad. J. Am. Chem. Soc. **133**, 11847 (2011)
71. Yano, M., Matsuhira, K., Tatsumi, M., Kashiwagi, Y., Nakamoto, M., Oyama, M., Ohkubo, K., Fukuzumi, S., Misaki, H., Tsukube, H.: "ON–OFF" switching of europium complex luminescence coupled with a ligand redox process. Chem. Commun. **48**, 4082–4084 (2012)
72. Kanazawa, K., Nakamura, N., Kobayashi, N.: High-contrast electroswitching of emission and coloration based on single–molecular fluoran derivatives. J. Phys. Chem. A **118**, 6026–6033 (2014)

73. Kanazawa, K., Nakamura, N., Kobayashi, N.: Electroswitchable optical device enabling both luminescence and coloration control consisted of fluoran dye and 1,4-benzoquinone. Sol. Energy Mater. Sol. Cells **145**, 42–53 (2016)
74. Hohenberg, P., Kohn, W.: Inhomogeneous electron gas. Phys. Rev. **136**, B864–B871 (1964)
75. Kohn, W., Sham, L.J.: Self-consistent equations including exchange and correlation effects. Phys. Rev. **140**, 1133–1138 (1965)
76. Salahub, D.R., Zerner, M.C. (eds.): The Challenge of d and f Electrons. ACS, Washington, DC (1989)
77. Parr, R.G., Yang, W.: Density Functional Theory of Atoms and Molecules. Oxford University Press, Oxford, UK (1989)
78. Bauernschmitt, R., Ahlrichs, R.: Treatment of electronic excitations within the adiabatic approximation of time dependent density functional theory. Chem. Phys. Lett. **256**, 454–464 (1996)
79. Casida, M.E., Jamorski, C., Kasida, K.C., Salahub, D.R.: Molecular excitation energies to high-lying bound states from time-dependent density-functional response theory: characterization and correction of the time-dependent local density approximation ionization threshold. J. Chem. Phys. **108**, 4439–4449 (1998)
80. Stratmann, R.E., Scuseria, G.E., Frisch, M.J.: An efficient implementation of time-dependent density-functional theory for the calculation of excitation energies of large molecules. J. Chem. Phys. **109**, 8218–8224 (1998)
81. Frisch, M.J., Trucks, G.W., Schlegel, H.B., Scuseria, G.E., Robb, M.A., Cheeseman, J.R., Scalmani, G., Barone, V., Mennucci, B., Petersson, G.A. et al.: Gaussian 09, rev A.02. Gaussian, Inc., Wallingford (2009)
82. Becke, A.D.: Density-functional exchange-energy approximation with correct asymptotic behavior. Phys. Rev. A **38**, 3098–3100 (1988)
83. Lee, C.T., Yang, W.T., Parr, R.G.: Development of the Colle-Salvetti correlation-energy formula into a functional of the electron-density. Phys. Rev. B **37**, 785–789 (1988)
84. Miehlich, B., Savin, A., Stoll, H., Preuss, H.: Results obtained with the correlation energy density functionals of Becke and Lee, Yang and Parr. Chem. Phys. Lett. **157**, 200–206 (1989)
85. Becke, A.D.: Density-functional thermochemistry. III. The role of exact exchange. J. Chem. Phys. **98**, 5648–5652 (1993)
86. Rassolov, V.A., Pople, J.A., Ratner, M.A., Windus, T.L., Windus, T.L.: 6-31G* basis set for atoms K through Zn. J. Chem. Phys. **109**, 1223–1229 (1998)
87. Miertus, S., Scrocco, E., Tomasi, J.: Electrostatic interaction of a solute with a continuum. A direct utilization of AB initio molecular potentials for the prevision of solvent effects. J. Chem. Phys. **55**, 117–129 (1981)
88. Tomasi, J., Mennucci, B., Cammi, R.: Quantum mechanical continuum solvation models. Chem. Rev. **105**, 2999–3093 (2005)
89. Yang, L., Ren, A.-M., Feng, J.-K., Wang, J.-F.: Theoretical investigation of optical and electronic property modulations of π-conjugated polymers based on the electron-rich 3,6-dimethoxy-fluorene unit. J. Org. Chem. **70**, 3009–3020 (2005)
90. Zhang, X., Chi, L., Ji, S., Wu, Y., Song, P., Han, K., Guo, H., James, T.D., Zhao, J.: Rational design of d-pet phenylethynylated-carbazole monoboronic acid fluorescent sensors for the selective detection of α-hydroxyl carboxylic acids and monosaccharide. J. Am. Chem. Soc. **131**, 17452–17463 (2009)
91. Saita, K., Nakazono, M., Zaitsu, K., Nanbu, S., Sekiya, H.: Theoretical study of photophysical properties of bisindolylmaleimide derivatives. J. Phys. Chem. A **113**, 8213–8220 (2009)
92. Wu, Y., Guo, H., Zhang, X., James, T.D., Zhao, J.: Chiral donor photoinduced-electron-transfer (d-pet) boronic acid chemosensors for the selective recognition of tartaric acids, disaccharides, and ginsenosides. Chem. –Eur. J. **17**, 7632–7644 (2011)

93. Prasanna de Silva, A., Nimal Gunarantne, H.Q., Rice, T.E.: Proton-controlled switching of luminescence in lanthanide complexes in aqueous solution: pH sensors based on long-lived emission. Angew. Chem. Int. Ed. **35**, 2116–2118 (1996)
94. Parker, D.: Luminescent lanthanide sensors for pH, pO_2 and selected anions. Coord. Chem. Rev. **205**, 109–130 (2000)
95. Keefe, M.H., Benkstein, K.D., Hupp, J.T.: Luminescent sensor molecules based on coordinated metals: a review of recent developments. Coord. Chem. Rev. **205**, 201–228 (2000)
96. Montalti, M., Prodi, L., Zaccheroni, N., Charbonniere, L., Douce, L., Ziessel, R.: A luminescent anion sensor based on a europium hybrid complex. J. Am. Chem. Soc. **123**, 12694–12695 (2001)
97. Hasegawa, Y., Yamamuro, M., Kanehisa, N., Kai, Y., Yanagida, S.: Luminescent polymer containing the Eu(III) complex having fast radiation rate and high emission quantum efficiency. J. Phys. Chem. A **107**, 1697–1702 (2003)
98. Bünzli, J.-C.G., Piguet, C.: Taking advantage of luminescent lanthanide ions. Chem. Soc. Rev. **34**, 1048–1077 (2005)
99. De Bettencourt-Dias, A.: Lanthanide-based emitting materials in light-emitting diodes. Dalton Trans., 2229–2241 (2007)
100. Nakamura, K., Hasegawa, Y., Kawai, H., Yasuda, N., Kanehisa, N., Kai, Y., Nagamura, T., Yanagida, S., Wada, Y.: Enhanced lasing properties of dissymmetric Eu(III) complex with bidentate phosphine ligands. J. Phys. Chem. A **111**, 3029–3037 (2007)
101. Binnemans, K.: Lanthanide-based luminescent hybrid materials. Chem. Rev. **109**, 42843–4374 (2009)
102. De Bettencourt-Dias, A., Barber, P.S., Viswanathan, S.: Aromatic N-donor ligands as chelators and sensitizers of lanthanide ion emission. Coord. Chem. Rev. **273–274**, 165–200 (2014)
103. Mortimer, R.J., Reynolds, J.R.: An in situ colorimetric measurement study of electrochromism in the di-n-heptyl viologen system. Displays **29**, 424–431 (2008)
104. Itaya, K., Uchida, I., Neff, V.D.: Electrochemistry of polynuclear transition metal cyanides: Prussian blue and its analogues. Acc. Chem. Res. **19**, 162–168 (1986)
105. Nassar, E.J., Goncalves, R.R., Ferrari, M., Messaddeq, Y., Ribeiro, S.J.L.: Titania-based organic–inorganic hybrid planar waveguides. J. Alloys Compd. **344**, 221–225 (2002)
106. Tan, M., Wang, G., Ye, Z., Yuan, J.: Synthesis and characterization of titania-based monodisperse fluorescent europium nanoparticles for biolabeling. J. Lumin. **117**, 20–28 (2006)
107. Rocha, L.A., Cluffi, K.J., Sacco, H.C., Nassar, E.J.: Influence on deposition speed and stirring type in the obtantion of titania films. Mater. Chem. Phys. **85**, 245–250 (2004)
108. Wang, X.-L., Yan, B.: Ternary luminescent lanthanide-centered hybrids with organically modified titania and polymer units. Colloid Polym. Sci. **289**, 423–431 (2011)
109. Celedon, S., Quiroz, C., Gonzalez, G., Sotomayor Torres, C.M., Benavene, E.: Lanthanides–clay nanocomposites: Synthesis, characterization and optical properties. Mater. Res. Bull. **44**, 1191–1194 (2009)
110. Lezhnina, M., Benavente, E., Bentlage, M., Echevarria, Y., Klumpp, E., Kynast, U.: Luminescent hybrid material based on a clay mineral. Chem. Mater. **19**, 1098–1102 (2007)
111. Ma, Y., Wang, H., Liu, W., Wang, Q., Xu, J., Tang, Y.: Microstructure, luminescence, and stability of a europium complex covalently bonded to an attapulgite clay. J. Phys. Chem. B **113**, 14139–14145 (2009)
112. Wada, Y., Sato, M., Tsukahara, Y.: Fine control of red–green–blue photoluminescence in zeolites incorporated with rare-earth ions and a photosensitizer. Angew. Chem. Int. Ed. **45**, 1925–1928 (2006)
113. Wang, Y., Li, H., Gu, L., Gan, Q., Li, Y., Calzaferri, G.: Thermally stable luminescent lanthanide complexes in zeolite L. Microporous Mesoporous Mater. **121**, 1–6 (2009)

114. Li, P., Zhang, Y., Wang, Y., Wang, Y., Li, H.: Luminescent europium(III)-β-diketonate complexes hosted in nanozeolite L as turn-on sensors for detecting basic molecules. Chem. Commun. **50**, 13680–13682 (2014)
115. Wang, Y., Li, H.: Luminescent materials of zeolite functionalized with lanthanides. CrystEngComm **16**, 9764–9778 (2014)
116. O'Regan, B., Grätzel, M.: A low-cost, high-efficiency solar cell based on dye–sensitized colloidal TiO_2 films. Nature **353**, 737–740 (1991)
117. Chen, X., Mao, S.S.: Titanium dioxide nanomaterials: synthesis, properties, modifications, and applications. Chem. Rev. **107**, 2891–2959 (2007)
118. Ardo, S., Meyer, G.J.: Photodriven heterogeneous charge transfer with transition-metal compounds anchored to TiO_2 semiconductor surfaces. Chem. Soc. Rev. **38**, 115–164 (2009)
119. Grätzel, M.: Recent advances in sensitized mesoscopic solar cells. Acc. Chem. Res. **42**, 1788–1798 (2009)
120. Imahori, H., Umemiya, T., Ito, S.: Large π-aromatic molecules as potential sensitizers for highly efficient dye-sensitized solar cells. Acc. Chem. Res. **42**, 1809–1818 (2009)
121. Hagfeldt, A., Boschloo, G., Sun, L., Kloo, L., Pettersson, H.: Dye-sensitized solar cells. Chem. Rev. **110**, 6595–6663 (2010)
122. Cummins, D., Boschloo, G., Ryan, M., Corr, D., Rao, S.N., Fitzmaurice, D.: Ultrafast electrochromic windows based on redox-chromophore modified nanostructured semiconducting and conducting films. J. Phys. Chem. B **104**, 11449–11459 (2000)
123. Bach, U., Corr, D., Lupo, D., Pichor, F., Ryan, M.: Nanomaterials-based electrochromics for paper-quality displays. Adv. Mater. **14**, 845–848 (2002)
124. Corr, D., Bach, U., Fay, D., Kinsella, M., McAtamney, C., O'Reilly, F., Rao, S.N., Stobie, N.: Coloured electrochromic "paper-quality" displays based on modified mesoporous electrodes. Solid State Ionics **165**, 315–321 (2003)
125. Ma, C., Taya, M., Xu, C.: Flexible electrochromic device based on poly (3,4-(2,2-dimethylpropylenedioxy) thiophene). Electrochim. Acta **54**, 598–605 (2008)
126. Freitag, M., Galoppini, E.: Cucurbituril complexes of viologens bound to TiO_2 films. Langmuir **26**, 8262–8269 (2010)
127. Kanazawa, K., Nakamura, K., Kobayashi, N.: Electrochemical luminescence modulation in a Eu(III) complex-modified TiO_2 electrode. J. Mater. Chem. C **3**, 7135–7142 (2015)
128. Grätzel, M.: Photoelectrochemical cells. Nature **414**, 338–344 (2001)
129. Szacilowski, K.: Digital information processing in molecular systems. Chem. Rev. **108**, 3481–3548 (2008)

Theoretical Insights in ECL

Alexander Oleinick, Oleksiy V. Klymenko, Irina Svir and Christian Amatore

Abstract ECL reaction mechanisms involved in all experimental approaches are characterized by a sequence of very fast second-order reactions taking place in the diffusion layer between transient species that, according to the exact process of ECL generation, may be generated at a single electrode, a pair of electrodes (anode-cathode), or through electron transfers between activated reactants and co-reactants. These extremely fast second-order reactions generate the extremely short-lived electronically excited state species, S*, which are deactivated through a rapid first-order emissive decay giving rise to the emission of light. These crucial species are therefore always generated and consumed within a narrow layer of solution of very small size compared to diffusion layers. Moreover, they are present within this layer at vanishingly small concentrations. This creates kinetic situations termed "reaction fronts" exact numerical treatment of which is almost impossible by classical numerical approaches. This chapter presents a series of numerical approaches based on the concepts developed by the authors to circumvent these severe complications and allow a precise and fast simulation of ECL reaction mechanisms. These are illustrated taking advantage of experimental examples featuring each main method of ECL generation.

A. Oleinick · O.V. Klymenko · I. Svir (✉) · C. Amatore (✉)
Departement de Chimie, CNRS UMR 8640 PASTEUR, Ecole Normale Superieure-PSL Research University, Sorbonne Universites-UPMC Paris 6, CNRS, 24 rue Lhomond, 75005 Paris, France
e-mail: irina.svir@ens.fr

C. Amatore
e-mail: christian.amatore@ens.fr

O.V. Klymenko
Department of Chemical and Process Engineering, University of Surrey, Guildford GU2 7XH, UK

F. Miomandre and P. Audebert (eds.), *Luminescence in Electrochemistry*,
DOI 10.1007/978-3-319-49137-0_7

1 Introduction

Electrochemiluminescence (ECL) has enjoyed considerable interest for fundamental reasons rooted in photochemistry and physical electrochemistry as well as for various analytical purposes including important biological analyses. ECL belongs to the general class of chemiluminescence in which an electronically excited state, S*, is generated by kinetically shunting a would-be highly exergonic electron transfer (ET) by a much less exergonic one. The excited species then decays to a lower energy level through the emission of a photon the energy of which is close to that of the difference in enthalpies. Therefore, ECL is very well suited for probing the Marcus inverted region by invoking competition between ET leading to fundamental states and ET resulting in excited product states followed by radiative decay (see Fig. 1).

Annihilation ECL was the first type of ECL reaction mechanisms to be investigated since it helped improve the fundamental understanding of highly energetic outer-sphere ET reactions [1–3]. In this mechanism, an excited state emitting light upon the return to its ground state is formed in solution from an oxidized and a reduced species that are sequentially generated at the same working electrode surface by fast specific potential modulation. Alternatively, the two oxidized/reduced species could be generated at two closely positioned microelectrodes, thus leading to a steady state ECL flux [4–8].

In the late 1970s, another way of ECL generation was discovered in which an inert parent luminophore molecule (or its reduced or oxidized form) reacts with a (oxidized or reduced, respectively) co-reactant leading ultimately to an emissive decay of the excited state via a cascade of reactions [9–13]. This quickly found applications for analytical detection of minute quantities of analytes that modulate ECL light intensity through interference in the reaction mechanism. The particularly low limits of ECL detection stem from extremely low noise levels in the optical

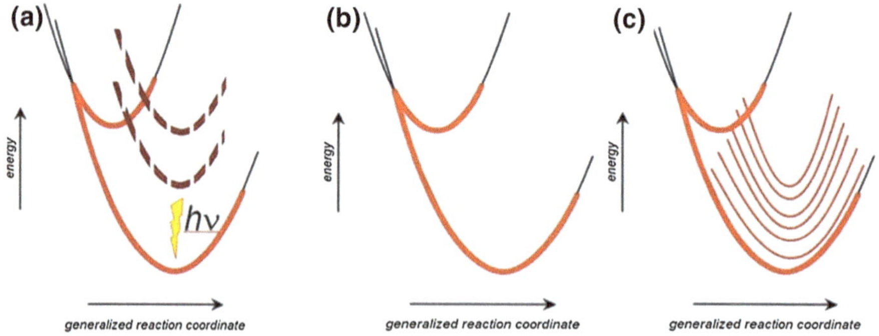

Fig. 1 Different reaction pathways for highly exergonic electron transfers in the Marcus inverted region. **a** Efficient ECL system involving a small density of vibronic excited states of the system of product(s), **b** classical situation corresponding to the Marcus inverted region leading to no significant ECL due to the absence of any vibronic excited state of the system of product(s), and **c** when the density of vibronic excited states of the system of product(s) is too high the system proceeds through a cascade of states leading to emission of infrared photons, i.e. no ECL generation

output (i.e. the ECL yield) because the light signal is practically uncoupled from the electrical input (i.e. the electrochemical trigger) thus precluding the transmission of significant electrical noise. Many reactant/co-reactant couples have been reported for different specific applications. In fact, an excellent review about many possible solutions involving this strategy has been published recently [13], and we wish to refer readers to this review for an overview of the various reactant/co-reactant couples that have been investigated. Indeed, in the present chapter we rather wish to focus on the mechanistic aspects of the various general kinetic situations that define ECL generation independently of the very nature of the species involved, though we will exemplify them using actual experimental situations.

Reaction mechanisms of both experimental approaches are characterized by a sequence of very fast second-order reactions taking place in the diffusion layer and generating the extremely short-lived electronically excited state(s), S*, which are deactivated through a rapid first-order emissive decay. Because of this, the emitters are confined to a narrow reaction zone and their concentrations are exceedingly small. This presents severe complications for the mathematical modelling of ECL reaction mechanisms and, in particular, for the evaluation of the ECL light intensity which depends on the small emitter concentration variations in space and time.

Furthermore, other peculiarities of this nontrivial electrochemical problem such as flux discontinuities (arising from rapid stepwise potential changes at the electrode/solution interface) or its intrinsically multiscale nature in both space and time (e.g. locally high reaction rates coupled with smooth diffusion patterns and rates) render its theoretical study rather challenging. Therefore, with certain exceptions (limiting cases, see below), analytical solutions are not obtainable, and numerical methods must be applied. The numerical treatments for such complex problems require exceptional precision and accuracy in both space and time to allow the prediction of almost negligible emitter concentrations. An overview of simulation approaches and techniques that have been employed over the years to model ECL is presented below including that developed by the authors and implemented in general purpose electrochemical software (KISSA) [14–20]. The advantages of this latter approach are illustrated by considering the three main reaction mechanisms leading to ECL under typical conditions [21].

Subsequently, we consider recently studied applications of our simulation approach involving some of the most complex mechanisms leading to ECL through the oxidation of alkyl amines in the presence of transition metal(II) complexes [22]. Following their seminal discovery by Bard et al., these systems found many analytical and biomedical applications. Yet, the design and optimisation of novel reaction pathways in search of increased efficiency in comparison with the original $Ru(bpy)_3^{2+}$/TPrA (tri-*n*-propylamine) system typically relies on the well-known and widely accepted mechanism of this classical couple. Variations in ECL intensities are then attributed to expected changes in thermodynamic and kinetic rate constants or to the stability of intermediates, though this is hard or impossible to verify independently. The resulting model may be far from the true case but still provides acceptable agreement with experimental data over a range of conditions upon fitting based on alleged system parameters. In this context, we have demonstrated on the basis of numerical simulation that one important parameter, namely the ratio of co-reactant diffusivities, had been overlooked in the interpretation of ECL behaviour. Thus, when the diffusion coefficients of the metal complex species decrease the intensity of the first ECL wave observed at the level of the amine oxidation peak greatly increases with respect to the second wave occurring around the formal potential of the metal(II) complex. This illustrates that the effects of electrochemical reactivity of the amine and the metal complex are sometimes overshadowed by much stronger effects imposed by local transport properties.

Finally, the solution to one of the most challenging problems tackled by us (from the point of view of both theory and experiment) will be presented. This inverse problem was formulated to unravel various optical distortions interfering with imaging ECL light emission using confocal microscopy in an electrochemical system consisting of two parallel band microelectrodes under steady state conditions [23]. A comprehensive mathematical model specifically developed for this situation accounts for experimental distortions arising from (a) the specific finite shape of the sampling volume in confocal microscopy, (b) the light transmitted through the diaphragm of a confocal microscope directly from out-of-focus emitters and (c) that transmitted upon reflection from polished platinum electrode surfaces. The results presented are based on a thorough analysis of optical and physicochemical effects simulated using carefully crafted mathematical and numerical techniques. This enables the real spatial distribution of ECL light emission to be recovered from confocal microscopy measurements by deconvoluting it from experimental distortions [24].

2 General Theoretical Approach

Over the years, a number of numerical approaches and software have been applied to simulate ECL systems. One notable example is DigiSim [25], which is a versatile program for simulating current–potential–time dependences for complex electrochemical reaction sequences. It employs a finite-difference approach and exponential spatial grid compression near an electrode surface. The latter allows for thin kinetic layers adjacent to the electrode to be represented with sufficient resolution to enhance the accuracy of the electrochemical current evaluation. While exponential grid expansion in general is certainly suitable for EC-type mechanisms with fast chemical steps, it is not appropriate for approximating very thin reaction layers or fronts occurring away from the electrode surface and changing their position with time as is the case in EE-type mechanisms and ECL annihilation. In addition, DigiSim does not provide a means for the direct computation of ECL light intensity. Nonetheless, this software has been used to approximately simulate ECL emission intensity by treating photons as an extremely quickly diffusing electroactive species [26]. The ECL output is then proportional to the computed flux of the photon pseudo-species through the electrode surface.

An alternative to using software specifically designed for electrochemical problems is to employ multipurpose simulation environments such as COMSOL Multiphysics [27] based on the finite-element method (FEM), which has found a variety of applications in computational electrochemistry. As long as the model is defined correctly, an appropriate solver is chosen and its parameters (namely, finite-element grid density and/or adaptation criteria and convergence goals.) are tuned accordingly, perfectly acceptable predictions can be obtained with such nonspecialized multipurpose software. However, it should be noted that setting up a nontrivial COMSOL Multiphysics simulation requires a substantial level of expertise in order for the program to effectively resolve specific electrochemical boundary conditions and travelling reaction fronts. Specifically, the option to automatically adapt the FEM simulation grid based on the estimated local approximation error is available, but it may fail in several classical situations encountered in electrochemistry [28, 29], particularly whenever drastically different spatial length scales must be resolved simultaneously. In the majority of such situations, an experienced user is able to aid and direct the grid adaptation to recover the expected convergence behaviour of the solver. However, such tweaking may be difficult to achieve by an insufficiently skilled user and normally leads to extended CPU times and memory occupation (see below). Therefore, even though very accurate results may be obtained with COMSOL Multiphysics in most circumstances, this requires solid user expertise and hence, there is a danger of obtaining invalid predictions by novice users who may fail to spot potential sources of inaccuracy in the numerical simulation.

A lack of specialized electrochemical software enabling direct simulation of any reaction mechanism leading to ECL has forced researchers to combine the capabilities of specialized and general purpose software. For example, a successor of DigiSim called DigiElch was used by Sartin et al. [30] to simulate voltammetric responses while concentration distributions of two light-emitting species and ensuing ECL intensity were computed with the help of COMSOL Multiphysics [27].

Our first attempt to address the inherent difficulties of numerical simulation of ECL resulted in the creation in 2001 of scientific software ECL-Package [31] specifically targeting the basic mechanism of annihilation-type ECL occurring at working (micro)electrodes of different geometries (planar, spherical, band, channel double band and disc). This work aimed at the development of accurate numerical approaches to the solution of the corresponding nonlinear problems primarily for microelectrodes such as disc and band that are characterized by the presence of edge effects. The latter have a severe effect on the distribution of current density over the electrode surface which affects local concentrations and in turn drastically impacts local rates of bimolecular reactions and as a consequence the ECL photon flux. These numerical features were addressed by implementing specifically designed approaches ranging from coordinate transformations to nonuniform finite-difference grids tailored to satisfy the two somewhat conflicting objectives of resolving the edge-effect problem while trying to minimize the overall number of grid nodes and hence the computational cost. Therefore, the annihilation ECL reaction mechanism and electrode geometries with different mass transport regimes served as a good test environment for perfecting numerical methods and grid-generation algorithms [31–34], including conformal mappings for some electrode shapes (sphere and disc) [31, 32], applicable to such problems [33].

Ten years later, in 2011, we developed a novel general approach for the numerical simulation of electrochemical problems which was subsequently implemented within KISSA software [14–20] for systems that can be described in one or two spatial dimensions. The software is capable of accurate and efficient simulation of reaction mechanisms consisting of any sequence of heterogeneous and homogeneous ET steps and first- or second-order homogeneous reactions with rate constants up to and beyond realistically achievable values (i.e. a few 10^{12} s^{-1} and 10^{11} M^{-1} s^{-1}, respectively). Later on, additional features were built into KISSA enabling it to account for the natural convection [20, 35–39] and pre-scan conditioning of the electrode [20]. Furthermore, kinetically controlled adsorption–desorption of reactants ("dynamic" adsorption) as well as chemical and electrochemical reactions in the adsorbed state (including generalized Laviron's mechanisms [40]) were also implemented thus significantly expanding the range and potential complexity of mechanisms [41] amenable to the latest version (1.2) of KISSA.

The efficiency of this simulation approach stems from a combination of two principal components. The first one consists in the use of conformal mappings of the space developed for most typical microelectrode and nanoelectrode geometries (see for example Tables 1 and 2, including (hemi)sphere, disc, cylinder, band and double bands) [17, 42–54]. The second component of the approach is the adaptivity of computational grids driven by a novel and original kinetic criterion, which allows swift concentration changes and/or (travelling) reaction fronts in the bulk of the solution to be identified for subsequent grid refinement in respective areas of the diffusion layer [14–17]. No user intervention is required at any stage of the solution process, so that only the problem formulation must be supplied to the program (i.e. a reaction mechanism, its kinetic parameters, electrode geometry and electrochemical technique to be simulated). Owing to this approach, both electrochemical currents and concentration distributions are simulated with high accuracy. Moreover, this is true for concentrations of all species regardless of their spatial localization so that even extremely minute distributions of highly reactive species restricted to exceedingly narrow reaction fronts or kinetic layers are computed with high precision.

Clearly, the described above complex reaction pathways that lead to ECL emission are fully within the scope of the adaptive simulation approach at the core of KISSA software. A specialized version of this original method aimed at ECL systems is presented below. Its high precision and ability to deal with any numerically challenging situations delineated previously are complemented with outstanding performance in terms of minimal computational times and memory occupation without the need for specific skills on the part of potential users. The validity, precision and accuracy of this approach are established against analytical limits obtained for seminal ECL mechanisms under extreme kinetic conditions putting the numerical simulation under the toughest conditions. The advantages of the approach are then illustrated through typical simulations of the three main ECL generation mechanisms [21].

As mentioned above, there are two strategies that underpin the novel simulation approach, neither of which depends on the expertise of or specific intervention from

Table 1 Computational grids and spatial transformations for major types of currently employed metallic solid or liquid microelectrode/nanoelectrodes

Electrode geometry/system	Coordinate transform	Computational grid in real space and conformal simulation area
Disc [42, 43]	$R=\sin\left(\frac{\pi}{2}\xi\right)/\cos\left(\frac{\pi}{2}\eta\right)$ $Z=\cos\left(\frac{\pi}{2}\xi\right)\ \tan\left(\frac{\pi}{2}\eta\right)$	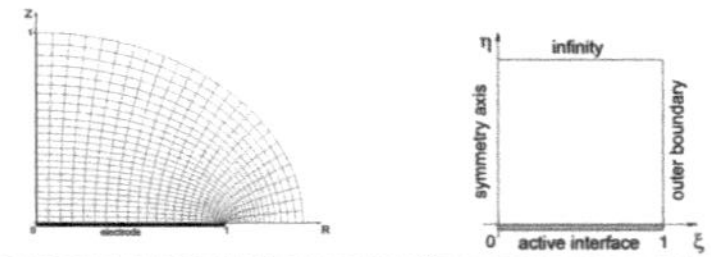
Ring [44]	$X=\sqrt{(1-a^2)\ \sin^2\left[\frac{\pi}{2}\omega\right]+a^2}$	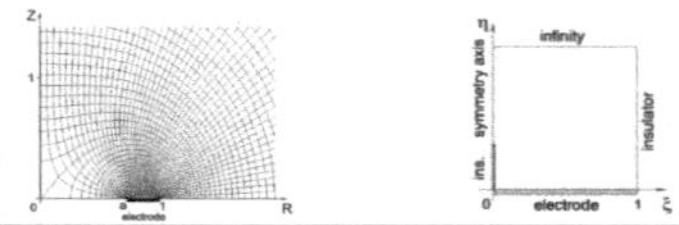
Nanotube cavity [45]	$X=1+\frac{(-\zeta_3)^{1-\beta}}{\pi\sqrt{\zeta_2}}\int_{\ln\zeta_2}^{\omega}\frac{\sqrt{e^{\omega}-\zeta_2}}{(e^{\omega}-\zeta_3)^{1-\beta}}\,d\omega$	
Channel double Band [46]	$Z=\frac{i}{\pi}\arccos\left[a-b\ \cos\left(\pi\left\{\xi+i\frac{\eta}{1-\eta}\right\}\right)\right]$	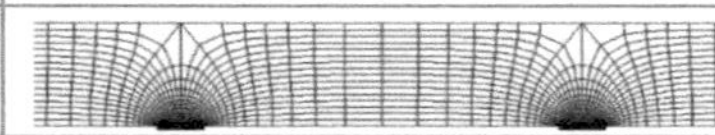
Droplet-modified electrode [48]	$R=-\frac{\sinh\left[\frac{\pi}{2}\tan\left(\frac{\pi}{2}\eta\right)\right]}{\cosh\left[\frac{\pi}{2}\tan\left(\frac{\pi}{2}\eta\right)\right]-\sin\left(\frac{\pi}{2}\xi\right)}$; $Z=\frac{\cos\left(\frac{\pi}{2}\xi\right)}{\cosh\left[\frac{\pi}{2}\tan\left(\frac{\pi}{2}\eta\right)\right]-\sin\left(\frac{\pi}{2}\xi\right)}$	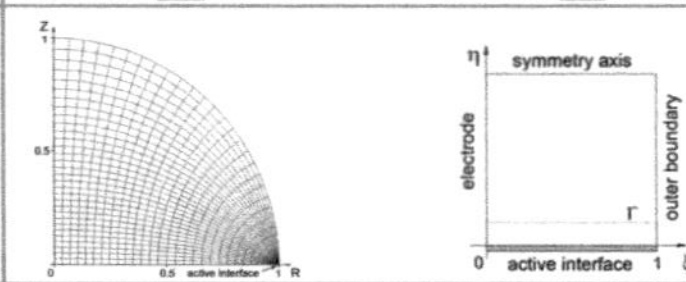

(continued)

Table 1 (continued)

Electrode geometry/system	Coordinate transform	Computational grid in real space and conformal simulation area
Two hemicylinders, neuron-electrode arrays [47, 49]	$X = \frac{a e^{-\xi} - (a-b)\cos\eta + b e^{\xi}}{2(\cosh\xi - \cos\eta)}$; $Y = \frac{(a-b)\sin\eta}{2(\cosh\xi - \cos\eta)}$	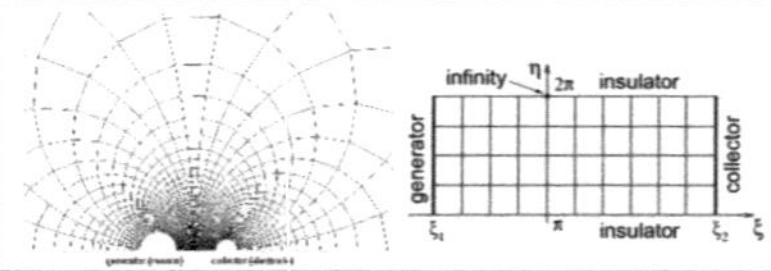
Nanopore, Recessed disc arrays [50, 51]	$X = -iH_p + C\int\limits_0^{\omega} \frac{1}{\left\lvert\sin\left(\frac{\pi}{2}\omega\right)\right\rvert}\left[\frac{\sin^2\left(\frac{\pi}{2}\omega\right) - u_3}{\sin^2\left(\frac{\pi}{2}\omega\right) - 1}\right]^{1-\alpha/\pi} d\omega$	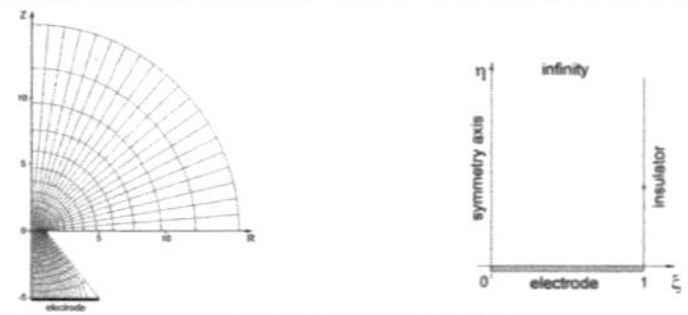
Two bands assembly [52]	$X = \sin\left(\frac{\pi}{2}\xi\right)\cosh\left(\frac{\pi}{2}\eta\right)$; $Y = \cos\left(\frac{\pi}{2}\xi\right)\sinh\left(\frac{\pi}{2}\eta\right)$	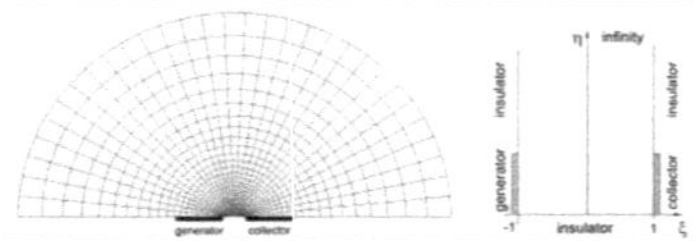
Vesicle (or partially masked spherical body) [53, 54]	$\omega^* = K\int\limits_0^{X}\prod_{i=1}^{m}\left(1 - \frac{\zeta}{X_i}\right)^{\gamma_i - 1} d\zeta$; $\omega = \xi + i\eta = \xi^* + f(\eta^*)$	

Table 2 Coordinate transformations used in KISSA-1D and KISSA-2D [17]

Electrode	Coordinate transformation	Transformed Laplacian, $\tilde{\Delta}$	Simulation area	Transformed coordinates
Sphere	$\rho = 1 - \frac{r_s}{r}$ $\tau_s = D_{\max} t / r_s^2$	$(1-\rho)^4 \frac{\partial^2}{\partial \rho^2}$	$\rho \in [0,\ 1]$	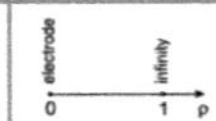
Cylinder	$\zeta = \frac{2}{\delta} \ln \frac{r_c}{r}$ $\tau_c = t / t_{\max}$	$\exp(-\delta\zeta) \frac{\partial^2}{\partial \zeta^2}$	$\zeta \in [0,\ \zeta_{\max}]$ $\zeta_{\max} = \frac{2}{\delta} \ln(1 + 3\delta)$ $\delta = 2\sqrt{D_{\max} t_{\max}} / r_c$	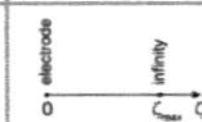
Disc	$R = \frac{r}{r_d} = \frac{\sin(\pi\xi/2)}{\cos(\pi\eta/2)}$ $Z = \frac{z}{r_d} = \cos\left(\frac{\pi\xi}{2}\right) \tan\left(\frac{\pi\eta}{2}\right)$ $\tau_d = D_{\max} t / r_d^2$	$\frac{4\left(\frac{\partial^2}{\partial \xi^2} + \cos^2\left(\frac{\pi\eta}{2}\right)\frac{\partial^2}{\partial \eta^2} + \frac{\pi}{2}\cot\left(\frac{\pi\xi}{2}\right)\frac{\partial}{\partial \xi}\right)}{\pi^2\left(\sec^2\left(\frac{\pi\eta}{2}\right) - \sin^2\left(\frac{\pi\xi}{2}\right)\right)}$	$\xi \in [0,\ 1]$ $\eta \in [0,\ 1]$	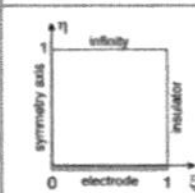
Band	$X = \frac{2x}{w} = \cos\left(\frac{\beta\varphi}{2}\right) \cosh\left(\frac{\beta\psi}{2}\right)$ $Y = \frac{2y}{w} = \sin\left(\frac{\beta\varphi}{2}\right) \sinh\left(\frac{\beta\psi}{2}\right)$ $\tau_b = t / t_{\max}$	$\frac{1}{4[\sinh^2(\beta\psi/2) + \sin^2(\beta\varphi/2)]}\left(\frac{\partial^2}{\partial \varphi^2} + \frac{\partial^2}{\partial \psi^2}\right)$	$\varphi \in [0,\ \pi/\beta]$ $\psi \in [0,\ \psi_{\max}]$ $\psi_{\max} = \frac{2}{\beta} \operatorname{arccosh}\left(1 + \frac{5\beta}{4}\right)$ $\beta = 8\sqrt{D_{\max} t_{\max}} / w$	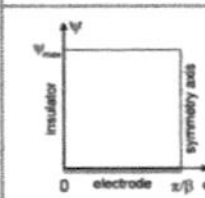

the user, allowing them to function in a completely automatic fashion within the software. The first of them relies on such specific (quasi)conformal mappings of the simulation space as to generate computational grids commensurate with diffusion patterns imposed by the geometric configuration of the working electrode(s) [42–54]. Conformal mappings (see Table 1) covering the major types of currently employed metallic solid or liquid micro/nanoelectrodes such as planar, spherical, cylindrical, disc and band electrodes, including those purposely or accidentally recessed or protruding with respect to the insulator, and their arrays [42, 43, 50, 51] (e.g. Fig. 2, Table 1) have been reported by us in recent years. The use of coordinate transformations is coupled with the second strategy designed to automatically identify reaction fronts and rapid concentration variations at each time step by detecting spatial locations where reaction term exceed a predefined numerical threshold. Whether such spatial regions are adjacent to the electrode surface (classical kinetic layers) or located away from it within the diffusion layer (reaction fronts), the second strategy commands appropriate spatial grid refinement (with respect to the magnitude and width of a kinetic zone/reaction front) in order to resolve the local kinetic terms with sufficient precision as to obtain accurate concentration values at the next time step (Table 2).

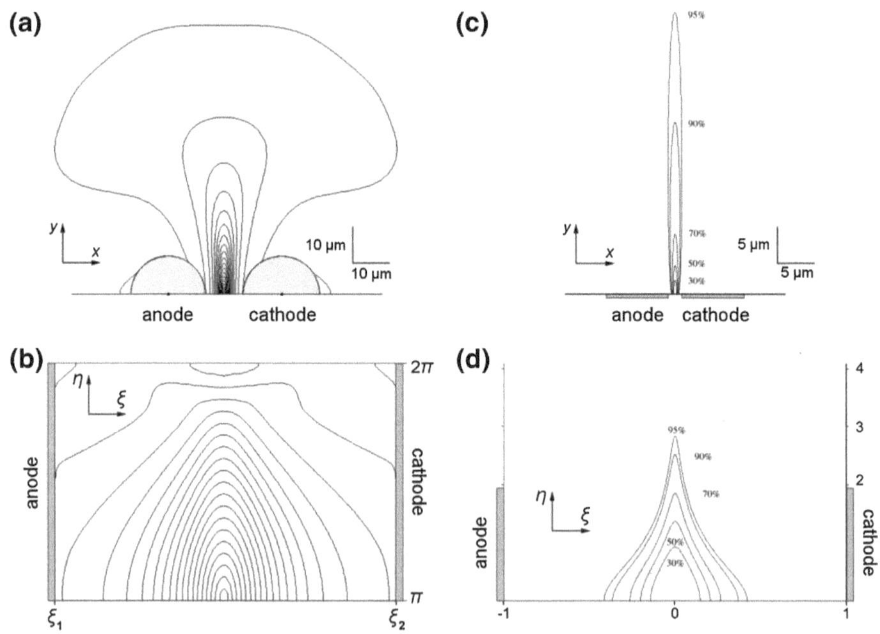

Fig. 2 Simulated concentration distribution of the light-emitting ECL species S^* in the real **a** and conformal space **b** for a two hemicylinder electrode assembly [49]. For comparison, computed concentration distribution of the light-emitting species S^* for the same ECL mechanism at two band microelectrodes in the real **c** and conformal **d** coordinates [24]. See also Table 1 for the analytically derived conformal transforms. Adapted from Refs. [24, 49]

In terms of computational grid adaptivity, to the best of our knowledge, the majority of previously proposed strategies (including that in COMSOL Multiphysics) rely on local concentration gradients numerically estimated on potentially inappropriate computational grids (although perhaps fitted for the preceding time step) necessitating an iterative process to achieve acceptable grid point distribution which may result in long CPU times. On the other hand, the method developed for KISSA (both 1D and 2D versions, see Table 2) employs the notion of "pure kinetics (KP)" introduced by Saveant's group [55–57]. In this context, computational grids perfectly suited to a kinetic problem at hand are generated based only on local chemical reaction rate terms. The combination of these two approaches ensures that a low-density finite difference grid is used where a conformal map is able to effectively resolve the diffusional problem at hand while grid point density is increased in the areas of strong reaction fronts (see, e.g. Fig. 3) [14]. Grid adaptation criteria are checked at each time step, and mesh point density adjustment is performed as required. This means that any sharp reaction fronts within the diffusion layer, whether stationary or travelling, are accurately tracked both in space and time with appropriate grid density changes (see, for example,

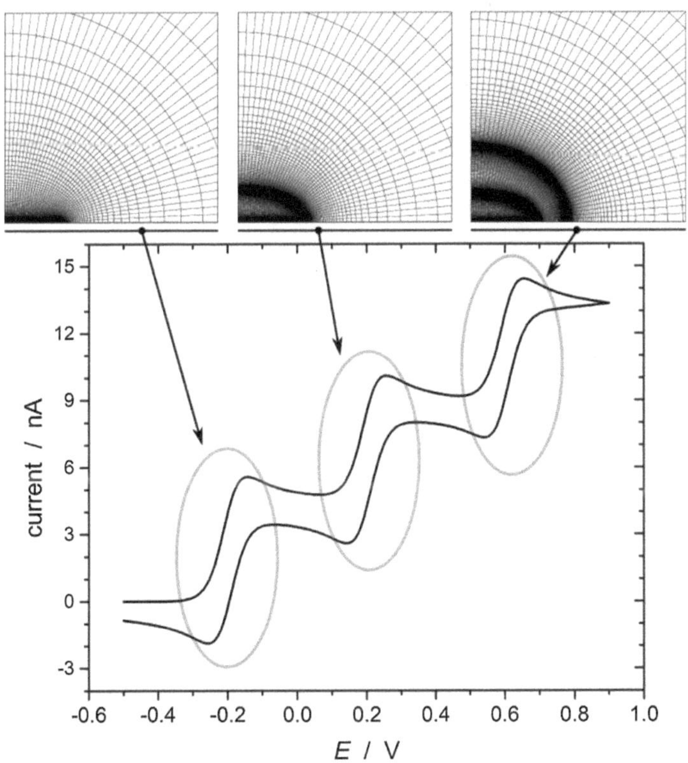

Fig. 3 Snapshots of the computational grid for the EEE mechanism [17] at a **disc** microelectrode illustrating its automatic adjustment with changing electrode potential under cyclic voltammetry as generated by KISSA-2D software

Fig. 4). Finally, to maintain the resulting precision, time grid adaptation may also be necessary with its compression required around sharp temporal variations of, e.g. the electrode potential (that is, at scan direction inversion in voltammetry or potential steps in chronoamperometry) [17]. This helps reduce computational burden and, last but not least, avoid numerical oscillations when discontinuous (potential steps) or nonsmooth (e.g. at inversion potentials in voltammetry) variations of the electrode potential are involved.

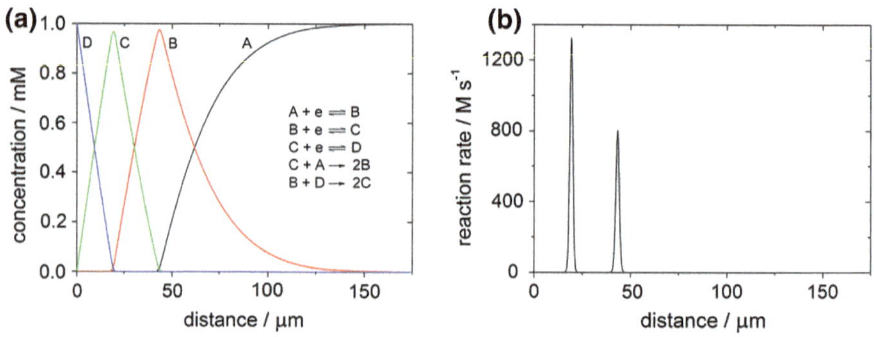

Fig. 4 Simulated normalized concentration profiles **a** and overall reaction rate **b** for the EEE reaction sequence during a chronoamperometric experiment at a **planar** electrode obtained using KISSA-1D. Adapted from Ref. [17]

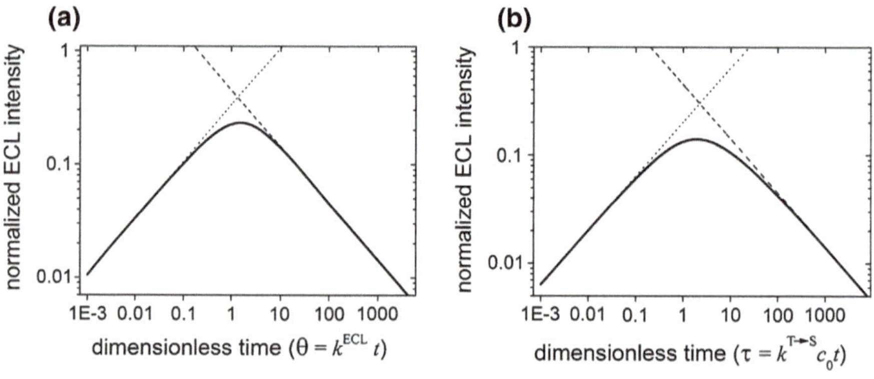

Fig. 5 **a** *Solid curve*: dimensionless ECL intensity variations predicted for **Scheme A** (see below) as a function of $k^{\mathrm{ECL}}t$ obtained through simulations by KISSA-1D. *Dashed line*: analytical limit at infinite times; *dotted line*: analytical limit at short times. **b** ECL intensity for **Scheme B** (see below) including the reaction sequence in Eqs. (8b) and (9b) plotted as a function of $k^{\mathrm{T}\rightarrow\mathrm{S}}c_0t$, i.e. when the triplet bimolecular decay in Eqs. (8a) and (8b) is the *rds*. *Dashed line*: analytical limit at infinite times; *dotted line*: analytical limit at short times. In both cases, the ECL intensity was evaluated for $k^{\mathrm{ET}}c_0t \geq 10^3$. Adapted from Ref. [21]

3 Simulation of "Typical" Complex ECL Mechanisms

Given a reaction mechanism of arbitrary complexity, our novel simulation approach yields very accurate concentration distributions *and* their gradients at the electrode surface (fluxes) with modest computational time and memory requirements. As will be established below, it is thus perfectly suited for the simulation of ECL mechanisms and dealing with their specific features imposing strict requirements on the numerical solver. To illustrate this, **Schemes A** and **B** representing two main ECL generation sequences are considered with annihilation ECL in **Scheme A** and a luminophore and co-reactant pathway in **Scheme B**. A plethora of other ECL sequences may exist as combinations of such basic routes of excited species generation. However, we focus on **Schemes A** and **B** as typical examples of experimental situations producing reactants taking part in the homogeneous ET steps (3) and (7) whether these reactants derive from the same precursor [as in Eq. (3)] or two distinct precursors perhaps at different bulk concentrations [as in (Eq. 7)]. Further below, we consider a different situation in which the oxidation of a single compound (e.g. TPrA) eventually leads to ECL emission as in **Scheme C** (see p. 234).

Scheme A

$$A - e \rightleftharpoons A^{+} \tag{1}$$

$$A + e \rightleftharpoons A^{-} \tag{2}$$

$$A^{+} + A^{-} \xrightarrow{k^{ET}} \varepsilon A^{*} + (2-\varepsilon)A \tag{3}$$

$$A^{*} \xrightarrow{k^{ECL}} A + \gamma h\nu \tag{4}$$

Scheme B

$$A - e \rightleftharpoons A^{+} \tag{5}$$

$$B + e \rightleftharpoons B^{-} \tag{6}$$

$$A^{+} + B^{-} \xrightarrow{k^{ET}} \varepsilon A^{3*} + (1-\varepsilon)A + B \tag{7}$$

$$2\,A^{3*} \xrightarrow{k^{T\to S}} \chi A^{*} + (2-\chi)A \tag{8a}$$

$$A^{*} \xrightarrow{k^{ECL}} A + \gamma h\nu \tag{9a}$$

or:

$$2\,A^{3*} \xrightarrow{k^{T\to S}} A_2^{*} \tag{8b}$$

$$A_2^{*} \xrightarrow{k^{ECL}} 2A + \gamma h\nu \tag{9b}$$

$$2\,A^{3*} \xrightarrow{k^{T\to S}} \chi A^* + (2-\chi)A \quad (8a)$$

$$A^* \xrightarrow{k^{ECL}} A + \gamma\, h\nu \quad (9a)$$

or:

$$2\,A^{3*} \xrightarrow{k^{T\to S}} A_2^* \quad (8b)$$

$$A_2^* \xrightarrow{k^{ECL}} 2A + \gamma\, h\nu \quad (9b)$$

Both in Scheme A and B, a species responsible for ECL emission is the exceedingly short-lived excited state A^*. Accurate numerical simulation of its concentration distribution is crucial for the evaluation of the total photon flux (Fig. 5) defined as:

$$I_{ECL}(t) = \phi_{ECL} k^{ECL} \int_V [A^*](t)\, dV \quad (10)$$

where ϕ_{ECL} is the overall quantum yield, t is the time and V is the volume of the light collection optical pathway across the cell. $\phi_{ECL} = \Gamma\gamma\varepsilon$ for **Scheme A** and **Scheme Bb**, and $\phi_{ECL} = \Gamma\gamma\varepsilon\chi$ for **Scheme Ba**, where Γ is a geometrical capture and collection efficiency factor the value of which is determined by the experimental setup, optical path and efficiency of photomultiplier used.

All the numerical complications that may be encountered in ECL simulations are essentially contained in Eq. (10). Indeed, when k^{ECL} is large, the magnitude of $[A^*](t)$ becomes very small while experiencing steep spatial gradients, i.e. decaying extremely rapidly away from the reaction front producing this species. Because the integral in (10) is taken along the whole optical pathway, it is prone to substantial numerical errors if $[A^*](t)$ is evaluated with insufficient resolution, which ultimately results in erroneous predictions of $I_{ECL}(t)$. Therefore, the accuracy and precision of simulated $[A^*](t)$ are key to the successful solution especially when this concentration is extremely small (i.e. at long times), making its correct numerical evaluation substantially more problematic. For this reason, our simulation outcomes have been tested versus analytical solution that may be derived under such drastic conditions to stress their validity.

3.1 Simulation of ECL from the Thianthrene-2, 5-Diphenyl-1,3,4-Oxadiazole System

The first example involves a reaction cascade initiated by the downhill homogeneous ET taking place between electrogenerated thianthrene (TH) cation radicals

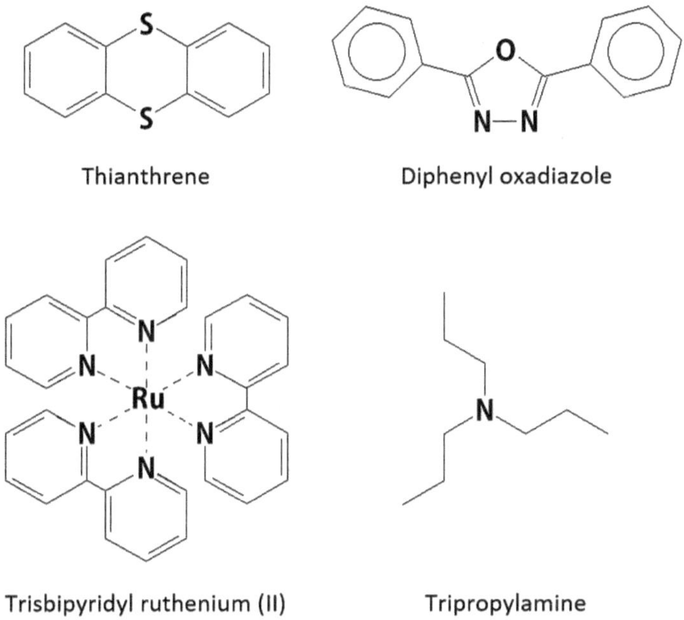

Chart 1 Chemical structures of the main species cited in this chapter

and 2,5-diphenyl-1,3,4-oxadiazole (PPD) anion radicals (see Chart 1 for the structures of the main chemical species cited in this chapter) [58]. This mechanism falls exactly into the framework of **Scheme B** (Eqs. (5)–(7), (8a), (9a) with A = TH, B = PPD and $\chi = 1$) and has been used by Bard and Faulkner in their textbook [9] as a typical experimental system for mixed ECL:

$$\mathrm{TH} - \mathrm{e} \rightleftharpoons \mathrm{TH}^{+} \tag{11}$$

$$\mathrm{PPD} + \mathrm{e} \rightleftharpoons \mathrm{PPD}^{-} \tag{12}$$

$$\mathrm{TH}^{+} + \mathrm{PPD}^{-} \xrightarrow{k_1} \mathrm{TH}^{3*} + \mathrm{PPD} \tag{13}$$

$$2\mathrm{TH}^{3*} \xrightarrow{k_2} \mathrm{TH}^{*} + \mathrm{TH} \tag{14}$$

$$\mathrm{TH}^{*} \xrightarrow{k_3} \mathrm{TH} + h\nu \tag{15}$$

Experimental ECL intensity in this system was studied by Michael and Faulkner [58] using a special sequence of four potential steps shown in the inset of Fig. 6a. The resulting ECL intensity dependences were represented as functions of the dimensionless time $\tau = (t - t_f)/t_f$, where t_f stands for the duration of the initial cathodic step generating PPD^{-}. The potential of the second step of duration $(t_0 - t_f)$ was such

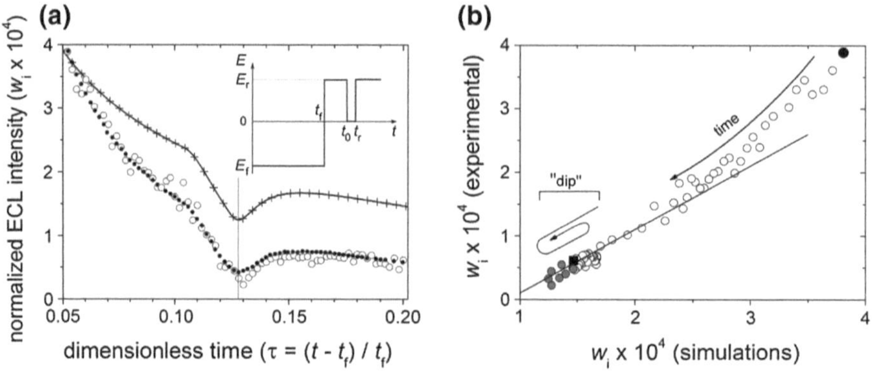

Fig. 6 **a** Normalized ECL intensity simulated with different approaches (*solid curve*—KISSA-1D, *crosses*—Comsol) and reproduced from Ref. [58] (*open circles*—experimental data, *dots*—fitted simulations), and **b** experimental ECL intensity data (*circles*) from Ref. [58] against simulations by KISSA-1D where the starting point ($\tau = 0.05$) is shown with a *filled circle,* the end point ($\tau = 0.2$) denoted by a *filled square* and points corresponding to the dip in w_i are filled with grey. See Ref. [21] for the values of thermodynamic and kinetic parameters used in the simulations. Figures adapted from Ref. [21]

as to generate TH^+ which subsequently reacted with PPD^- clearing it from the vicinity of the working electrode. The potential of the third pulse lasting $(t_r - t_0)$ seconds was chosen between the formal potentials of TH and PPD to reduce TH^+ at the electrode while continuing to oxidize PPD^-. The fourth and last potential step was anodic again to regenerate TH^+ at the electrode surface. Such an unusual potential step sequence was devised to prove that the ECL emission was generated away from the electrode surface through homogeneous reactions within the diffusion layer. The use of this potential sequence enabled to reveal the diffusional delay between concentration changes occurring at the electrode surface and the time at which they reach the thin zone of ECL generation where TH^+ and PPD^- annihilate each other. This is manifested in a smooth dip in the ECL intensity delayed with respect to the time interval beginning at t_0 and ending at t_r.

The present simulation approach perfectly reproduces this phenomenology and the ECL intensity dip as shown in Fig. 6. Although the computed ECL intensity correlates well with the experimental data, two problems can be identified (see Fig. 6b). The first one consists in noticeably smoother simulated ECL intensity transients compared with the experimental data. Such discrepancies are not unusual in ECL experiments and are not surprising. In the case at hand, they may be attributed to frequently occurring reactions with impurities [3] that may scavenge a fraction of the electrogenerated PPD^- and TH^+. A more serious problem is observed at longer times where, despite a good linear correlation with unit slope between experimental data and simulations, there is a constant shift between them in magnitude. We could not offer any feasible explanation for this puzzling discrepancy and thus resorted to using independent simulation software, COMSOL

Multiphysics (v. 4.3) [27], in order to establish the validity of our approach through a comparison of their numerical predictions. It turned out that the results obtained using COMSOL Multiphysics and our approach were identical (although the former took 15 times longer than KISSA-1D to reach similar accuracy with CPU times of 149 s versus 9.5 s) as shown in Fig. 6. Therefore, the shift of experimental data with respect to computational predictions is most probably related to a wrongly assigned origin of the vertical axis in the published figure.

3.2 Effect of Natural Convection on Annihilation-Type ECL

In the preceding sections, we implicitly assumed that, in the absence of hydrodynamic flow (i.e. when the solution is macroscopically still), diffusion layers grow indefinitely with time (this is a typical assumption made in theoretical studies). It is thus normally accepted in electrochemical models that only diffusion-reaction processes are responsible for the electrochemical behaviours observed. However, this assumption is not true in typical experiments as was discovered by Nernst who proposed the notion of the "Nernst diffusion layer" to explain discrepancies from ideal behaviour. This concept was re-examined by us some time ago invoking the seminal contribution by Levich on spontaneous hydrodynamics in solution near a solid wall. A sound model of the "natural convection layer" was thus derived [35] and its physicochemical correctness thoroughly verified experimentally [35–39]. It was recently incorporated into KISSA software to ensure that the effects of natural convection are predicted correctly in situations involving pre-scan or pre-step rest times and/or prolonged experiments when diffusion layer thickness would otherwise have exceeded the thickness of the stagnant layer (which takes a few seconds in classical electrolytes under typical electrochemical conditions) [20].

Natural convection can obviously be negligible in short (with the overall duration not exceeding a few seconds) ECL experiments because diffusion-reaction processes occur essentially in the viscously still solution layer. While this is typically the case for ECL experiments involving stepped potentials, it may not be so when using voltammetric techniques. Indeed, regarding the ET steps in Eqs. (3) or (7), to increase the exergonicity required to generate visible photons, one needs to oppose two redox species with the largest difference in standard potentials (i.e. generally around a few volts). Therefore, scanning voltammetrically one wave and then the other may take more than a few seconds, and the resulting diffusion layer may reach the boundary of the stagnant layer during the experiment. Even if this is not the case, the general formulation of diffusion-natural convection transport is still valid (practically reducing in this case to the diffusion-only limit), and our simulation approach covers every possible situation (although the option to take natural convection into account can be switched off in the program). In the following, this effect is illustrated on the basis of a voltammetric study of an annihilation-type ECL system involving electron–donor-substituted phenylethynylanthronitriles, in which parent molecules may be oxidized or reduced at different sites [59]. The reaction mechanism reproduced in Eqs. (16)–(21) was found to fit the ECL behaviour of four out of six compounds synthesized and examined by the authors of Ref. [59]:

$$(\mathrm{A-D}) + \mathrm{e} \rightleftharpoons (^{\cdot-}\mathrm{A-D}) \tag{16}$$

$$(\mathrm{A-D}) - \mathrm{e} \rightleftharpoons (\mathrm{A-D}^{\cdot+}) \tag{17}$$

$$(\mathrm{A-D}^{\cdot+}) + (^{\cdot-}\mathrm{A-D}) \xrightarrow{k^{\mathrm{ET}}} {}^{3}(\mathrm{A-D})^{*} + (\mathrm{A-D}) \tag{18}$$

$${}^{3}(\mathrm{A-D})^{*} + {}^{3}(\mathrm{A-D})^{*} \xrightarrow{k^{\mathrm{T}\to\mathrm{S}}} (\mathrm{A-D})_2^{*} \tag{19}$$

$$(\mathrm{A-D})_2^{*} \xrightarrow{k^{\mathrm{ECL}}} (\mathrm{A-D})_2 + h\nu \tag{20}$$

$$(\mathrm{A-D})_2 \xrightarrow{k^{\mathrm{cleav}}} 2(\mathrm{A-D}) \tag{21}$$

This reaction mechanism is equivalent to the generic one in Scheme Bb upon replacing A and B with (A–D) and noting that the reaction in Eq. (9b) is represented with two sequential elementary steps in Eqs. (20) and (21). It should be mentioned that the above reaction mechanism is undoubtedly oversimplified. This conclusion is corroborated by published in Ref. [59] experimental data with considerably smaller and irreversible oxidation waves compared to the reversible reduction ones of the same chemical compound. However, despite this, we use this mechanism to illustrate the ability of KISSA-1D to accurately track the concentration distribution of excimer species $(\mathrm{A-D})_2^{*}$.

Similarly to the experimental results in Ref. [59], we assume that voltammetric scans begin in the cathodic direction. Then $(^{\cdot-}\mathrm{A-D})$ is first generated at the electrode surface and diffuses into the bulk of the solution where a part of it may survive while the scan is reversed towards the potential of the oxidation wave to produce $(\mathrm{A-D}^{\cdot+})$.

With the onset of $(\mathrm{A-D}^{\cdot+})$ generation, it reacts immediately with the remaining $(^{\cdot-}\mathrm{A-D})$ still present near the electrode surface to form the excimer species $(\mathrm{A-D})_2^{*}$. As more $(\mathrm{A-D}^{\cdot+})$ is generated at the electrode, it must diffuse farther and farther into the solution to encounter and react with $(^{\cdot-}\mathrm{A-D})$ according to Eq. (18). As a result, the front of reaction (18) moves away from the electrode surface together with the concentration distribution of $(\mathrm{A-D})_2^{*}$, which remains extremely narrow owing to the very fast decay of the excimer compared with its rate of diffusion.

Figure 7b demonstrates this over a time range corresponding to scanning the oxidation wave under diffusion-only conditions. The respective advancement of the automatically enhanced computational grid density around the reaction front is shown in Fig. 7a. This permits the narrow and extremely minute concentration distribution of the excimer to be precisely followed at all times to ensure it is approximated with sufficient resolution and accuracy. On the other hand, the same reaction sequence results in rather different variations of the excimer concentration distribution when natural convection effects are taken into account as shown in Fig. 7c assuming the convection-free layer thickness of 200 μm [20, 35–39]. Natural convection does not tangibly affect the system while the reduction wave is

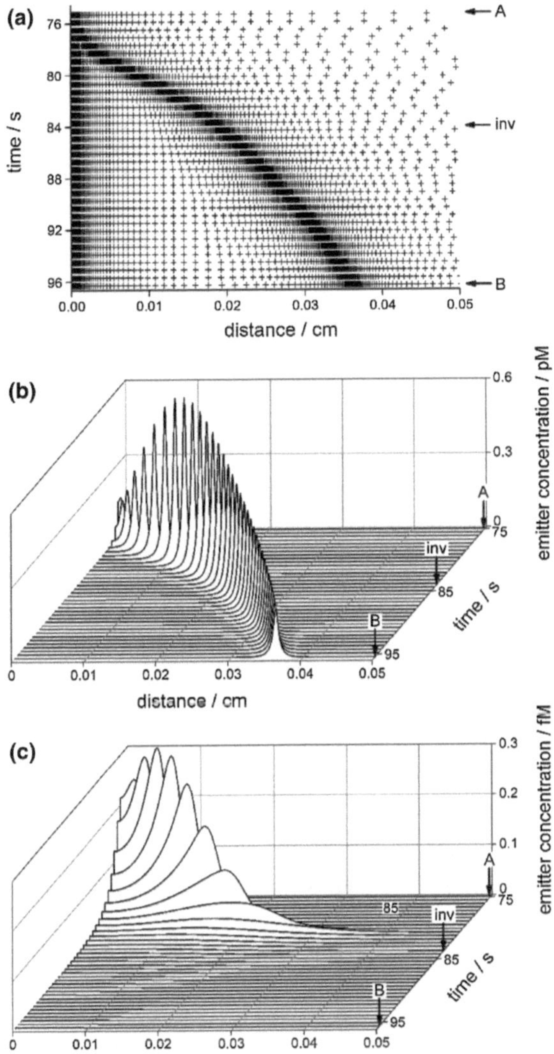

Fig. 7 Adaptive computational grid **a** and concentration profiles of the excimer $(A - D)_2^*$ (**b** and **c**) generated in the sequence of Eqs. (16)–(21) during a voltammetric scan ($v = 50$ mV/s) in the absence **a**, **b** or in the presence **c** of natural convection considering a convection-free layer of 200 μm. **a** Spontaneous evolution of the simulation grid with time to track automatically the reaction front created by Eq. (18) so the $(A - D)_2^*$ concentration profile in **b** may be simulated with excellent precision and accuracy. See Ref. [21] for the values of thermodynamic and kinetic parameters and experimental conditions used in the simulations. Figures adapted from Ref. [21]

scanned because the diffusion layer does not have time to reach the boundary of the stagnant layer. However, it has a highly pronounced effect on the distribution of $(^{\cdot-}A-D)$ within the important part of the diffusion layer. This is rationalized by considering that $(^{\cdot-}A-D)$ is both oxidized at the electrode and drawn from the stagnant layer towards well-mixed bulk solution at a significantly higher rate than in the absence of natural convection throughout the reverse potential sweep following scanning the reduction wave. Predictably, this results in drastically diminished concentration of $(A-D)_2^*$, and its relatively quicker decay. Overall though, $(A-D)_2^*$ is simultaneously confined to a thinner layer of solution adjacent to the electrode surface.

3.3 *Simulation of ECL in the $Ru(bpy)_3^{2+}$/Tri-n-Propylamine (TPrA) System*

The discovery of ECL systems involving $Ru(bpy)_3^{2+}$ and substituted alkyl amines such as tripropylamine (TPrA) has been the subject of extensive mechanistic and development studies [9–13] owing to their important applications as a highly sensitive method for immunoassays and DNA analyses. Herein, we do not intend to revisit the question of the validity of the published reaction mechanism. This sequence of reactions is used here to illustrate the ease with which our new approach is able to solve electrochemical problems of such complexity. Thus, we rely hereafter on the mechanism that incorporates all the mechanistic paths considered in a recent study by Bard et al. [60] (Eqs. 22–32):

Scheme C

$$\mathrm{TPrAH^+} \rightleftharpoons \mathrm{TPrA} + \mathrm{H^+}, \quad \mathrm{p}K_\mathrm{a} = 10.4 \tag{22}$$

$$\mathrm{TPrA} - \mathrm{e} \rightleftharpoons \mathrm{TPrA^{\cdot +}}, \quad k^0_{\mathrm{TPrA0/+}}, E^0_{\mathrm{TPrA0/+}} \tag{23}$$

$$\mathrm{TPrA^{\cdot +}} \rightleftharpoons \mathrm{TPrA^{\cdot}} + \mathrm{H^+}, \quad k_f^{\mathrm{TPrA^{\cdot}}}, K_{\mathrm{eq}}^{\mathrm{TPrA^{\cdot}}} \tag{24}$$

$$\mathrm{TPrA^{\cdot}} + \mathrm{Ru(bpy)_3^{2+}} \rightleftharpoons \mathrm{Im^+} + \mathrm{Ru(bpy)_3^{+}}, \quad k_f^1, K_{\mathrm{eq}}^1 \tag{25}$$

$$\mathrm{TPrA^{\cdot +}} + \mathrm{Ru(bpy)_3^{+}} \rightleftharpoons \mathrm{TPrA} + \mathrm{Ru(bpy)_3^{2+*}}, \quad k_f^2, K_{\mathrm{eq}}^2 \tag{26}$$

$$\mathrm{Ru(bpy)_3^{2+*}} \rightarrow \mathrm{Ru(bpy)_3^{2+}} + h\nu, \quad k_f^{\mathrm{ECL}} \tag{27}$$

$$\mathrm{Ru(bpy)_3^{2+}} - \mathrm{e} \rightleftharpoons \mathrm{Ru(bpy)_3^{3+}}, \quad k^0_{\mathrm{Ru2+/3+}}, E^0_{\mathrm{Ru2+/3+}} \tag{28}$$

$$Ru(bpy)_3^{+} - e \rightleftharpoons Ru(bpy)_3^{2+}, \quad k^0_{Ru+/2+}, E^0_{Ru+/2+} \tag{29}$$

$$Ru(bpy)_3^{3+} + Ru(bpy)_3^{+} \rightarrow Ru(bpy)_3^{2+} + Ru(bpy)_3^{2+*}, \quad k^{ET} \tag{30}$$

$$TPrA^{\cdot} + Ru(bpy)_3^{3+} \rightleftharpoons Im^{+} + Ru(bpy)_3^{2+*}, \quad k_f^3, K_{eq}^3 \tag{31}$$

$$Im^{+} \xrightarrow{+H_2O} (CH_3)_2CO + DPrA, \quad k_{hydrol} \tag{32}$$

In **Scheme C**, Im^{+} stands for the iminium product and DPrA for the dipropylamine. Note also that, with respect to Ref. [60], we have added reaction (31) which was not considered by the authors of the original work but is essential because Ru (III) is a stronger oxidant than Ru(II).

This system has a fairly unusual ECL response featuring two distinct "ECL waves" originating from three different mechanistic regimes activated as a function of the electrode potential, although in all the cases the only photon-emitting species is $Ru(bpy)_3^{2+*}$. The more anodic ECL wave appears at electrode potentials corresponding to considerable oxidation of $Ru(bpy)_3^{2+}$ (Eq. 28), the product of which then reacts with $Ru(bpy)_3^{+}$ (Eq. 30) formed in a reaction between $TPrA^{\cdot}$ and $Ru(bpy)_3^{2+}$ (Eq. 25) as well as with the $TPrA^{\cdot}$ radical itself (Eq. 31) to yield the photon emitting species $Ru(bpy)_3^{2+*}$ [60]. The second less anodic ECL wave owes its existence to two mechanistic pathways primarily depending on the ratios of the concentrations of $TPrA^{\cdot}$ and $TPrA^{\cdot+}$ radicals to that of $Ru(bpy)_3^{2+}$ [60].

The complexity of this reaction mechanism necessitates highly accurate numerical simulation to both precisely compute the concentration distributions of all intermediates regardless of their magnitude and spatial localization (as illustrated above) and track the interplay between the different mechanistic pathways depending on the electrode potential and bulk concentrations of $Ru(bpy)_3^{2+}$ and TPrA. Therefore, this complex system represents a perfect test case for our simulation approach. As illustrated in Fig. 8a, the results of computations demonstrate all the features of the experimental behaviour reported by Bard et al., i.e. the ECL response with two waves of correct shape and magnitude with the first small-amplitude wave flattening out before $Ru(bpy)_3^{2+}$ can be significantly oxidized, and the second one having a CV-like shape at higher potentials. The analysis of concentration variations of the four primary reactants participating in the generation of $Ru(bpy)_3^{2+*}$ and leading to the first and the second ECL intensity waves (Fig. 8b, c) also confirms the mechanistic interpretation of this process proposed previously [60]. The ability to accurately compute such concentration distributions in order to validate any tentative mechanistic interpretation is another valuable advantage of our approach.

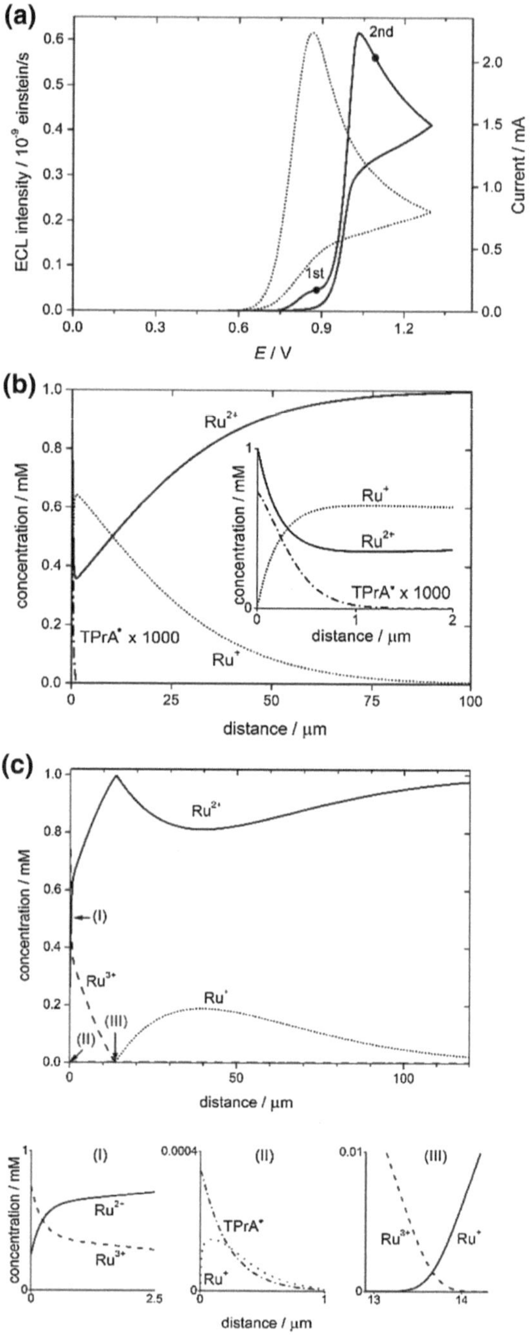

Fig. 8 **a** Simulated ECL intensity (*solid curve*) and voltammogram (*dotted curve*) for the $Ru(bpy)_3^{2+}$/TPrA system at $v = 0.1\ V\,s^{-1}$, $[Ru(bpy)_3^{2+}] = 1$ mM and [TPrA] = 0.1 M under conditions (pH buffered at 8.5) similar to those reported in Ref. [60]. **b and c** Concentration profiles at the first ECL wave, $E = 0.88$ V **b**, or at the second one, $E = 1.09$ V **c**: $Ru(bpy)_3^{2+}$ (*solid curve*), $Ru(bpy)_3^{3+}$ (*dashed curve*), $Ru(bpy)_3^{+}$ (*dotted curve*) and $TPrA^{\cdot}$ (*dashed-dotted curve*). See Ref. [21] for the values of thermodynamic and kinetic parameters used in the simulations. Figures adapted from Ref. [21]

3.4 Conclusions

The main merit of our novel simulation approach with respect to modelling any electrochemical reaction mechanism resulting in ECL emission lies in the ability to accurately compute concentration distributions in both time and space of all reacting species regardless of how short-lived they are and how small and spatially localized their concentrations are. Indeed, concentrations of photon emitting species are typically extremely low owing to immense rates of their formation and decay which outstrip most usual reaction rates. Despite this, these concentrations are crucial for the evaluation of the ECL intensity which is a key quantity of interest in a variety of modern physicochemical or bioanalytical applications. The new approach owes its remarkable properties to a successful combination of specific conformal maps resolving numerical artefacts due to diffusional transport and automatic adjustment of the computational grid in the conformal space [14–20] led by a novel criterion [14, 21] tracking any rapid kinetic variations within the diffusion layer in order to resolve them with higher mesh point density and hence accuracy. This ensures that any reactions, even with rate constants considerably higher than those physically possible, can be successfully treated in a completely automatic fashion without the need for any decisions from the user. The kinetic grid adaptation criterion is formulated on the basis of the magnitude of reaction rate terms, and therefore, its evaluation does not require a preliminary integration step or iterative estimation typical for common finite-difference methods designed to deal with rapid linear or nonlinear chemical reactions. This is essential for the efficient numerical treatment of most kinetic situations encountered in ECL systems.

We believe that all these features present important advantages to most electrochemical users without advanced expertise in mathematics. This becomes evident in comparison with other existing software and approaches. Thus, even though general purpose simulation software such as COMSOL Multiphysics [27] is capable of solving this type of problem with accuracy similar to that of our approach (as illustrated in Fig. 6), this requires considerably longer CPU times while the problem has to be fully defined in the form of rigorous mathematical equations (including reaction terms and specific electrochemical boundary conditions for a particular reaction mechanism) instead of simply entering a reaction scheme in a conventional chemical format.

4 Effects of Diffusion Rates on ECL Generation by Amine/Transition Metal(II) Systems

There have been manifold applications of electrochemically induced chemiluminescence generated through the oxidation of alkyl amines in the presence of transition metal(II) complexes since its discovery by Bard with co-workers. The mechanism of the classical $Ru(bpy)_3^{2+}$/TPrA (tri-*n*-propylamine) couple is often considered in the design and optimization of other systems, in which ECL intensity changes are explained by factors such as changes in thermodynamic and kinetic rate constants or the stability of intermediates. However, another important parameter is the ratio of the diffusivities of the two co-reactants. The effects of these different factors can be analyzed through our simulation approach without any requirement for specific syntheses [22].

4.1 Generalized Scheme of ECL

In biomedical applications, the most frequently used ECL system is shown in **Scheme C1** (derived from that presented by Bard et al. [60]), which involves a Metal(II) complex (M, typically a *d*-block transition metal from group 8 or 9, viz., Ru or Ir) and an aliphatic amine (AH) [21, 60]. For biomedical assays, the mechanism in **Scheme C1** is typically operated at the AH oxidation potential so that the ECL-emitting species M^{2+*} is generated by reaction (37).

$$AH + H^{+} \rightleftharpoons HAH^{+} \tag{33}$$

$$AH - e \rightleftharpoons AH^{\circ +} \tag{34}$$

$$AH^{\circ +} \rightleftharpoons A^{\circ} + H^{+} \tag{35}$$

$$A^{\circ} + M^{2+} \rightarrow A^{+} + M^{+} \tag{36}$$

$$M^{+} + AH^{\circ +} \rightarrow AH + (1 - \varepsilon_5)M^{2+} + \varepsilon_5 M^{2+*} \tag{37}$$

$$M^{+} - e \rightarrow M^{2+} \tag{38}$$

$$M^{2+*} \rightarrow M^{2+} + \varphi h\nu \tag{39}$$

$$M^{2+*} + e \rightarrow M^{+} \tag{40}$$

$$M^{2+*} - e \rightarrow M^{3+} \tag{41}$$

$$M^{+} + M^{3+} \rightarrow M^{2+*} + M^{2+} \tag{42}$$

$$M^{3+} + e \rightleftharpoons M^{2+} \tag{43}$$

$$M^{3+} + A^{\circ} \rightarrow (1 - \varepsilon_{12})M^{2+} + \varepsilon_{12} M^{2+*} + A^{+} \tag{44}$$

$$A^{+} \rightarrow \text{etc.} \tag{45}$$

Scheme C1 General mechanism of a second-generation ECL system deduced on the basis of the $Ru(bpy)_3^{2+}$ (M^{2+}) and TPrA (AH) couple. Note that the solution ETs in Eqs. (37) and (44) may not produce M^{2+*} quantitatively (hence ε_5 and ε_{12} may differ from unity); likewise, the photon emission reaction (39) proceeds with a quantum yield $\varphi < 1$; the values of φ, ε_5 and ε_{12} depend on a particular system of reactants and determine the fraction of ETs yielding photons. Note also that the iminium derivative of AH, A^+, and the corresponding imine radical, A°, may experience several follow-up or parallel reactions not included in **Scheme C1** (see **Scheme C2** for the corresponding reactions). Reaction (43) is written in the direction followed before M^{2+} becomes directly oxidizable at the electrode surface; its direction is inverted afterwards and thus serves to generate the second ECL wave

Our analyses here are restricted to the mechanistic sequence in **Scheme C1** [22, 60]. However, the chemical fates of the imine radical A° and the iminium cation A^+ (see **Scheme C2**) are not considered within the framework of this mechanism [61–63]; these are determined by the exact AH structure and could complicate the mechanistic sequence somewhat. It is presumed that, if such pathways were operating in the canonical $Ru(bpy)_3^{2+}$/TPrA system, their contributions would be indirectly included in the derived kinetic and thermodynamic constants. Therefore, the values of these constants may differ from the real case, but still provide good agreement with experimental voltammetric and ECL results under usual conditions [21, 60]. Because of this, we use the reaction mechanism in **Scheme C1** and take the kinetic and thermodynamic constants reported in Ref. [60], disregarding the reactions in **Scheme C2**.

$$A^\circ - e \rightarrow A^+ \quad (46a)$$

and/or:

$$A^\circ + AH^{\circ+} \rightarrow A^+ + AH \quad (46b)$$

$$A^+ + AH \rightarrow A_2H^+ \rightarrow \text{etc.} \quad (47a)$$

$$A^+ + Nuc \rightarrow A - Nuc^+ \rightarrow \text{etc.} \quad (47b)$$

$$2A^\circ \rightarrow A_2' \quad (48)$$

$$A^\circ + O_2 \rightarrow A^+ + O_2^{\circ-} \rightarrow \text{etc.} \quad (49a)$$

and/or:

$$A^\circ + O_2 \rightarrow A\text{-}OO^\circ \rightarrow \text{etc.} \quad (49b)$$

Scheme C2 Potential additional reactions involving A^+ (iminium) and A° (imine radical) [61–63] not considered in the general mechanistic **Scheme C1**; however, their interference may alter the ECL yields predicted on the basis of **Scheme C1**. Nuc represents any nucleophile different from AH, e.g. water leading to a ketone and another amine (viz., a dipropylamine for TPrA); this reaction is already included in **Scheme C1** as step (45), as was described in Refs. [21, 60]. Species A_2H^+ and A_2' are different dimers formed by a carbon–nitrogen or a carbon–carbon coupling, respectively. Note that, apart from consuming A° and A^+, the formation of A-Nuc^+, A_2H^+ and A_2' amounts to the production of new amine derivatives which may then take part in the ECL generation process. Lastly, it should be remarked that bioanalytical applications of ECL are often performed in non-thoroughly deaerated aliquots so that dioxygen may also interfere [64], presumably through spontaneous ET or coupling with A°, protonation by $AH^{\circ+}$, etc., thus leading to reactive oxygen species (ROS) and subsequent degradation of co-reactants. O_2 may also be involved in ET from the reduced form M^+ of the metal complex thus generating ROS and decreasing the ECL yield, although this is certainly a minor pathway relative to the reduction by/coupling with A° owing to the low M^{2+} concentration compared to that of AH

In the first ECL wave, the ET between $AH^{\circ+}$ and M^+ (formed by reduction of M^{2+} by A°) is the critical process. The continuous regeneration of M^{2+} occurs through three routes: M^+ oxidation (Eq. 38) and M^{3+} reduction (Eq. 43) at the electrode surface and M^{2+*} returning to the fundamental state in the solution bulk (Eq. 39). Therefore, even if the electrode potential is insufficiently high for direct oxidation of M^{2+}, its concentration is maintained (except for a tiny fraction stored dynamically as M^+ and M^{3+}), which allows significant ECL to be achieved even if the M^{2+} concentration is much smaller than that of AH. Thus, the mechanism in **Scheme C1** can be considered to depend essentially on the concentration fluxes of $AH^{\circ+}$ and A°. M^{2+} plays a vital part, but because of its constant regeneration, it acts rather like a catalyst. Consequently, the maximum intensity of the first ECL wave, and hence, the maximal production of M^{2+*} in a potential range sufficiently low that M^{2+} is not oxidizable at the electrode surface (Eq. 43), occurs near the AH oxidation peak. Beyond this point, the rate of $AH^{\circ+}$ and A° production decreases diffusionally with increasing potential, so the intensity of the first ECL wave also decreases in an almost diffusion-controlled manner.

This is seen in Fig. 9b, which shows the ECL wave of Fig. 9a split into its two components. Likewise, Fig. 9c, d highlight the origins of the ECL contributions in terms of the M^{2+*} concentration profiles, i.e. the reaction fronts giving rise to each ECL emission component (note that such splitting is impossible experimentally, but is readily achieved in numerical simulations). From Fig. 9c, it is seen that the first ECL wave originates from a bell-shaped emitter distribution, which remains adjacent to the electrode surface. The depletion of M^+ and M^{2+*} at the electrode surface, compensated by the production of M^{2+*}, which decreases diffusionally with distance from the electrode owing to the diffusion-reaction depletion of $AH^{\circ+}$ and A°, lead to this characteristic bell shape, perfectly explaining the reported effect of scan rate on the intensity of the first ECL wave. Indeed, an increase in scan rate, v, leads to an increase in the $AH^{\circ+}$ and A° fluxes because of the decrease in the diffusion layer thickness; this is also reflected in the $v^{1/2}$ dependence of the voltammetric current amplitude.

Despite all this, one cannot conclude that M^{2+} concentration is unimportant in this process. Its almost continuous recycling makes its diffusion practically irrelevant over this potential range, but the ECL intensity is controlled by its *local* concentration through the rate of Reaction (36). Note, however, that M^{2+} is not fully recycled owing to dynamic "storage" of its small fraction in the form of M^+ and some M^{3+}. Yet, the global balance equation for **Scheme C1** over this potential range is given by (γ is the ratio of bulk concentrations of M^{2+} and AH):

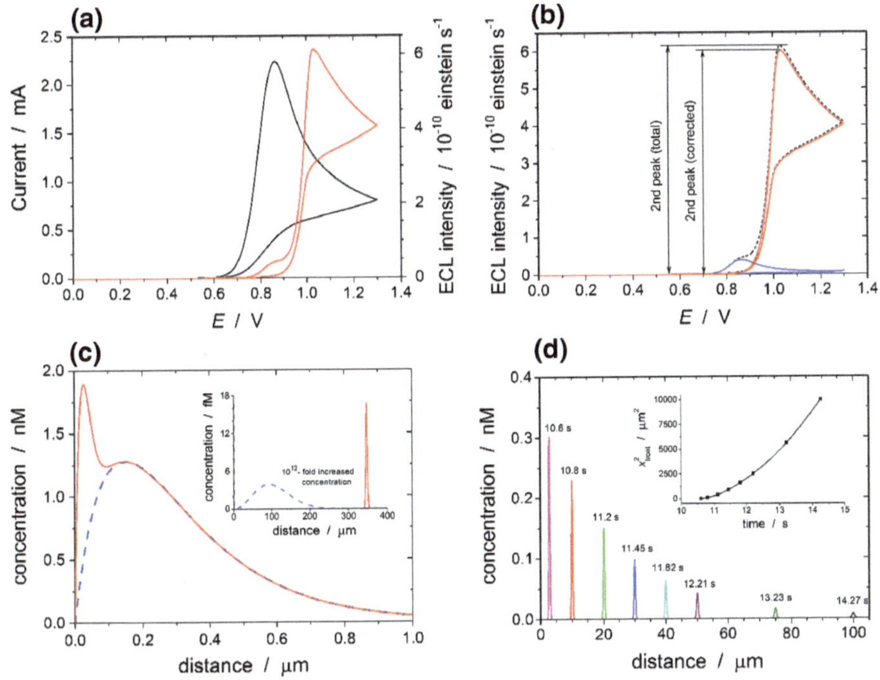

Fig. 9 **a** Current and ECL intensity for **Scheme C1** simulated for thermodynamic and kinetic parameters reported in Refs. [21, 60] for the classical TPrA/Ru(bpy)$_3^{2+}$ ECL system; $[AH]_{bulk} = 0.1$ M, $[M^{2+}]_{bulk} = 10^{-3}$ M, scan rate $v = 0.1$ V/s. **b** Two-wave decomposition of the ECL voltammogram from (**a**) (*dashed line*) into its components produced by Reaction (37) (*blue line*) and Reactions (42) and (44) (*red line*); *arrows* show the second ECL wave and indicate how its peak intensity can be evaluated from the baseline ("total") or from the "diffusion tail" of the first ECL wave ("corrected"); see also Fig. 10c. **c** Temporal and spatial variations of the two $M^{2+}*$ concentration profiles generated near the electrode surface and sustaining part of the total ECL emission; the contribution corresponding to the first ECL wave is extracted from the global profile as the blue curve; data taken at $E = 0.8645$ V under the same conditions as in (**a, b**). The *inset* shows the second reaction front (*sharp red curve*) propagating diffusionally into the solution when M^{3+} is produced at the electrode surface at the end of the forward scan under the same conditions as in (**a, b**); note that the features shown in the main figure (**c**) are seen compressed onto the ordinate axis owing to the change of abscissa scale. **d** Diffusional propagation of the reaction front ($M^{2+}*$ concentration profile) sustaining part of the second ECL wave intensity; times elapsed from the beginning of voltammetric scan are indicated above each curve; the *inset* in **d** represents the variations with time of the position of the maxima (squared abscissa values) of the diffusionally propagating $M^{2+}*$ concentration profiles; this provides evidence of the progressive diffusional displacement of the front towards the solution bulk. Adapted from Ref. [22]

$$AH - 2e \xrightarrow{[\gamma M^{2+}]} A^+ + H^+ + \varepsilon_5 \varphi h\nu \quad (E < E^{1/2}_{M^{2+/3+}}) \tag{50}$$

Thus, it is clear that the AH diffusion rate (and that of its derivatives) is the controlling factor in this system, and M^{2+} formally acts as a catalyst.

The situation changes drastically when the electrode potential is such that M^{2+} is oxidizable at the electrode surface. Now, rather than M^{3+} being reduced at the electrode surface, the opposite happens, and M^{2+} is oxidized to yield M^{3+}; Eq. (42) can therefore proceed efficiently, leading to the second ECL wave. The global balance equation is then given by:

$$\gamma\, M^{2+} + AH - (2+\gamma)e \rightarrow \gamma M^{3+} + A^{+} + H^{+} + \varepsilon_5 \varphi h\nu \quad (E > E^{1/2}_{M^{2+/3+}}) \quad (51)$$

in this potential range (note that ε is a function of ε_5 and ε_{12} which can be represented as $\varepsilon_5 + \varepsilon_{12}$ in a first approximation). Thus, the second ECL wave depends greatly on the speed at which M^{2+} is replenished, by diffusion, near the electrode surface where M^{3+} is generated. A second reaction front is then initiated by the latter species to produce M^{2+*} near the electrode surface through reaction (42) (see Fig. 9c). However, owing to the relative excess of M^{3+} versus M^{+} a significant fraction of the former is able to diffuse deeper into the solution, where it encounters a flux of M^{+} formed through reaction (36), the magnitude of which decays over time owing to the diffusion-reaction depletion of A°. Therefore, reaction (42) also takes place in the solution, within a reaction zone which travels diffusionally away from the electrode surface (Fig. 9d). Thus, the second ECL wave encompasses two components that depend crucially on the diffusion rate of M^{2+} and its related species, both featuring reaction (42), the role of which is practically negligible at potentials lower than that at which M^{2+} may be oxidized at the electrode surface. Therefore, the intensity of the second ECL wave should be influenced greatly by the diffusivity of M^{2+} and related species. The diffusion rate of AH and its derivatives also plays a role (in relation to the production of A°), but it is a minor role as the concentration of AH is generally around one hundred times higher than that of M^{2+}.

The fact that there are two ECL intensity waves if M^{2+} can be oxidized within the experimental potential window is now thoroughly explained by this analysis. Even more interesting is the insight that different diffusional conditions are involved in the formation of each wave since they are produced by different mechanistic sequences. It is thus evident that high diffusional fluxes of AH (first ECL wave) or of AH and M^{2+} (second ECL wave) should be sustained, which is in accordance with the dependence of the wave amplitude on the potential sweep rate [60]. On the other hand, the above analysis also establishes the strong but distinct influence on the ECL wave intensity from the values of diffusivities D_a (applying to all AH-derived species) and D_m (corresponding to all M^{2+}-related compounds) [65, 66]. This seems to be a significant finding (although typically ignored) since when M^{2+} is bound to large biomolecules in biomedical applications of ECL [67–71] the resulting compounds may be far less diffusive than it is usually assumed in typical voltammetric experiments rationalized by invoking the classical analyses of the TPrA/$Ru(bpy)_3^{2+}$ system. For instance, recent studies have suggested that the local concentration of M^{2+} may be enhanced by binding it to a soluble dendrimeric scaffold [72] or by depositing them onto nanoparticles (e.g. Ru(II) complexes on silica as done by Zanarini et al. [73]). Despite different aims behind them, both strategies resulted in significant enhancement of the first ECL wave occurring around $E_{\mathrm{AH}}^{1/2}$. Clearly, such dendrimers or nanoparticles are decorated with multiple M^{2+} complexes, which implies considerably lower diffusion coefficients than those pertaining to individual metal complexes and therefore lower than those of the amines. Because of this, these two experimental systems perfectly highlight the significance of being able to manipulate relative diffusional fluxes when optimising a system's ECL response.

Consequently, the following sections focus on the quantification of the effect of different relative values of D_a and D_m on the two-wave light emission response corresponding to the two mechanistically distinct ECL generation sequences.

4.2 *Effect on Each ECL Wave of Reducing D_m While Keeping D_a Constant*

A variety of approaches to changing the diffusivities of AH and/or M^{2+} and of species deriving from them can be envisaged in order to shift the positions and strength of the reaction fronts and hence alter the important fluxes presented in Fig. 9c, d. This can be achieved by joining AH and/or M^{2+} (as well as their related species) to large molecules without affecting the chemical structure of their active sites which determine the kinetics and thermodynamics of reactions in **Scheme C1**. Because of this, the published thermodynamic and kinetic parameters for the classical $Ru(bpy)_3^{2+}$/TPrA system [22] were adopted throughout the subsequent quantitative evaluation accomplished on the basis of numerical computations by KISSA [14–22]. However, at the beginning, we present some qualitative considerations with respect to the dependence of each ECL wave intensity on the values of D_a and D_m. These are based on the deliberation of the reaction fronts that essentially shape the ECL response of the system defined in **Scheme C1**.

Because the product of AH oxidation effectively plays the role of a redox catalyst for the reduction of M^{2+}, one does not anticipate to see a significant effect of decreasing D_m with D_a kept at the same value. The efficiency of this redox catalysis depends almost exclusively on the diffusional transport of AH, particularly given that the concentration of AH is approximately one hundred times higher than that of M^{2+}. This also clarifies the fact that the oxidation of M^{2+} does not reveal itself in experimental voltammograms in such circumstances. On the other hand, a considerable enhancement of the first ECL wave intensity is expected upon decreasing D_m while keeping D_a constant. Indeed, in this case, owing to its reduced diffusivity, most of M^+ remains close to the electrode, which boosts the rate of reaction (37). Therefore, the first ECL wave intensity is expected to rise when D_m is reduced until it becomes mainly governed by the value of D_a.

A qualitative analysis of the second ECL wave intensity proves to be not as straightforward. This is partly due to the fact that the second ECL wave is superimposed onto the diffusional "tail" of the increasing first wave (Fig. 9b and inset of Fig. 10b). Additionally, when D_m is low, M^{2+} cannot be replenished quickly enough near the electrode surface where it is consumed through reaction (43). Because of this, the intensity of the second ECL wave (measured from the diffusional tail of the first one) is expected to decrease with diminishing D_m/D_a ratio. These opposite effects preclude any definitive qualitative conclusions about the absolute peak ECL intensity of the second wave (i.e. as measured from the baseline; see Fig. 9b and inset of Fig. 10b). Therefore, numerical simulations by KISSA [14–20] are employed to confirm these qualitative inferences in a quantitative manner.

Thus, the above conclusion that D_m/D_a ratio does not tangibly affect the electrochemical current is verified by Fig. 10a. Indeed, reducing D_m by over three orders of magnitude results in a current variation of just about 5%. On the contrary, decreasing D_m/D_a ratio leads to a dramatic amplification of the first ECL peak, as envisioned from qualitative considerations, which sees it develop from an insignificant pre-wave at $D_m = 5 \times 10^{-6}$ cm^2 s^{-1} (Fig. 9a, b) to the main ECL feature at $D_m = 10^{-9}$ cm^2 s^{-1} (Fig. 10b). At the same time, the amplitude of the second ECL peak (as measured from the diffusional extrapolation of the first wave) drops considerably as predicted above. On the other hand, the absolute amplitude of the second wave (i.e. measured from the baseline) stays unaffected by the value of D_m except for extremely minute values of D_m/D_a resulting in its sharp decrease eventually leading to its complete disappearance so that only the first ECL wave remains (this limiting case is not illustrated as this requires unfeasibly small values of D_m unless M^{2+} is immobilized on a molecular structure that is "transparent" to the diffusion of amines [74–77]). These variations of the ECL intensity are presented in Fig. 10c.

Fig. 10 Simulated current **a** and ECL **b** intensities under the same conditions as in Fig. 9 as a ▶ function of the diffusion coefficient of all metal complexes: $D_m = 5 \times 10^{-6}$ (*black*), 5×10^{-7} (*red*), 10^{-7} (*green*), 5×10^{-8} (*magenta*), 10^{-8} (*cyan*) and 10^{-9} (*blue*) cm^2 s^{-1}; $D_a = 5 \times 10^{-6}$ cm^2 s^{-1}, $v = 0.1$ V s^{-1}, $[AH]_{bulk} = 0.1$ M, $[M^{2+}]_{bulk} = 10^{-3}$ M. The *arrows* in **a** and **b** show the directions of variation upon decreasing the metal complex diffusion coefficient (see text). **c** Variations of peak ECL intensity versus the ratio of diffusion coefficients. Corrected peak values (*green curve*) are determined as schematized in the inset and Fig. 9b. See Ref. [22] for the method used in **c** to separate the different components due to each ECL wave, and for the values of the kinetic and thermodynamic parameters used in simulations based on **Scheme C1**. Adapted from Ref. [22]

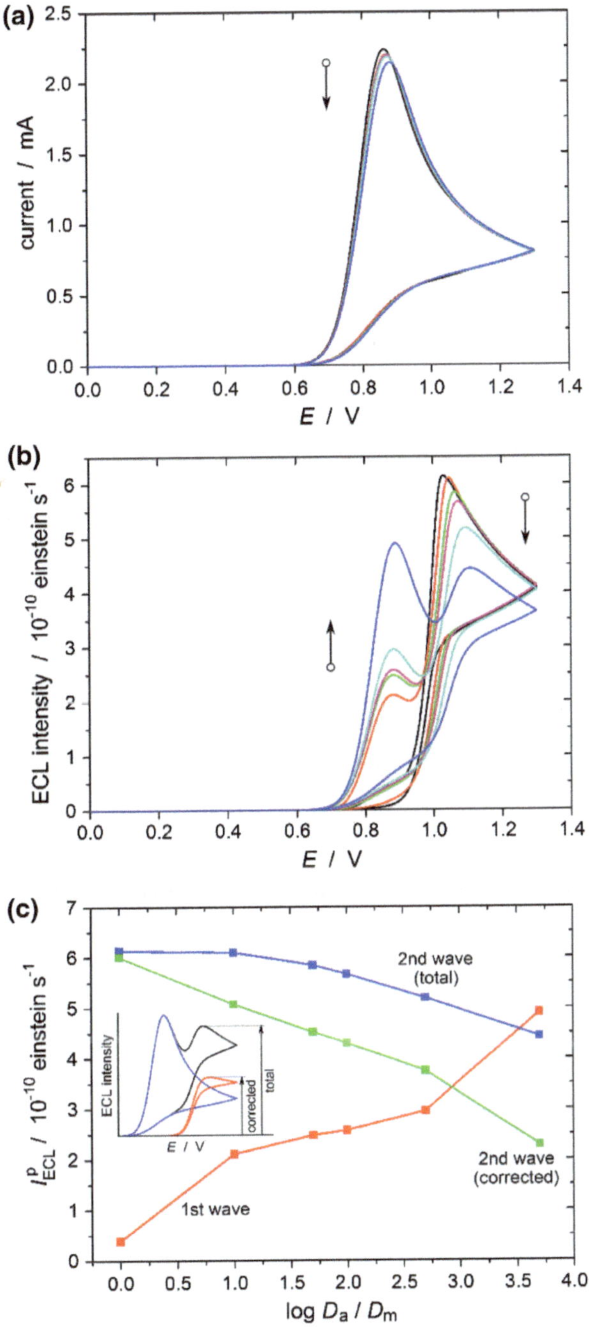
(a)
current / mA
E / V
(b)
ECL intensity / 10^-10 einstein s^-1
E / V
(c)
I^p_ECL / 10^-10 einstein s^-1
2nd wave
(total)
2nd wave
(corrected)
1st wave
ECL intensity
E / V
total
corrected
log Da / Dm

4.3 Effect of Reducing D_a While Keeping D_m Constant on Each ECL Wave

In this case, both the voltammetric current and the ECL intensity ought to decrease according to the global balance Eq. (50) (before M^{2+} can be oxidized at the electrode) and Eq. (51) (when M^{2+} is oxidizable at the electrode) because both depend on the diffusion rates of amine-related species which are in great excess ($\sim$100-fold) with respect to M^{2+} under typical experimental conditions. Computational results reported in Fig. 11a, b are in perfect agreement with this reasoning. Remarkably, the voltammetric current and second ECL wave amplitudes reduce at the same rate but that of the first ECL wave drops considerably faster so a moderate decrease in D_a/D_m results in this wave becoming negligible. This is another indication of the role the diffusion rate of amine-related species plays in the formation of the first wave ECL flux originating from the reaction front adjacent to the electrode surface (Fig. 9c). On the other hand, the intensity of the second ECL wave decreases to a lesser extent (Fig. 11c) since it is mainly commanded by the diffusion of the metal complex M^{2+} and species deriving from it. Thus, since AH is in 100-fold excess, the reaction front (Fig. 9d) travelling away from the electrode surface and leading to an observable second ECL wave can be sustained even for values of D_a as low as to confine the oxidation product of AH and the related intermediates to the very vicinity of the electrode.

4.4 Conclusions

The presented computational results have revealed that decreasing D_m without altering D_a leads to a substantial increase in the first ECL wave intensity and a simultaneous decline of the second wave. This essentially indicates that the amine oxidation is the main driver of the first ECL wave while M^{2+} is constantly recirculated. On the contrary, M^{2+} is the primary promoter of the second ECL wave, which disappears as soon as D_m becomes too small to ensure continuous production of M^{3+} at the electrode.

These insights point out that for a candidate ECL system to show better performance than the canonical $Ru(bpy)_3^{2+}$/TPrA case any structural alterations of the amine (e.g. to avoid parasitic reactions in **Scheme C2**, or to adjust the kinetics of the cation radical $AH^{\circ+}$ to optimize the system) should not lead to a considerable decrease in D_a. On the other hand, the first ECL wave can be enhanced by decreasing D_m, for instance, by binding the transition metal complexes onto large dendrimeric formations or particles [72, 73]. Owing to the fact that this is the important wave in the majority of bioanalytical or biomedical ECL applications this latest observation is highly significant.

Finally, it can be concluded that, in view of the essential influence of diffusivities on the outcome of bioanalytical ECL detection, analysis of solution voltammetric

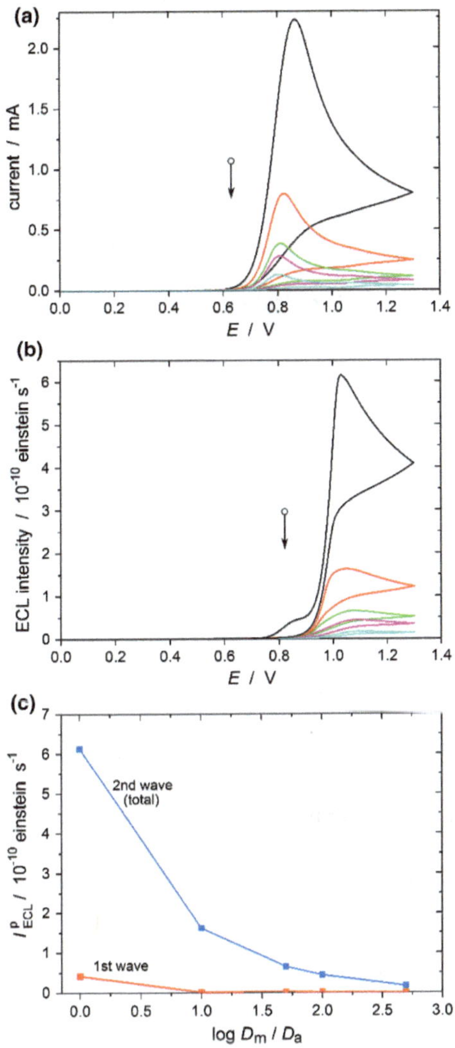

Fig. 11 Simulated current **a** and ECL **b** intensities under the same conditions as in Fig. 9 as a function of the diffusion coefficient of all amine-related species: $D_a = 5 \times 10^{-6}$ (*black*), 5×10^{-7} (*red*), 10^{-7} (*green*), 5×10^{-8} (*magenta*), and 10^{-8} (*cyan*) $cm^2\ s^{-1}$; $D_m = 5 \times 10^{-6}\ cm^2\ s^{-1}$, $v = 0.1\ V\ s^{-1}$, $[AH]_{bulk} = 0.1$ M, $[M^{2+}]_{bulk} = 10^{-3}$ M. The *arrows* in **a** and **b** show the direction of variation upon decreasing the metal complex diffusion coefficient (see text). **c** Variations of the peak ECL intensity with the ratio of diffusion coefficients. See Ref. [22] for the method used in **c** to separate the different components due to each ECL wave, and for the values of the kinetic and thermodynamic parameters used in simulations of **Scheme C1**. Adapted from Ref. [22]

experiments requires extra caution. Indeed, experimental setup for bioanalytical detection typically involves freely diffusing amines while the transition metal complex is immobilized using biological recognition methods involving, e.g. an "electrode-antibody-analyte-antibody-M^{2+}" assembly [67–72]. Thus, M^{2+} becomes very concentrated in a narrow range at a distance of several tens of nanometres from the electrode surface. Its diffusion coefficient, D_m, is then practically zero and any ECL response is made possible by simultaneous presence of $AH^{\circ+}$ and A° (both ensuing from the oxidation of amine at the electrode surface and diffusing away from it) at appropriate concentrations for reactions (35)–(37) to proceed. Work is ongoing in our team in order to examine theoretically this peculiar situation.

5 Conclusion and Perspectives

In this chapter, we have tried to evidence the big theoretical difficulties arising when one wants to predict ECL flux intensities correctly. These difficulties are not specific to ECL but are certainly made extremely acute under situations leading to efficient ECL generation.

Indeed, in any ECL mechanism, the excited species responsible for the ECL light emission results from the fast endergonic reaction of two reactants and decays rapidly through a photon emission. Under most frequent situations, because of the dynamics of diffusion-reaction pathways, the two reactant species are segregated in to two adjacent zones of the diffusion layer near the electrode. Hence, they may react only within a thin solution strip separating these two solution zones, thus creating a sharp reaction front in which the photon-emitting species is formed through a high production flux and is consumed almost as fast. The most significant difficulty is that, because of the interplay between diffusion and convection, this sharp reaction front moves into the diffusion layer, so its very position at any moment is hard, or impossible, to predict. As a result, predicting or rationalizing experimental ECL fluxes requires high computational precision to determine the time-dependent sharp concentration profile of the emitting species, so as to obtain excellent accuracy in the determination of its spatial integral, the value of which is proportional at any time to the measured ECL flux [21, 22].

This strong difficulty inherent in ECL situations is complicated further by the high curvatures of the concentration profiles of electroactive species' imposed by diffusional transport near the edges of electrodes of finite sizes. In other words, the correct prediction or rationalization of the outcome of such electrochemical experiments requires the use of extremely dense simulation grids both in space and in time. This ideal scope is naturally impossible to fulfil with present computers unless prohibitive computational times and memory occupation are allowed. It is clear that this cannot be the case in common practice.

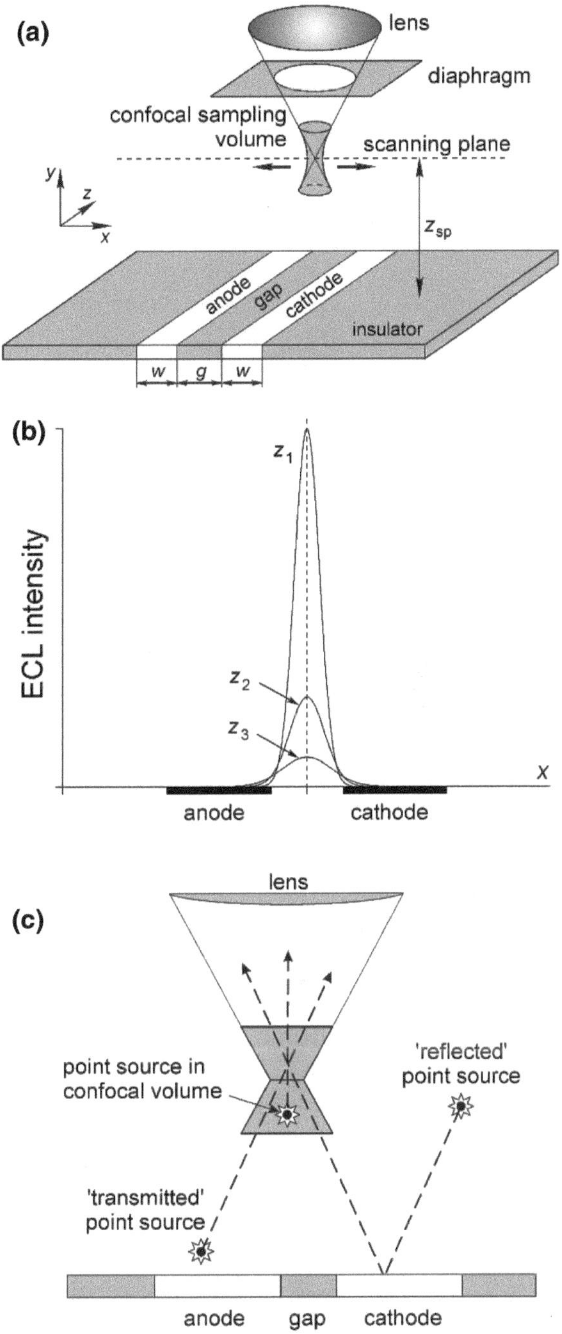

Fig. 12 Principle of confocal microscopy imaging of ECL emitted at a double band microelectrode assembly. **a** General principle of the optical arrangement used when sampling at a constant height z from the electrodes/insulator plane, **b** schematic predicted pattern of ECL intensity expected under the conditions shown in **a** at three different constant heights $z_1 < z_2 < z_3$ [23, 24], and **c** schematized representation of the two additional distortions incurred in confocal microscopy probing of the spatial distribution of ECL point sources besides that due to the large aperture. Adapted from Ref. [24]

We have shown in this chapter that the general computational approaches that we derived in devising our software KISSA automatically adapt to these local spatial–temporal difficulties and circumvent them. This powerful ability rests on the combination of calculations in conformal spaces in order to eliminate and smooth any difficulties created by the existence of strongly curved flux lines near any spatial discontinuity, with an automatic enhancement of spatial precision through the dynamic refinement of spatial simulation grids over solution ranges traversed by reaction fronts, so that the simulation grids automatically adapt to treat the local competition between extremely fast reactions.

The validity and the success of this approach, even compared to calculations performed using COMSOL Multiphysics, has been established in several instances in this chapter, with a particular focus on ECL reactions, but also for more classical kinetic schemes in molecular electrochemistry. Nonetheless, all these presentations were cast within a classical electrochemical perspective, i.e. when it is essential to predict the precise outcome of actual experiments either to rationalize or to optimize them. Basically, in ECL, this requires highly precise predictions of the time variations of the integrals of the concentration profile of the ECL-emitting species, since this is the only experimental measurement available under classical electrochemical conditions.

However, thanks to the development of confocal spectroscopies [23, 24, 78, 79], one may think of obtaining direct access to the spatial measurement of the very extent of the micrometric spatial volumes within which light is emitted, which is tantamount to accessing the spatial distribution of the emitting species. However, it is clear that, despite the micrometric resolution of confocal microscopy, such measurements cannot provide an exact mapping of the light generation areas. Indeed, they are necessarily distorted owing to the fact that the emitting species experiences drastic spatial gradients over submicrometric distances. Furthermore, any light beam emitted from solution domains located far from the confocal microscope focal point but, owing to geometrical optical paths, able to pass through the microscope diaphragm will be detected. This evidently includes direct light beams but also those reflected by surfaces such as those of polished electrodes (see Fig. 12c). It should thus be easily conceived that despite their importance, measurements based on confocal microscopy are not a panacea, but may only report distorted information.

In this final and prospective section we wish to illustrate how our present approach can be used to reconstruct more precise data about the spatial repartition of the emitting species based on measured confocal microscopic data. For this purpose, to illustrate the performance of our numerical approaches for solving such "inverse problems," we rely on ECL generation based on **Scheme A** as disclosed in Sect. 3, in which the oxidation and reduction of the substrate A occur simultaneously at two different parallel band electrodes (see Fig. 2c, d) [4–8, 24, 52, 80]. Evidently, one should also conceive that other distortions may affect the confocal pattern, such as self-interferences due to the beams emitted by the point source and mirage effects due to solution composition gradients when they are sufficient to induce significant optical index gradients [78, 79], but those, being specific to each set of given experimental conditions, will not be considered in the following.

Ideally, if light samples were collected from infinitely small volumes, the confocal picture would report the pointwise density of photon emission, viz.:

$$I_{\text{ECL}}^{\text{point}} = N_A \Phi_{\text{ECL}} k_f c^*, \tag{52}$$

where N_A is Avogadro's number, and c^* is the local concentration of the singlet state species A^* formed in Eq. (3), which is the source of the photon emission in Eq. (4). Indeed, for a point sampling, the transmitted and reflected intensities would be negligible owing to the two small solid angles concerned (compare Fig. 12c). Conversely, when the sampling volume is not negligible, as occurs experimentally, the confocal microscope integrates the photons emitted collectively by all points located in the confocal sampling volume along the directions that may be collected through the confocal diaphragm aperture. Hence, the importance of the directionality of photon emission means that the confocal intensity does not result from a simple integration of the point source intensities such as in Eq. (52), but that each point source intensity has to be weighted statistically by its solid angle, as imposed by its position within the confocal sampling volume. One can easily appreciate the complexity of the problem and the fact that the corresponding prediction of the spatial distribution of c^* is highly critical. In addition, as evident from Fig. 12c, the confocal aperture also collects those photons transmitted or reflected by point sources located far away from the confocal volume. In other words, contrary to classical confocal measurements, the sampling volume may be viewed as being nearly infinite and limited only by the extent of the solid angles through which the confocal diaphragm "neck" is "seen" by each point source present in this near-infinite volume (see Fig. 13).

Figure 14 compares the effect of such distortions as measured experimentally (Fig. 14b) to the pattern predicted theoretically (Fig. 14a) in the absence of such distortions as it would be recorded by an ideal confocal microscope with a point aperture, i.e. according to Eq. (52). One can easily observe that even at the closest heights the distortions are very substantial. Conversely, Fig. 14c represents the theoretically predicted pictures at different sampling heights when taking into account the above distortions. It is readily seen that up to a height of ca. 40 μm, there is a very good concordance between the patterns observed experimentally in Fig. 14b and reconstructed in Fig. 14c, and that both differ from the real pattern in Fig. 14a. Above 40 μm, the experimental recording (Fig. 14b) deviates significantly from that reconstructed in Fig. 14c upon considering the above distortions, essentially by showing a lower intensity in the symmetry plane and the presence of additional "wings" apparently located above the insulating materials. The reduction in intensity at a large distance from the electrode plane is expected since the reconstruction in Fig. 14c does not take into account the effect of natural convection [20, 35–39], the effect of which is to eliminate a significant flux of A^+ and A^-

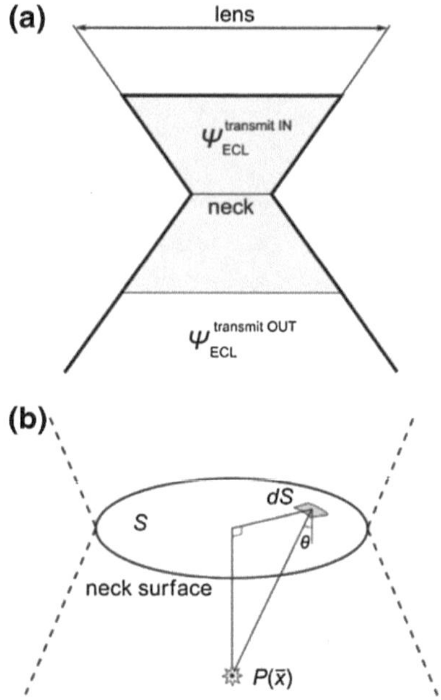

Fig. 13 **a** Schematic representation of the spatial origin of the two transmitted types of light beams detected by a confocal microscope exploring ECL; note that, to refer to classical confocal sampling, the ECL intensities due to point sources located within the confocal sampling volume ($\Psi_{ECL}^{transmit\ IN}$) and outside it but nonetheless collected ($\Psi_{ECL}^{transmit\ OUT}$) are indicated. Note that $\Psi_{ECL}^{transmit\ OUT}$ represents the sum of the light intensity transmitted directly or after reflection from the mirror like surfaces of polished electrodes (compare Fig. 12c). **b** Geometric description of the elementary solid angle $d\Xi(\bar{x})$ subtended by a differential surface element dS of the confocal volume neck defined in (a). The "neck" surface is defined by the cross-section of smallest area within the confocal sampling volume. Adapted from Ref. [24]

versus that predicted based on diffusion alone. Hence, the formation of A* by the ET annihilation of A^+ and A^-, which have travelled along excessively long diffusion paths from the electrode surfaces [20, 35–39], is impeded. The presence of the "wings" is more cumbersome but is certainly due for the most part to light reflected by the insulating white material flanking the two band electrodes, an additional distortion not considered in the model.

However, for our purpose in this published work, it was sufficient to establish the performance of our numerical approaches by reconstructing with a high precision the light intensity pattern monitored experimentally through confocal microscopy. In other words, as evidenced in Sects. 2 and 3 of this chapter, our approaches not only provide exceptional precision in predicting experimental ECL fluxes emitted during the course of classical ECL voltammetric or amperometric experiments as embodied in KISSA [14–20], but also in rationalizing and integrating the effect of

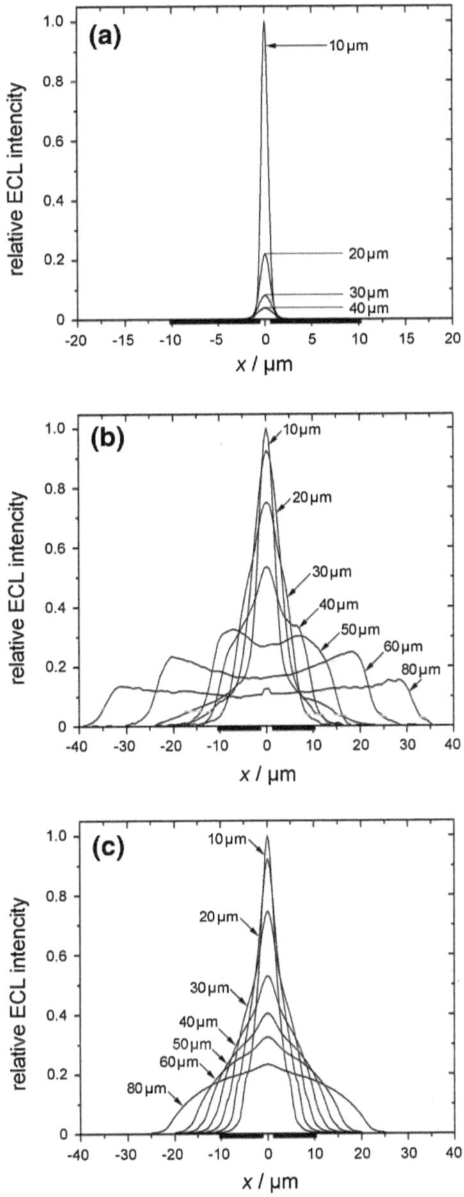

Fig. 14 **a** Predicted ECL intensities recorded at different sampling heights (see Fig. 12a) according to the mechanism shown in **Scheme A** by an ideal confocal microscope. **b** Experimental measurements performed with a confocal microscope [23, 24]. **c** Reconstructed effect of the distortions due to the confocal sampling as described in Figs. 12 and 13 based on the intensity patterns shown in (a) [23]. Adapted from Ref. [24]

optical distortions when one tries to map experimentally the very locations of ECL emissions [21, 22].

Acknowledgements This work was supported by CNRS UMR8640 PASTEUR (CNRS, ENS and UPMC) and the French Ministry of Research and Higher Education as well as by ANR Chaires d'excellence (CHEX) 2010, ANR-10-CHEX-0012, MicroNanoChem.

References

1. Santhanam, K.S.V., Bard, A.J.: J. Am. Chem. Soc. **87**, 139–140 (1965)
2. Feldberg, S.W.: J. Am. Chem. Soc. **88**, 390–393 (1966)
3. Faulkner, L.R., Bard, A.J..: Techniques of electrogenerated chemiluminescence in electroanalytical chemistry. In: Bard, A.J. (eds.), vol. 10, pp. 1–95. Marcel Dekker, New York (1977)
4. Amatore, C.: Electrochemistry at ultramicroelectrodes in physical electrochemistry: principles, methods and applications. In: Rubinstein, I. (ed.), chap. 4, pp. 131–208. Marcel Dekker, New York (1995)
5. Bartelt, J.E., Drew, S.M., Wightman, R.M.: J. Electrochem. Soc. **139**, 70–74 (1992)
6. Amatore, C., Fosset, B., Maness, K.M., Wightman, R.M.: Anal. Chem. **65**, 2311–2316 (1993)
7. Collinson, M.M., Pastore, P., Maness, K.M., Wightman, R.M.: J. Am. Chem. Soc. **116**, 4095–4096 (1994)
8. Pastore, P., Magno, F., Collinson, M.M., Wightman, R.M.: J. Electroanal. Chem. **397**, 19–26 (1995)
9. Bard, A.J., Faulkner, L.R.: Electrochemical Methods: Fundamentals and Applications. John Wiley & Sons, New York (2002)
10. Miao, W.: Chem. Rev. **108**, 2506–2553 (2008)
11. Yin, X.-B., Dong, S., Wang, E.: TrAC-Trend Anal. Chem. **23**, 432–441 (2004)
12. Pyati, R., Richter, M.M.: Annu. Rep. Prog. Chem., Sect. C: Phys. Chem. **103**, 12–78 (2007)
13. Liu, Z., Qi, W., Xu, G.: Chem. Soc. Rev. **44**, 3117–3142 (2015)
14. Amatore, C., Klymenko, O.V., Svir, I.: Electrochem. Commun. **12**, 1170–1173 (2010)
15. Amatore, C., Klymenko, O.V., Svir, I.: Electrochem. Commun. **12**, 1165–1169 (2010)
16. Klymenko, O.V., Oleinick, A., Svir, I., Amatore, C.: Rus. J. Electrochem. **48**, 593–599 (2012)
17. Klymenko, O.V., Svir, I., Oleinick, A., Amatore, C.: ChemPhysChem **13**, 845–859 (2012)
18. Amatore, C., Klymenko, O.V., Svir, I.: Electrochim. Acta **56**, 4422–4423 (2011)
19. Klymenko, O.V., Svir, I., Amatore, C.: Electrochem. Commun. **12**, 1378–1382 (2010)
20. Amatore, C., Klymenko, O.V., Svir, I.: Anal. Chem. **84**, 2792–2798 (2012)
21. Klymenko, O.V., Svir, I., Amatore, C.: ChemPhysChem **14**, 2237–2250 (2013)
22. Svir, I., Oleinick, A., Klymenko, O.V., Amatore, C.: ChemElectroChem **2**, 811–818 (2015)
23. Amatore, C., Pébay, C., Servant, L., Sojic, N., Szunerits, S., Thouin, L.: ChemPhysChem **7**, 1322–1327 (2006)
24. Amatore, C., Oleinick, A., Klymenko, O.V., Thouin, L., Servant, L., Svir, I.: ChemPhysChem **8**, 1664–1676 (2007)
25. Rudolph, M., Reddy, D.P., Feldberg, S.W.: Anal. Chem. **66**, 589A (1994)
26. Ketter, J.B., Forry, S.P., Wightman, R.M., Feldberg, S.W.: Electrochem. Solid. St. **7**, E18–E22 (2004)
27. Comsol Multiphysics, COMSOL, Inc., Burlington, MA
28. Cutress, I.J., Dickinson, E.J.F., Compton, R.G.: J. Electroanal. Chem. **638**, 76–83 (2010)
29. Dickinson, E.J.F., Ekström, H., Fontes, E.: Electrochem. Commun. **40**, 71–74 (2014)
30. Sartin, M.M., Shu, C., Bard, A.J.: J. Am. Chem. Soc. **130**, 5354–5360 (2008)
31. Svir, I.B., Oleinick, A.I., Klimenko, A.V.: J. Electroanal. Chem. **513**, 19–125 (2001)
32. Svir, I.B., Oleinick, A.I.: J. Electroanal. Chem. **499**, 30–38 (2001)
33. Kukoba, A.V., Bykh, A.I., Svir, I.B.: Fresenius' J. of Anal. Chem. **368**, 439–442 (2000)
34. Svir, I.B., Klimenko, A.V., Compton, R.G.: Radiotekhnika **118**, 92–101 (2001)
35. Amatore, C., Szunerits, S., Thouin, L., Warkocz, J.-S.: J. Electroanal. Chem. **500**, 62–70 (2001)
36. Amatore, C., Szunerits, S., Thouin, L., Warkocz, J.S.: Electroanalysis **13**, 646–652 (2001)
37. Amatore, C., Pebay, C., Scialdone, O., Szunerits, S., Thouin, L.: Chem. Eur. J. **7**, 2933–2939 (2001)
38. Baltes, N., Thouin, L., Amatore, C., Heinze, J.: Angew. Chem. Int. Ed. **43**, 1431–1435 (2004)
39. Amatore, C., Knobloch, K., Thouin, L.: J. Electroanal. Chem. **601**, 17–28 (2007)

40. Klymenko, O.V., Svir, I., Amatore, C.: J. Electroanal. Chem. **688**, 320–327 (2013)
41. Klymenko, O.V., Svir, I., Amatore, C.: Mol. Phys. **112**, 1273–1283 (2014)
42. Oleinick, A., Amatore, C., Svir, I.: Electrochem. Commun. **6**, 588–594 (2004)
43. Amatore, C., Oleinick, A.I., Svir, I.: J. Electroanal. Chem. **597**, 69–76 (2006)
44. Amatore, C., Oleinick, A., Svir, I.: J. Electroanal. Chem. **564**, 245–260 (2004)
45. Amatore, C., Oleinick, A., Klymenko, O.V., Delacôte, C., Walcarius, A., Svir, I.: Anal. Chem. **80**, 3229–3243 (2008)
46. Amatore, C., Oleinick, A., Svir, I.: Electrochem. Commun. **6**, 1123–1130 (2004)
47. Amatore, C., Oleinick, A.I., Svir, I.B.: J. Electroanal. Chem. **553**, 49–61 (2003)
48. Amatore, C., Oleinick, A., Svir, I.: J. Electroanal. Chem. **575**, 103–123 (2005)
49. Amatore, C., Oleinick, A., Svir, I.: Electrochem. Commun. **5**, 989–994 (2003)
50. Amatore, C., Oleinick, A.I., Svir, I.: J. Electroanal. Chem. **597**, 77–85 (2006)
51. Amatore, C., Oleinick, A.I., Svir, I.: Anal. Chem. **81**, 4397–4405 (2009)
52. Fosset, B., Amatore, C.A., Bartelt, J.E., Michael, A.C., Wightman, R.M.: Anal. Chem. **63**, 306–314 (1991)
53. Amatore, C., Oleinick, A.I., Svir, I.: ChemPhysChem **11**, 149–158 (2010)
54. Amatore, C., Oleinick, A.I., Svir, I.: ChemPhysChem **11**, 159–174 (2010)
55. Amatore, C., Savéant, J.-M.: J. Electroanal. Chem. **85**, 27–46 (1977)
56. Amatore, C.: Principles and methods. Basic concepts in organic electrochemistry. In: Baizer, M., Lund, H. (eds.), chap. 2, pp. 11–119. M. Dekker, New York (1991)
57. Amatore, C.: Organic electrochemistry, fifth edition: revised and expanded. In: Hammerich, O., Speiser, B. (eds.), chap. 1, pp. 3–96; chap. 10, pp. 371–394. CRC Press, Taylor & Francis Group, Boca Raton (2016)
58. Michael, P.R., Faulkner, L.R.: J. Am. Chem. Soc. **99**, 7754–7761 (1977)
59. Elangovan, A., Kao, K.-M., Yang, S.-W., Chen, Y.-L., Ho, T.-I., Su, Y.O.: J. Org. Chem. **70**, 4460–4469 (2005)
60. Miao, W., Choi, J.-P., Bard, A.J.: J. Am. Chem. Soc. **124**, 14478–14485 (2002)
61. Mariano, P.S., Stavinoha, J., Bay, E.: Tetrahedron **37**, 3385–3393 (1981)
62. Andrieux, C.P., Savéant, J.M.: J. Electroanal. Chem. **26**, 223–235 (1970)
63. Ames, J.R., Brandange, S., Rodriguez, B., Castagnoli Jr., N., Ryan, M.D., Kovacic, P.: Biorg. Chem. **14**, 228–241 (1986)
64. Zheng, H., Zu, Y.: J. Phys. Chem. B **109**, 12049–12053 (2005)
65. Scott, A.M., Pyati, R.: J. Phys. Chem. B **105**, 9011–9015 (2001)
66. Zheng, L., Chi, Y., Dong, Y., Zhang, L., Chen, G.: J. Phys. Chem. C **112**, 15570–15575 (2008)
67. Adcock, J.L., Barrow, C.J., Barnett, N.W., Conlan, X.A., Hogan, C.F., Francisa, P.S.: Drug Test. Analysis **3**, 145–160 (2011)
68. Yuan, L., Zhou, L., Li, J., Shi, L., Chen, L., Huang, C., Yanac, Z., Cai, Q.: Anal. Methods **5**, 3626–3630 (2013)
69. Joshi, T., Barbante, G.J., Francis, P.S., Hogan, C.F., Bond, A.M., Gasser, G., Spiccia, L.: Inorg. Chem. **51**, 3302–3315 (2012)
70. Xu, X.H.N., Jeffers, R.B., Gao, J., Logan, B.: Analyst **126**, 1285–1292 (2001)
71. Bergmann, F, Cysewski, R., De Cola, L., Dziadek, S., Fernandez Hernandez, J.M., Josel, H., Seidel, C.: Patents Numbers (ROCHE DIAGNOSTICS GMBH): WO2014019709-A2; WO2014019709-A3; WO2014019707-A2; WO2014019707-A3; WO2014019711-A1; WO2014019708-A1
72. DeCola, L.: P. Abstract of ECL meeting 2014. Bertinorro (Italy)
73. Zanarini, S., Rampazzo, E., Della Ciana, L., Marcaccio, M., Marzocchi, E., Montalti, M., Paolucci, F., Prodi, L.: J. Am. Chem. Soc. **131**, 2260–2267 (2009)
74. Zeng, Q., McNally, A., Keyes, T.E., Forster, R.J.: Electrochem. Commun. **10**, 466–470 (2008)
75. Devadoss, A., Spehar-Délèze, A.M., Tanner, D.A., Bertoncello, P., Marthi, R., Keyes, T.E., Forster, R.J.: Langmuir **26**, 2130–2135 (2010)
76. Brennan, J.L., Keyes, T.E., Forster, R.J.: J. Electroanal. Chem. **662**, 30–35 (2011)

77. Edwards, D.A.: J. Phys. Chem. C **117**, 6747–6751 (2013)
78. Amatore, C., Bonhomme, F., Bruneel, J.L., Servant, L., Thouin, L.: Electrochem. Commun. **2**, 235–239 (2000)
79. Amatore, C., Bonhomme, F., Bruneel, J.L., Servant, L., Thouin, L.: J. Electroanal. Chem. **484**, 1–17 (2000)
80. Svir, I.B., Oleinick, A.I., Compton, R.G.: J. Electroanal. Chem. **560**, 117–126 (2003)

Applications of Electrogenerated Chemiluminescence in Analytical Chemistry

Neso Sojic, Stéphane Arbault, Laurent Bouffier and Alexander Kuhn

Abstract The great success of electrogenerated chemiluminescence (ECL) in analytical chemistry can be measured by the widespread use of the technology in different fields, ranging from basic research to commercial clinical and biological applications. Indeed, this remarkable readout method offers intrinsic advantages by comparison with other transduction methods: high sensitivity, extremely wide dynamic range, and insensitivity to matrix effects. In addition, its versatility allows exploiting various types of biomolecular interactions and therefore to detect specifically targeted analytes of biological interest such as proteins, nucleic acids, and enzymatic substrates. Numerous assay formats, biosensors, or analytical strategies with new ECL labels or with label-free approaches have been proposed by using nanostructured materials: carbon nanotubes, metal or doped nanoparticles, graphene, carbon dots, quantum dots, or ultrathin films. The development of analytical ECL has also been fueled by discovering novel luminophores and efficient co-reactants and also by deciphering the complexity of the ECL mechanisms at the minute scale. The combination of ECL with microfluidics, paper-based materials, bipolar electrochemistry, and portable miniaturized devices has led to various intriguing and promising analytical applications.

1 Introduction

Electrogenerated chemiluminescence, also referred to as electrochemiluminescence (ECL), belongs to the family of luminescence phenomena in which the transition of an excited state of a luminophore (an atom or a molecule) to a state of lower energy is accompanied by the emission of photon [1, 2]. Such light-emitting events are classified according to the nature of the process providing the energy to reach the excited state. For example, in photoluminescence (i.e., fluorescence or phosphorescence),

---

N. Sojic (✉) · S. Arbault · L. Bouffier · A. Kuhn
Institut des Sciences Moléculaires, CNRS UMR 5255, University of Bordeaux,
Bordeaux INP ENSCBP, Pessac, France
e-mail: Neso.Sojic@enscbp.fr

F. Miomandre and P. Audebert (eds.), *Luminescence in Electrochemistry*,
DOI 10.1007/978-3-319-49137-0_8

photoexcitation occurs by the absorption of photons by the luminophore. In chemiluminescence, the energy source is provided by the homogeneous chemical reactions between at least two reactive species. ECL is a specific case where these reactive intermediates are produced electrochemically at the electrode surface. Then, they undergo a chemical step characterized by a highly exergonic electron-transfer reaction generating the excited state of the luminophore. Light emission occurs during the relaxation to the ground state. So ECL is initiated by an electrochemical step, continues with a homogeneous chemical reaction, and ends with a photochemical step. It differs from electroluminescence which is the radiative recombination of electrons and holes in a material, usually a semiconductor. In addition, ECL can be considered as the competition between a light-emitting and a non-radiative transition in the Marcus inverted region [3]. Indeed, the energetic electron-transfer reaction occurring during the second step can actually favor excited-state formation over the ground-state formation. Therefore, ECL evidences the existence of an inverted Marcus region for electron-transfer reactions.

Since ECL combines electrochemical addressing and orthogonal optical detection modalities, it has attracted large interest in different fields such as photochemistry, electrochemistry, and analytical chemistry [4–9]. Due to its interdisciplinary nature and also to its remarkable characteristics, ECL has shown great achievements in fundamental research but also for practical applications, mainly in the diagnostic market with immunoassays [10–13]. For analytical applications, the information is contained in the light signal, and its intensity is directly proportional to the concentration of a limiting reactant involved in the ECL process. For example, the luminophore could be used as a label in a biorecognition chain, such as in immunoassays. Modern detectors can easily measure light intensity at very low levels which allows the development of ultrasensitive analytical methods based on ECL technology.

Compared with other sensing methods used in bioassays, ECL obviously does neither need any radio-isotope tracer nor optical excitation (photoluminescence-based methods), which avoids background from scattered light or sample auto-fluorescence. It just requires a photodetector and a potentiostat, which is a fairly light instrumentation favorable for its integration with fluidic approaches within high-throughput systems. From an analytical point of view, it offers many intrinsic advantages. First, extremely low detection limits can be reached for biomolecules (picomoles), in particular since ruthenium-based labels can be regenerated in situ at the electrode surface. As a result, many photons are produced per measurement cycle, which greatly enhances the sensitivity of the ECL technique. The most extensively investigated ECL luminophore is tris(2,2′-bipyridyl) ruthenium(II), $Ru(bpy)_3^{2+}$, which can be considered as an ECL standard. Second, if the luminophore is used as a label in the presence of an excess of co-reactant [14], an extremely wide dynamic range greater than six orders of magnitude may be achieved [15]. Third, since ECL is initiated by the application of a suitable potential at an electrode surface, it allows the time, the duration (application of the potential), and position (localized near the electrode surface) of the light-emitting reactions to be precisely controlled, which can improve the signal-to-noise ratio. In addition, the control of the luminescence process in space enables multiplexed analyses using electrode arrays and imaging techniques. Fourth, ECL-active labels such as $Ru(bpy)_3^{2+}$ derivatives can be easily conjugated to biomolecules (e.g., proteins, peptides, and oligonucleotides) by coupling their amino reactive groups to the bipyridine—N-hydroxysuccinimide (NHS) ligands of the label. Fifth, ECL allows the simultaneous measurement of two experimental parameters versus potential (i.e., light intensity and Faradaic current), and thus a great selectivity and control over the light-emitting reactions. Combining both parameters yields also important insights in the ECL mechanisms. Sixth, it provides a fast time response, typically a few seconds. Finally, the reagents are very stable, and the reactive intermediates are electrogenerated in situ.

2 Principles of ECL

As already mentioned, ECL is the light emitted from an electronically excited state which is produced by energetic electron-transfer reactions. The earliest ECL reactions occurred by the annihilation reaction between species electrogenerated by pulsing the electrode potential [16–18]. However, they were limited to aprotic solvents. After the initial reports with aromatic hydrocarbons, the study of the remarkable electrochemical and ECL properties of the water-soluble $Ru(bpy)_3^{2+}$ luminophore was reported by Bard et al., and it represented a major breakthrough in the field [19, 20]. The reasonable stability of the oxidized form, $Ru(bpy)_3^{3+}$, in water paved the way to the discovery of ECL in aqueous phase. In the early 1980s, ECL was generated in aqueous solutions containing both $Ru(bpy)_3^{2+}$ and a co-reactant such as oxalate by applying a single anodic potential [21]. It opened a new area for ECL. Indeed, the large majority of ECL bioanalytical applications are based on a mechanism involving the reaction of a sacrificial co-reactant species with the luminophore. After a first report on chemiluminescence emission obtained by mixing $Ru(bpy)_3^{3+}$ and tri-*n*-propylamine (TPrA) [22], Leland and Powell described the use of TPrA as a very efficient co-reactant that generates strong ECL signals in aqueous solutions [23]. In the early 1990s, ECL bioassays were reported that led to the development of commercial ECL systems for immunoassays [10, 11, 24]. Considering all these developments, ECL mechanisms can be separated into two dominant pathways: the annihilation pathway and the co-reactant pathway.

2.1 Annihilation Pathway

In the annihilation pathway, an oxidized and a reduced species are electrochemically produced at the electrode surface by applying alternating pulsed potentials. Then, these two species react together according to an annihilation reaction to generate the electronically excited state of the luminophore. It relaxes to the ground state and emits a photon. Depending on the available enthalpy of the annihilation reaction and of the energy levels of the luminophore, the possible mechanism for the ECL annihilation pathway can be presented as follows:

$$A - e^- \rightarrow A^{\cdot +} \quad \text{(oxidation at the electrode)} \tag{1}$$

$$A + e^- \rightarrow A^{\cdot -} \quad \text{(reduction at the electrode)} \tag{2}$$

$$A^{\cdot +} + A^{\cdot -} \rightarrow (2 - \alpha)A + \alpha {}^1A^* \quad \text{(formation of the singlet excited state)} \tag{3}$$

or

$$A^{\cdot +} + A^{\cdot -} \rightarrow (2 - \beta)A + \beta {}^3A^* \quad \text{(formation of the triplet excited state)} \tag{4}$$

$${}^3A^* + {}^3A^* \rightarrow (2 - \gamma)A + \gamma {}^1A^* \quad \text{(triplet-triplet reaction)} \tag{5}$$

$${}^1A^* \rightarrow A + h\nu \quad \text{(ECL emission)} \tag{6}$$

If the homogeneous electron-transfer reaction between the electrogenerated species is exergonic enough, the annihilation process leads directly to the emitting singlet excited state of the luminophore (Eqs. 1–3). In this case, the system has sufficient available energy and the emitted ECL follows the singlet route, also called "S-route." But, if the enthalpy of the annihilation reaction is insufficient to populate directly the singlet state, the system may follow the "T-route" where the triplet state is first produced (Eq. 4). For such an "energy deficient system," triplet–triplet annihilation may eventually yield the singlet emitting state in a second step (Eq. 5). Both routes have been extensively studied with various systems [25, 26].

2.2 *Co-reactant Pathways ("Reductive–Oxidative," "Oxidative–Reductive")*

An alternative mode of ECL generation with more complex mechanistic pathways is based on the use of sacrificial co-reactants (e.g., oxalate, TPrA, peroxydisulfate, benzoyl peroxide (BPO),…). A co-reactant can be defined as a chemical species that, upon electrochemical oxidation or reduction, produces very reactive intermediates capable to react with the oxidized or reduced luminophore to generate the desired excited state. The corresponding mechanisms are often referred to as "oxidative–reductive" ECL and "reductive–oxidative" ECL, respectively [4]. The method is very simple, because it requires just the application of a single potential step. An example of an "oxidative–reductive" co-reactant is the oxalate ion that was the first ECL co-reactant discovered in water [21]. The luminophore $Ru(bpy)_3^{2+}$ and the co-reactant $C_2O_4^{2-}$ are both oxidized at the electrode surface. Then, upon bond cleavage, $C_2O_4^{\cdot-}$ forms a strong reducing species ($CO_2^{\cdot-}$) that reduces $Ru(bpy)_3^{3+}$ and generates $Ru(bpy)_3^{2+*}$, which emits light. Thus, ECL mechanism of oxalate is often referred to as an "oxidative–reductive" route because the sequences of the electron-transfer reactions are as follows: (1) The electrochemical oxidation at the electrode surface and (2) the reduction of the oxidized luminophore by the radical. Another example of an "oxidative–reductive" system is the model $Ru(bpy)_3^{2+}$/TrPA system which is considered as a standard in ECL. ECL mechanism of $Ru(bpy)_3^{2+}$ with amine-based co-reactants [14, 27] depends on a great variety of parameters: nature of the co-reactant, electrode material, solvent, pH, presence of surfactant in the solution, respective concentrations of the co-reactant and of the ruthenium complex, and hydrophobicity of the electrode surface. For example, at high concentrations of $Ru(bpy)_3^{2+}$ (> 0.1 mM), the "catalytic route" (also called "EC" route) is the dominant process for ECL [28–30]. Along this path, the catalytic oxidation of TPrA occurs by a reaction with electrogenerated $Ru(bpy)_3^{3+}$. In the commercialized bead-based immunoassays, the luminophore is used as a label and is immobilized on the bead surface. In this case, ECL follows the "revisited" route where only TPrA is oxidized at the electrode surface [31]. The resulting radicals, $TPrA^{\cdot+}$ and $TPrA^{\cdot}$, diffuse to react with the luminophore [32], as detailed in the Sect. 4 and also in the Chapter "Theoretical Insights in ECL" by Amatore et al. Most of the co-reactant ECL reports are based on the "oxidative–reductive" scheme because it operates efficiently in water. Indeed, it just requires imposing a simple anodic potential and also oxygen evolution is very slow in aqueous solutions. Therefore, it does not interfere significantly with the ECL generation.

For the "reductive–oxidative" route, the situation is completely different in aqueous solutions because very cathodic potentials are required. Thus, hydrogen evolution dominates and the electrogenerated species decompose too rapidly to obtain a stable ECL process. However, in pure organic solvents or mixed acetonitrile/water solutions, strong ECL intensity may be generated by using peroxydisulfate or BPO as co-reactants. First, the co-reactant is reduced simultaneously with the luminophore at the surface of the electrode. After dissociation, the peroxydisulfate or BPO co-reactants produce strong oxidants $SO_4^{\cdot}$ or $Ph\text{-}CO_2^{\cdot}$, respectively, which oxidize the reduced luminophore to generate the excited state. Finally, radiative de-excitation generates the ECL emission.

To select efficient co-reactants giving strong ECL intensity, a number of criteria should be fulfilled: solubility, low oxidation or reduction potentials, stability, toxicity, kinetics, quenching effects, low ECL background, adequate redox potentials, and sufficient lifetimes of the radicals.

3 Analytical Strategies

As stated in the previous section, ECL is intrinsically based on a sequential combination of electron transfers, chemical steps, and light emission. Three key elements can be potentially tuned in order to achieve ECL: the nature of the luminescent moiety, the interface where the electron-transfer steps take place, and the chemical mechanism allowing the formation of the excited state. This section will focus on the main analytical possibilities that were historically reported to design ECL-sensing strategies.

3.1 Co-reactant ECL (Co-reactant is the Analyte)

ECL could be promoted according to either annihilation or co-reactant pathways. By employing the same luminophore, various co-reactants can be used in order to generate ECL following reductive–oxidative or oxidative–reductive mechanisms. The amount of light generated at the electrode–solution interface directly depends on the concentration of the dye [33] but also of the co-reactant [23]. Therefore, any ECL-active co-reactant can be potentially detected and quantified as its concentrations will directly influence ECL intensity.

For example, the determination of oxalate in aqueous solutions was reported as early as in 1983 with an ECL intensity varying linearly over a concentration range between 1 and 100 μM [34]. Such concentration values are compatible with the level typically found in normal human blood or urine samples. In a more recent study, the quantification of oxalate was directly performed from urine and plasma samples by combining reversed-phase ion pair HPLC with an ECL readout [35]. The detection limit was found to be below 1 μM corresponding to an injection of only 25 pmol. The calibration curve for oxalate was linear throughout the entire clinical range, and the concentration values were compared with a conventional enzymatic method with a discrepancy lower than 1%.

Peroxydisulfate concentration was also measured accurately when used as a co-reactant for $Ru(bpz)_3^{2+}$ cathodic ECL (with bpz = 2,2′-bipyrazine). The method is highly specific with an impressive linear range over six orders of magnitude of concentration between 1 nM and 1 mM when a rotating disk is used as the working electrode [36]. Pyruvate was equally determined in a comparable manner with $Ru(bpy)_3^{2+}$ anodic ECL in the presence of cerium nitrate, increasing the diversity of possible ECL analytes [37]. The limit of detection (LOD) was found to be about 0.3 μM (equivalent to 27 ppb), and the interference of various chemicals such as EDTA and citrate was eliminated because of their presence in potential samples from fermentation media.

Tertiary and secondary amines can be quantified without chemical modification or derivatization by taking advantage of ECL reaction, typically with $Ru(bpy)_3^{2+}$

used as an effective luminophore. However, many tertiary amine compounds are capable to produce intense ECL, whereas others do not generate significant light emission. In that context, a remarkable structure/activity relationship study was published a decade ago by Knight and Greenway [38]. The key chemical features have now been clearly identified in order to rationalize the corresponding ECL proficiency. First, the presence of a hydrogen atom on the carbon in position α is essential for the radical formation regardless the mechanistic pathway. Electron-withdrawing groups such as carbonyl, hydroxyl, or halogen located on the amine will destabilize the radical intermediate and therefore drop down ECL emission. The presence of electron-donating substituents will contrarily favor the radical stabilization and increase the resulting ECL activity. This is in fact a very classic reactivity trend based on the stabilization of the radical intermediates which hinders subsequent reactivity and thus reduces ECL accordingly.

Relevant examples illustrating this structure/activity relationship could be briefly mentioned. A direct comparison of the ECL emission recorded with mono-, di-, and tri-propylamine reveals a marked increase from 4 to 48 and 660, respectively (given in arbitrary units) [39]. This implies that changing from a tertiary to a primary amine could potentially lower ECL by a factor up to 165. Also, the following trend Phenyl > Hydroxy > Chloro in the ethylamine series is well explained by the electron-withdrawing ability. Indeed, phenylethylamine is almost four times more efficient than chloroethylamine which directly correlates with the Hammett substituent constant [40]. Another more quantitative method to predict ECL efficiency of various alkylamines was also previously reported based on the redox potential value of the reducing intermediate involved in the charge-transfer steps leading to the excited state [22]. The reader should be aware that the comparison between co-reactant and the prediction of relative ECL intensity based on the corresponding molecular structure should be made with caution. Indeed, several effects that may be antagonist may influence the ECL strength like the respective oxidizing/reducing power of the reactive intermediates as well as their relative stability. This is why a comparison between structurally related molecules seems only reasonable especially with complicated co-reactants exhibiting several functional groups. For example, the superior ECL capability of 2-(dibutylamino)ethanol compared to TPrA was nicely reported by Liu et al. [27], and the enhancement is due to the promotion of amine oxidation thanks to the presence of the hydroxyethyl group.

The most common anesthetic drugs exhibit a chemical structure combining a tertiary amine and an aromatic moiety connected together either by an amide or an ester bonding (Fig. 1). Thus, the presence of the amine moiety makes them suitable for ECL determination. The ECL activity directly matches the chemical structure of the anesthetics. Indeed, bupivacaine (also known under the brand name Marcaine) exhibits the higher ECL intensity (100% a.u.) thanks to its long alkyl groups, whereas the signal recorded for lidocaine and tetracaine, in the same conditions, steadily decreases to 63 and 16% because of the shorter ethyl and methyl groups, respectively [38].

It is also noteworthy that many NADH-dependent enzymatic processes have been measured by ECL because the NADH coenzyme contains an amine group

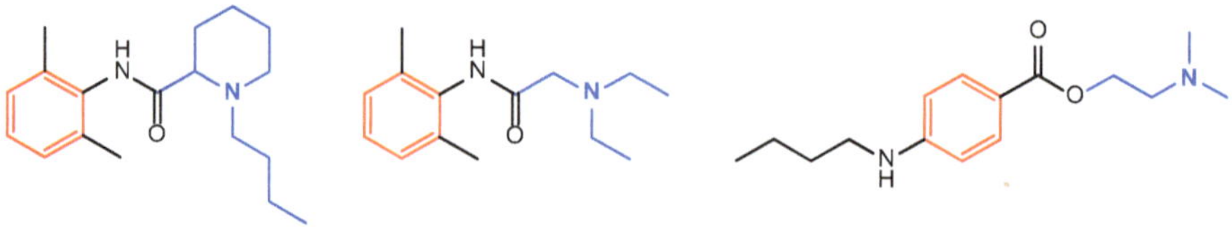

Fig. 1 Chemical structures of local anesthetics with the aromatic ring drawn in *red* and the tertiary amine in *blue* [38]

which can act as a co-reactant for $Ru(bpy)_3^{2+}$ ECL, whereas NAD^+ does not behave as a co-reactant [41]. Such strategies will be detailed below in Sect. 4.

3.2 *Luminophore ECL (ECL Labels)*

It was already mentioned that the ECL intensity depends on the concentrations of both emitter and co-reactant. However, almost all practical applications in biomedical diagnostics are performed under experimental conditions where the ECL intensity is directly proportional to the luminophore amount. Also, the only ECL dye used in commercial assays is the $Ru(bpy)_3^{2+}$ complex which exhibits a strong ECL efficiency. It is readily available and can be further chemically functionalized. In the latter cases, $Ru(bpy)_3^{2+}$ is covalently attached to one of the biochemical species involved in the recognition process which is the so-called ECL labeling strategy. Then, the ECL is promoted in the presence of a large excess of co-reactant (generally TPrA) in order to be only sensitive to the dye content either in solution or surface-immobilized structures. In this strategy, the emitter is chemically bound to the analyte which allows a proper quantification. ECL is a very useful toolbox for molecular diagnostics which necessitates low-concentration detection of various bioanalytes and markers, especially when diseases are at an early stage of development. Many different biomolecules need to be detected including DNA, RNA, peptides, proteins, or specific antibodies. Therefore, the two most classic chemical labeling processes include the presence of an activated ester such as NHS linked to the ruthenium complex or a phosphoramidite conjugate (Fig. 2). The former reacts with primary amines to form an amide bond, which is useful in the case of biomolecules containing amino acids, whereas the latter allows incorporating the label into oligonucleotides.

The detection of DNA with ECL labels as well as the development of immunoassays has been established in practical clinical diagnostic more than 25 years ago [11]. In fact, the use of ECL labeling based on ester bond formation through NHS activation was first reported in DNA assays back in 1991 [42]. This opened the door to numerous applications for the rapid detection of polymerase chain reaction–amplified products from specific genes or viruses. More recently,

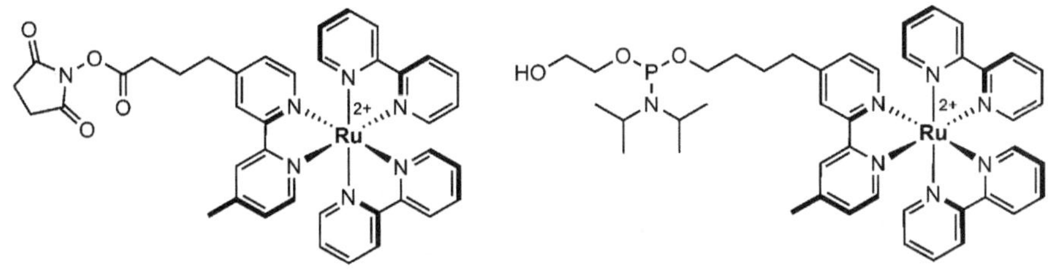

Fig. 2 Chemical structures of $Ru(bpy)_3{}^{2+}$ luminophore conjugated with an activated ester (*left*) or a phosphoramidite (*right*) for ECL labeling of various biomolecules [7]

ECL labeling was also used to evaluate the surface coverage of biosensors, especially for immobilized DNA and protein [43]. The strategy adopted in some recent contributions in the field of DNA biosensors combines ECL labeling and detection with supramolecular assembly such as hairpin [44] or aptamer folding [45] for the selective detection of small molecular targets.

The transduction of antigen/antibody recognition is by far the largest application of ECL. This strategy used generally the classic $Ru(bpy)_3{}^{2+}$/TPrA system where the dye is used as a label to report the immune affinity. This was applied to a wide variety of assays which relate the emitting label to analyte concentration such as β-amyloid peptide, insulin, interleukins, or various cancer antigen markers. This field was reviewed in detail by several authoritative reviews from Richter [7], Miao [4], and Liu et al. [9] and will be discussed in Sect. 4.

3.3 NP-Based ECL

The promotion of ECL at the surface of nanoparticles is an emerging and very promising field. The general idea is to position semi-conducting particles at the electrode surface to allow the injection of excitons either in the valence band or conduction band of the quantum dots (QDs). Several mechanistic pathways can be followed by using appropriate co-reactants in order to generate either anodic or cathodic ECL (Fig. 3). This area of ECL has been extensively reviewed recently by the group of Wu et al. [46] and Zhao et al. [47]. Therefore, in this subsection, the attention will be solely focused on the historical publications.

The very first contribution was introduced by Bard with silicon nanocrystals with a size ranging between 2 and 4 nm in diameter [48]. Indeed, the electrochemical gap between the onset of electron and hole injection expands with decreasing particle size. The ECL spectra recorded in solution feature a significant redshift compared to photoluminescence (PL) with a maximum centered at 640 nm against 420 nm, respectively. This first seminal work was followed by a series of contributions exemplifying the versatility with plenty of different QDs such as 3 nm in diameter CdSe prepared from Cd-acetate and Se powder [49]. The same typical

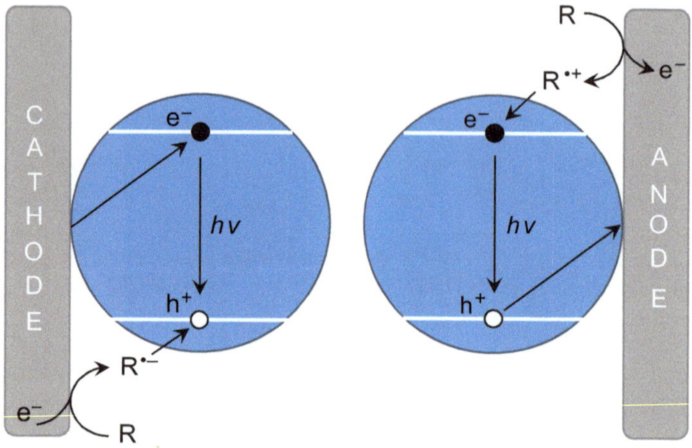

Fig. 3 Scheme of ECL promotion with QDs. The excitons (*holes and electrons*) are injected by electrode and co-reactant following the cathodic or anodic pathways [46]. Reprinted with permission from Ref. [46]. Copyright 2014 American Chemical Society

redshift of ~200 nm with respect to PL spectrum was assigned to the implication of surface states in the emission process. CdSe was also used later as an ECL-active nanocrystal film employed for sensing applications in aqueous solution. This was illustrated for the detection of H_2O_2 co-reactant with a typical LOD of 10^{-7} M [50]. ZnSe [51], CdTe [52] or CdS [53] have also been reported. These QDs are either used alone or in combination with other classes of materials in order to prepare composites with tunable properties or enhanced surface area (see next section). QDs have then been used to prepare modified electrodes and tested as ECL biosensors for low-density lipoprotein (LDL). The LDL concentration was measured through the decrease in ECL signal resulting from the binding to a specific receptor with an outstanding 6 ng L^{-1} LOD [53]. Another analytical strategy involving ECL energy transfer to analytes was also demonstrated [54]. In the latter case, the energy transfer takes place from the excited CdTe QDs to quencher catechols such as dopamine or adrenalin as model analytes with a linear response of 4 orders of magnitude from μM to nM range, typically.

Apart from these transition metal-based QDs, carbon-based semiconducting systems have also been described. ECL spectra of carbon nanocrystals released from graphite were measured with a maximum wavelength of both anodic and cathodic processes centered at 535 nm [55]. Most of the time, expensive techniques such as laser ablation or proton beam irradiation are employed, as well as harsh chemical degradation approaches from various carbon sources. However, it is noteworthy that a facile microwave-assisted pyrolysis synthesis of fluorescent and ECL-active carbon nanoparticles has been performed in a matter of minutes [56]. The size is adjusted by the synthesis duration and reasonable monodispersity was achieved with particles exhibiting between 2 and 4 nm in diameter. Two-color graphene QDs with a greenish-yellow luminescence were reported for the very first time recently [57]. The ECL mechanistic pathway was studied together with an application to Cd^{2+} sensing based on ECL quenching by cysteine as masking agent.

Noble metal clusters constituted by a few atoms have also been recently investigated by Hesari et al. [58]. Such ultrasmall chemical entities exhibit properties that are close to a single molecule behavior. Several clusters such as Au_8, Au_{25}, and Au_{38} have been discussed in the literature. In particular, the ECL emission is remarkable in the near-infrared region with a fundamental mechanistic study which is now well-established [59]. It is also noticeable that, for the Au_{38}/TPrA system, the ECL yield is 3.5 times higher than the reference $Ru(bpy)_3^{2+}$/TPrA one [60]. The cathodic ECL of Au_{25} nanoclusters with $K_2S_2O_8$ as co-reactant deposited on an ITO electrode was used for the detection of dopamine, and this contribution opened the door to new promising material platforms for ECL biosensors [61]. Finally, the photophysical properties of silver nanoclusters consisting of only a few metal atoms were also reported. They exhibit both tunable photo- and electro-chemiluminescence such as solvatochromism because of a dramatic sensitivity to the local chemical environment [62].

3.4 Nanomaterials ECL

The development of efficient and sensitive ECL platforms for chemical or biochemical sensing became regularly more sophisticated thanks to the design of complex modified surfaces and architectures [63]. This area is currently rapidly expanding as it combines and takes benefit of surface-immobilized structures based on nanoparticles, nanotubes, and polymer or metal complexes thin films.

The initial works were simply done by immobilizing $Ru(bpy)_3^{2+}$ alone [64] or inside Nafion on an electrode surface for the preparation of a regenerable ECL-active layer [40, 65]. Later, the combination of Nafion-doped sols was found to show even better properties. The amount of cationic luminophore exchanged into these films strongly depends on the proportion of each chemical ingredient in the mixture. The ECL response was studied with TPrA or sodium oxalate as co-reactant, and greater ECL emission was recorded in the composite Nafion/silica films when compared to Nafion only [66]. Such a strategy was generalized from SiO_2 to TiO_2 sol/gel with enhanced ECL capability [67]. The optimal composition ratio for ECL was found to be 50% Nafion content and the corresponding modified electrode was studied for co-reactant titration with typically μM (oxalate) and sub-μM LOD (TPrA). The ECL sensor was also combined with HPLC for the determination of erythromycin in human urine samples. Direct incorporation of $Ru(bpy)_3^{2+}$ dye inside SiO_2 NPs casted onto electrode surfaces provides a new generation of multilayer ECL-active films [68]. The corresponding films showed improved linearity of the signal as a function of the analyte concentration and enhanced stability due to the multilayer architecture and high-active surface area.

This sensor was then further optimized to achieve up to 3 nM LOD for TPrA which is 3 orders of magnitude lower than the benchmark Nafion-based ECL sensor with long-term stability [69].

QDs have been extensively developed in combination with several forms of nanomaterials in order to obtain synergetic effects. For example, a novel strategy was developed by combining CdSe nanocrystals, carbon nanotubes, and chitosan biopolymer [70]. A label-free ECL sensor for the detection of human immunoglobine antibody (IgG) was built up. All these ingredients put together allow biocompatibility and more importantly a large surface area generating a 20-fold higher ECL intensity for enhanced sensitivity. The interaction between surface plasmons of Au metal nanoparticles and ECL-active semiconducting particles such as CdS QDs was also evidenced. This was illustrated with CdS thin films for the ultrasensitive detection of thrombin aptamer [71]. The multifunctional system showed up to a 5-fold ECL signal increase when compared to that without the incorporation of Au NPs with an outstanding linearity range from 0.1 to 100 fM. The amplification of ECL based on a QDs platform for the sensing of antioxidants was achieved with graphene oxide [72]. In fact, the latter facilitated the promotion of excitons inside CdTe QDs and triggered the generation of reactive oxygen species, thus leading to a large ECL enhancement. Glutathione peptide was selectively detected at μM concentration, even in the presence of thiol-containing competitors such as cysteine. Another system constituted by a polymer-protected graphene with Au and CdSe nanoparticles composite was successfully combined for the detection of IgG [73]. The advantage of plasmonic particles with an improved surface area offered an impressive linearity response over 6 orders of magnitude with a LOD of 5 pg L^{-1} for ultrasensitive protein detection. The same trend was equally observed with functionalized graphene and gold nanorods multilabeled with an enzyme such as glucose oxidase (GOx) and secondary antibody [74]. Luminol was used as an ECL-active luminophore, and the whole system was applied to the detection of prostate specific antigen (PSA) protein, a biomarker of prostate cancer. The multifunctional platform works through a cascade of amplifications of luminol ECL in the presence of glucose and oxygen and pg mL^{-1} of PSA is detectable in such conditions.

To summarize, the design of surface-modified electrodes combining intimately nanomaterials and plasmonic effects [75–77] is very promising for point-of-care diagnostics, especially in the context of clinical screening of trace biomarkers.

3.5 Bipolar ECL Detection

Bipolar electrochemistry (BPE) has attracted in the last few years an exponentially increasing interest in the scientific community due to some unique features, which can also be of primary importance in the context of ECL. The concept is less well known than classic electrochemistry where a conventional three-electrode setup, composed of a working, counter, and reference electrodes, is used [1]. In such a

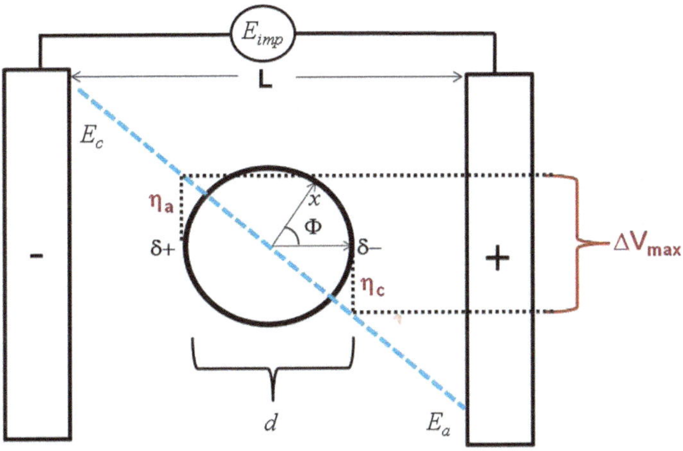

Fig. 4 Polarization of a conducting object. Scheme illustrating the 2D projection of the polarization of a spherical conducting object located between two feeder electrodes in a solution

normal electrochemical setup, the electrode of interest is the working electrode, where, depending on its polarization with respect to the solution, either an oxidation or a reduction reaction occurs. For BPE, things are slightly different, because both oxidation and reduction reactions occur simultaneously on the same electrode, which is not physically connected to a power supply through an electric contact.

A straightforward way to generate a bipolar electrode is to place a (semi)conducting object in a solution in which an electric potential gradient exists, generated by two feeder electrodes which are not in physical contact with the immersed object. As the conducting object is by definition equipotential, it will experience an inhomogeneous potential difference with respect to the solution. This will automatically lead to a situation where some parts of the object are more likely to undergo an oxidation reaction, whereas other sections will become preferential sites for a reduction reaction (Fig. 4).

As a consequence of the presence of this electric field, individual anodic and cathodic polarisation potentials η_a and η_c will be established at every point of the object. Depending on the localisation x on the surface of the object, these potentials vary and can be calculated as:

$$\eta_x = \mathrm{E}\frac{d}{2}\cos\Phi \qquad (7)$$

for a spherical object, with d being the object diameter and $\mathrm{E} = E_{imp}/L$ the electric field. As shown in Fig. 4, the maximum polarization potential difference occurs between the extremities of the object, and its value, $\eta_a - \eta_c = \Delta V_{max}$, can be calculated as:

$$\Delta V_{max} = \varepsilon d \tag{8}$$

This value directly induces the reactivity at the extremities of the polarized interface which can be used for many different applications [78, 79], and among others also to generate ECL. In terms of application of this concept to the detection of analytes, three main routes can be followed:

- either the analyte is simultaneously also a co-reactant of the ECL reaction (direct bipolar ECL reporting via oxidation of the analyte)
- or the analyte is detected by a reduction reaction, which is coupled through the bipolar electrode to the oxidative ECL reaction (indirect bipolar ECL reporting)
- or the analyte is indirectly involved in the ECL process such as, for example, by quenching the excited state

Manz and co-workers explored for the first time the possibility to use BPE for triggering ECL by following the direct detection route [80]. In their experiments, a photodetector measured remotely the ECL emission generated at a bipolar electrode, and it was shown that the electrochemical activity is directly related to the ECL intensity (Fig. 5a, b). This fits perfectly the requirements for performing analytical BPE, since an electrochemical reaction occurring on the bipolar object is correlated with, and transduced into, an optical signal. The second indirect route has been mainly investigated by Crooks and co-workers during the last decade and

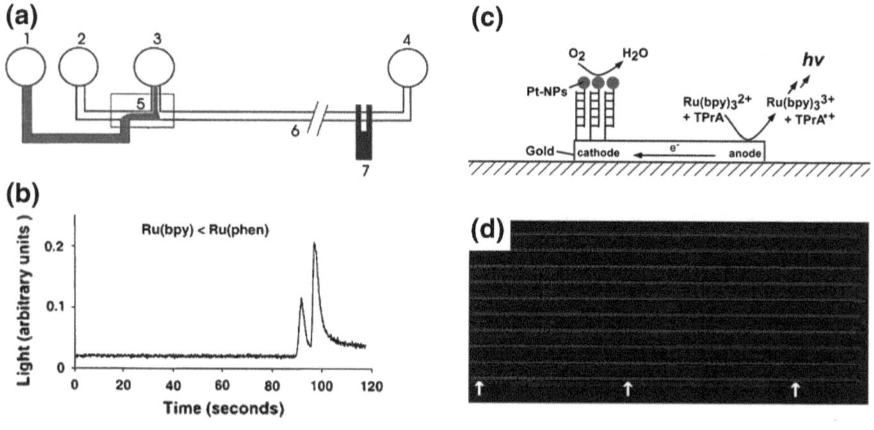

Fig. 5 Experimental illustration of the difference between direct and indirect bipolar ECL reportings. **a** Layout of a microfluidic device: *1* sample reservoir; *2* buffer reservoir; *3* sample waste; *4* buffer waste; *5* double-T injector; *6* separation channel; *7* bipolar detection electrode. **b** Electropherograms for $Ru(bpy)_3^{2+}$ and $Ru(phen)_3^{2+}$ separation with direct bipolar ECL detection [80]. **c** Scheme of indirect bipolar ECL reporting of DNA hybridization based on the coupling of the Pt-catalyzed reduction of oxygen on one side with the oxidation of the ECL ingredients on the other side of a bipolar electrode [81]. **d** A typical array of bipolar ECL electrodes with a density of 2000 recording elements per cm^2 [82]. Adapted with permission from Refs. [80–82]. Copyright 2009 American Chemical Society

established them as the leader of BPE-based sensing in general due to their pioneering work describing various bipolar configurations for such applications [78]. In the reported experiments, they used reduction reactions for the sensing event and oxidations as reporter reactions, many of them being based on ECL, simply because it offers an easy visual readout (see Fig. 5c, d). The seminal contributions of this group demonstrated that ECL generation at the vicinity of a bipolar electrode is a very simple and useful way to detect and quantify a spatially delocalized reaction, and therefore the concept has subsequently also been extensively exploited by many other groups in the last few years [83–94].

From a more theoretical point of view, some constraints are related to the use of the bipolar ECL approach. In order to observe ECL emission on a bipolar object, an adequate potential difference between the solution and its two poles has to be induced. Considering the classic $Ru(bpy)_3^{2+}$/TPrA system, its oxidation at the anodic side of the object requires at least a potential of E_{ox} = 1 V versus Ag/AgCl. On the other hand, a reduction reaction has to occur simultaneously at the cathodic pole with equal intensity. For the case of an aqueous system (pH 7.4), this will involve the reduction of protons, which takes place, for example, on carbon typically at $E_{red} = -1.1$ V versus Ag/AgCl. Therefore, ECL can only be generated if the polarization potential difference ΔV between both extremities of the bipolar electrode is at least equal to $|E_{red} - E_{ox}| = 2.1$ V. Taking into account Eq. 8 from above, this means that a minimum global electric field between the feeder electrodes of $\varepsilon = 2.1\ V/d$ is needed to observe ECL emission on an object with a characteristic dimension d. For very small objects, this can be a drawback, because very high absolute voltages (of the order of kV or more) might be necessary to trigger the ECL reaction for a given experimental setup. On the other hand, one of the undeniable advantages is that ECL can be generated on an object without direct electrical connection, and thus it is also possible to address thousands or millions of objects simultaneously with just two feeder electrodes. These latter aspects open the door to some completely new applications of ECL [83, 90], like the emission of light from moving objects [95, 96], or the generation of light in the bulk of a solution (Fig. 6) [97, 98].

All these very appealing features allow envisioning the development of low-cost high-throughput screening methods based on various already well-established analytical concepts such as DNA (de)hybridization [94] or enzymatic reactions [87, 99] when combining them with a bipolar electrochemical setup.

4 Analytical Applications

The main applications of all the methodological and chemical systems described above are dedicated to bioanalyses. These applications can be divided into three main types and objectives: the immunoassays, DNA analyses, and more recent aptamers and DNAzymes detections. The immunoassays are the major ones, not

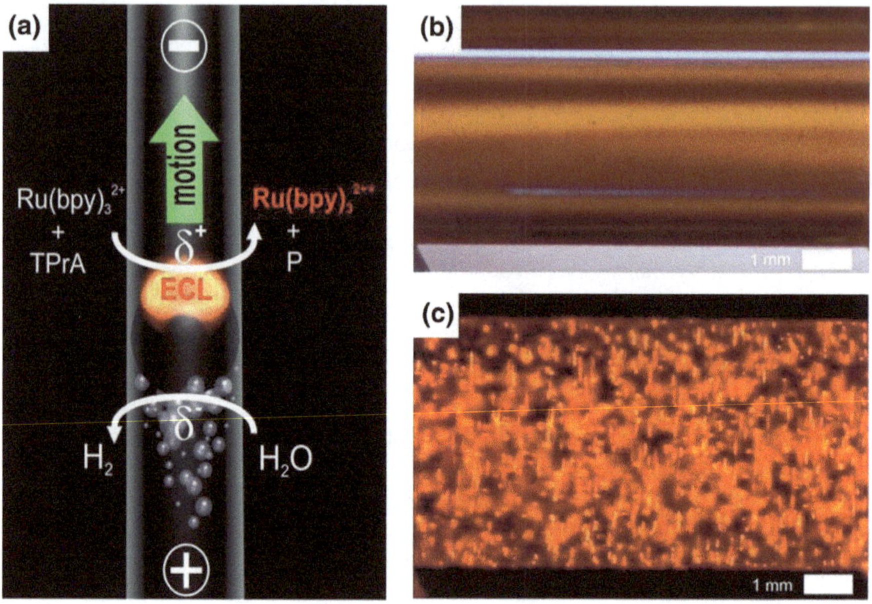

Fig. 6 **a** Scheme of a light-emitting bipolar electrochemical swimmer. Concomitant reduction of H_2O at the cathodic pole (*bottom of the bead*) and oxidation of ECL reagents at the anodic pole (*top of the bead*) induces simultaneous motion and light emission in a glass capillary [95]. **b** Bulk bipolar ECL emission from a suspension of carbon nanotubes (CNTs) in a capillary. Images of the suspension of CNTs **b** under *white light* without electric field and **c** in the *dark* of the bulk ECL emitted by the CNTs in the presence of electric field [97]. Reprinted with permission from Ref. [95]. Copyright 2012 Wiley. Reprinted with permission from Ref. [97]. Published by The Royal Society of Chemistry

only from the number of literature reports, but rather because of massive daily use for clinical diagnostics.

4.1 ECL Immunoassays

The pioneer works in ECL immunoassay (ECLIA) was done in particular by Bard and Whitesides [24], owing to a newly synthesized ruthenium chelate having a NHS residue. Thus, the availability of an efficient ruthenium label for bio-macromolecules (proteins, nucleic acids) has rendered possible the development of highly sensitive ECL bioassays [11, 42].

ECL immunosensing is mainly carried out in solid-phase heterogeneous formats, including direct, competitive, and sandwich immunoassay modes; the most common mode being the sandwich-type immunoassay [4, 13, 43]. In all types of heterogeneous immunoassay, the primary antibody (Ab1) is immobilized on the solid phase. In this case, the antibody must remain reactive after being adsorbed or linked to a solid surface and must retain its structural integrity. Several approaches have been reported to immobilize this primary antibody: the covalent binding through glutaric dialdehyde [100], via L-cysteine [101, 102], via silanization [103] or 3,4,9,10-perylene-tetracarboxylic acid [104]. Besides, the sandwich-type ECLIA involves the conjugation of a secondary antibody (Ab2) labeled with the luminophore, typically a ruthenium complex.

Most of the commercial ECL-based systems for bioassay reported so far are based on the capture at the electrode surface of magnetic beads (a few microns in diameter), on which the immuno-complex formation occurs (Fig. 7). The major advantages of beads as a support are their greater surface area than the flat electrode surface, and the easy concentration of reacting beads by a magnet positioned below the electrode. In addition, it avoids the covalent immobilization of the biomolecules on the electrode surface which allows analyzing sequentially different samples with a high throughput. In the commercial Elecsys systems (Roche Diagnostics Inc.), conventional antigen–antibody reactions are achieved on the surface of streptavidin-coated paramagnetic microbeads owing to biotinylated antibodies. The ECL reaction then occurs on the bead surface between the TPrA co-reactant oxidized at the electrode, and the immobilized ruthenium label. More than 80

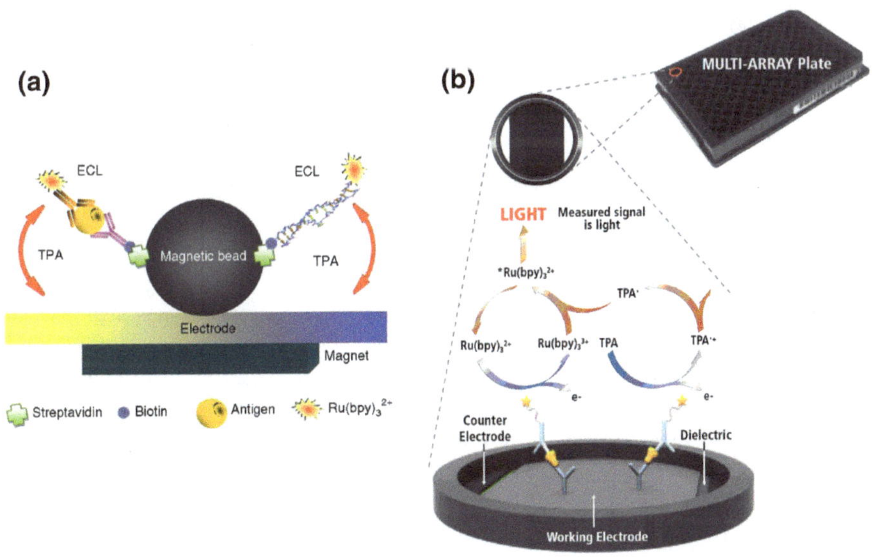

Fig. 7 Two major methodological approaches involved in commercial ECL bioassays. **a** Scheme of the sandwich immuno- and nucleic acid assays achieved on the surface of magnetic beads modified with either a target antibody or nucleic acid (streptavidin–biotin linkage) [12]. The secondary antibody or nucleic acid probes are labeled with $Ru(bpy)_3^{2+}$ moieties. ECL occurs owing to the immobilization of the microbead on the electrode surface under the effect of a magnet, and following addition of TPrA. **b** Scheme of the multiarray system integrated in a titer plate, based on sandwich immunoassays at the surface of a carbon electrode. Several sensors exist within each well allowing multi-ECLIA. Adapted from image by Meso Scale Diagnostics Inc.

bioassays are currently available with these systems, including cardiac markers, tumor markers, bone markers, infectious disease, thyroid function tests, anemia, and fertility tests (Roche Diagnostics Inc.). Recently, a new class of imaging-based ECL devices has been commercialized by Meso Scale Discovery Inc. (MSD). MSD systems are based on the immobilization of antibodies on electrodes and afford a multiarray technology. It uses the carbon-type electrode surface for bioaffinity binding assays, integrated into the bottom of 96-well microtiter plates, since the carbon surface has a 10-fold greater binding capacity than polystyrene. The binding reagent is immobilized on the carbon electrode plate surface, the ECL label (NHS-Ester tagged) is captured on the electrode via the binding complex, and finally the light is collected by a CCD camera. MSD ECLIA can be multiplexed by spotting up to 10 different antigens onto discrete areas of the carbon electrode within each well of the plate.

Understanding the excellent sensitivity of the ECL bead-based immunoassays is essential for further applications and it requires deciphering the ECL mechanisms of the model $Ru(bpy)_3^{2+}$/TPrA system. Different competitive mechanistic pathways have been proposed by Miao et al. [31]. They involve cascades of reactions of short-lived radicals with different redox potentials and reactivity at the minute scale [105] which is challenging to simulate accurately in solution phase (see Chapter "Theoretical Insights in ECL" for a detailed description). They can be classified into two main groups depending on how $Ru(bpy)_3^{2+}$ is oxidized. The first one requires the heterogeneous oxidation of $Ru(bpy)_3^{2+}$ at the electrode surface. However, these mechanistic routes cannot account for the ECL features reported at low oxidation potentials and also for the sensitivity of the bead-based assays [106–108]. Indeed, in this case, only the $Ru(bpy)_3^{2+}$ labels located on the bead within electron tunnelling distance from the electrode surface would be directly oxidized, meaning that an infinitesimal fraction of labels would contribute to the ECL signal following these paths. A second "revisited" route involving the mediated oxidation of $Ru(bpy)_3^{2+}$ by the cation radical ($TPrA^{\cdot+}$) has been proposed to explain how ruthenium centres located at micrometric distances from the electrode might generate ECL [31]. In this path, only the co-reactant TPrA is oxidized at the electrode, and the resulting radicals, $TPrA^{\cdot+}$ and $TPrA^{\cdot}$, diffuse over short distances and react with the luminophore to generate its excited state. Maximum ECL intensity occurs in the micrometric region where concentrations of $TPrA^{\cdot+}$ and $TPrA^{\cdot}$ radicals are locally the highest. Only the luminophores located in the 3-μm region next to the electrode contribute to the ECL signal, and this finite reaction layer defines the optimal size of the functionalized beads for the bioassays [32]. This "revisited" route is essential in bead-based ECL assays and also to propose novel strategies for ECL immunosensing [109].

Many ECL-based immunoassays have been reported in the literature and are commercially available. Overall, they were first dedicated to the detection of cancer biomarkers, then for cardiovascular dysfunctions, for hormones and drugs, and for major immune antibodies such as the IgG [13]. Cancer markers detected by ECLIA include the α-fetoprotein (AFP), the carbohydrate antigen, the carcinoma embryonic antigen, the human chorionic gonadotrophin antigen, and the prostate protein antigen. Heart dysfunctions are analyzed owing to platelet-derived growth factor, Human Cardiac Troponin I marker [110]; the human hormone glucagon [111], IgG [112], and drugs like clenbuterol [113] can be analyzed from blood samples. Extremely sensitive detections of these biomarkers have been achieved owing to the rapid developments of nanomaterials [63, 114–117] notably over the last decade, since they act (see Sect. 3.4 for detailed properties) as: (1) architectures to load large amounts of ECL labels; (2) energy acceptors to quench ECL; (3) novel labels for ECL detection; (4) electrocatalysts of ECL reactions.

Quantum dots (QDs) are increasingly used in the development of ECL bioassays. As explained above, they have appeared as very promising ECL luminophores since they offer a broad excitation spectrum, a narrow and size-tunable luminescence spectrum, a high resistivity towards photobleaching and a high-quantum yield [47–49, 118]. Most of the ECL immunoassays with QDs are based on the quenching, inhibition, or enhancement of the ECL intensities in co-reactant-depending ECL systems, using either $S_2O_8^{2-}$, SO_3^{2-}, or H_2O_2 as co-reactant [119]. Let us mention in particular the work by Peng et al. who reported the first label-free ECL immunosensor based on CdSe QDs. These were electrodeposited directly on a gold electrode from an electrolyte solution (containing cadmium sulfate, ethylenediamine tetra-acetate, and selenium dioxide) by cycling the potential between 0 and −1.2 V versus SCE [120]. The antibody was directly immobilized onto the Au electrode modified by CdSe, which had enough binding sites though being electrodeposited. The specific immunoreaction of AFP with anti-AFP resulted in the decrease of ECL intensity. Another striking result was reported by Tian et al. [121]. They showed an ECL immunoassay method for ultrasensitive detection of prostate protein antigen, by remarkably efficient energy-transfer-induced ECL quenching from the CdS QDs sensitized TiO_2 nanotube array to the activated CdTe QDs functionalized multi-walled carbon nanotubes (MWCNTs) composite. The ECL intensity decrement was logarithmically related to the concentration of the prostate protein antigen (PSA) in the range of 1.0 fg mL^{-1} to 10 pg mL^{-1} with a LOD of 1 fg mL^{-1}. Using CdSe QDs as ECL emitters, Ju et al. were able to monitor the carbohydrate expression on living cells by combining the specific recognition of lectin to carbohydrate groups with the functionalization of immobilized QDs [122].

However, it must be pointed out that the limited functionalization of ligands and variable efficiency in the bioconjugation to antibodies are the main drawbacks of QDs. Several approaches are proposed for the conjugation of QDs with Ab2: (1) the direct conjugation of amino/carboxyl groups using actived esters; (2) a direct conjugation to the QDs surface through the antibody thiol groups; (3) indirect conjugations using bridge proteins such as a modified protein G or avidin reacting with biotinylated antibodies. Ultimately, any employed QDs should be previously modified in order to obtain water solubility and the required functional groups, preferably carboxylate or amino groups.

Noble metal particles structured at the nanoscale are also favorable NPs for ECL with improved mass transport and electron conductivity [5, 58, 115, 123–125]. For instance, nanoporous gold has attracted considerable attention in recent years due to its high surface-to-volume ratio, high in-plane conductivity, good stability, and biocompatibility. These nanoporous materials have been used for electrochemical sensors, while they are basically used as ECL labels [126]. The primary antibody was immobilized on the Au NPs modified electrode through L-cysteine and glutaraldehyde, and then the antigen and the functionalized ruthenium-modified silica nanoporous gold composite-labeled secondary antibody were conjugated successively to form a sandwich-type immunocomplex through the specific interaction. Nanoporous Pt–Ru alloy has also been widely studied because of its uniform structural dimensions [102, 126].

Besides, nanosized-carbon particles, materials, and nanoelectrodes constitute a very promising category [109, 115, 123, 127, 128]. Especially, graphene sheets, with their two-dimensional structure, unique mechanical properties, excellent conductivity, large surface area, unique graphitized basal plane structure, as well as good biocompatibility might be excellent supporting material for signal amplification in ECL sensors. Chitosan-functionalized graphene-modified glassy-carbon electrodes have been employed to increase the loading of Ab1 and catalyzed the cathodic ECL reaction, which was further amplified by gold nanorods and an enzyme-catalyzed reaction [129]. Another striking example is an ECL immunosensor prepared from graphene-CdS QDs-alginate as the immobilizing support and CdSe/ZnS QDs as the label. CdSe/ZnS QDs labeled on the Ab2 could absorb emitted photons in the system, providing a possible application in bioassays of QDs [121]. In parallel, CNTs are attractive as building blocks for ECL sensors because they possess a unique combination of excellent electrical and electrochemical properties, they can be easily modified with different functional groups for the attachment of biomolecules [130, 131], and they possess a very high surface-area-to-weight ratio. Paolucci et al. exploited these remarkable properties of CNTs to build a highly sensitive ECL biosensor for the detection of palytoxin, a potent marine toxin which contaminates seafood [132]. They obtained very reproducible and reliable responses in complex matrices, such as mussels and microalgae, reaching the LOD of 0.07 ng mL^{-1}. Recently, an interfusion of graphene and CNT in a Nafion film was coated on an electrode for $Ru(bpy)_3^{2+}$ adsorption in order to demonstrate the co-reactant-based approach of the $Ru(bpy)_3^{2+}$/poly(ethylenimine) system for human chorionic gonadotrophin ECLIA [104].

All these different strategies based on ever more and more complex architectures lead currently to exceptionally sensitive ECLIA. They are all solely based on the high specificity of the antigen–antibody interaction. However, diagnostic applications would ultimately require the simultaneous analysis of several antibodies from the same sample (especially in the blood). This task should be performed using either multilabel, or multianalyte (on an array) strategies [106, 133, 134], both constituting specific research areas dedicated to the analytical methodology [135–139].

4.2 DNA Assays

DNA is the second main biological target of ECL bioassays. Indeed, the sensitive detection of DNA is of great importance for medical diagnosis, environmental investigations, gene expression analysis, pharmaceutical studies, forensic analyses, etc. [140, 141]. ECL DNA assays include mainly two types of label-free and label-based ECL detection strategies. Meanwhile, various amplification techniques have been used to improve the sensitivity of DNA detection [43, 142, 143]. These include nanomaterials with multiple grafting sites to increase the number of tags, enzyme-assisted signal-amplification processes, and DNA-amplification techniques such as rolling-circle amplification (RCA), target-induced repeated primer extension, hybridization chain reaction (HCR), and loop-mediated amplification [144].

First, metal or semiconductor NPs, owing to their unique optical and electrical properties, were widely employed as labels for the amplified detection of DNA. Since a single NP (e.g., Au NP) can be loaded with thousands of DNA molecules, the ECL intensity is typically enhanced by 2–3 orders of magnitude on an Au NP self-assembled electrode compared with a bare gold electrode. Jiang et al. developed a novel ECL DNA sensor for *Mycobacterium tuberculosis* (MTB), by using luminol-functionalized Au NPs [124, 125] as labels [145]. Owing to these novel ECL labels, the amplification by Au NPs and the biotin–streptavidin system, extremely high sensitivity (LOD of 170 pg L^{-1}) for a single-stranded DNA (ssDNA) was obtained.

Then, among the techniques for DNA amplification, RCA is a powerful, though simple, biochemical method based on the use of a DNA polymerase to generate long, linear, tandem, repetitive, ssDNA under isothermal conditions (Fig. 8). As the synthesized long DNA molecules contain many repetitive sequence motifs, this improves the biosensor analytical properties. For instance, Jiang et al. reported a highly sensitive strategy for the ECL detection of DNA of *Clostridium perfringens* [146]. In the presence of target DNA, the RCA primer was attached to the electrode and the RCA reaction was executed isothermally. The products of RCA were incubated with hemin to form hemin/G-quadruplex DNAzymes, which competitively consumed oxygen in the detection buffer and thus quenched the ECL emission of the $O_2/S_2O_8{}^{2}$ system. The decrease of ECL emission was related to the quantity of the target DNA, so that the ECL sensor provided both sensitivity to detect the amount of *Clostridium perfringens* target DNA, and also the selectivity to discriminate target DNA from non-target sequences even with a difference of only one base.

DNA nanoassembly via target-probe hybridization is a simple and effective approach to build ECL DNA sensors. If properly designed, DNA can assemble into

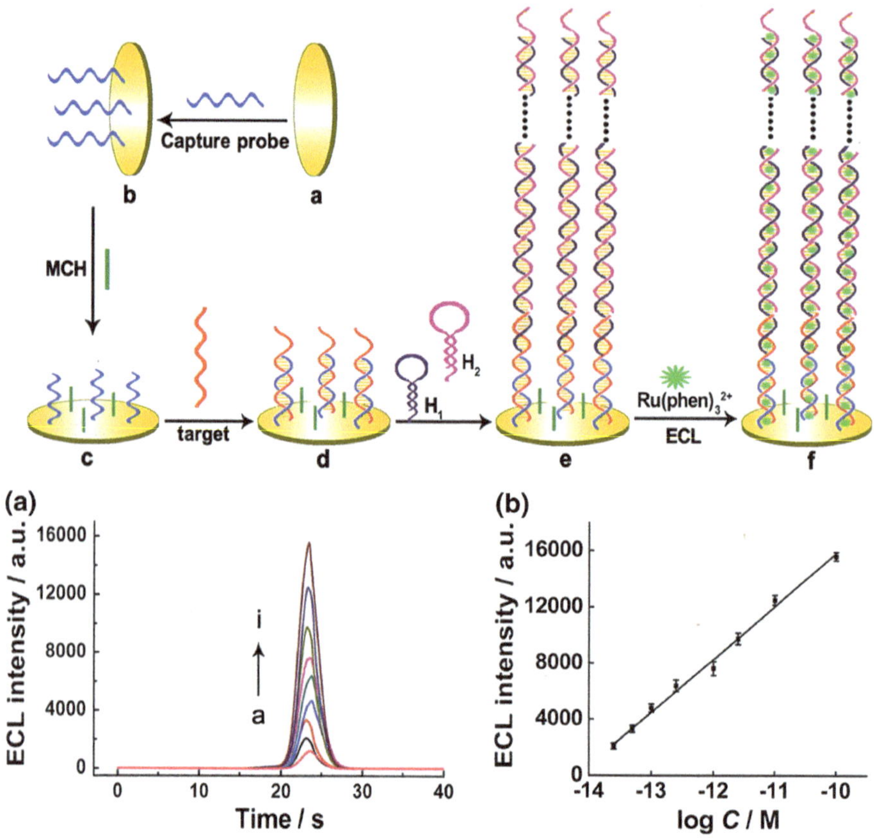

Fig. 8 ECL detection of DNA based on a HCR strategy. *Top* ECL measurements are obtained after immobilizing the capture DNA, adding 2.5 pM of target DNA and making hairpin probes reacting with the hybridized system. Subsequently, 2 mM $Ru(phen)_3^{2+}$ is added so as to intercalate with the DNA assembly and provide ECL in 0.1 M phosphate buffer solution (pH 7.5) containing 20 mM TPrA. *Below* **a** ECL responses of the sensor to different concentrations of target DNA (*a*) 0 fM, (*b*) 25 fM, (*c*) 50 fM, (*d*) 100 fM, (*e*) 250 fM, (*f*) 1 pM, (*g*) 2.5 pM, (*h*) 10 pM, (*i*) 100 pM. **b** The resulting calibration plot of log c versus ECL intensity [147]. Reprinted with permission from Ref. [147]. Copyright 2012 American Chemical Society

different structures in two- or three dimensions. Based on long-range, self-assembled, DNA nanostructures as carriers for ECL signal amplification, Liu et al. fabricated a label-free, ultrasensitive ECL biosensor for detection of microRNA-21, a biomarker for cancers and other diseases (Fig. 9) [148]. Due to the high sensitivity (1 fM LOD) and high selectivity, wide linear range (1 fM–1 pM) and simplicity of the methodology, this approach is of particular interest.

HCR is another DNA-nanoassembly method for DNA detection; this is an enzyme-free process, where a hybridization event is triggered by an initiator (target) and leads to the polymerization of oligonucleotides into a long nicked

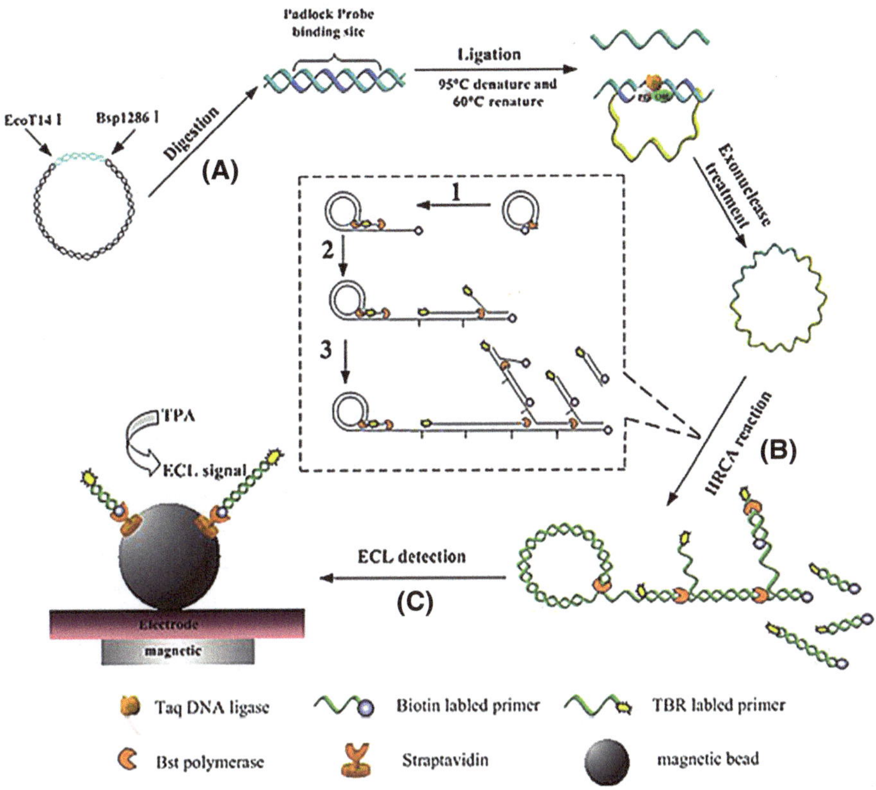

Fig. 9 Principle of magnetic bead-based ECL hyperbranching rolling circle amplification (HRCA) assay for DNA (from Listeria monocytogenes herein). (A) Genomic DNA digestion and padlock-probe circularization. (B) HRCA reaction with a biotin-labeled primer and a $Ru(bpy)_3^{2+}$-labeled primer. (C) ECL detection [148]. Reprinted with permission of Springer from Ref. [148]

double-stranded DNA (dsDNA). The HCR-based ECL strategy not only enables low femtomolar detection of sequence-specific DNA, but also shows high selectivity against single-base mismatch sequences. As an example, a highly sensitive, universal strategy for ECL detection of a DNA sequence specific to *Escherichia coli* was reported based on a HCR amplification approach [147]. Authors obtained a linear response between 25 fM and 100 pM with an estimated LOD of 15 fM for a DNA sequence, which is a copy of partial region of the *E. coli* 16S rRNA gene, specific to urinary tract infections.

Besides, ECL starts to be used to assess DNA damages, which include the oxidation, alkylation, or hydrolysis of bases, adducts' formation, and mismatch of bases caused by environmental hazards. Many biological methods such as the comet assay and the unscheduled DNA-synthesis test have been used to detect DNA damages. However, these methods are complex, time consuming, and have low efficiency and sensitivity. Recently was reported an ECL biosensor to detect specific sequences of DNA by using CdTe@SiO_2 as nanoprobes for signal amplification [149, 150]. Based on the ECL detection of CdTe@SiO_2 labeled on probe DNA, a sensitive assay for the sequence-specific DNA detection was developed. The "sandwich-type" DNA complexes were fabricated by self-assembly of an aminated capture DNA on a glassy-carbon electrode and hybridized with one end of target DNA, the other end of which was recognized with signal DNA labeled with CdTe@SiO_2. Under optimum conditions, the peak current value increased with the concentration of target DNA within a wide dynamic range (0.1 nM–2 μM) and an excellent LOD (30 pM). An ECL signal corresponding to a single-base-mismatch was also detected. The experiments indicated that the sensing system could differentiate single-base-mismatched DNA from complementary DNA. Strikingly, the biosensor could successfully be applied to study the DNA damage induced by several genotoxic chemicals (styrene, aflatoxin, ochratoxin, methanol, and ethanol) [150].

In recent years, isothermal amplification reactions based on DNA machines attracted much attention in designing sensitive analysis of DNA and single-nucleotide polymorphism (SNP) for their autonomous circular amplification and simple operations [151]. These isothermal amplification reactions use enzyme-assisted replication or scission reactions as the DNA machines. A new ECL approach for detection of SNPs was based on an isothermal cycle-assisted, triple-stem probe labeled with gold and CdTe NPs [152]. With its simplicity, selectivity, and sensitivity, this molecular tool holds great promise for SNP discovery and analysis, and might be extensively applied in the future for clinical research and diagnostics.

As already mentioned, the detection of DNA damages by oxidative stress processes or by chemicals is also a central biomedical issue. Rusling et al. demonstrated in a series of articles that ECL can be obtained in ultrathin metallopolymer films containing either osmium or ruthenium complexes from oxidized DNA without using a sacrificial co-reactant [133, 134, 153]. The guanine bases present in oligonucleotides generate the ECL emission by reacting with the organometallic complexes. ECL signals are sensitive to oligonucleotide hybridization and chemical damage of dsDNA [153]. Such an approach allowed them to detect DNA oxidation and nucleobase adducts from chemical damage, such as 8-oxoguanine [133]. ECL arrays were thus reported to screen the relative propensity of different enzymes (e.g., cytochrome P450 enzymes) to produce genotoxic metabolites [134].

4.3 Aptamers ECL Sensors

A large number of functional oligonucleotides including aptamers and DNAzymes have been synthesized and used for the development of biosensors [154, 155]. Consequently, they have attracted increasing interest recently for ECL bioanalytical applications. Aptamers are single-stranded nucleic acids (DNA or RNA) with special structures that allow high recognition ability to a broad range of molecular targets including small molecules, ions, proteins, and even whole cells. Aptamers exhibit many advantages when compared to traditional molecular recognition elements, including simple synthesis, good stability, easy storage, high affinity and specificity, and possible further chemical modifications for their processing.

Various protocols of aptamer-based ECL biosensing for the determination of proteins and small molecules have been reported [156, 157]. A striking example for the ultrasensitive ECL detection of thrombin was reported by Chen et al. [158]. The sensor is based on layer-by-layer assembly of CdTe QDs onto polystyrene microbeads as ECL signal amplification labels (Fig. 10). Upon using these polystyrene-$(CdTe)_2$ labels, and playing with the number of QD layers, about 2–4 orders of magnitude improvement in the LOD for thrombin (350 fM) was obtained when compared with other signal amplification routes. ECL aptasensors for the detection of small molecules have also been widely reported. ECL quenching of $Ru(bpy)_3{}^{2+}$-labeled silica NPs by ferrocene was used to analyze adenosine [159]. First, complementary DNA strands, labeled with the $Ru(bpy)_3{}^{2+}$-silica NPs, were immobilized onto the gold electrode. Then, ferrocene-labeled aptamer was provided to hybridize the complementary DNA on the electrode surface, resulting in the quenching of ECL from $Ru(bpy)_3{}^{2+}$. In the presence of adenosine, the ferrocene-labeled aptamer preferentially forms an adenosine–aptamer complex and prevents DNA hybridization, which induces an ECL signal recovery. By using this approach, a LOD of 31 pM for adenosine was achieved.

Though being extremely efficient and sensitive, the above approaches involve complicated modification procedures. Thus, label-free ECL detection is also extensively investigated according to its advantages of simplicity. In particular, label-free and sensitive ECL aptasensors for cations were developed. In a first example, dedicated to the detection of potassium ions, Dang et al. have reported that a free-state G-rich DNA aptamer can enhance the ECL signal of chitosan/$Ru(bpy)_3{}^{2+}$/silica NPs on an electrode more effectively than its G-quadruplex structure which binds K^+ [160]. By analyzing the decrease of ECL signal, a LOD of 300 pM for K^+ was achieved. ECL probe intercalation in DNA aptamers was used for a label-free and "turn-on" sandwich ECL biosensor for Hg^{2+} ions; this was based on the thymidine–Hg^{2+}–thymidine coordination [161]. In this protocol, two functional ssDNA were designed with a thymidine-rich section for the recognition of mercury ions and could partially hybridize with each other forming an intermolecular duplex for the intercalation of $Ru(phen)_3{}^{2+}$ luminophore. Owing to the specificity of interaction between Hg^{2+} and multithymidine, and high sensitivity of super sandwich ECL detection, a LOD of 250 pM for mercury was obtained.

Besides small molecules and proteins, ECL aptasensors have also been used for the detection of cell markers [92, 162, 163]. An ultrasensitive hybrid bipolar electrode–ECL biosensor for the detection of cancer cell surface proteins was recently proposed [162] by using a ferrocene-labeled aptamer as a signal recognition and amplification probe. The hybrid bipolar electrode (see part 3.5 for details on ECL in bipolar electrochemistry) is composed of an ITO-coated glass as the anode, and Au NPs as the

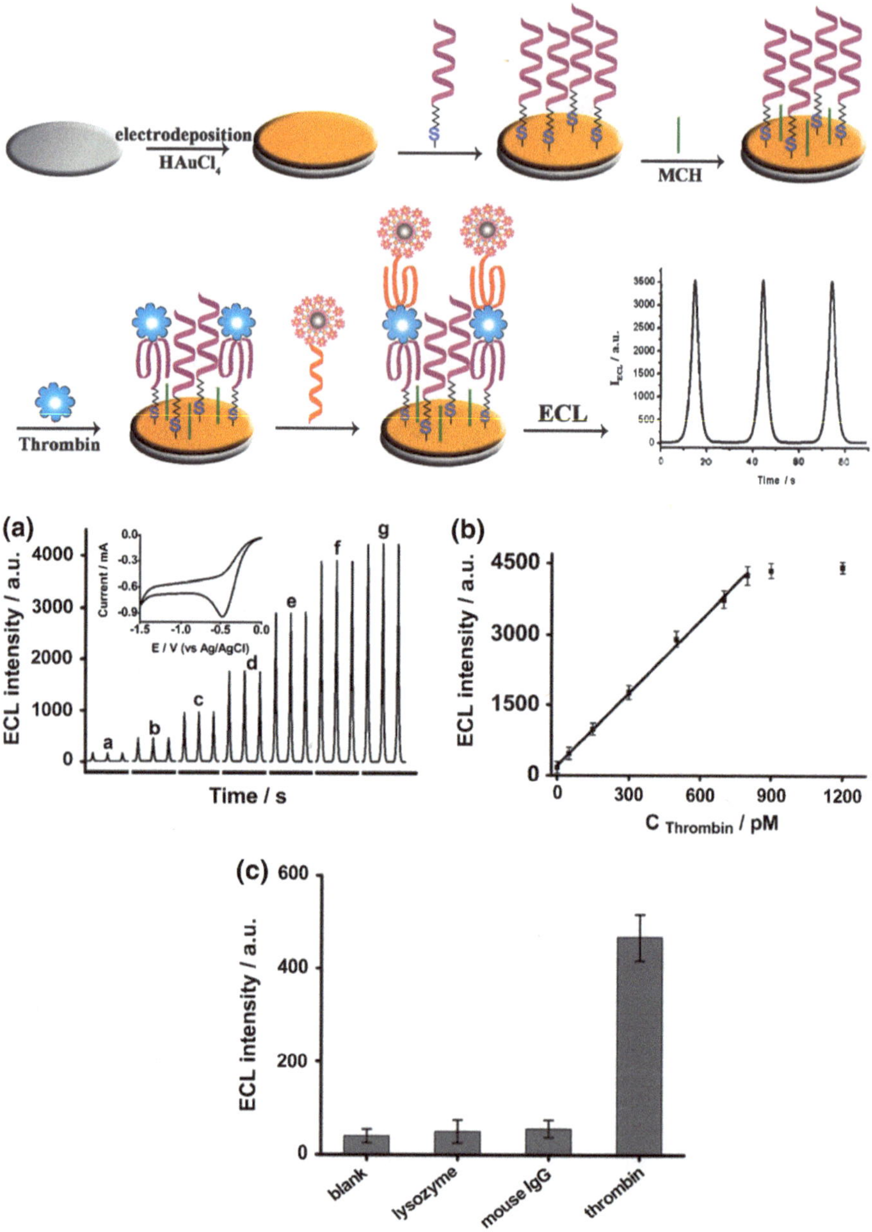

cathode. Owing to its conductivity and large surface area, this system provides large cathodic currents resulting in enhanced ECL of the $Ru(bpy)_3{}^{2+}$–TPrA system on the anode. Then, a ferrocene-labeled aptamer was introduced to hybridize the capture DNA immobilized on the cathode NPs. Ferrocene quenched the ECL response,

◀ **Fig. 10** *Top* Schematic of the different steps for the construction of an ECL sensor for thrombin. It consists in the deposition of gold nanoparticles (Au NPs) on a glassy-carbon electrode, the self-assembly of the primary thrombin-binding aptamer on the Au NP-modified electrode, the surface blocking with 6-mercapto-1-hexamine (MCH), the association of the thrombin analyte with the primary thrombin-binding aptamer and the secondary thrombin binding aptamer-conjugated PS bead-$(CdTe)_2$ QDs. *Down* **a** ECL profiles for thrombin detection based on PS-$(CdTe)_2$ assembly label at different concentrations: (*a*) 0.5 pM, (*b*) 50 pM, (*c*) 150 pM, (*d*) 300 pM, (*e*) 500 pM, (*f*) 700 pM and (*g*) 800 pM. *Inset* Typical voltammogram corresponding to the presence of 500 pM thrombin. **b** Calibration plot of ECL intensity versus thrombin concentration. **c** Selectivity of the ECL detection of thrombin (50 pM) against interference proteins, including lysozyme (500 pM) and mouse IgG (500 pM) [158]. Reprinted with permission from Ref. [158]. Published by The Royal Society of Chemistry

which can recover in the presence of cancer cells, since labeled aptamers can combine with target proteins expressed on the cell surface. The coupling of the cathodic ECL at Au nanoparticles with the anodic ferrocene induced signal quenching amplification, allowed a LOD of 20 cells with this approach.

4.4 Enzymatic Assays

Also for the detection of enzymatic reactions, ECL has some distinct advantages because the, in many cases highly selective, enzymatic activity can be tracked in time and space if at least one of the reaction products can act as a co-reactant in one or the other of the classic ECL systems. Depending on which compound is produced by the enzyme, there are mainly three options that have been largely explored and also extensively reviewed in the recent literature [4, 41, 164].

The first very common possibility is to use enzymes belonging to the family of dehydrogenases. The "by-product" of their enzymatic activity is the transformation of the coenzyme NAD^+ into NADH. The latter one can play the role of an ECL co-reactant for the $Ru(bpy)_3{}^{2+}$ system because, exactly like TPrA, it has a tertiary amine function. Therefore, and under the condition that $Ru(bpy)_3{}^{2+}$ is present in excess, the recorded ECL signal is directly proportional to the production of reduced coenzyme, which in turn is correlated with the concentration of the enzyme's substrate in solution. The family of dehydrogenases contains hundreds of enzymes belonging to the group of oxidoreductases which oxidize a substrate and transfer one or more hydrides to an electron acceptor such as NAD^+. Therefore, this gives access to ECL analysis of classic analytes such as glucose [165] or ethanol [166], but in principle also many other biologically relevant substrates can be detected based on this concept.

A second approach is based on the fact that another family of oxidoreductases, namely oxidases uses oxygen as an electron acceptor instead of NAD^+, and therefore produces hydrogen peroxide as a side product. This obviously allows working with the luminol/H_2O_2 ECL system as an optical readout of the enzymatic activity, again either for the most frequent analytes like glucose [167], but also for analytes which were not accessible with the $Ru(bpy)_3^{2+}$/NAD^+ system [168–170].

The third possibility which has been widely studied to couple an enzymatic reaction with an ECL producing system involves the use of quantum dots. ECL with QDs has been investigated, among others, for co-reactants which can be correlated with enzymatic activity (e.g., H_2O_2, O_2). Cathodic and anodic ECL processes have been explored in this context. For cathodic ECL, QDs and H_2O_2 are simultaneously reduced at the electrode, and then the produced hydroxyl radical can act as a strong oxidizing agent and inject a hole into the QD, which can further react to generate an excited state by annihilation. In this way, most enzymes producing H_2O_2 are in principle able to enhance QD ECL [171]. It is also possible to develop enzymatic ECL biosensors utilizing anodic QD ECL [172]. However, a drawback can be in some cases the quite high positive potentials which are necessary to trigger the process.

These listed reaction schemes are usually carried out in conventional three-electrode electrochemical setups, but as already mentioned in Sect. 3.5, one quite original and promising extension of these approaches is to trigger ECL in a bipolar electrochemistry configuration. Recently, a couple of enzymatic systems were employed for analyte detection based on bipolar electrochemistry. Oxidation of blood glucose by means of glucose oxidase yielding H_2O_2 was analyzed in a three-compartment closed bipolar sensing platform allowing a straightforward wireless readout of the chemical information based on $Ru(bpy)_3^{2+}$ ECL [85]. The luminol system has also been explored in a bipolar configuration for detecting glucose [87]. In this case, the H_2O_2 generated by the enzymatic reaction was not directly involved in the ECL generation as a co-reactant, but was reduced at the bipolar cathode with the help of a catalytically active Prussian Blue layer (Fig. 11a). This reduction reaction was intrinsically coupled with the luminol/H_2O_2 oxidation at the anodic side of the bipolar electrode, leading finally to a light signal which is proportional to the glucose concentration. Very recently, it has been possible to extend this approach to a dual enzymatic detection [173] based on the concept of 3D ECL described in Sect. 3.5 [97, 98].

Another interesting extension based on the synergistic combination of enzymatic ECL detection with bipolar electrochemistry is based on the fact that the latter one can also be used in a very simple way to generate motion [174, 175]. If such a motion is coupled with ECL emission, this leads to a light-emitting moving object [95]. If in addition the ECL reaction is triggered by an enzymatic step, this results in a roving biosensor [99]. As a consequence, this gives access to spatio-temporal detection of varying concentrations of an analyte based on the light emission from a single object cruising through the solution (Fig. 11b).

Although these last examples are for the moment at the stage of academic research, it can be envisioned that such new concepts might help in the future to map enzymatically the analyte concentration in inhomogeneous samples either by 3D ECL or by the roving single particle approach. Naturally, all the studies mentioned in this paragraph can also be extended towards the elaboration of

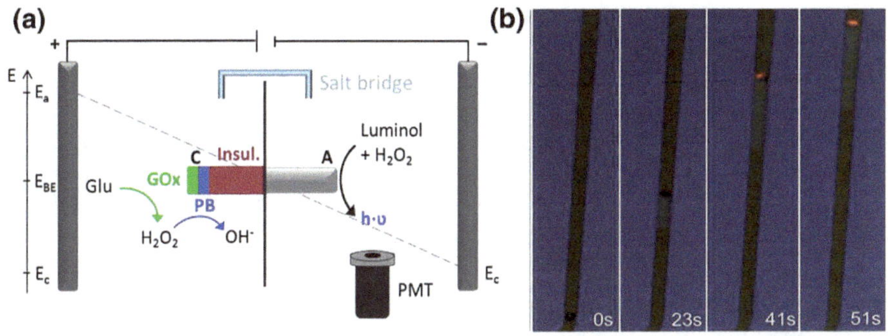

Fig. 11 **a** Schematic illustration of the measurement concept for the bipolar enzymatic ECL detection of glucose with the luminol/H_2O_2 system [87]. **b** Enzymatic ECL sensing of glucose concentration with a moving particle. Series of optical images showing a GC bead emitting ECL at different times during its motion. The bead is positioned in a capillary filled with 100 mM PBS solution (pH = 7.4) containing 9.28 U mL^{-1} GDH, 1.5 mM $Ru(bpy)_3^{2+}$, 10 mM NAD^+ and a few drops of surfactant. When it reaches a glucose-rich region, while moving towards the top of the capillary, light emission starts [99]. Reprinted with permission of Elsevier from Ref. [87]. Reprinted with permission from Ref. [99]. Published by The Royal Society of Chemistry

immunosensors and DNA sensing, as enzymes can be easily used as labels in these kinds of systems (see Sects. 4.1 and 4.2). This again underlines the tremendous application potential of these concepts.

5 Conclusion and Perspectives

Since the first reports in the 1960s, ECL has evolved from an interesting laboratory phenomenon to a powerful ultrasensitive analytical technique. It is a very versatile method because it combines the principles and the tools of both electrochemistry and photochemistry. Therefore, more factors are involved in the global ECL process which expands the interplay possibilities. In addition to the developments in its own area, progresses in each of these fields may also lead to advances in the ECL domain. Its remarkable characteristics and the discovery of co-reactant ECL in aqueous solutions make such readout mechanisms particularly attractive for bioanalytical applications. Thus, ECL is currently commercialized in the clinical diagnostic market for a wide range of immunoassays measuring specific biomarkers involved in different pathologies, such as cardiac and infectious diseases, thyroid disorders, and tumors.

Looking to the future, it seems obvious that ECL will find more analytical applications due to the demand in medical diagnostics for lower LODs, in biodefense, and in environmental analysis. However, fundamental research on ECL remains also essential for continuing to push the barriers which we believed to be unsurpassable at a given time, as usual in science. For example, compared to chemiluminescence, one of the main restrictions of ECL is the confinement of the process to the electrode surface. To overcome such a limitation, 3D bulk ECL has been generated at the level of millions of micro- and nano-emitters dispersed in solution [97]. Addressing the electrodes in a wireless manner by bipolar electrochemistry gave thus various original applications [78, 83]. Another area that is expected to grow rapidly is the paper-based microfluidic sensors based on ECL detection and portable multiplexed systems with a high degree of integration [84, 176, 177]. The research on new efficient co-reactants [27], of luminophores emitting at different wavelengths [177–181] and of new nano-objects (e.g., Au NPs [58–60], QDs [48, 49], and doped silica NPs [182, 183]), offers the opportunity to develop new analytical detection strategies and assays, but also to continue to study fundamental issues in chemistry, biology or physics.

References

1. Bard, A.J., Faulkner, L.R.: Electrochemical Methods: Fundamentals and Applications, 2nd edn. Wiley, New York (2001)
2. Bard, A.J.: Electrogenerated Chemiluminescence. M. Dekker, New York (2004)
3. Kapturkiewicz, A.: In: Alkire, R.C., Gerischer, H., Kolb, D.M., Tobias, C.W. (eds.): Advances in Electrochemical Science and Engineering, vol. 5, p. 1. Wiley-VCH
4. Miao, W.: Chem. Rev. **108**, 2506 (2008)
5. Bertoncello, P., Stewart, A.J., Dennany, L.: Anal. Bioanal. Chem. **406**, 5573 (2014)
6. Forster, R.J., Bertoncello, P., Keyes, T.E.: Ann. Rev. Anal. Chem. **2**, 359 (2009)
7. Richter, M.M.: Chem. Rev. **104**, 3003 (2004)
8. Rampazzo, E., Bonacchi, S., Genovese, D., Juris, R., Marcaccio, M., Montalti, M., Paolucci, F., Sgarzi, M., Valenti, G., Zaccheroni, N., Prodi, L.: Coord. Chem. Rev. **256**, 1664 (2012)
9. Liu, Z., Qi, W., Xu, G.: Chem. Soc. Rev. **44**, 3117 (2015)
10. Yang, H., Leland, J.K., Yost, D., Massey, R.J.: Nat. Biotechnol. **12**, 193 (1994)
11. Blackburn, G.F., Shah, H.P., Kenten, J.H., Leland, J., Kamin, R.A., Link, J., Peterman, J., Powell, M.J., Shah, A., Talley, D.B., Tyagi, S.K., Wilkins, E., Wu, T.G., Massey, R.J.: Clin. Chem. **37**, 1534 (1991)
12. Zhou, X., Zhu, D., Liao, Y., Liu, W., Liu, H., Ma, Z., Xing, D.: Nat. Protoc. **9**, 1146 (2014)
13. Muzyka, K.: Biosens. Bioelectron. **54**, 393 (2014)
14. Yuan, Y., Han, S., Hu, L., Parveen, S., Xu, G.: Electrochim. Acta **82**, 484 (2012)
15. Ege, D., Becker, W.G., Bard, A.: J. Anal. Chem. **56**, 2413 (1984)
16. Hercules, D.M.: Science **145**, 808 (1964)
17. Visco, R.E., Chandross, E.A.: J. Am. Chem. Soc. **86**, 5350 (1964)
18. Santhanam, K.S.V., Bard, A.J.: J. Am. Chem. Soc. **87**, 139 (1965)
19. Tokel-Takvoryan, N.E., Hemingway, R.E., Bard, A.J.: J. Am. Chem. Soc. **95**, 6582 (1973)
20. Tokel, N.E., Bard, A.J.: J. Am. Chem. Soc. **94**, 2862 (1972)
21. Rubinstein, I., Bard, A.J.: J. Am. Chem. Soc. **103**, 512 (1981)
22. Noffsinger, J.B., Danielson, N.D.: Anal. Chem. **59**, 865 (1987)
23. Leland, J.K., Powell, M.J.: J. Electrochem. Soc. **137**, 3127 (1990)
24. Bard, A.J., Whitesides, G.M.: US Patent 5,221,605, 22 June 1993
25. Faulkner, L.R., Bard, A.J.: J. Am. Chem. Soc. **90**, 6284 (1968)
26. Glass, R.S., Faulkner, L.R.: J. Phys. Chem. **85**, 1160 (1981)
27. Liu, X., Shi, L., Niu, W., Li, H., Xu, G.: Angew. Chem. Int. Ed. **46**, 421 (2007)
28. Zu, Y., Bard, A.: J. Anal. Chem. **73**, 3960 (2001)

29. Kanoufi, F., Zu, Y., Bard, A.J.: J. Phys. Chem. B **105**, 210 (2001)
30. Zu, Y., Bard, A.: J. Anal. Chem. **72**, 3223 (2000)
31. Miao, W., Choi, J.-P., Bard, A.J.: J. Am. Chem. Soc. **124**, 14478 (2002)
32. Sentic, M., Milutinovic, M., Kanoufi, F., Manojlovic, D., Arbault, S., Sojic, N.: Chem. Sci. **5**, 2568 (2014)
33. Cruser, S.A., Bard, A.J.: Anal. Lett. **1**, 11 (1967)
34. Rubinstein, I., Martin, C.R., Bard, A.: J. Anal. Chem. **55**, 1580 (1983)
35. Skotty, D.R., Nieman, T.A.J.: Chromatogr. B: Biomed. Sci. Appl. **665**, 27 (1995)
36. Yamashita, K., Yamazaki-Nishida, S., Harima, Y., Segawa, A.: Anal. Chem. **63**, 872 (1991)
37. Knight, A.W., Greenway, G.M.: Analyst **120**, 2543 (1995)
38. Knight, A.W., Greenway, G.M.: Analyst **121**, 101R (1996)
39. Downey, T.M., Nieman, T.A.: Anal. Chem. **64**, 261 (1992)
40. Brune, S.N., Bobbitt, D.R.: Anal. Chem. **64**, 166 (1992)
41. Chen, X.-M., Su, B.-Y., Song, X.-H., Chen, Q.-A., Chen, X., Wang, X.-R.: Trac-Trend Anal. Chem. **30**, 665 (2011)
42. Kenten, J.H., Casadei, J., Link, J., Lupold, S., Willey, J., Powell, M., Rees, A., Massey, R.: Clin. Chem. **37**, 1626 (1991)
43. Miao, W., Bard, A.: J. Anal. Chem. **75**, 5825 (2003)
44. Zhang, J., Qi, H., Li, Y., Yang, J., Gao, Q., Zhang, C.: Anal. Chem. **80**, 2888 (2008)
45. Li, Y., Qi, H., Peng, Y., Yang, J., Zhang, C.: Electrochem. Commun. **9**, 2571 (2007)
46. Wu, P., Hou, X., Xu, J.-J., Chen, H.-Y.: Chem. Rev. **114**, 11027 (2014)
47. Zhao, W.-W., Wang, J., Zhu, Y.-C., Xu, J.-J., Chen, H.-Y.: Anal. Chem. **87**, 9520 (2015)
48. Ding, Z., Quinn, B.M., Haram, S.K., Pell, L.E., Korgel, B.A., Bard, A.J.: Science **296**, 1293 (2002)
49. Myung, N., Ding, Z., Bard, A.J.: Nano Lett. **2**, 1315 (2002)
50. Zou, G., Ju, H.: Anal. Chem. **76**, 6871 (2004)
51. Myung, N., Bac, Y., Bard, A.J.: Nano Lett. **2003**, 3 (1053)
52. Bae, Y., Myung, N., Bard, A.J.: Nano Lett. **4**, 1153 (2004)
53. Jie, G., Liu, B., Pan, H., Zhu, J.-J., Chen, H.-Y.: Anal. Chem. **79**, 5574 (2007)
54. Liu, X., Jiang, H., Lei, J., Ju, H.: Anal. Chem. **79**, 8055 (2007)
55. Zheng, L., Chi, Y., Dong, Y., Lin, J., Wang, B.: J. Am. Chem. Soc. **131**, 4564 (2009)
56. Zhu, H., Wang, X., Li, Y., Wang, Z., Yang, F., Yang, X.: Chem. Commun. **5118** (2009)
57. Li, L.-L., Ji, J., Fei, R., Wang, C.-Z., Lu, Q., Zhang, J.-R., Jiang, L.-P., Zhu, J.-J.: Adv. Funct. Mater. **22**, 2971 (2012)
58. Swanick, K.N., Hesari, M., Workentin, M.S., Ding, Z.: J. Am. Chem. Soc. **134**, 15205 (2012)
59. Hesari, M., Workentin, M.S., Ding, Z.: Chem. Sci. **5**, 3814 (2014)
60. Hesari, M., Workentin, M.S., Ding, Z.: ACS Nano **8**, 8543 (2014)
61. Li, L., Liu, H., Shen, Y., Zhang, J., Zhu, J.-J.: Anal. Chem. **83**, 661 (2011)
62. Diez, I., Pusa, M., Kulmala, S., Jiang, H., Walther, A., Goldmann, A.S., Mueller, A.H.E., Ikkala, O., Ras, R.H.A.: Angew. Chem. Int. Ed. **48**, 2122 (2009)
63. Bertoncello, P., Forster, R.J.: Biosens. Bioelectron. **24**, 3191 (2009)
64. Obeng, Y.S., Bard, A.J.: Langmuir **7**, 195 (1991)
65. Rubinstein, I., Bard, A.J.: J. Am. Chem. Soc. **102**, 6641 (1980)
66. Khramov, A.N., Collinson, M.M.: Anal. Chem. **72**, 2943 (2000)
67. Choi, H.N., Cho, S.-H., Lee, W.-Y.: Anal. Chem. **75**, 4250 (2003)
68. Guo, Z., Shen, Y., Wang, M., Zhao, F., Dong, S.: Anal. Chem. **76**, 184 (2004)
69. Zhang, L., Dong, S.: Anal. Chem. **78**, 5119 (2006)
70. Jie, G., Zhang, J., Wang, D., Cheng, C., Chen, H.-Y., Zhu, J.-J.: Anal. Chem. **80**, 4033 (2008)
71. Wang, J., Shan, Y., Zhao, W.-W., Xu, J.-J., Chen, H.-Y.: Anal. Chem. **83**, 4004 (2011)
72. Wang, Y., Lu, J., Tang, L., Chang, H., Li, J.: Anal. Chem. **81**, 9710 (2009)
73. Li, L.-L., Liu, K.-P., Yang, G.-H., Wang, C.-M., Zhang, J.-R., Zhu, J.-J.: Adv. Funct. Mater. **21**, 869 (2011)

74. Xu, S., Liu, Y., Wang, T., Li, J.: Anal. Chem. **83**, 3817 (2011)
75. Zhang, J., Gryczynski, Z., Lakowicz, J.R.: Chem. Phys. Lett. **393**, 483 (2004)
76. Yuk, J.S., O'Reilly, E., Forster, R.J., MacCraith, B.D., McDonagh, C.: Chem. Phys. Lett. **513**, 112 (2011)
77. Devadoss, A., Dickinson, C., Keyes, T.E., Forster, R.: J. Anal. Chem. **83**, 2383 (2011)
78. Fosdick, S.E., Knust, K.N., Scida, K., Crooks, R.M.: Angew. Chem. Int. Ed. **52**, 10438 (2013)
79. Loget, G., Zigah, D., Bouffier, L., Sojic, N., Kuhn, A.: Acc. Chem. Res. **46**, 2513 (2013)
80. Arora, A., Eijkel, J.C.T., Morf, W.E., Manz, A.: Anal. Chem. **73**, 3282 (2001)
81. Chow, K.-F., Mavré, F., Crooks, R.M.: J. Am. Chem. Soc. **130**, 7544 (2008)
82. Chow, K.-F., Mavré, F., Crooks, J.A., Chang, B.-Y., Crooks, R.M.: J. Am. Chem. Soc. **131**, 8364 (2009)
83. Qi, W., Lai, J., Gao, W., Li, S., Hanif, S., Xu, G.: Anal. Chem. **86**, 8927 (2014)
84. Feng, Q.-M., Pan, J.-B., Zhang, H.-R., Xu, J.-J., Chen, H.-Y.: Chem. Commun. **50**, 10949 (2014)
85. Zhang, X., Li, J., Jia, X., Li, D., Wang, E.: Anal. Chem. **86**, 5595 (2014)
86. Zhang, J.-D., Yu, T., Li, J.-Y., Xu, J.-J., Chen, H.-Y.: Electrochem. Commun. **49**, 75 (2014)
87. Eßmann, V., Jambrec, D., Kuhn, A., Schuhmann, W.: Electrochem. Commun. **50**, 77 (2015)
88. Wu, S.Z., Zhou, Z.Y., Xu, L.R., Su, B., Fang, Q.: Biosens. Bioelectron. **53**, 148 (2014)
89. Shi, H.W., Wu, M.S., Du, Y., Xu, J.J., Chen, H.Y.: Biosens. Bioelectron. **55**, 459 (2014)
90. Wu, M.-S., Yuan, D.-J., Xu, J.-J., Chen, H.-Y.: Chem. Sci. **4**, 1182 (2013)
91. Zhang, X., Chen, C., Li, J., Zhang, L., Wang, E.: Anal. Chem. **85**, 5335 (2013)
92. Wu, M.-S., Yuan, D.-J., Xu, J.-J., Chen, H.-Y.: Anal. Chem. **85**, 11960 (2013)
93. Wu, M.-S., Xu, B.-Y., Shi, H.-W., Xu, J.-J., Chen, H.-Y.: Lab Chip **11**, 2720 (2011)
94. Wu, M.-S., Qian, G.-S., Xu, J.-J., Chen, H.-Y.: Anal. Chem. **84**, 5407 (2012)
95. Sentic, M., Loget, G., Manojlovic, D., Kuhn, A., Sojic, N.: Angew. Chem. Int. Ed. **51**, 11284 (2012)
96. Bouffier, L., Zigah, D., Adam, C., Sentic, M., Fattah, Z., Manojlovic, D., Kuhn, A., Sojic, N.: ChemElectroChem **1**, 95 (2014)
97. Sentic, M., Arbault, S., Bouffier, L., Manojlovic, D., Kuhn, A., Sojic, N.: Chem. Sci. **6**, 4433 (2015)
98. de Poulpiquet, A., Diez-Buitrago, B., Milutinovic, M., Goudeau, B., Bouffier, L., Arbault, S., Kuhn, A., Sojic, N. ChemElectroChem **3**, 404–409 (2016)
99. Sentic, M., Arbault, S., Goudeau, B., Manojlovic, D., Kuhn, A., Bouffier, L., Sojic, N.: Chem. Commun. **50**, 10202 (2014)
100. Guo, Z., Hao, T., Wang, S., Gan, N., Li, X., Wei, D.: Electrochem. Commun. **14**, 13 (2012)
101. Zhang, M., Dai, W., Yan, M., Ge, S., Yu, J., Song, X., Xu, W.: Analyst **137**, 2112 (2012)
102. Zhang, M., Ge, S., Li, W., Yan, M., Song, X., Yu, J., Xu, W., Huang, J.: Analyst **137**, 680 (2012)
103. Guo, Z., Hao, T., Duan, J., Wang, S., Wei, D.: Talanta **89**, 27 (2012)
104. Liao, N., Zhuo, Y., Chai, Y., Xiang, Y., Cao, Y., Yuan, R., Han, J.: Chem. Commun. **48**, 7610 (2012)
105. Klymenko, O.V., Svir, I., Amatore, C.: ChemPhysChem **14**, 2237 (2013)
106. Deiss, F., LaFratta, C.N., Symer, M., Blicharz, T.M., Sojic, N., Walt, D.R.: J. Am. Chem. Soc. **131**, 6088 (2009)
107. Miao, W., Bard, A.: J. Anal. Chem. **76**, 7109 (2004)
108. Miao, W., Bard, A.: J. Anal. Chem. **76**, 5379 (2004)
109. Habtamu, H.B., Sentic, M., Silvestrini, M., De Leo, L., Not, T., Arbault, S., Manojlovic, D., Sojic, N., Ugo, P.: Anal. Chem. **87**, 12080 (2015)
110. Shen, W., Tian, D., Cui, H., Yang, D., Bian, Z.: Biosens. Bioelectron. **27**, 18 (2011)
111. Sloan, J.H., Siegel, R.W., Ivanova-Cox, Y.T., Watson, D.E., Deeg, M.A., Konrad, R.J.: Clin. Biochem. **45**, 1640 (2012)
112. Wang, G., Jin, F., Dai, N., Zhong, Z., Qing, Y., Li, M., Yuan, R., Wang, D.: Anal. Biochem. **422**, 7 (2012)

113. Yao, X., Yan, P., Tang, Q., Deng, A., Li, J.: Anal. Chim. Acta **798**, 82 (2013)
114. Bertoncello, P.: Front. Biosci.-Landmark **16**, 1084 (2011)
115. Deng, S., Ju, H.: Analyst **138**, 43 (2013)
116. Lei, J., Ju, H.: Trac-Trend Anal. Chem. **30**, 1351 (2011)
117. Lisdat, F., Schaefer, D., Kapp, A.: Anal. Bioanal. Chem. **405**, 3739 (2013)
118. Zhou, H., Gan, N., Li, T., Cao, Y., Zeng, S., Zheng, L., Guo, Z.: Anal. Chim. Acta **746**, 107 (2012)
119. Lin, D., Wu, J., Yan, F., Deng, S., Ju, H.: Anal. Chem. **83**, 5214 (2011)
120. Peng, S., Zhang, X.: Microchim. Acta **178**, 323 (2012)
121. Tian, C.-Y., Zhao, W.-W., Wang, J., Xu, J.-J., Chen, H.-Y.: Analyst **137**, 3070 (2012)
122. Han, E., Ding, L., Lian, H., Ju, H.: Chem. Commun. **46**, 5446 (2010)
123. Li, J., Guo, S., Wang, E.: Rsc Adv. **2**, 3579 (2012)
124. Chai, Y., Tian, D., Wang, W., Cui, H.: Chem. Commun. **46**, 7560 (2010)
125. Tian, D., Zhang, H., Chai, Y., Cui, H.: Chem. Commun. **47**, 4959 (2011)
126. Zhang, Y., Ge, S., Wang, S., Yan, M., Yu, J., Song, X., Liu, W.: Analyst **137**, 2176 (2012)
127. Chikkaveeraiah, B.V., Bhirde, A.A., Morgan, N.Y., Eden, H.S., Chen, X.: ACS Nano **6**, 6546 (2012)
128. Xu, Y., Liu, J., Gao, C., Wang, E.: Electrochem. Commun. **48**, 151 (2014)
129. Wang, T., Zhang, S., Mao, C., Song, J., Niu, H., Jin, B., Tian, Y.: Biosens. Bioelectron. **31**, 369 (2012)
130. Konry, T., Bouhifd, M., Cosnier, S., Whelan, M., Valsesia, A., Rossi, F., Marks, R.S.: Biosens. Bioelectron. **22**, 2230 (2007)
131. Le Goff, A., Holzinger, M., Cosnier, S.: Electrochim. Acta **56**, 3633 (2011)
132. Zamolo, V.A., Valenti, G., Venturelli, E., Chaloin, O., Marcaccio, M., Boscolo, S., Castagnola, V., Sosa, S., Berti, F., Fontanive, G., Poli, M., Tubaro, A., Bianco, A., Paolucci, F., Prato, M.: ACS Nano **6**, 7989 (2012)
133. Dennany, L., Forster, R.J., White, B., Smyth, M., Rusling, J.F.: J. Am. Chem. Soc. **126**, 8835 (2004)
134. Hvastkovs, E.G., So, M., Krishnan, S., Bajrami, B., Tarun, M., Jansson, I., Schenkman, J.B., Rusling, J.F.: Anal. Chem. **2007**, 79 (1897)
135. Liu, Z., Zhou, C., Zheng, B., Qian, L., Mo, Y., Luo, F., Shi, Y., Choi, M.M.F., Xiao, D.: Analyst **136**, 4545 (2011)
136. Sardesai, N.P., Barron, J.C., Rusling, J.F.: Anal. Chem. **83**, 6698 (2011)
137. Li, L., Chen, Y., Lu, Q., Ji, J., Shen, Y., Xu, M., Fei, R., Yang, G., Zhang, K., Zhang, J.-R., Zhu, J.-J.: Sci. Rep. **3** (2013)
138. Li, W., Li, L., Li, M., Yu, J., Ge, S., Yan, M., Song, X.: Chem. Commun. **49**, 9540 (2013)
139. Li, Z., Wang, Y., Kong, W., Li, C., Wang, Z., Fu, Z.: Biosens. Bioelectron. **39**, 311 (2013)
140. Burton, P.R., Clayton, D.G., Cardon, L.R., Craddock, N., Deloukas, P., Duncanson, A., Kwiatkowski, D.P., McCarthy, M.I., Ouwehand, W.H., Samani, N.J., Todd, J.A., Donnelly, P., Barrett, J.C., Davison, D., Easton, D., Evans, D., Leung, H.-T., Marchini, J.L., Morris, A.P., Spencer, C.C.A., Tobin, M.D., Attwood, A.P., Boorman, J.P., Cant, B., Everson, U., Hussey, J.M., Jolley, J.D., Knight, A.S., Koch, K., Meech, E., Nutland, S., Prowse, C.V., Stevens, H.E., Taylor, N.C., Walters, G.R., Walker, N.M., Watkins, N.A., Winzer, T., Jones, R.W., McArdle, W.L., Ring, S.M., Strachan, D.P.; Pembrey, M., Breen, G., St. Clair, D., Caesar, S., Gordon-Smith, K., Jones, L., Fraser, C., Green, E.K., Grozeva, D., Hamshere, M.L., Holmans, P.A., Jones, I.R., Kirov, G., Moskvina, V., Nikolov, I., O'Donovan, M.C., Owen, M.J., Collier, D.A., Elkin, A., Farmer, A., Williamson, R., McGuffin, P., Young, A.H., Ferrier, I.N., Ball, S.G., Balmforth, A.J., Barrett, J.H., Bishop, D.T., Iles, M.M., Maqbool, A., Yudasheve, N., Hall, A.S., Braund, P.S.; Dixon, R.J., Mangino, M., Stevens, S., Thompson, J.R., Bredin, F., Tremelling, M., Parkes, M., Drummond, H., Lees, C.W., Nimmo, E.R., Satsangi, J., Fisher, S.A., Forbes, A., Lewis, C.M., Onnie, C.M., Prescott, N.J., Sanderson, J., Mathew, C.G., Barbour, J., Mohiuddin, M.K., Todhunter, C.E., Mansfield, J.C., Ahmad, T., Cummings, F.R., Jewell, D.P.: Nature **447**, 661 (2007)

141. McCarthy, J.J., Hilfiker, R.: Nat. Biotechnol. **18**, 505 (2000)
142. Sun, B., Qi, H., Ma, F., Gao, Q., Zhang, C., Miao, W.: Anal. Chem. **82**, 5046 (2010)
143. Wang, S., Milam, J., Ohlin, A.C., Rambaran, V.H., Clark, E., Ward, W., Seymour, L., Casey, W.H., Holder, A.A., Miao, W.: Anal. Chem. **81**, 4068 (2009)
144. Ding, C., Zhang, W., Wang, W., Chen, Y., Li, X.: Trac-Trend Anal. Chem. **65**, 137 (2015)
145. Jiang, J., Chai, Y., Cui, H.: Rsc Adv. **1**, 247 (2011)
146. Jiang, D., Liu, F., Liu, C., Liu, L., Li, Y., Pu, X.: Anal. Meth. **6**, 1558 (2014)
147. Chen, Y., Xu, J., Su, J., Xiang, Y., Yuan, R., Chai, Y.: Anal. Chem. **84**, 7750 (2012)
148. Liu, T., Chen, X., Hong, C.-Y., Xu, X.-P., Yang, H.-H.: Microchim. Acta **181**, 731 (2014)
149. Wang, J., Zhao, W.-W., Zhou, H., Xu, J.-J., Chen, H.-Y.: Biosens. Bioelectron. **41**, 615 (2013)
150. Wei, W., Zhou, J., Li, H., Yin, L., Pu, Y., Liu, S.: Analyst **138**, 3253 (2013)
151. Zhou, H., Liu, J., Xu, J.-J., Chen, H.-Y.: Chem. Commun. **47**, 8358 (2011)
152. Zhou, H., Liu, J., Xu, J.-J., Chen, H.-Y.: Anal. Chem. **83**, 8320 (2011)
153. Dennany, L., Forster, R.J., Rusling, J.F.: J. Am. Chem. Soc. **125**, 5213 (2003)
154. Lee, J.H., Yigit, M.V., Mazumdar, D., Lu, Y.: Adv. Drug Deliv. Rev. **62**, 592 (2010)
155. Zhou, W., Huang, P.-J.J., Ding, J., Liu, J.: Analyst **139**, 2627 (2014)
156. Liu, Z., Zhang, W., Hu, L., Li, H., Zhu, S., Xu, G.: Chem. Eur. J. **16**, 13356 (2010)
157. Hu, L., Bian, Z., Li, H., Han, S., Yuan, Y., Gao, L., Xu, G.: Anal. Chem. **81**, 9807 (2009)
158. Chen, Y., Jiang, B., Xiang, Y., Chai, Y., Yuan, R.: Chem. Commun. **47**, 7758 (2011)
159. Chen, L., Cai, Q., Luo, F., Chen, X., Zhu, X., Qiu, B., Lin, Z., Chen, G.: Chem. Commun. **46**, 7751 (2010)
160. Dang, J., Guo, Z., Zheng, X.: Anal. Chem. **86**, 8943 (2014)
161. Yuan, T., Liu, Z., Hu, L., Zhang, L., Xu, G.: Chem. Commun. **47**, 11951 (2011)
162. Shi, H.-W., Wu, M.-S., Du, Y., Xu, J.-J., Chen, H.-Y.: Biosens. Bioelectron. **55**, 459 (2014)
163. Tian, C.-Y., Xu, J.-J., Chen, H.-Y.: Chem. Commun. **48**, 8234 (2012)
164. Hu, L., Xu, G.: Chem. Soc. Rev. **39**, 3275 (2010)
165. Wilson, R., Kremeskötter, J., Schiffrin, D.J., Wilkinson, J.S.: Biosens. Bioelectron. **11**, 805 (1996)
166. de Poulpiquet, A., Diez-Buitrago, B., Milutinovic, M., Sentic, M., Arbault, S., Bouffier, L., Kuhn, A., Sojic, N.: (2016, submitted)
167. Marquette, C.A., Blum, L.J.: Anal. Chim. Acta **1** (1999)
168. Marquette, C.A., Degiuli, A., Blum, L.J.: Biosens. Bioelectron. **19**, 433 (2003)
169. Leca, B., Blum, L.J.: Analyst **125**, 789 (2000)
170. Wang, X.F., Zhou, Y., Xu, J.J., Chen, H.Y.: Adv. Funct. Mater. **19**, 1444 (2009)
171. Dai, H., Wu, X., Xu, H., Wei, M., Wang, Y., Chen, G.: Electrochem. Comm. **11**, 1599 (2009)
172. Liu, X., Ju, H.: Anal. Chem. **80**, 5377 (2008)
173. Jameison, F., Sanchez, R.I., Dong, L., Leland, J.K., Yost, D., Martin, M.T.: Anal. Chem. **68**, 1298 (1996)
174. Loget, G., Kuhn, A.: Nat. Commun. **2**, 535 (2011)
175. Roche, J., Carrara, S., Lannelongue, J., Loget, G., Bouffier, L., Kuhn, A., Sanchez, J., Fischer, P.: Sci. Rep. **4**, 6705 (2014)
176. Delaney, J.L., Hogan, C.F., Tian, J., Shen, W.: Anal. Chem. **83**, 1300 (2011)
177. Doeven, E.H., Barbante, G.J., Kerr, E., Hogan, C.F., Endler, J.A., Francis, P.S.: Anal. Chem. **86**, 2727 (2014)
178. Swanick, K.N., Ladouceur, S., Zysman-Colman, E., Ding, Z.: Angew. Chem. Int. Ed. **51**, 11079 (2012)
179. Doeven, E.H., Barbante, G.J., Hogan, C.F., Francis, P.S.: Chempluschem **80**, 456 (2015)
180. Valenti, G., Rampazzo, E., Bonacchi, S., Khajvand, T., Juris, R., Montalti, M., Marcaccio, M., Paolucci, F., Prodi, L.: Chem. Commun. **48**, 4187 (2012)

181. Zanarini, S., Felici, M., Valenti, G., Marcaccio, M., Prodi, L., Bonacchi, S., Contreras-Carballada, P., Williams, R.M., Feiters, M.C., Nolte, R.J.M., De Cola, L., Paolucci, F.: Chem. Eur. J. **17**, 4640 (2011)
182. Zanarini, S., Rampazzo, E., Ciana, L.D., Marcaccio, M., Marzocchi, E., Montalti, M., Paolucci, F., Prodi, L.: J. Am. Chem. Soc. **131**, 2260 (2009)
183. Zanarini, S., Rampazzo, E., Bonacchi, S., Juris, R., Marcaccio, M., Montalti, M., Paolucci, F., Prodi, L.: J. Am. Chem. Soc. **131**, 14208 (2009)

Electrochemically Driven Luminescence in Organometallic and Inorganic Systems

Andrea Fiorani, Giovanni Valenti, Elena Villani, Massimo Marcaccio, Enrico Rampazzo, Luca Prodi and Francesco Paolucci

Abstract This chapter analyses the literature appeared within the decade that follows the publication of M. Richter's exhaustive chapter dedicated to *metal chelates* in the comprehensive ECL monograph edited by A. J. Bard in 2004. In this chapter, we have attempted to cover, although somehow selectively, the published work on the application of metal chelates in ECL, organizing the material, similarly to Richter's choice, according to the main metal. Perhaps not surprisingly, among the metal chelate systems, $Ru(bpy)_3^{2+}$ (bpy = 2,2′-bipyridine) has still been, over the last decade, the main star in the ECL sky as previously, in view in particular of its outstanding role in bioanalytical research and commercial applications. Nonetheless, the importance of other coordination and organometallic systems, especially those based on iridium, has grown in the recent research literature because of their photophysical and electrochemical properties that may offer great advantages in the technical development of ECL. A variety of reviews pertaining to particular aspects of metal chelates application in ECL, in particular for (bio)analytical purposes but also covering many other aspects of this fascinating area, are available to which the reader is directed for further information.

1 Introduction

Electrochemically generated chemiluminescence or electrogenerated chemiluminescence (ECL) is a redox-induced light emission in which species generated at electrodes undergo high-energy electron transfer reaction to form light-emitting excited states [1–7].

The ECL emission is initiated and controlled by the electrode potential. As ECL is a method of producing light at an electrode, it represents an ideal marriage between electrochemical and spectroscopic methods [2, 3]. A chemical species to

A. Fiorani · G. Valenti · E. Villani · M. Marcaccio · E. Rampazzo · L. Prodi · F. Paolucci (✉)
Department of Chemistry Giacomo Ciamician, Alma Mater Studiorum—University of Bologna, Via Selmi 2, 40126 Bologna, Italy
e-mail: francesco.paolucci@unibo.it

F. Miomandre and P. Audebert (eds.), *Luminescence in Electrochemistry*,
DOI 10.1007/978-3-319-49137-0_9

be an efficient ECL emitter has to fulfil not only various electrochemical and photophysical requirements, but also other criteria depending on the envisaged application. In the perspective of its use in ECL-based biological assays, for example, the luminophore needs to have a number of important features, besides an intense and stable ECL signal in physiological conditions, such as the presence of suitable binding sites for bioconjugation, and good solubility in aqueous buffers to allow the labelling reaction. Inorganic coordination compounds and, in particular, ruthenium [8–10] and iridium [11–17] complexes play a privileged role in such a context since they display a rich and highly stable redox behaviour, although not always accessible in aqueous media, and long-lived, luminescent excited states. However, most of the reported ECL luminophores display limited water solubility, problems in the introduction of binding sites without affecting their pristine photophysical/electrochemical properties; however, they lack stability during electrochemical switching. All this restricts very much the number of possible candidates, and ruthenium polypyridine complexes (in particular, $Ru(bpy)_3{}^{2+}$) are practically the only luminophores so far employed in the commercial applications of ECL. Among the new approaches to innovative ECL luminophores that would increase the available colours, also allowing multiplexing ECL analysis, one of the most promising is that involving nanostructured dyes, e.g. doped nanoparticles [18, 19].

At the basis of ECL is chemiluminescence, i.e. the light emission phenomenon taking place when a highly exergonic reaction occurs between an electron acceptor *A* and an electron donor *D*, thus generating the excited state of one of the two products (either oxidized *D* or reduced *A*).

$$D + A \rightarrow D^{+*} + A^- \quad (1)$$

$$D^{+*} \rightarrow D^+ + h\nu \quad (2)$$

The existence of a class of redox processes that are kinetically highly unfavoured because of their very high exergonicity (i.e. the inverted region, originally predicted by Marcus [20, 21]) is the mechanistic basis explaining why reaction (1) would lead to the generation of an excited state (D^{+*}) rather than the more stable ground state (D^+). This is shown in Fig. 1 where the free energy curve for reactants intersects with that of products in their ground states at higher energy than with that where one of the products is electronically excited.

Electrochemistry provides a very convenient way to ignite a chemiluminescence reaction through the generation, at the electrode surface and upon the application of the electrochemical pulse, of a pair of high-energy reactants, thus leading to electrogenerated chemiluminescence.

The simplest mechanism to generate electrochemically an emitting excited state is the so-called *annihilation mechanism*. Herein, such a mechanism is illustrated in Fig. 2, in the case of the most popular fluorophore ever used in ECL, namely $Ru(bpy)_3^{2+}$. Such a species, dissolved in a proper electrochemical solution,

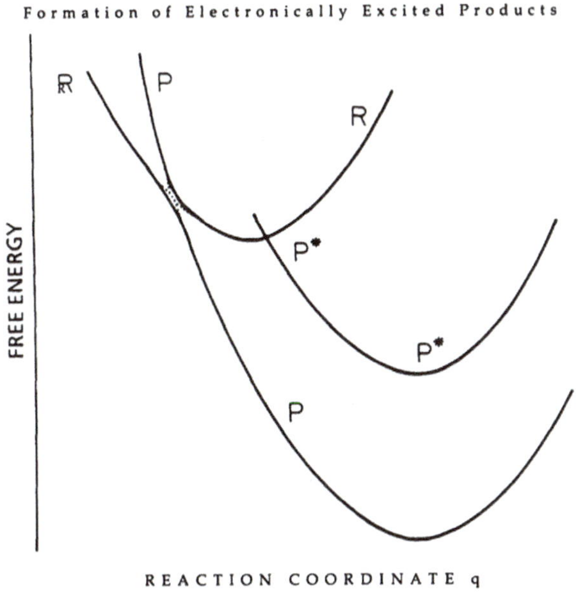

Fig. 1 The favoured formation of an electronically excited state of the products of a highly exergonic reaction (Reproduced from Nobel Lecture by Rudolph A. Marcus, "Electron transfer reactions in chemistry: theory and experiment" From Nobel Lectures, Chemistry 1991–1995, Editor Bo G. Malmström, World Scientific Publishing Co., Singapore, 1997. © The Nobel Foundation 1992)

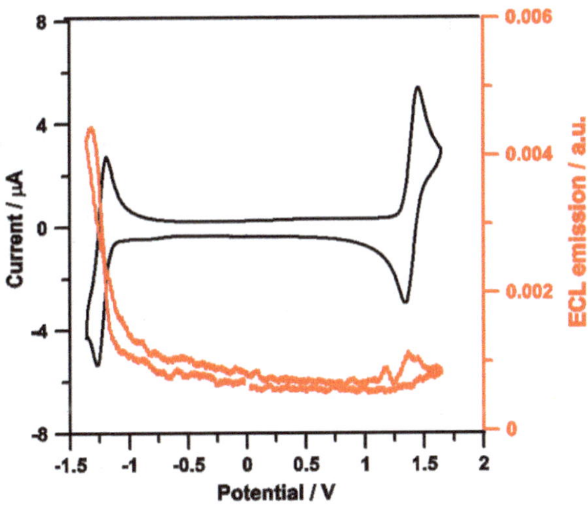

Fig. 2 Cyclic voltammetric curves and the ECL intensity profile of $Ru(bpy)_3^{2+}$ (0.5 mM) in $TBAPF_6/CH_3CN$ (0.07 M). Working electrode: Pt disc (2 mm diameter); reference electrode: SCE; scan rate = 1 V s^{-1}; $T = 25$ °C; PMT bias 750 V

typically acetonitrile and tetrabutylammonium hexafluorophosphate as supporting electrolyte, is first reduced and then oxidized upon the application of a potential rapidly changing from highly reductive to highly oxidative values, thus bringing about the enrichment of the diffusion layer in close proximity to the electrode surface with both the reduced and oxidized forms of the precursor.

The high-energy species generated at the electrode may then react with one another, in analogy with the pair of reactants *A*/*D* of the chemiluminescent reaction (see Eqs. 1 and 2), to finally produce light emission

$$Ru(bpy)_3{}^{2+} + e^- \rightarrow Ru(bpy)_3{}^{+} \tag{3}$$

$$Ru(bpy)_3{}^{2+} \rightarrow Ru(bpy)_3{}^{3+} + e^- \tag{4}$$

$$Ru(bpy)_3{}^{+} + Ru(bpy)_3{}^{3+} \rightarrow Ru(bpy)_3{}^{2+*} + Ru(bpy)_3{}^{2+} \tag{5}$$

$$Ru(bpy)_3{}^{2+*} \rightarrow Ru(bpy)_3{}^{2+} + light \tag{6}$$

The opposite reaction scheme, where oxidative conditions are first applied followed by reductive ones, may alternatively be used and usually results in a similar emission.

The question of which of the two parent species in reaction (5), either $Ru(bpy)_3{}^{+}$ or $Ru(bpy)_3{}^{3+}$, form the excited state is of some interest [22]. On the basis of simple orbital overlap arguments, the formation of $Ru(bpy)_3{}^{2+*}$ from the oxidized parent (involving ligand-to-ligand electron transfer) should be kinetically favoured over formation from the reduced parent (which requires metal-to-metal electron transfer). Insights into the mechanism of formation of excited state were recently obtained upon investigating mixed inorganic ECL systems involving iridium complexes in which the electron density of their HOMO is largely delocalized over their ligands as well as on the metal [23]. For example, the HOMO of $Ir(ppy)_3$ has been estimated to be less than 50% metal based, on the basis of DFT calculations. Therefore, the HOMO/HOMO electron transfer route ought to be relatively less disfavoured when an iridium complex is the oxidant compared to the case where a ruthenium complex with a purely metal-based HOMO is used.

The ECL emission is spatially located in a region of few tens of μm from the electrode surface (the reaction layer), where both reactants reach sufficiently high concentrations at the same time to allow the bimolecular annihilation process to occur at a finite rate. Chemical stability of the electrochemically generated *A* and *D* species is therefore a very important issue, since parasitic reactions that may remove one (or both) species from the reaction layer would affect drastically the overall efficiency of the annihilation reaction and would result in a lower ECL intensity.

In order to lead to the generation of the emitting excited state, the annihilation reaction needs to satisfy energy requirements that are strictly related to both the redox and photophysical properties of the precursor. In fact, the free energy for the generation of the excited state (ΔG_{ES}) can be calculated according to the following equations [1].

$$\Delta G_{\mathrm{ES}} \approx \Delta G_{\mathrm{annihil}} + \frac{1240.8}{\lambda(\mathrm{nm})} \tag{7}$$

$$\Delta G_{\mathrm{annihil}} \approx -\mathrm{e}\left(\mathrm{E}_{\frac{1}{2},\mathrm{Ox}} - \mathrm{E}_{\frac{1}{2},\mathrm{Red}}\right) + \Delta G_{\mathrm{el}} \tag{8}$$

Equation (8) includes the ΔG_{el} term that accounts for the electrostatic interactions between the partners of the annihilation reaction when they are both charged. Such a term may represent a significant fraction of the free energy involved in the process and should therefore be taken into account in the calculation of ΔG_{ES}.

ECL has been widely exploited over the last 20 years as a highly sensitive and selective detection technique, in particular for bio-related applications such as microchip capillary electrophoresis to detect amino acids, oxalate, NADH (nicotinamide adenine dinucleotide), alkylamines and nucleic acids [24–26]. ECL systems are commercialized, for example, by Roche Diagnostics Corp. and by MesoScale Diagnostic [27] for a number of assays with a particular focus on immunoassay to detect many important biological molecules and macromolecules such as DNA [28–30], cancer biomarker proteins [31, 32], cytochrome [33], cocaine [34], toxins or viruses [35, 36].

The development of coreactant approach was a crucial point for the implementation of ECL as analytical technique; in fact, the majority of commercially available ECL-based instrumentations employs this strategy. Unlike the typical annihilation procedure, in this kind of ECL the excited state is generated through the reaction between two different precursors, an *emitter* and a *coreactant*, and the oxidation (or reduction) of these species is first carried out. The coreactant can be chosen among different classes of chemicals, and its fundamental feature is to present a highly unstable oxidized (or reduced) form that undergoes a fast chemical degradation, thus forming a high-energy radical. Two different mechanisms are then possible, depending on the nature of the coreactant, i.e. the *"oxidative-reduction"* or the *"reductive-oxidation"* one. In the former case, the potential is increased in order to oxidize both the emitter and the coreactant. The radical (directly formed at the electrode through the degradation of the oxidized coreactant) acts as a strong reductant towards the emitter generating its reduced form. This process takes place in the diffusion layer where the two species finally undergo the usual annihilation reaction that gives rise to the ECL emission. In the *"reductive-oxidation"* mechanism, the reduced coreactant (generated at the electrode) undergoes a rapid further reaction generating a strong oxidizing radical, which reacts with the luminophore generating its oxidized form. Finally, a last mechanism, the "*mixed annihilation*" [1, 37] one, is also possible, in which the luminophore and the coreactant act as independent precursors to generate the high-energy pair for the annihilation reaction (1).

Coreactant ECL presents unique advantages that explain its prevalence with respect to the annihilation procedure in all bioanalytical applications. First of all, a high stability of both the oxidized and reduced forms of the luminophore is no longer required, thus increasing the number of potential ECL reporters that may be used in the analysis. Moreover, also solvents with a limited potential window can be used. The solvent discharge (i.e. its oxidation or reduction) would in fact prevent the generation of at least one of the two annihilation partners. This is the case, for instance, of aqueous buffer solutions, where hydrogen evolution may strongly compete with the generation of the reduced luminophore, thus inhibiting the annihilation procedure. Finally, the possibility to operate only in the region of positive potentials (*oxidative-reduction mechanism*) avoids any possible interference of oxygen (that is a reducible species) in the generation of ECL partners, thus permitting in principle to perform analysis in air.

2 Ruthenium

2.1 Mononuclear Complexes

Since the first ECL experiment involving ruthenium was proved, by using the complex $Ru(bpy)_3^{2+}$ [38], many chelates of this metal were synthesized for ECL studies and applications, though this complex remains the most popular [3, 6, 7].

Here, we focus our attention on ruthenium complexes over the last decade, as for the previous years they were already reviewed [1].

One of the main objectives when designing a ruthenium complex is to enhance the ECL efficiency compared to the standard $Ru(bpy)_3^{2+}$ or tuning the emission wavelengths, usually for analytical applications prospect. This was achieved by means of two strategies, such as the functionalization of the bpy ligand or replacing it with a different ligand to obtain other homo- or heteroleptic complexes.

Examples concerning the former approach see the functionalization of bpy as shown in Fig. 3 [39].

The PL quantum yields of the complexes were found to be higher in the buffer solutions than in the deaerated acetonitrile solutions, contrary to general perception. The integrated ECL intensity (buffer solution/TPA as coreactant) for 8 s did not follow the PL quantum yields. Complexes with the bpy moiety functionalized with

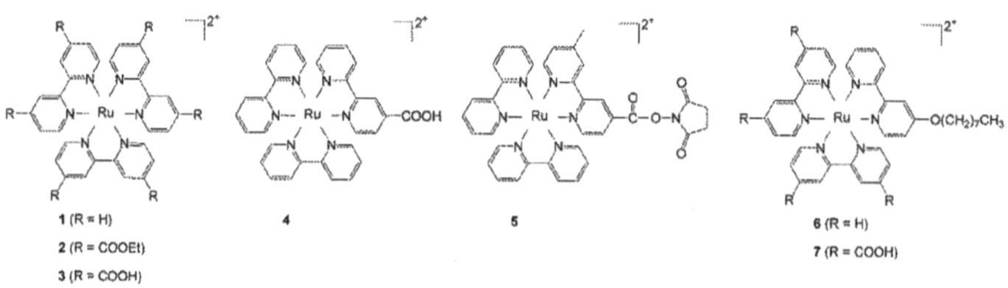

Fig. 3 Ruthenium complexes with carboxyl-, NHS ester- and alkoxy-functionalized bpys (Reproduced from [39]. Copyright 2005 American Chemical Society)

carboxyl, N-hydroxysuccinimide ester group showed higher ECL intensity than $Ru(bpy)_3{}^{2+}$ although PL quantum yields are lower compared to the standard.

Similar ligands, bpy derivatives, were synthesized in order to tune the ECL emission wavelength for a series of heteroleptic ruthenium complexes in annihilation ECL [40]. Furthermore, functionalized ruthenium complexes were designed for coupling with other molecules or surfaces for ECL-sensing applications. Electron-withdrawing (EW) or electron-donating (ED) groups on the ligands of $Ru(bpy)_3{}^{2+}$ cause shifts in the emission spectrum, by lowering the π^* energy levels or destabilising the d (HOMO) orbital in the ground-state species.

DFT calculations showed that the LUMO is dominated by the substituted ligand for the complexes containing EW substituents and by the unsubstituted bpy ligand for the complexes containing ED groups. Electron-withdrawing groups, such as ester functionality, had a significant effect, resulting in a large positive shift in the first reduction potential. There was also a small positive shift in the formal potential of the Ru^{2+}/Ru^{3+} couple. By contrast, the ligand reductions remained evenly spaced and virtually unaffected by the methyl substitution. The carboxylate functionality significantly decreased the reduction potential relative to the unsubstituted complex, but only moderately increased the oxidation potential, with the ester producing a larger shift than the amide. In the case of alkyl substituent, the ligand-based reductions were unaffected by the electron-donating effect, but the complexes are slightly more easily oxidized. Moreover, neither the alkyl chain length nor its end group had any effect on the redox potentials

The electrochemical HOMO–LUMO gap is a good predictor of the spectroscopic properties, and excellent correlation was observed between the difference in ground-state redox potentials, $\Delta E^{\circ}_{\text{II-I}}$, and the energy corresponding to absorption or emission maxima.

In general, the PL quantum yield (Φ_p) is a reasonable predictor of the annihilation ECL yield (Φ_{ECL}), within related sets of compounds such as this, despite the range of factors that may degrade the efficiency of population of the excited state.

Tuning the ECL properties of ruthenium complexes was also achieved by choosing different ligands. For example, the tetrazolate ring was largely investigated as an alternative ligand for ruthenium, in mono- and bimetallic complexes [41, 42].

The PL and ECL spectra of tetrazolate complexes revealed a red-shifted emission with respect to the $Ru(bpy)_3{}^{2+}$. ECL intensities (Fig. 4) for mononuclear compounds are lower than for $Ru(bpy)_3{}^{2+}$ but are of the same order of magnitude. The dinuclear Ru complex, Ru(4-TBN)Ru, shows an emission about 3 times more intense than the reference one. Moreover, the emission intensity can be tuned by either including or not the second reduction and oxidation steps in the potential scan. Considering the four different mononuclear complexes, the higher efficiency was for the ligands containing tetrazolate bound with pyridine and benzene. Another interesting point is the change in ECL efficiency between Ru(4-TBN) and Ru(4-TBN)Ru. The addition of a second Ru(bpy)(tpy) centre causes an increase of ECL efficiency by 6 times, with the two Ru centres in Ru(4-TBN)Ru behaving

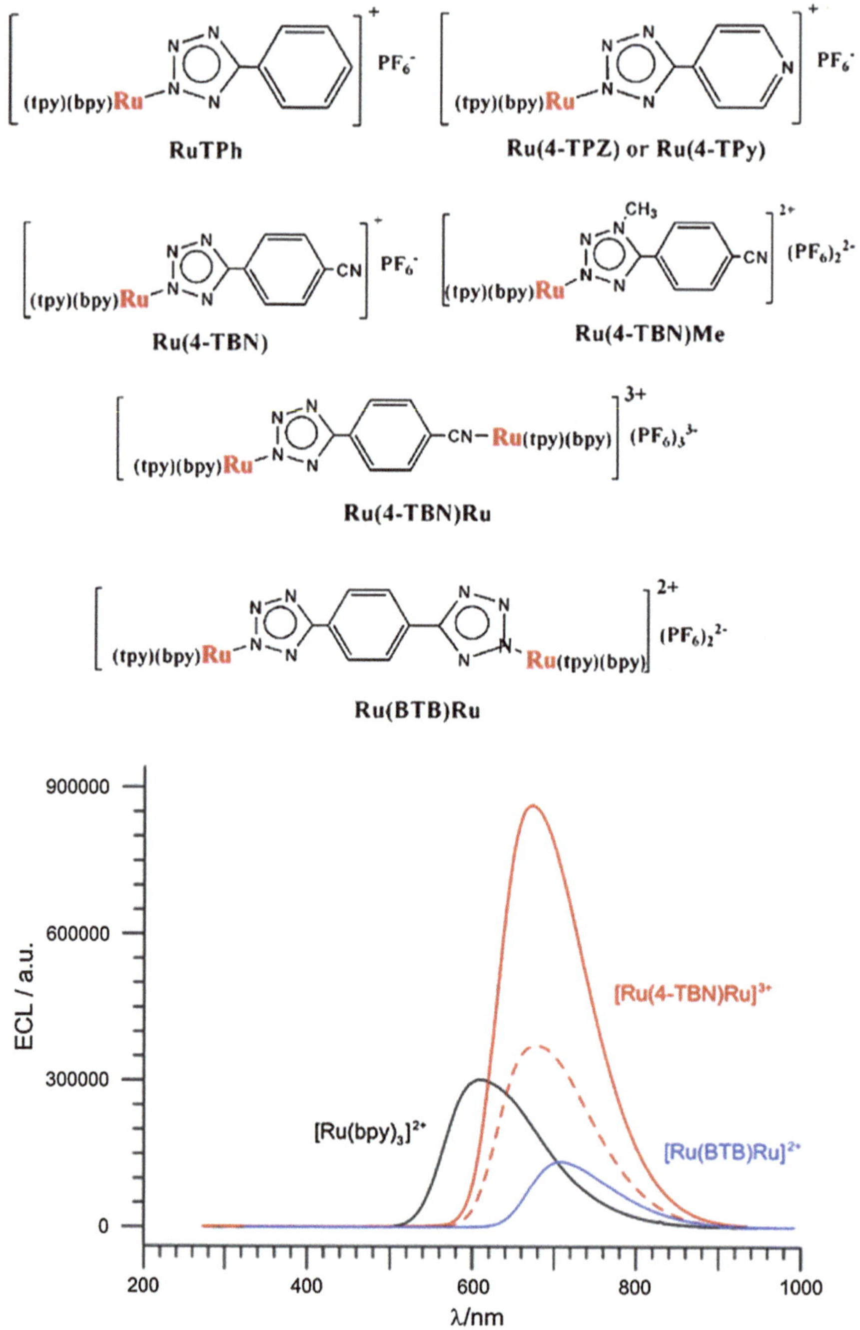

Fig. 4 Tetrazolate ruthenium complexes and ECL spectra of bimetallic complexes compared to $Ru(bpy)_3^{2+}$ (Reproduced from [42] and [8]. Copyright 2006 American Chemical Society)

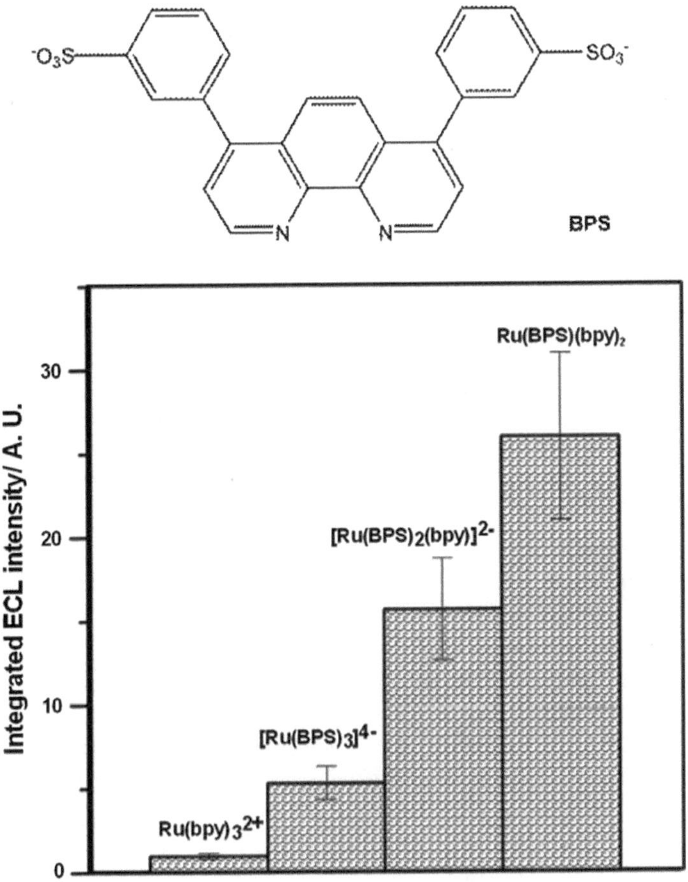

Fig. 5 The BPS ligand and related complexes integrated intensity in coreactant ECL with TPA (Reproduced from [44]. Copyright 2010 American Chemical Society)

independently [43]; thus, the inclusion of the second oxidation and reduction localized on the second Ru centre causes the ECL intensity to double. The moderate presence of oxygen has been found to affect the ECL efficiency of $Ru(bpy)_3^{2+}$ much more than for tetrazolate complexes.

The intense ECL emission of Ru(4-TBN)Ru suggested its use for the preparation of light-emitting solid devices, which show emission intensity comparable to that of $Ru(bpy)_3^{2+}$ based devices. In addition, Ru(4-TBN)Me can be also considered as a luminophore for the design of NIR (780 nm) light-emitting solid devices.

Other homo- and heteroleptic ruthenium complexes were synthesized with the bis-chelating disulphonate bathophenanthroline (BPS) ligand, and their properties were investigated in both annihilation ECL and coreactant method with TPA (Fig. 5) [9, 44].

The complexes showed lower emission intensity compared to $Ru(bpy)_3^{2+}$ in annihilation ECL, while the intensity was higher for coreactant ECL with TPA. The cause of the reduced ECL quantum yield of $[Ru(BPS)_3]^{4-}$ in both annihilation and coreactant-assisted methods appears to be the formation of a filming product at the electrode during oxidation, while this does not occur when the negative charge is reduced or eliminated.

N-heterocyclic carbene (NHC) ligands where used to synthesize a series of four ruthenium complexes of the form $[Ru(bpy)_2(C^\wedge N)]^{2+}$ ($C^\wedge N$ is a bidentate pyridine-functionalized imidazolylidene- or benzimidazolylidene-based NHC ligand) [45]. The maximum emission wavelength, for both the PL and the ECL spectra, showed a red-shift for imidazolylidene-based complexes, while the benzimidazolylidene-based ligands are blue-shifted compared to $Ru(bpy)_3^{2+}$. The red-shift in emission occurs as a result of a narrowing of the HOMO–LUMO gap caused by destabilization of the HOMO, while the blue-shift in the emission maxima is in agreement with the stabilization of the HOMO, compared to $Ru(bpy)_3^{2+}$. The influence of the benzimidazolylidene-based NHC ligand can be related to π back-bonding, where back-donation of electron density from the metal to the ligand mitigates the strong sigma donating properties commonly found for NHCs. The complexes exhibit moderately intense PL and compare well with $Ru(bpy)_3^{2+}$ in terms of their ECL intensity.

Metal complexes containing acetylacetonate (acac) ligands were demonstrated viable, diverse and tunable redox-active tags for DNA modification [46]. Therefore, $(bpy)_2Ru(acac)(PF_6)$ complexes were studied for ECL generation feasible for sensors applications [47]. ECL efficiencies with TPA as a coreactant for the $(bpy)_2Ru(acac)^+$ complexes are lower than a $Ru(bpy)_3^{2+}$ standard and may reflect the instability of the complexes upon electrochemical oxidation and subsequent interaction with the coreactant.

Despite the fact that it is not a proper complex, electrogenerated chemiluminescence of a $(bpy)_2Ru(II)$–porphyrin conjugate was investigated [48]. This complex is quite unstable; in fact, the shoulder at 464 nm (the MLCT $Ru^{II}(d\pi) \rightarrow$ bpy (π^*) transition) decreases with time and it may reflect decomposition of $(bpy)_2Ru(II)$–porphyrin complex upon prolonged exposure to solution and loss of the $Ru(bpy)_2Cl^+$ moiety or substitution of the chloride for a solvent molecule and subsequent loss of the MLCT transition. ECL was produced with coreactant TPA in acetonitrile, and the ECL emission spectrum of the complex peaks at 656 nm and is similar to the PL emission of the porphyrin Soret emission at 655 nm, while no ECL spectrum matching the MLCT state of the ruthenium moiety is observed. The ECL efficiencies of porphyrin and $(bpy)_2Ru(II)$–porphyrin complex were 0.099 and 0.14, respectively, much lower than the $Ru(bpy)_3^{2+}$ standard ($\Phi/ecl = 1$).

Very unusual ligands and complexes were synthesized for intra- or self-enhanced ECL. In this case, the complex is the combination of the luminophore (metal and ligands) and the coreactant, directly bound to the ligands. The coreactants used are amine derivate [49] or a Schiff base [50] (Fig. 6).

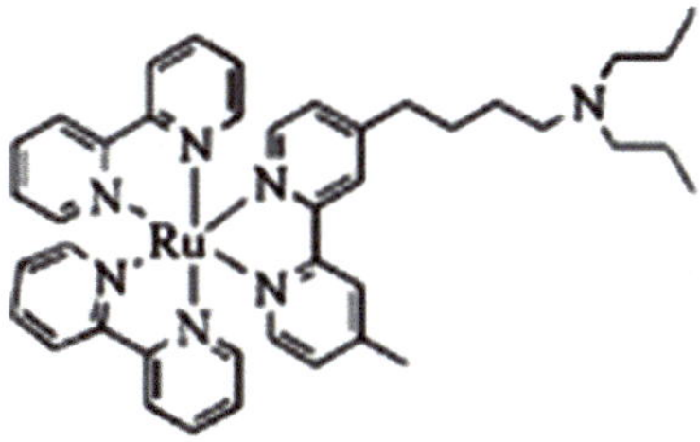

1

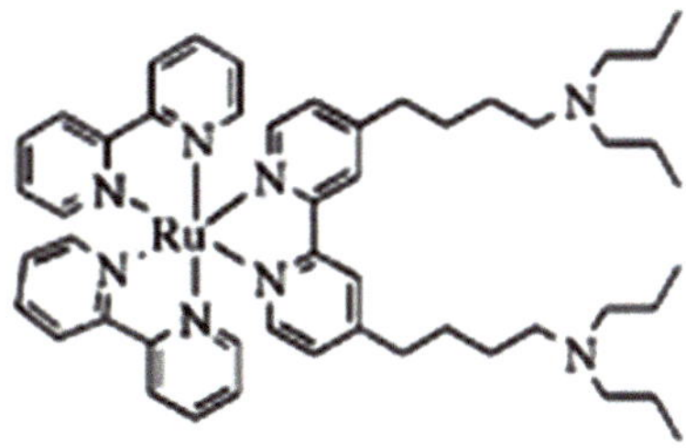

2

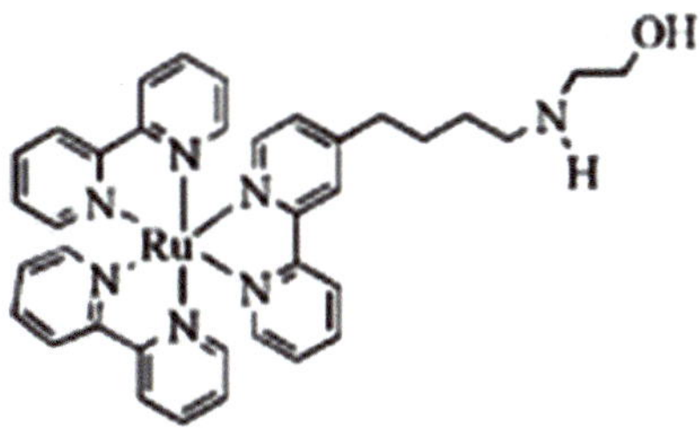

3

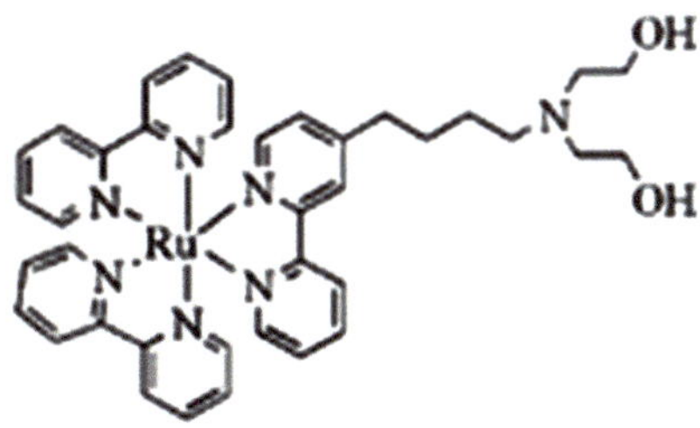

4

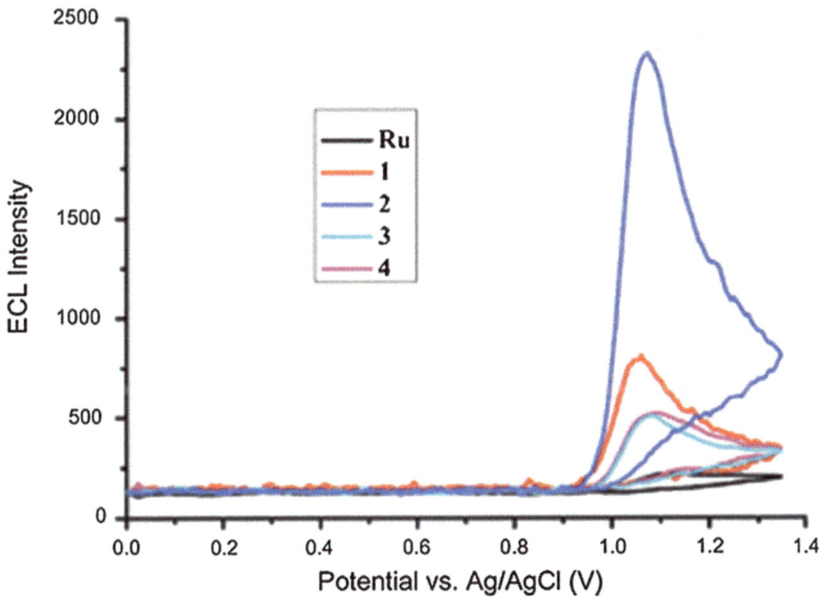

Fig. 6 Structure of conjugated ruthenium(II) tri-bypyridyl complexes and ECL intensity during CV of ruthenium complexes (Reproduced from [49] by permission of The Royal Society of Chemistry)

When considering the amine derivate complexes, the ECL emission is generated with the classical "oxidative-reductive" mechanism for amines.

The ECL mechanism for Schiff base ruthenium complexes is similar to "oxidative-reductive"; in fact, imino group receives electrons from the oxidized phenolic radicals and then electron-rich imines react with Ru(III) leading to the self-enhancement of ECL.

2.2 Metallopolymers and Surface Films

To develop sensing application, ruthenium complexes were immobilized on electrode surfaces by using polymers or coatings. Usually, ruthenium complexes are bound by side groups to the polymer backbone. A commonly used polymer is poly (4-vinylpyridine) (PVP) [51] where the ruthenium complex is coordinated by the nitrogen of the pyridine. Poly(4-vinylpyridine)-Ru metallopolymer was used to tune the emission wavelength in coreactant ECL. The efficient energy transfer from the excited-state $[Ru(dpp)_2]^{2+*}$ centres, created by mediated reduction of oxalate to the $[Ru(dpp)_2Cl]^+$ centres, allowed the wavelength of maximum emission to be shifted from approximately 600 to 710 nm.

Electrodes modified with thin films of the redox polymer $[Ru\text{-}(bpy)_2(PVP)_{10}]^{2+}$ produced ECL of greater efficiency and higher intensity than was observed for the same species freely diffusing in solution. This result highlights a key advantage of polymeric luminophores, that is the matrix may decrease the rate of radiative decay and block access of quenchers, e.g. molecular oxygen. Significantly, transient luminescence measurements reveal that Ru^{3+} acts as an efficient electron transfer quencher of the emissive Ru^{2+*} state, suggesting that optimum ECL intensity will be obtained at intermediate luminophore loadings [52].

Interestingly, metallocomplex $[Ru\text{-}(bpy)_2(PVP)_{10}]^{2+}$ compared to $[Ru(bpy)_2(PPyBBIM)_{10}]^{2+}$ (PPyBBIM, poly[2-(2-pyridyl)-bibenzimidazole) displayed slower rates of charge transfer, while the ECL emission intensity, with oxalate coreactant, was significantly higher than that obtained for $[Ru(bpy)_2(PPyBBIM)_{10}]^{2+}$ [53]. The ECL may be quenched by energy or electron transfer to the conducting polymer backbone, or the benzimidazole film may be more compact limiting access of the coreactant.

Nafion is another polymer used to incorporate $Ru(bpy)_3^{2+}$ into films, where the interactions with ruthenium complex are electrostatic [54]. When the film is rather thick, the ECL response is low because electron hopping between oxidized and reduced ruthenium centres which regenerates $Ru(bpy)_3^{3+}$ at the film–solution interface is limiting. In this case, the intensity scales as the inverse of the film thickness, which represents the diffusion length. For thinner films, the generation of the ECL reagent occurs rapidly and the limitation switches to diffusion in solution. This transition appears to occur at a film thickness of approximately 90 nm.

Instead of polymers, surface grafting can be applied, where pyrrole-functionalized bpy of ruthenium complex was electropolymerized on glassy carbon electrode and the ECL response was investigated both in annihilation and with coreactants (TPA and oxalate) [55].

In addition, the direct casting on ITO of $Ru(bpy)_3^{2+}$ immobilized in SiO_2-coated MWNTs was used. The ECL properties were demonstrated with the coreactant TPA [56].

The formation of $[Ru(bpy)_2dcb]Cl_2$ (dcb, 4,4′-dicarboxy-2,2′-bipyridine) monolayer on fluorine-doped tin oxide (FTO) electrodes was used for the detection of proline in a wide linear range, with low detection limit and good reproducibility [57].

Another method of monolayer formation was proved with $[Ru(bpy)_2(4,4'\text{-}(CH_3(CH_2)_{14}COO)_2\text{-bpy})](ClO_4)_2$, (Ru-LC) [58]. Ru-LC is totally insoluble in water, and precipitation from an acetonitrile solution leads to self-assembled nanobelts. ECL was obtained from the nanobelts dispersed in water containing TPA (Fig. 7).

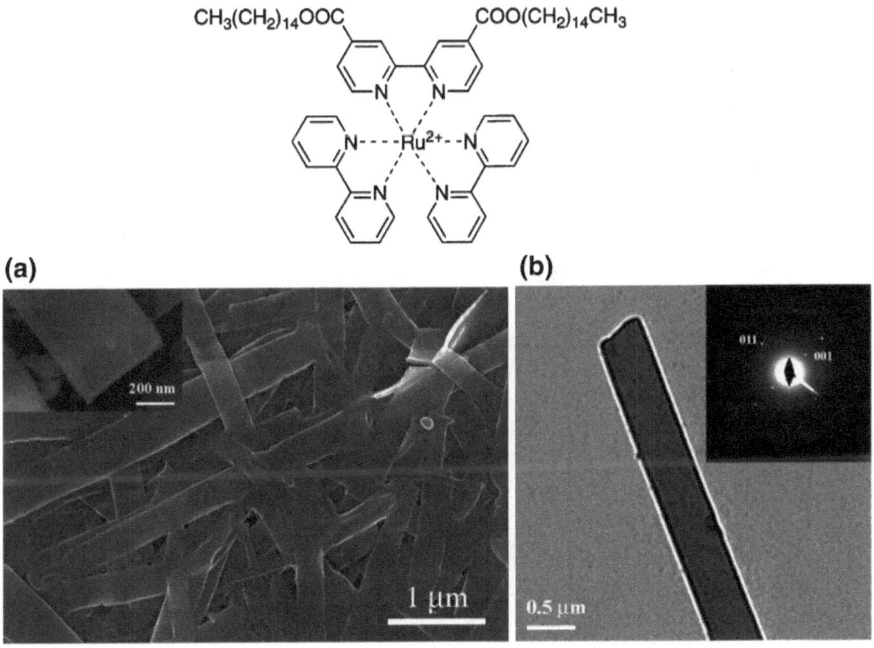

Fig. 7 Structure of Ru-LC EM and SEM (**a**) and TEM (**b**) images of the product. *Insets* in panels **a** and **b** show side-face SEM image and SAED pattern of single nanobelt, respectively (Reproduced from [58]. Copyright 2008 American Chemical Society)

2.3 Polynuclear Complexes

Multimetallic ruthenium complexes were synthesized to enhance the ECL intensity by adding more Ru centres in one molecule. Bimetallic Ru complexes [59, 60] (Fig. 8) are reported to give higher ECL emission compared to monometallic $Ru(bpy)_3{}^{2+}$, with TPA coreactant. Considering that two ruthenium active centres exist in the dinuclear complexes, their ECL intensity was also compared with that of 2 μM monometallic $Ru(bpy)_3{}^{2+}$ as a reference. Figure 8 illustrates the ECL emission of 1 μM dinuclear complexes with different aliphatic chain lengths and 2 μM $Ru(bpy)_3{}^{2+}$ with addition of 5 mM TPA as coreactant. The result demonstrated that the ECL intensity is affected by the length of the spacer, with the highest ECL intensity for the C_{10} alkyl chain. A similar trend in the ECL emission was observed with 2-(dibutylamino)ethanol (DBAE) coreactant.

Ruthenium multimetallic dendrimers were synthesized with different bridging ligands, such as polyamidoamine (PAMAM) [61], polyamine (PA) [62] and ester linkage [63]. The ECL emission with TPA coreactant is generally higher than the respective monometallic complexes, but not if compared to equimolar ratio of Ru.

3 Iridium

3.1 Mononuclear Complexes

Compared to the large amount of orange-red-emitting metal complexes used in ECL, mainly Ru complexes, there are relatively few blue emitters available. Among others, Ir(III) complexes are the most effective to design blue emitters for ECL and the green fluorophore $Ir(ppy)_3$ is their progenitor (λ_{max} ca. 520 nm).

The emission spectrum of $Ir(ppy)_3$ and related complexes evidences the substantial MLCT character of the emitting state [64]. From green to blue, hypsochromic shift is usually achieved by incorporating electron-withdrawing substituents on the Ir complex ligand, which stabilizes the HOMO to a much greater extent than the LUMO [11, 65]. Furthermore, the ECL efficiency of Ir(III) complexes was found higher than that of $Ru(bpy)_3{}^{2+}$ under identical conditions [66]. ECL efficiency of $Ir(ppy)_3$ was investigated with various solvent mixtures (30% trifluoroethanol, 50% acetonitrile and pure aqueous 0.1 M H_2PO_4). The ECL efficiency is higher in 50% acetonitrile, then 30% trifluoroethanol and lower in aqueous buffer [67].

Among the various ancillary ligands, most of those that provide the appropriate steric and electronic environment around the central element are related to nitrogen aromatic compounds. The common 2-phenylpyridine (ppy) was replaced by aryltriazole, tetrazolate, acetylacetonate and amide or thiophene derivates, just to name a few.

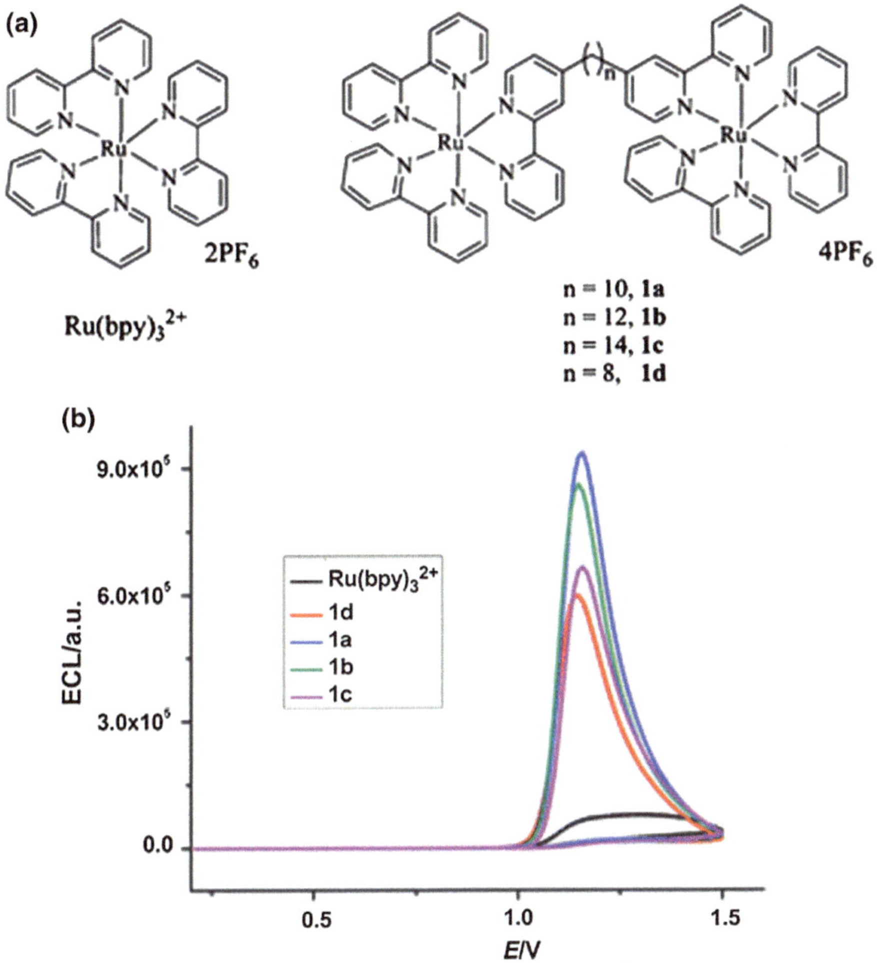

Fig. 8 **a** Ruthenium bimetallic complexes and **b** ECL emission with TPA coreactant (Reproduced from [59] by permission of The Royal Society of Chemistry)

Neutral and cationic cyclometalated iridium complexes (Fig. 9) containing ppy or bpy ligands and bidentate triazole-derived ligands were investigated by several groups. Moreover, the ligands were further functionalized to allow the tuning of emission wavelength. ECL properties of click chemistry-customizable iridium complexes were reported [11]. The introduction of a different moiety such as adamantane or cyclodextrin on the triazole–pyridine ligand did not affect the oxidation potential, while fluorination of the ppy ligand was suitable to modulate the emission energy. ECL emission was obtained both by annihilation and by coreactant (2-dibutylamino ethanol) methods with bright green and blue ECL emission generated by the coreactant approach in phosphate buffer.

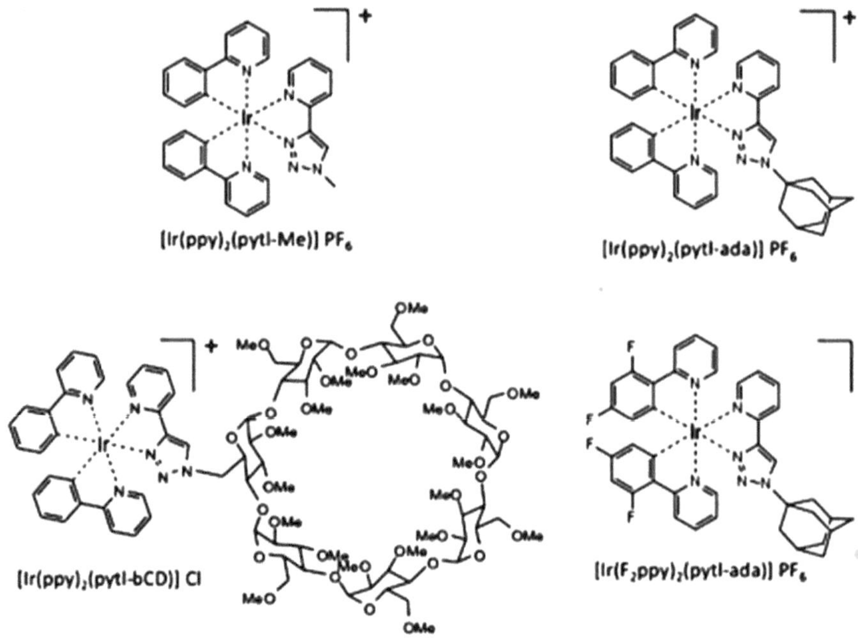

Fig. 9 Chemical formulae of the iridium complexes (Reproduced from [11] by permission of John Wiley & Sons, Inc.)

Iridium complexes with aryltriazole cyclometalated and bipyridyl ligands are reported to give very high ECL emission intensity, up to four times higher than $Ru(bpy)_3^{2+}$ with coreactant benzoyl peroxide (BPO). Emission wavelength ranges from yellow (587–579 nm) to blue-green (504–503 nm) for fluorinated aryltriazole [68]. When the ligand 4,4′-dimethylamino-2,2′-bipyridine (dmabpy) was used, a further enhancement of ECL emission was observed (Fig. 10). The ECL efficiency in the apparent annihilation process was increased to 550% of that of $Ru(bpy)_3^{2+}$ when dimethylamino was oxidized, and light emission was enhanced 16 times. This dramatic increase in ECL is attributed to the unique architecture of the dmabpy ligand acting as a self-coreactant [69]. Fluorine and t-Bu substituents on the ligands increase the electrochemical gap, thus promoting blue-shifts in the ECL emission; however, they decrease the ECL intensity [70].

Iridium(III) complex $[Ir(df\text{-}ppy)_2(ptb)]^+$ (ptb = 1-benzyl-1,2,3-triazol-4-ylpyridine) was attractive as a ECL blue emitter, featuring a considerable hypsochromic shift (ca. 60 nm) and over 16-fold greater coreactant (TPA) ECL intensities than the parent Ir $(ppy)_3$.

Furthermore, different ligands such as difluorophenylpyridine (df-ppy) and/or an ancillary triazolylpyridine ligand [3-phenyl-1,2,4-triazol-5-ylpyridinato (ptp) or 1-benzyl-1,2,3-triazol-4-ylpyridine (ptb)] were used to tune the emission wavelength of Ir(III) complexes (Fig. 11) [71].

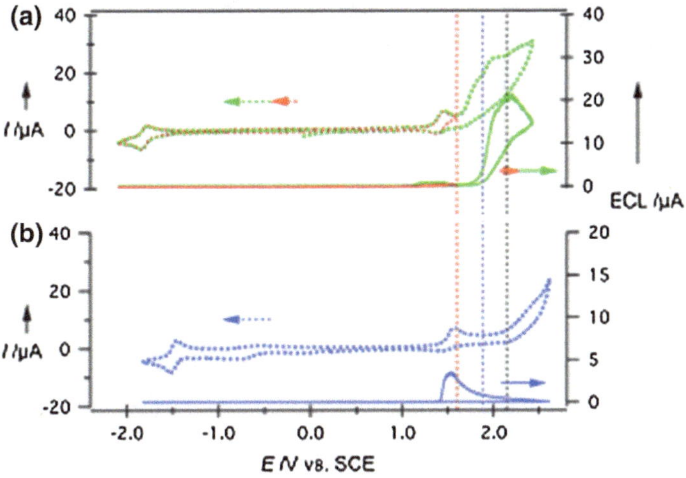

Fig. 10 Cyclic voltammograms (*dotted*) overlaid with corresponding ECL–voltage curves (*solid*) in acetonitrile of **a** $[(dFphtl)_2Ir(dmabpy)]PF_6$ and **b** $[(dFphtl)_2Ir(bpy)]PF_6$, both 0.6 mM (Reproduced from [69] by permission of John Wiley & Sons, Inc.)

Heteroleptic Ir(III) complexes of the general form Ir(ppy)2(C^C:) have been reported, where C^C represents a bidentate cyclometalated phenyl-substituted imidazolylidene ligand. The complexes were from the cyclometalated phenyl ring of the N-heterocyclic carbenes (NHCs) ligand being unsubstituted or having electron-donating (OMe and Me) or electron-withdrawing (Cl and F) groups. The complexes gave moderate to intense annihilation ECL; however, only the fluorinated complex produced significant coreactant ECL. The ECL emission profiles are essentially identical with the PL spectra, but fluorine substituents at the 2- and 4-positions of this ligand bring about a 50 nm blue-shift in emission wavelength; furthermore, the complex exhibits the highest ECL emission intensity in annihilation, slightly higher than the standard $Ru(bpy)_3^{2+}$ [72].

Like ruthenium, a widely investigated set of ligands for iridium are acetylacetonates [73]. Early studies reported extremely efficient ECL systems with TPA coreactant (Fig. 12) [66]. Acetylacetonates are especially useful to design new multicolour complexes from fairly green to deep red regions of UV–Vis spectra when they are combined with different organic ligands able to coordinate Ir(III) ions by means of the covalent metal–nitrogen and the ionic metal–carbon bonds [13]. The possibility to obtain stable ECL emissions in aqueous solutions from completely insoluble neutral $Ir(pq)_2acac$ (pq = 2-phenylquinoline) was demonstrated by encapsulating the Ir(III) complex inside silica-PEG nanoparticles with DBAE as coreactant [74].

Avobenzone [1-(4-methoxyphenyl)-3-(4-tert-butylphenyl)propane-1,3-dione], an acetylacetonate derivative, with 2-phenylbenzo[d]thiazole showed extremely higher ECL efficiency than $Ru(bpy)_3^{2+}$ when using TPA as a coreactant [75].

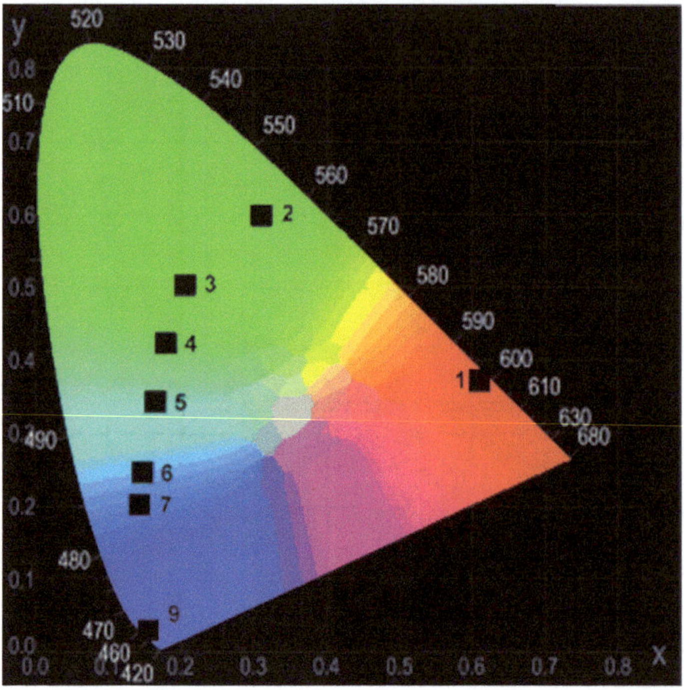

Fig. 11 CIE 1931 (two-degree observer) chromaticity characterization of $Ru(bpy)_3^{2+}$ (*1*), $Ir(ppy)_3$ (*2*), $[Ir(ppy)_2(ptp)]$ (*3*), $[Ir(ppy)_2\text{-}(ptb)]^+$ (*4*), $[Ir(df\text{-}ppy)_3]$ (*5*), $[Ir(df\text{-}ppy)_2(ptp)]$ (*6*), $[Ir(df\text{-}ppy)_2(ptb)]^+$ (*7*) and $Ir(pmi)_3$ (*9*) (Reproduced from [71] by permission of John Wiley & Sons, Inc.)

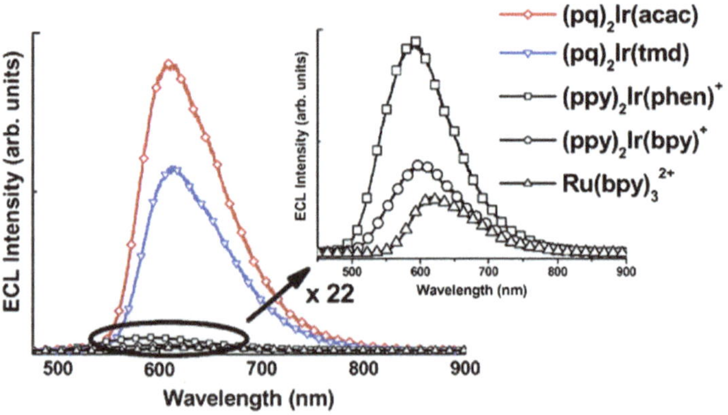

Fig. 12 ECL spectra of $Ru(bpy)_3^{2+}$, $[(ppy)_2Ir(bpy)]^+$, $[(ppy)_2Ir(phen)]^+$, $(pq)_2Ir(acac)$ and $(pq)_2Ir(tmd)$ (tdm = 2,2′,6,6′-tetramethylhepta-3,5-dione) in acetonitrile solution. Inset: 22 times magnified ECL data for $Ru(bpy)_3^{2+}$, $[ppy2Ir(bpy)]^+$ and $[ppy2Ir(phen)]^+$ (phen = phenanthroline) (Reproduced from [66]. Copyright 2005 American Chemical Society)

Other carbonyl moieties such as phenylacetamide ligands were compared to acac in a series of bis-cyclometalated iridium(III), where acac showed the highest ECL efficiency [76]. Similar complex with N-phenyl methacrylamide ligand was proved for ECL emission at directly modified glassy carbon electrode with MWNTs/complex film, and steady ECL in the presence of tri-n-propylamine as coreactant was produced [77].

Another comparison of acac and its derivates (tdm = 2,2′,6,6′-tetramethylhepta-3,5-dione and dibenzoylmethane anions) was made with picolinate-like ligands (3-isoquinaldate, picolinate, quinaldate anions) [78]. A series of six heteroleptic complexes were studied in annihilation and coreactant (TPA and persulphate) ECL. The gap potential ($E^0_{Ox} - E^0_{Red}$) is not greatly influenced by the different ligands, which affected mainly the ECL emission. Picolinate complex resulted in the highest ECL emission in annihilation, while complexes with acac and tdm showed the highest ECL emission with coreactant, TPA and persulphate, respectively.

A class of air-stable cationic-functionalized bipyridyl-cyclometalated iridium (III) complexes were obtained with sulphur derivate ligands. Cyclometalated thiophene-based iridium(III) complexes showed an ECL emission quantum yield by annihilation of 0.5–0.1% (with respect to $Ru(bpy)_3{}^{2+}$ standard) and a trend very similar to that of the luminescence ones [12]. Tetrathiafulvalene-fused phenanthroline ligand was used in $[Ir(ppy)_2(L)]$ for TPA coreactant ECL [79]. Such complexes showed a red-shifted PL and ECL emission around 600 nm.

An example of iridium(III)-cored dendrimer was investigated both in solution and in solid-state ECL with coreactant TPA [80].

The solution-phase ECL of iridium(III)-cored dendrimer was found to be enhanced compared to the simple parent complex $Ir(ppy)_3$ although the dendrimer displays lower PL quantum yield. However, the solid-state ECL of $Ir(ppy)_3$ is moderately intense and the enhancement seen in solution for the dendrimer was not seen in the solid state. This has been attributed to poorer charge transport in the dendrimer film due to the dilution of the iridium(III) complex redox sites, the lipophilicity of the dendrimer and amorphous nature of the film, meaning that ECL can only occur near the surface of the film.

3.2 *Metallopolymers*

ECL properties of Ir(III) complexes bound to polymers were investigated by using Nafion, poly-9-vinylcarbazole (PVK) and block copolymers. Nafion-Ir complex $[(pqcm)_2Ir(bpy)](PF6)$ (pqcmH = 2-phenyl-quinoline-4-carboxylic acid methyl ester) drop-cast on glassy carbon electrode showed enhanced ECL emission compared to Ir complex with oxalate as coreactant in aqueous solution [81].

ECL properties of three different Ir complexes with ppy, acac and picolinate ligands were compared when bound to Nafion or PVK and coated on Pt mesh electrode. ECL emission was investigated in phosphate buffer and acetonitrile with TPA as coreactant. Nafion-Ir picolinate resulted in the higher ECL emission intensity in phosphate buffer, while PVK-Ir acac and ppy showed the highest intensity in acetonitrile [82].

$Ir(ppy)_2(bpy)^+$ complex was incorporated into a block copolymer composed of three main blocks and a biotin label. The hydrophobic blocks contained butyl chains and the iridium complexes, while the poly(ethylene glycol) (PEG44) portion imparted water solubility. Micelles formation is enabled in aqueous solution, while reopening is achieved in acetonitrile. ECL emission with TPA as coreagent was studied both in acetonitrile and in buffer solutions. In the former solvent, Ir-copolymer showed intense ECL emission; however, in the latter, no ECL was detected because Ir complexes were buried within the polymer coil. Formation and reopening of the self-assembled micelles allowed recovery of the polymer and near-complete retention of its original ECL intensity [83].

4 Mixed Ru–Ir Systems

The search for new systems for multicolour ECL has focussed in recent years on mixed annihilation systems combining either more than one organic compound [1, 19] or a transition metal chelate with a non-emissive organic compound [13, 14, 84, 85]. Several studies on the ECL from cyclometalated iridium(III) chelates with emission maxima spanning the entire visible region have been reported in recent years [11, 13, 64, 71, 74, 85, 86]. Mixed metal chelate coreactant ECL systems with ruthenium(II) and iridium(III) complexes have also been reported, where emission is resolved by selective excitation at different potentials [65a, b, 87–89].

When comparing Ru(II) complexes with Ir(III) cyclometalates, the relative HOMO/LUMO orbitals distribution has to be taken into account in order to explain their different behaviours in ECL. While in Ru complexes, the HOMO is mostly localized on the metal centre and the LUMO on the ligands, in the Ir cyclometalates the HOMO is largely delocalized over their ligands as well as on the metal. Schmittel and coworkers [90] described two azacrown ether-appended iridium(III) complexes as probes for metal cations operating on oxidative-reduction ECL with tri-*n*-propylamine as coreactant in acetonitrile. They showed a notable ECL sensing for Ba^{2+} and Ag^+, respectively, through a ninefold emission intensity enhancement and emission red-shift. Such a behaviour was not observed in the structurally analogous azacrown ether-appended ruthenium(II) complexes as a consequence of the different distributions of the orbitals involved in the ECL generation and cation binding.

Schmittel et al. also showed that the use of some non-Kekulé-structured heteronuclear Ru–Ir systems could lead to different emissions in PL and ECL and that the ratio of emissive transitions was dependent upon the applied potential, so that the wavelength of maximum ECL intensity could be tuned from 649 to 611 nm with increasing anodic scan range [91].

Ding and coworkers reported the first example of ECL from the heterometallic soft salt $[Ru(dtbubpy)_3][Ir(ppy)_2(CN)_2]_2$ (dtbubpy = 4,4′-di-tert-butyl-2,2′-bipyridine; ppyH = 2-phenyl-pyridine), [Ir][Ru][Ir], which is a 2:1 stoichiometric mixture of complexes containing a cationic $Ru(dtbubpy)_3^{2+}$ and an anionic $[Ir(ppy)_2(CN)_2]$ [92]. Such a species shows significant ion-pairing interaction under both annihilation and coreactant conditions. Interestingly, ECL was generated through the annihilation mechanism involving radicals from both complexes, in a potential window from 1.16 to 1.48 V. $[Ru]^{2+}$ is reduced to its radical cation $[Ru]^{+}$ at −1.36 V, and [Ir] is oxidized to its radical cation $[ir]^{+}$ at 1.07 V. The excited species $[Ru]^{2+*}$ is generated by electron transfer from the HOMO of reduced $[Ru]^{+}$ to the SOMO of oxidized $[ir]^{+}$. Coreactant ECL (with tri-n-propylamine) displayed a unique light emission at 634 nm due to the $[Ru]^{2+*}$. No ECL signal was generated from [Ir]* due to the electrocatalytic reduction of $[Ru]^{2+}$ by [Ir].

A series of heterodinuclear complexes $[(bpy)_2Ru(bpy)(CH_2)n(bpy)Ir(df\text{-}ppy)_2]^{3+}$ (where bpy is 2,2′-bipyridyl, df-ppy is 2-(2,4-difluorophenyl)pyridine, and $n = 10$, 12, 14) were investigated. The ECL colour was shown (for $n = 12$) to be switchable between green and red using the electrode potential [93]. By changing the electrode potential from 0.55 to 0.90 V, a good linear calibration curve between the ECL intensity ratio (I_{Ru}/I_{Ir}) and the scanning potential was obtained.

Hogan and coworkers demonstrated the possible use of electrode potential to switch the colour of the ECL emission from a binary mixture of luminophores, between green and red or green and blue [65b]. This was made possible from the observed quenching of coreactant ECL of $Ir(ppy)_3$ at moderate to high potentials. The switch-off phenomenon was identified as the result of electron transfer quenching between TPA^{+} and the highly reducing excited-state $Ir(ppy)_3$*.

The same group recently reported a multiplexed electrogenerated chemiluminescence detection system, suitable for the development of low-cost, portable clinical diagnostic devices, where the distinct excitation and emission properties of concomitant electrochemiluminophores and the selective quenching of coreactant $Ir(ppy)_3$ ECL at high overpotentials were exploited [89]. A deep red emitter $Ru(bpy)_2(dm\text{-}bpy\text{-}dc)]^{2+}$, where dm-bpy-dc = dimethyl 2,2′-bipyridine-4,4′-dicarboxylate (λmax = 685 nm), was combined with a suitable blue emitter ($Ir(df\text{-}ppy)_3$, λmax = 495 nm) and a third concomitant electrochemiluminophore (the spectrally overlapping green emitter $Ir(ppy)_3$).

In the case of the mixed annihilation ECL of tris(2,2′-bipyridine) ruthenium(II) with various cyclometalated iridium(III) chelates, the existence of multiple, closely spaced reductions and oxidations of the metal chelates and the absence of the T-route pathways provided the possibility to fine-tune the energetics and predict the observed ECL emission colour based on the basis of simple estimations of the exergonicity of the reactions leading to excited-state production [94]. It was shown that, because of the delocalized nature of the HOMO in the iridium complexes, $Ru(bpy)_3^{+}$ would form an excited state by transfer of an electron from its HOMO to that of the Ir oxidized species (Fig. 13).

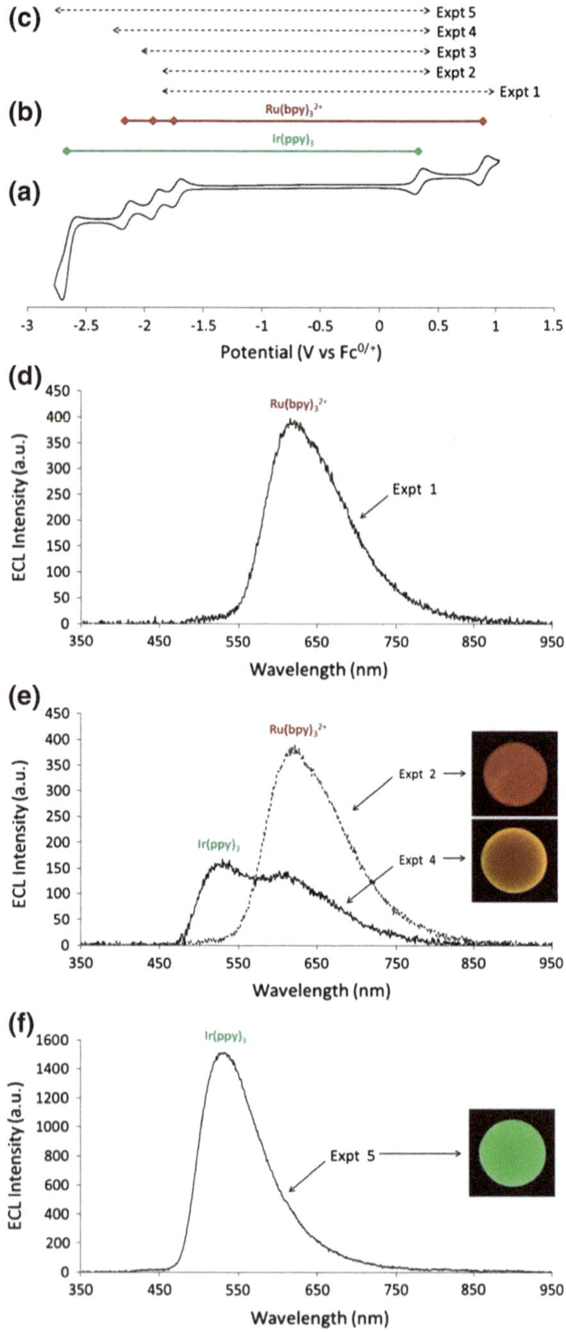

Fig. 13 **a** Cyclic voltammogram of 0.25 mM $Ru(bpy)_3^{2+}$ and 0.25 mM $Ir(ppy)_3$ in acetonitrile containing 0.1 M $TBAPF_6$. **b** Relevant reduction and oxidation potentials of the two metal chelates. **c** Illustration of potentials used in annihilation ECL experiments. **d–f** Spectra and photographs of the ECL at the working electrode of selected annihilation ECL experiments using 0.003 mM $Ru(bpy)_3^{2+}$ and 0.25 mM $Ir(ppy)_3$ in acetonitrile containing 0.1 M $TBAPF_6$ (Reproduced from [23]. Published by The Royal Society of Chemistry)

5 Platinum

Luminescent square-planar platinum (II) complexes have gained major attention because of their extensive application in many fields such as chemosensors, photocatalysis and organic light-emitting diodes (OLEDs) thanks to their robust PL properties [95, 96].

The strong spin–orbit coupling of the Pt heavy metal atom allows a strong efficient intersystem crossing between the singlet and triplet states which can lead to a high quantum yield of the emission from the triplet state. In addition, the electrochemical property of this class of complex generally shows well-defined redox behaviour. Moreover, one of the main motivations behind the ECL investigation of Pt (II) complexes is that they are promising candidates for the OLED screen application. For example, Pt-based porphyrin shows intense ECL emission both with coreactant and with annihilation reactions [97].

The ECL from Pt (II) salophen complexes were recently reported by Hogan and coworkers [98]. This study was motivated by the need of new ECL-sensing materials possessing tunable emission wavelength. By methodologically varying the position of the OMe substituents on the phenoxy ring of the salophen ligand, they demonstrated the ability to systematically and independently modulate the energetics of the HOMO or LUMO. Although the facility to independently tune the HOMO and LUMO was used for modulating the ECL emission colours maintaining constant the oxidation potential, the Pt (II) salophen complex shows an ECL efficiency lower than the $Ru(bpy)_3{}^{2+}$ standard (Fig. 14).

On the other hand, Pt complexes also show intense ECL emission with reductive-oxidation coreactant such as peroxydisulphate ($S_2O_8{}^{2-}$). This strategy was used for generating intense ECL emission from Pt (II) alkynyl terpyridine complex. A bright orange luminescence could be observed by reducing this complex and $S_2O_8{}^{2-}$. In this case, the less negative reduction potential (−0.92 V vs. SCE) is significantly shifted by 650 mV towards more positive potential compared with the $Ru(bpy)_3{}^{2+}/S_2O_8{}^{2-}$ [99].

The main drawback of the ECL from Pt(II) complex is related to the strong effect of the microenvironments solution on the emission such as the interaction of the excited-state species with the solvent, electrolyte and coreactant molecules.

Many of the recent ECL studies on the Pt (II) complexes were focussed on complexes containing polypyridinic ligands, while only few papers have reported so far the ECL properties of heteroleptic (C^N) Pt (LX) complexes (where C^N represents the cyclometalated ligand and LX the ancillary ligand) [99, 100]. This class of cyclometalated complexes typically shows excellent PL properties and high emission quantum yields. Neutral cyclometalated Pt(II) complexes with a phenylpyridine ligand and β-diketone moiety (named acac) have been reported by Feng Li et al. [101]. This class of complexes has been investigated both for annihilation and by using coreactant and shows an ECL quantum efficiency comparable with

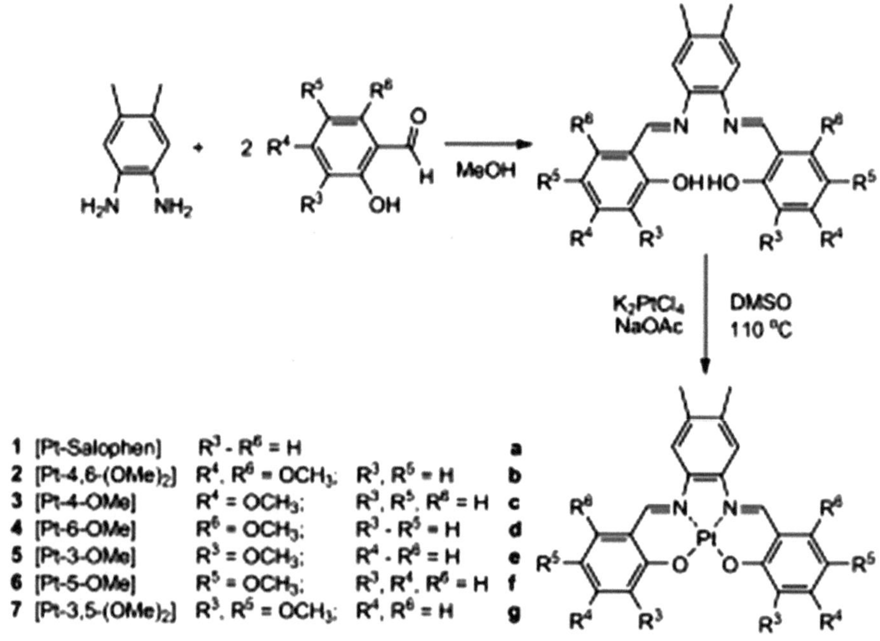

Fig. 14 Synthesis of the Schiff base ligand and platinum (II) complexes (Reproduced from [98] by permission of John Wiley & Sons, Inc.)

$Ru(bpy)_3{}^{2+}$ in addition to the advantage of a significant decrease of the oxidation potential.

6 Miscellaneous

Although $Ru(bpy)_3{}^{2+}$ is the most popular luminophore in ECL, it has relatively low ECL efficiency (Φ_{ECL} = photons emitted per redox event) of approximately 0.05 [102]. Aiming to improve the ECL performances by increasing the magnitude of emission, there is continuing interest in studying new metal and metal–ligand systems for a variety of ECL applications.

Osmium(II) complexes that incorporate combinations of phenanthroline and diphosphine or diarsine ligands are promising alternatives to traditional $Ru(bpy)_3{}^{2+}$ system for chemiluminescence and ECL detection [103]. These complexes provide convenient control of redox and luminescence properties using a variety of readily available ligands.

Pennington and coworkers examined the coreactant ECL of different osmium complexes of the form $[Os(N)_2(L)]^{2+}$, where N is a polypyridyl ligand and L is a bidentate arsine or phosphine ligand [104], reaching the ECL quantum efficiency eightfold greater than $Ru(bpy)_3^{2+}$ in aqueous solution.

Mixed metal supramolecular complexes such as $[((phen)_2\text{-}Ru^{II}(dpp))_2Rh^{III}Cl_2]^{5+}$ (where phen = 1,10-phenanthroline, dpp = 2,3-bis(2-pyridyl)pyrazine, Fig. 15) couple light-absorbing ruthenium centres to reactive metal centres that could offer the ability to control spectroscopic and redox properties important to photochemical applications [105].

The intense ECL of a bis-cyclometalated alkynylgold(III) complex using TPA as a coreactant at the fluorosurfactant-modified gold electrode (FSO-Au) was observed for the first time by Chen et al. [106]. The TPA oxidation current at the bare Au electrode (straight line, Fig. 16a) is suppressed mainly due to the coverage of the oxide layer, and a very weak ECL peak appears. However, in the presence of the adsorbed FSO layer (dashed line, Fig. 16a), the initial growth of the electrode

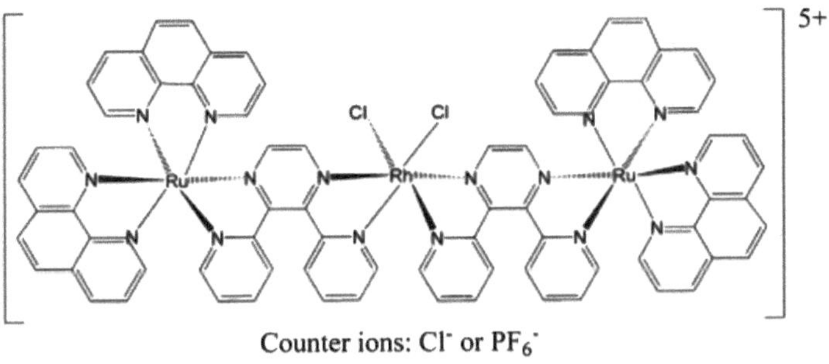

Fig. 15 Structure of $[((phen)_2Ru(dpp))_2RhCl_2]^{5+}$ (Reproduced from [107]. Copyright 2009 American Chemical Society)

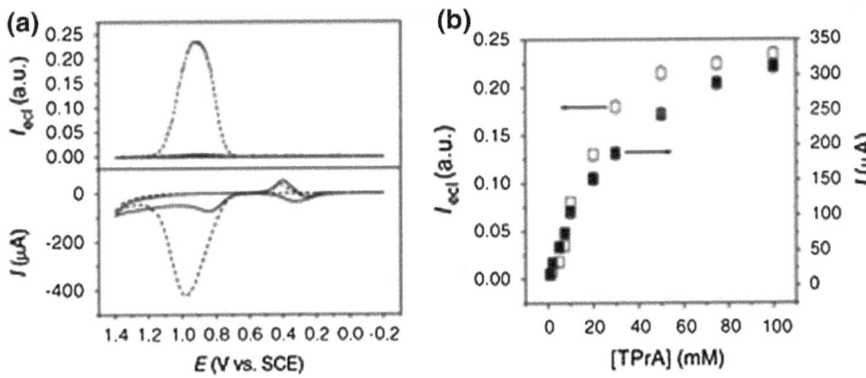

Fig. 16 **a** CVs and corresponding ECL curves at bare Au (−) and FSO-Au (- -) electrodes. **b** TPA oxidation current (*filled square*) and ECL intensity (*open square*) as a function of TPA concentration obtained at the FSO-Au electrode (Reprinted from 108 by permission of The Royal Society of Chemistry)

surface oxide is significantly suppressed and facile TPA oxidation is achieved showing a well-defined ECL peak. Compared with the emission at the bare Au electrode, a 55-fold enhanced ECL signal is observed when the electrode is pretreated with FSO. The relationship of the TPA oxidation current and the ECL intensity on TPA concentration at the FSO-Au electrode is shown in Fig. 16b. With increasing TPA concentration, both TPA oxidation current and ECL intensity increase with a similar trend.

The ECL of a metallopolymer containing $[Re(bpy)(CO)_3]^+$ moieties coordinated to a poly(4-vinylpyridine) polymer chain is generated upon reduction in the presence of benzoyl peroxide (BPO) as coreactant [107]. The ECL analysis shows a dual component, resulting from a state of chemical equilibria between the metallopolymer and the solvent product: the oxidation or reduction of the metal centre causes the Re–(4-vinylpyridine) bond to be cleaved, thus triggering the release of fluorophore, where the vacant coordination site is rapidly occupied by a solvent molecule that emits light at a different wavelength.

The ECL of the rhenium(I) carbonyl complexes was investigated in aqueous solution using the coreactant TPA in the presence of several surfactants [108]. The ECL intensity was observed to be enhanced upon addition of several kinds of surfactants when the surfactant concentration was lower than the critical micelle concentration (cmc).

Preliminary results from the ECL studies of the ionic fac-$Re(CO)_3$(L)(acetonitrile)$^+$ PF_6^- and the neutral fac-$Re(CO)_3$(MPBI)Cl complexes were briefly presented by Czerwieniec et al. [109]. The recorded ECL transients deviate considerably from that expected for simple annihilation of the electrochemically generated reduced and oxidized forms. Most probably, this is due to additional parasitic processes occurring during the electrochemical excitation.

A sensor fabricated with the zinc(II) complex composited with ctDNA (calf thymus DNA) exhibits sensitive ECL intensity and long-term stability [110]. This effect is explained by the fact that the amine groups on the ctDNA molecules can facilitate radical generation and decrease the excitation energy of the ECL reaction of the pure zinc(II) complex.

Novel 5,15-bis(9-anthracenyl)porphyrin derivatives show unexpected ECL properties, which revealed to be largely controlled by the electronic characteristics of the peripheral anthracenyl substituents [111].

Bisalicylideneethylenediamino (salen) metal complexes exhibit ECL properties in acetonitrile solutions [112]. Although the ECL emission efficiencies are weaker than the $Ru(bpy)_3^{2+}$ under identical conditions, the excited state formed electrochemically appears to be the same as that formed in PL.

7 Lanthanide(III) Metal Ions and Complexes

Many trivalent lanthanide metal complexes show quite high luminescence efficiencies (~5%), large Stokes shifts (~300 nm) and long excited-state lifetimes (up to several hundreds of μs). These interesting characteristics are normally obtained by intramolecular energy transfer processes from ligands to the lanthanide centre and can also be exploited for ECL. One of the goals that push to study the lanthanide(III) metal ions in ECL is their exploitation in novel-type immunoassay and DNA assay combining wavelength discrimination and a time-resolved approach to the detection of different labels. The most studied lanthanide(III) metal ions in this field are Tb(III), Dy(III) and secondly Eu(III).

An important role in the ECL processes involving lanthanide(III) metal ions is done by the chemical nature of the electrode, having a strong influence in defining the production of the reactive species involved in the ECL mechanism. In this framework, Hakansson et al. [113] studied the ECL emission properties of Tb(III) chelates by cathodic pulse polarization in aqueous solution, using oxide-covered Mg and n-ZnO:Al/MgO composite electrodes. This work was motivated to explore low-cost and relatively impure materials to produce disposable electrodes for ECL-based immunoassays using Tb(III) chelates as labels (ligands **1** and **2**, Fig. 17). The ECL emission of Tb(III) chelates derived from redox reactions in which hot electrons injected in the electrolyte solution were involved in one-electron steps exciting the ligand with subsequent sensitization of the central lanthanide ion. The excited states of Tb(III) chelates (radiative transitions

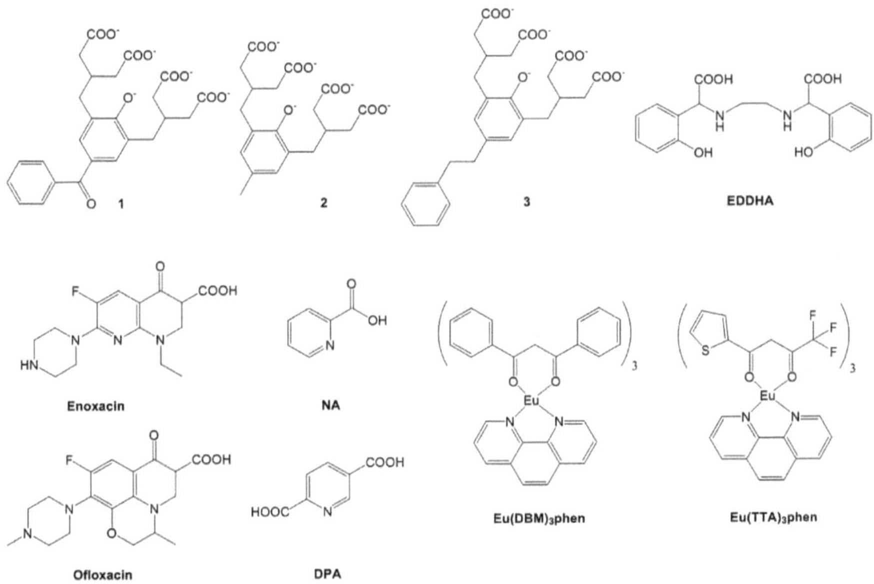

Fig. 17 Main chemical structures of ligands and analytes reported in this section

$^5D_4 \rightarrow {}^7F_J$) and their ECL emission were detected using time-resolved ECL (TR-ECL) thanks to their quite long luminescence lifetime. Oxide-covered Mg electrodes having a native layer of MgO were better than the ones produced by chemical oxidation in TR-ECL experiment in water solution containing peroxodisulphate ions. They showed linear calibration plots towards Tb(III) labels with a six order of magnitude concentration range, which seems promising for bioaffinity applications. In the case of composite electrodes fabricated by deposition of MgO on n-ZnO:Al/MgO, a 30-nm oxide film was the minimal thickness necessary to obtain a smooth ZnO surface on the electrode and to detect a TR-ECL signal. In a related study, the same group studied the ECL emission of hydratated and chelated Tb(III) ions with oxide-covered aluminium electrodes. In this case, a very low stability of the insulating aluminium oxide layer covering the electrode was observed (aluminium as cathode, 1.0 M $NaSO_4$ and 0.2 M borate buffer at pH 9.2) since aluminium oxide has an amphoteric nature and the layer dissolved in alkaline conditions. The detection limit of Tb(III) ions was 10^{-5} M for aquated ions and less than 10^{-6} M for Tb(III) in chelated form. More interesting results [114] were obtained preparing transparent electrodes by atomic layer deposition (ALD) of Y_2O_3 on ZnO:Al glasses. With Tb(III) chelates (ligands **1** and **3**, Fig. 17), these working electrodes showed a linear cathodic ECL emission for six orders of magnitude and a limit of detection down to 0.1 nM.

The ECL emission of Tb(III) and Dy(III) was studied by Staninski et al. [115] with cathodic pulse polarization using aluminium electrodes covered by a very stable oxide layer doped with lanthanide(III) chelates complexes. These experiments were carried out in the presence of aqueous electrolyte solutions containing the ligand EDDHA (ethylenediamine DL(*o*-hydroxy-phenylacetic acid, Fig. 17) and peroxodisulphate ion participating in the formation of highly oxidizing and reducing species such as hydroxyl and sulphate radicals and hydrated electrons. These species excited the lanthanide centres by a ligand to metal energy transfer with the following final emitting transitions: $^5D_4 \rightarrow {}^7F_J$ (Tb(III): $J = 6, 5, 4$ and 3, $\lambda \approx 488, 545, 582$ and 621 nm) and $^4F_{9/2} \rightarrow {}^6H_{15/2}$ $^4F_{9/2} \rightarrow {}^6H_{13/2}$ (Dy(III); $\lambda \approx 475$ and 570 nm). The same author measured the ECL emission of Tb(III)(acac)$_3$-phen complex adsorbed on the Al_2O_3 layer covering aluminium electrodes by pulse cathodic polarization of high amplitude, finding a linear relationship between the integral ECL intensity of Tb(III) ions and the thickness of the Al_2O_3 layer. This strategy can be useful also for the determination of electrode oxide layer porosity and thickness.

In some cases, analytes of pharmaceutical or biological interest can bind ECL-active lanthanide(III) ions, a behaviour that can be exploited for their ECL determination. The main limitation of this simple approach is the poor selectivity caused by the possible presence of interference, such as other chelating organic molecules or metal ions competitive with the lanthanide(III)–analyte binding process. The Tb(III) ECL emission in the presence of two fluorinated quinolones (enoxacin and ofloxacin, Fig. 17)—two synthetic antibacterial agents—was measured by Chen et al. [116] using the oxidation of Na_2SO_3 occurring at 0.72 V. Fluorochinolone can act as chelating moieties towards Tb(III) ions enhancing their luminescence, and this strategy offers the possibility of their analytical determination in aqueous solution by ECL. The ECL mechanism that was postulated considered the electrochemical production of SO_2* from the oxidation of Na_2SO_3, an intermolecular energy transfer from this high-energy intermediate to the ligand and then an intramolecular energy transfer to Tb(III). The ECL intensity was constant at 4.0–8.0 pH, while a decrease out of these limits was due to a destabilization of the chelates and to the formation of Tb(III) hydroxide, respectively. The feasibility of these measurements to determine the concentration of fluorinated quinolones was underlined considering the effects of some common excipients, ions and organic compounds. Calibration curves in the 10^{-10}–10^{-7} M concentration range were obtained.

The ECL emission of Tb(III) in the presence of nicotinic acid (NA, Fig. 17) was measured by Zhu et al. and proposed as a possible method for the quantitative determination of this molecule. Tb(III)$(NA)_6(H_2O)_4$ ECL emission was measured using $K_2S_2O_8$ as coreactant at GCE in the presence of acetic acid/acetate buffer (pH = 5.5), supplying a NA detection limit of 0.6 μM and a linear concentration range spanning from 2×10^{-6} M to 2.0×10^{-4} M. The ECL signal intensity from the complex was screened varying the concentration of the coreactant (0.0–10 mM) and the pH (3.5–6.0), with a maximum ECL intensity with $[K_2S_2O_8] = 8.0$ mM and pH = 5.5. The tolerance limit towards ions and common organic molecules was measured, indicating a low interference of many species, except for Fe^{3+}, Cu^{2+}, uric acid, ascorbic acid and dopamine.

In a contiguous study, Chen et al. [117] studied the ECL emission properties of Tb(III)$(DPA)_3(NO_3)_3(H_2O)_3$ (DPA = pyridine-2,6-dicarboxylate, Fig. 17) in water solution. In this case, the ECL intensity increased with pH, suggesting that OH^- ions played an important role in the ECL process. Tb(III)$(DPA)_3{}^{3+}$ was used to measure the concentration of free DPA molecules, since DPA has a high affinity towards this complex and is a biomarker and major constituent of bacterial spores. With GCE at pH 7.5 (TRIS buffer), the detection limit for DPA was around 6.0×10^{-8} M with a linear concentration range spanning from 2.0×10^{-7} to 3.4×10^{-5} M. Mention was made regarding the limited interference of various chemical species such as metal ions and small molecules (glucose, amino acids), but quantitative data were missing.

The ECL yield of systems containing Eu(III) complexes is few order of magnitude lower than similar systems containing Tb(III). The main reason for this is the difference in the redox properties of Tb(III) and Eu(III) ions ($E^0(Tb^{3+}/Tb^{2+}) = -3.6$ V; $E^0(Eu^{3+}/Eu^{2+}) = -0.35$ V). On Al/Al_2O_3 electrodes, only Tb(III) and Dy(III) hydrated ions show detectable emission [117]. With cathode-generated ECL using Al/Al_2O_3 electrodes and $K_2S_2O_8$ as coreactant, Lis et al. [118] measured ultraweak emission of Eu(III) hydrated ions ($^5D_0 \rightarrow {}^7F_1$; $\lambda = 594$ nm and $^5D_0 \rightarrow {}^7F_2$; $\lambda = 613$ nm) that was independent of the Eu(III) counterions (Cl^-, $SO_4{}^{2-}$, $ClO_4{}^-$).

H.-X. Yu et al. studied the cathodic ECL behaviour of acetonitrile, acetonitrile-1,10-phenantroline (phen) and acetonitrile Eu(III) ternary complexes ($Eu(DBM)_3$phen and $Eu(TTA)_3$phen, DBM = dibenzoylmethane, TTA = thenoyltrifluoroacetone, Fig. 17) at a gold electrode [119]. It was found that the secondary ligand phen can enhance the cathodic ECL emission of ternary Eu(III) complexes in acetonitrile. ECL experiments for these systems were carried out in detail using $TBABF_4$ and $(TBA)_2S_2O_8$. The main electrolysis products of acetonitrile were analysed by FT-IR and GC–MS (tributylamine and acetonitrile polymer), and an ECL mechanism was consequently proposed: it shows the formation by $SO_4^{\bullet-}$ of non-emissive ligand excited states that undergo intramolecular energy transfer to form the emitting excited state of the Eu(III) metal centre. In another work, Q. Jiang et al. [120] measured the ECL emission of multidentate phenolic Eu(III) complexes by TR-ECL.

From a general point of view, the use of lanthanide(III) ions in ECL seems quite promising. The key points to face in the next future will be the development of complexes with suitable functional groups for bioconjugation, a characteristic that is partially hampered by the presence of a crowded and not strictly defined coordination sphere for these metal centres. Even more, a quantitative comparison with the ECL performances of the more common Ru(III) and Ir(III) complexes is to better define the perspective of lanthanide(III) metal complexes in ECL applications.

8 Conclusion

This chapter reviewed the development of metal chelates used in electrogenerated chemiluminescence within the last decade. Among the wide range of metals available, ruthenium is confirmed the most used in fundamental electrochemical studied, also employed for analytical applications, while the interest for iridium is rapidly increasing.

Moreover, $Ru(bpy)_3{}^{2+}$ was the first complex used for ECL and it has been the standard in both aqueous and organic solvents, through the years until nowadays. Despite the great amount of research on Ru and Ir, other metals have drawn the attention for ECL applications, such as transition metals (Pt, Os, Rh, Re, etc.) and lanthanides (Tb, Dy and Eu).

References

1. Bard, A.J. (ed.): Electrogenerated Chemiluminescence. Marcel Dekker, New York (2004)
2. Richter, M.M.: Chem. Rev. **104**, 3003 (2004)
3. Miao, W.: Chem. Rev. **108**, 2506 (2008)
4. Forster, R.J., Bertoncello, P., Keyes, T.E.: Annu. Rev. Anal. Chem. **2**, 359 (2009)

5. Hu, L.Z., Xu, G.B.: Chem. Soc. Rev. **39**, 3275 (2010)
6. Liu, Z., Qi, W., Xu, G.: Chem. Soc. Rev. **44**, 3117 (2015)
7. Hesari, M., Ding, Z.: J. Electrochem. Soc. **163**, H3116 (2016)
8. Stagni, S., Palazzi, A., Zacchini, S., Ballarin, B., Bruno, C., Marcaccio, M., Paolucci, F., Monari, M., Carano, M., Bard, A.J.: Inorg. Chem. **45**, 695 (2006)
9. Zanarini, S., Della Ciana, L., Marcaccio, M., Marzocchi, E., Paolucci, F., Prodi, L.: J. Phys. Chem. B **112**, 10188 (2008)
10. Zanarini, S., Rampazzo, E., Bich, D., Canteri, R., Della Ciana, L., Marcaccio, M., Marzocchi, E., Montalti, M., Panciatichi, C., Pederzolli, C., Paolucci, F., Prodi, L., Vanzetti, L.:J. Phys. Chem. C **112**, 2949 (2008)
11. Zanarini, S., Felici, M., Valenti, G., Marcaccio, M., Prodi, L., Bonacchi, S., Contreras-Carballada, P., Williams, R.M., Feiters, M.C., Nolte, R.J.M., De Cola, L., Paolucci, F.: Chem. Eur. J. **17**, 4640 (2011)
12. Bandini, M., Bianchi, M., Valenti, G., Piccinelli, F., Paolucci, F., Monari, M., Umani-Ronchi, A., Marcaccio, M.: Inorg. Chem. **49**, 1439 (2010)
13. Kapturkiewicz, A., Nowacki, J., Borowicz, P.: Electrochim. Acta **50**, 3395 (2005)
14. Kapturkiewicz, A., Chen, T.M., Laskar, I.R., Nowacki, J.: Electrochem. Commun. **6**, 827 (2004)
15. Avilov, I., Minoofar, P., Cornil, J., De Cola, L.: J. Am. Chem. Soc. **129**, 8247 (2007)
16. Orselli, E., Kottas, G.S., Konradsson, A.E., Coppo, P., Frohlich, R., Frtshlich, R., De Cola, L., van Dijken, A., Buchel, M., Borner, H.: Inorg. Chem. **46**, 11082 (2007)
17. Kapturkiewicz, A., Nowacki, J., Borowicz, P., Phys, Z.: Chem. **220**, 525 (2006)
18. Rampazzo, E., Bonacchi, S., Genovese, D., Juris, R., Marcaccio, M., Montalti, M., Paolucci, F., Sgarzi, M., Valenti, G., Zaccheroni, N., Prodi, L.: Coordin. Chem. Rev. **256**, 1664 (2012)
19. Valenti, G., Rampazzo, E., Bonacchi, S., Khajvand, T., Juris, R., Montalti, M., Marcaccio, M., Paolucci, F., Prodi, L.: Chem. Commun. **48**, 4187 (2012)
20. Marcus, R.A., Siddarth, P.: In: Kochanski, E. (ed.) Photoprocesses in Transition Metal Complexes, Biosystems and Other Molecules: Experiment and Theory, p. 49. Kluwer, Norwall, Massachusetts (1992)
21. Valenti, G., Fiorani, A., Di Motta, S., Bergamini, G., Gingras, M., Ceroni, P., Negri, F., Paolucci, F., Marcaccio, M.: Chem. Eur. J. **21**, 2936 (2015)
22. Juris, A., Balzani, V., Barigelletti, F., Campagna, S., Belser, P., Von Zelewsky, A.: Coord. Chem. Rev. **84**, 85 (1988)
23. Kerr, E., Doeven, E.H., Barbante, G.J., Hogan, C.F., Bower, D.J., Donnelly, P.S., Connell, T.U., Francis, P.S.: Chem. Sci. **6**, 472 (2015)
24. Habtamu, H.B., Sentic, M., Silvestrini, M., De Leo, L., Not, T., Arbault, S., Manojlovic, D., Sojic, N., Ugo, P.: Anal. Chem. **87**, 12080 (2015)
25. Stewart, A.J., Hendry, J.: L. Dennany. Anal. Chem. **87**, 11847 (2015)
26. Bist, I., Song, B., Mosa, I.M., Keyes, T.E., Martin, A., Forster, R.J., Rusling, J.F.: Electrochemiluminescent array to detect oxidative damage in ds-DNA using [Os $(bpy)_2$(phen-benz-COOH)]$^{2+}$/Nafion/Graphene films. ACS Sens. (2016). doi:10.1021/acssensors.5b00189
27. Roche Diagnostic (© 2016 F. Hoffmann-La Roche Ltd) http://www.roche-diagnostics.co.in/Products/Pages/RocheElecsysSystems.aspx Accessed 10 Feb 2016
28. Shan, Y., Xu, J.J., Chen, H.Y.: Chem. Commun. **8**, 905 (2009)
29. Dennany, L., Forster, R.J., White, B., Smyth, M., Rusling, J.F.: J. Am. Chem. Soc. **126**, 8835 (2004)
30. Dennany, L., Forster, R.J., Rusling, J.F.: J. Am. Chem. Soc. **125**, 5213 (2003)
31. Wang, H., Chai, Y., Yuan, R., Cao, Y., Bai, L.: Anal. Chim. Acta **815**, 16 (2014)
32. Valenti, G., Rampazzo, E., Biavardi, E., Villani, E., Fracasso, G., Marcaccio, M., Bertani, F., Ramarli, D., Dalcanale, E., Paolucci, F., Prodi, L.: Faraday Discuss. **185**, 299 (2015)
33. Hvastkovs, E.G., So, M., Krishnan, S., Bajrami, B., Tarun, M., Jansson, I., Schenkman, J.B., Rusling, J.F.: Anal. Chem. **79**, 1897 (2007)

34. Sun, B., Qi, H.L., Ma, F., Gao, Q.A., Zhang, C.X., Miao, W.J.: Anal. Chem. **82**, 5046 (2010)
35. Shao, K., Wang, J., Jiang, X., Shao, F., Li, T., Ye, S., Chen, L., Han, H.: Anal. Chem. **86**, 5749 (2014)
36. Zamolo, V., Valenti, G., Venturelli, E., Chaloin, O., Marcaccio, M., Boscolo, S., Castagnola, V., Sosa, S., Berti, F., Fontanive, G., Poli, M., Tubaro, A., Bianco, A., Paolucci, F., Prato, M.: ACS Nano **6**, 7989 (2012)
37. Valenti, G., Bruno, C., Rapino, S., Fiorani, A., Jackson, E.A., Scott, L.T., Paolucci, F., Marcaccio, M.: J. Phys. Chem. C **114**, 19467 (2010)
38. Tokel, N.E., Bard, A.J.: J. Am. Chem. Soc. **94**, 2862 (1972)
39. Zhou, M., Robertson, G.P., Roovers, J.: Inorg. Chem. **44**, 8317 (2005)
40. Barbante, G.J., Hogan, C.F., Wilson, D.J.D., Lewcenko, N.A., Pfeffer, F.M., Barnett, N.W., Francis, P.S.: Analyst **136**, 1329 (2011)
41. Stagni, S., Orselli, E., Palazzi, A., De Cola, L., Zacchini, S., Femoni, C., Marcaccio, M., Paolucci, F., Zanarini, S.: Inorg. Chem. **46**, 9126 (2007)
42. Zanarini, S., Bard, A.J., Marcaccio, M., Palazzi, A., Paolucci, F., Stagni, S.: J. Phys. Chem. B **110**, 22551 (2006)
43. Stagni, S., Palazzi, A., Zacchini, S., Ballarin, B., Bruno, C., Marcaccio, M., Paolucci, F., Monari, M., Carano, M., Bard, A.J.: Inorg. Chem. **45**, 695 (2006)
44. Della Ciana, L., Zanarini, S., Perciaccante, R., Marzocchi, E., Valenti, G.: J. Phys. Chem. C **114**, 3653 (2010)
45. Barbante, G.J., Francis, P.S., Hogan, C.F., Kheradmand, P.R., Wilson, D.J.D., Barnard, P.J.: Inorg. Chem. **52**, 7448 (2013)
46. Weizman, H., Tor, Y.: J. Am. Chem. Soc. **124**, 1568 (2002)
47. Brooks, S.C., Vinyard, D.J., Richter, M.M.: Inorg. Chim. Acta **359**, 4635 (2006)
48. Vinyard, D.J., Swavey, S., Richter, M.M.: Inorg. Chim. Acta **360**, 1529 (2007)
49. Sun, S., Yang, Y., Liu, F., Fan, J., Peng, X., Kehr, J., Sun, L.: Dalton Trans. 7969 (2009)
50. Li, P., Jin, Z., Zhao, M., Xu, Y., Guoa, Y., Xiao, D.: Dalton Trans. **44**, 2208 (2015)
51. Spehar-Deleze, A.M., Pellegrin, Y., Keyes, T.E., Forster, R.J.: Electrochem. Commun. **10**, 984 (2008)
52. Dennany, L., Hogan, C.F., Keyes, T.E., Forster, R.J.: Anal. Chem. **78**, 1412 (2006)
53. O'Reilly, E.J., Keyes, T.E., Forster, R.J., Dennany, L.: Analyst **138**, 677 (2013)
54. Bertoncello, P., Dennany, L., Forster, R.J., Unwin, P.R.: Anal. Chem. **79**, 7549 (2007)
55. Ding, S.N., Cosnier, S., Shan, D., Sun, Y.M., Wang, Y.: Electrochem. Commun. **12**, 905 (2010)
56. Guo, S., Wang, E.: Electrochem. Commun. **9**, 1252 (2007)
57. Dennany, L., O'Reilly, E.J., Keyes, T.E., Forster, R.J.: Electrochem. Commun. **8**, 1588 (2006)
58. Yu, J., Fan, F.R.F., Pan, S., Lynch, V.M., Omer, K.M., Bard, A.J.: J. Am. Chem. Soc. **130**, 7196 (2008)
59. Sun, S., Li, F., Liu, F., Yang, X., Fan, J., Song, F., Sun, L., Peng, X.: Dalton Trans. **41**, 12434 (2012)
60. Sun, S., Yang, Y., Liu, F., Pang, Y., Fan, J., Sun, L., Peng, X.: Anal. Chem. **81**, 10227 (2009)
61. Kim, J.K., Park, H.S., Jun, Y.M., Hwang, R.Y., Lee, W.Y., Kim, B.H.: Synth. Met. **150**, 93 (2005)
62. Lee, D.N., Park, H.S., Kim, E.H., Jun, Y.M., Lee, J.Y., Lee, W.Y., Kim, B.H.: Bull. Korean Chem. Soc. **27**, 99 (2006)
63. Lee, D.N., Soh, B.K., Kim, S.H., Juna, Y.M., Yoon, S.H., Lee, W.Y., Kim, B.H.: J. Organomet. Chem. **693**, 655 (2008)
64. (a) Vogler, A., Kunkely, H.: High energy processes in organometallic chemistry. In: Suslick, K. S. (ed.) ACS Symposium Series, vol 333. American Chemical Society, Washington DC, p. 155 (1987) (b) Nishimura, K., Hamada, Y., Tsujioka, T., Shibata, K., Fuyuki, T.: Jpn. J. Appl. Phys. **40**, L945 (2001) (c) Bruce, D., Richter, M. M.: Anal. Chem. **74**, 1340 (2002)

65. (a) Muegge, B. D., Richter, M. M.: Anal. Chem. **76**, 73 (2004) (b) Doeven, E. H., Zammit, E. M., Barbante, G. J., Francis, P. S., Barnettb, N. W., Hogan, C. F.: Chem. Sci. **4**, 977 (2013)
66. Kim, J.I., Shin, I.S., Kim, H., Lee, J.K.: J. Am. Chem. Soc. **127**, 1614 (2005)
67. Robinson, W.D., Richter, M.M.: Luminescence **30**, 67 (2015)
68. Swanick, K.N., Ladouceur, S., Zysman-Colman, E., Ding, Z.: Chem. Commun. **48**, 3179 (2012)
69. Swanick, K.N., Ladouceur, S., Zysman-Colman, E., Ding, Z.: Angew. Chem. Int. Ed. **51**, 11079 (2012)
70. (a) Swanick, K. N., Ladouceur, S., Zysman-Colman, E., Ding, Z.: RSC Adv. **3**, 19961 (2013) (b) Ladouceur, S., Swanick, K. N., Gallagher-Duval, S., Ding, Z., Zysman-Colman, E.: Eur. J. Inorg. Chem. **2013**, 5329 (2013)
71. Barbante, G.J., Doeven, E.H., Kerr, E., Connell, T.U., Donnelly, P.S., White, J.M., Lópes, T., Laird, S., Wilson, D.J.D., Barnard, P.J., Hogan, C.F., Francis, P.S.: Chem. Eur. J. **20**, 3322 (2014)
72. Stringer, B.D., Quan, L.M., Barnard, P.J., Wilson, D.J.D., Hogan, C.F.: Organometallics **33**, 4860 (2014)
73. (a) Li, C., Lin, J., Yang, X., Wan, J., Organomet, J.: Chem. **696**, 2445 (2011) (b) Zhou, Y., Li, W., Yu, L., Liu, Y., Wanga, X., Zhou, M.: Dalton Trans. **44**, 1858 (2015) (c) Ni, X., Li, T., Song, Q.: J. Electroanal. Chem. **719**, 30 (2014)
74. Zanarini, S., Rampazzo, E., Bonacchi, S., Juris, R., Marcaccio, M., Montalti, M., Paolucci, F., Prodi, L.: J. Am. Chem. Soc. **131**, 14208 (2009)
75. Zhou, Y., Gao, H., Wang, X., Qi, H.: Inorg. Chem. **54**, 1446 (2015)
76. Zhu, S., Song, Q., Zhang, S., Ding, Y.: J. Mol. Struct. **1035**, 224 (2013)
77. Sun, S.Q., Song, Q.J., Yuan, H.F., Ding, Y.Q.: Chin. Chem. Lett. **19**, 1509 (2008)
78. (a) Shin, I. S., Kim, J. I., Kwon, T. H., Hong, J. I., Lee, J. K., Kim, H.: J. Phys. Chem. C **111**, 2280 (2007) (b) Shin, I. S., Yoon, S., Kim, J. I., Lee, J. K., Kim, T. H., Kim, H.: Electrochim. Acta **56**, 6219 (2011)
79. Qin, J., Deng, S.Y., Qian, C.X., Li, T.Y., Ju, H.-X.: J. Organomet. Chem. **750**, 7 (2014)
80. Reid, E.F., Burn, P.L., Lo, S.C., Hogan, C.F.: Electrochim. Acta **100**, 72 (2013)
81. Dong, Y.P., Ni, Z.Y., Zhang, J., Tong, B.H., Chu, X.F.: J. Lumin. **136**, 165 (2013)
82. Muegge, B.D., Richter, M.M.: Luminescence **20**, 76 (2005)
83. Tefashe, U. M., Metera, K. L., Sleiman, H. F., Mauzeroll, J.: Langmuir **29**, 12866
84. (a) Kapturkiewicz, A., Szrebowaty, P., Angulo, G., Grampp, G.: J. Phys. Chem. A. **106**, 1678 (2002) (b) Kapturkiewicz, A., Szrebowaty, P.: J. Chem. Soc. Dalton Trans. **3219** (2002)
85. Kapturkiewicz, A., Angulo, G.: Dalton Trans. **3907** (2003)
86. Stagni, S., Colella, S., Palazzi, A., Valenti, G., Zacchini, S., Paolucci, F., Marcaccio, M., Albuquerque, R.Q., De Cola, L.: Inorg. Chem. **47**, 10509 (2008)
87. Bruce, D., Richter, M.M.: Anal. Chem. **74**, 1340 (2002)
88. Doeven, E. H., Zammit, E. M., Barbante, G. J., Hogan, C. F., Barnett, N. W., Francis, P. S.: Angew. Chem. Int. Ed. **51**, 4354 (2012)
89. Doeven, E.H., Barbante, G.J., Kerr, E., Hogan, C.F., Endler, J.A., Francis, P.S.: Anal. Chem. **86**, 2727 (2014)
90. Lin, H., Cinar, M.E., Schmittel, M.: Dalton Trans. **39**, 5130 (2010)
91. Schmittel, M., Shu, Q., Cinar, M.E.: Dalton Trans. **41**, 6064 (2012)
92. Swanick, K.N., Sandroni, M., Ding, Z., Zysman-Colman, E.: Chem. Eur. J. **21**, 7435 (2015)
93. Sun, W., Sun, S., Jiang, N., Wang, H., Peng, X.: Organomet. **34**, 3385 (2015)
94. Kerr, E., Doeven, E.H., Barbante, G.J., Hogan, C.F., Bower, D.J., Donnelly, P.S., Connell, T.U., Francis, P.S.: Chem. Sci. **6**, 472 (2014)
95. Yu, C., Wong, K.M.C., Chan, K.H.Y., Yam, V.W.W.: Angew. Chem. Int. Ed. **44**, 791 (2005)
96. Shikhova, E., Danilov, E.O., Kinayyigit, S., Pomestchenko, I.E., Tregubov, A.D., Camerel, F., Retaileau, P., Ziessel, R., Castellano, F.N.: Inorg. Chem. **46**, 3038 (2007)

97. Long, T.R., Richter, M.M.: Inorg. Chim. Acta **358**, 2141 (2005)
98. Reid, E.F., Cook, V.C., Wilson, D.J.D., Hogan, C.F.: Chem. Eur. J. **19**, 15907 (2013)
99. Chen, Z., Wong, K.M.C., Kwok, E.C.H., Zhu, N., Zu, Y., Yam, V.W.W.: Inorg. Chem. **50**, 2125 (2011)
100. Strassert, C. A., Chien, C. H., Galvez Lopez, M. D., Kourkoulos, D., Hertel, D., Meerholz, K., De Cola, L.: Angew. Chem. Int. Ed. **50**, 946 (2011)
101. Li, C., Wang, S., Huang, Y., Zheng, B., Tian, Z., Wena, Y., Li, F.: Dalton Trans. **42**, 4059 (2013)
102. Luttmer, J.D., Bard, A.J.: J. Phys. Chem. **85**, 1155 (1981)
103. Zammit, E.M., Barbante, G.J., Carlson, B., Doeven, E.H., Barnett, N.W., Hogan, C.F., Richter, M.M., Francis, P.S.: Analyst **137**, 2766 (2012)
104. Pennington, N.S., Richter, M.M., Carlson, B.: Dalton Trans. **39**, 1586 (2010)
105. Wang, S., Milam, J., Ohlin, A.C., Rambaran, V.H., Clark, E., Ward, W., Seymour, L., Casey, W.H., Holder, A.A., Miao, W.: Anal. Chem. **81**, 4068 (2009)
106. Chen, Z., Wong, K. M. C., Au, V. K. M., Zu, Y., Yam, V. W. W.: Chem. Commun. **791** (2009)
107. Valenti, G., O'Reilly, E.J., McNally, A., Keyes, T.E., Marcaccio, M., Paolucci, F., Forster, R.J.: Chem. PlusChem. **78**, 55 (2013)
108. Li, M.J., Lin, M., Xie, R., Liu, X., Wei, Q.H., Chen, G.N.: Electrochim. Acta **56**, 9344 (2011)
109. Czerwieniec, R., Kapturkiewicz, A., Lipkowski, J., Nowacki, J.: Inorg. Chim. Acta **358**, 2701 (2005)
110. Yang, J., Liu, B., Wu, J., Zhang, S., Jin, B., Tian, Y.: New J. Chem. **39**, 1404 (2015)
111. Sooambar, C., Troiani, V., Bruno, C., Marcaccio, M., Paolucci, F., Listorti, A., Belbakra, A., Armaroli, N., Magistrato, A., De Zorzi, R., Geremia, S., Bonifazi, D.: Org. Biomol. Chem. **7**, 2402 (2009)
112. Schnuriger, M., Tague, E., Richter, M.M.: Inorg. Chim. Acta **379**, 158 (2011)
113. Håkansson, M., Jiang, Q., Helin, M., Putkonen, M., Niskanen, A.J., Pahlberg, S., Ala-Kleme, T., Heikkilä, L., Suomi, J., Kulmala, S.: Electrochim. Acta **51**, 289 (2005)
114. Håkansson, M., Helina, M., Putkonena, M., Jianga, Q., Kotirantaa, M., Suomia, J., Niskanenb, A.J., Ala-Klemec, T., Kulmala, S.: Anal. Chim. Acta **541**, 137 (2005)
115. Staninski, K., Lis, S., Komar, D.: Electrochem. Commun. **8**, 1071 (2006)
116. Chen, S., Ding, F., Liu, Y., Zhao, H.: Spectrochim. Acta, Part A **64**, 130 (2006)
117. Yang, Y., Zhang, Y., Shu, G., Dong, Q., Zou, L., Zhu, Y.: J. Lumin. **141**, 71 (2013)
118. Staninski, K., Lis, S.: Opt. Mater. **33**, 1540 (2011)
119. Yu, H.X., Cui, H., Guan, J.B.: Luminescence **21**, 81 (2006)
120. Jiang, Q., Håkansson, M., Spehar, A.M., Ahonen, J., Ala-Kleme, T., Kulmala, S.: Anal. Chim. Acta **558**, 302 (2006)

Light-Emitting Electrochemical Cells

Frédéric Dumur

Abstract During the last years, research on light-emitting electrochemical cells (LECs) has literally exploded. Interest for this kind of devices is supported by the specificity of LECs, namely that these devices are among the simplest electroluminescent devices known today. In their simplest form, LECs have the structure of early organic light-emitting diodes (OLEDs), with one layer of organic materials containing mobile ions sandwiched between two metal electrodes. Generally speaking, LECs have several advantages over OLEDs, such as a single-layer configuration, operating at low voltages and allowing the use of air-stable electrodes. Devices can also be entirely solution-processed, what is largely favorable to the development of low-cost and large-area lighting applications in the future. In this chapter, the recent advances and the main remaining challenges concerning LECs are discussed.

1 Introduction

Since the discovery of the first light-emitting electrochemical cells (LECs) by Heeger et al. in 1995 [1], research into LECs has been pursued intensively because of their potential applications, in among other things, flat panel displays, and lighting technologies [2]. LECs offer unique opportunities to replace today's state-of-the-art energy-saving lamps with devices of warmer light emission [3]. LECs can be thin, flexible, lightweight, robust, and promise energy-efficient operation and extremely low-cost production. Using solution-processable materials and flexible substrates, it also allows large-area devices to be created in a cost-effective way using inexpensive manufacturing technologies such as the newspaper-printing-like roll-to-roll process, electrospray deposition methods, or inkjet printing. If OLEDs and LECs are both thin-film electroluminescent devices,

F. Dumur (✉)
Institut de Chimie Radicalaire ICR, Aix Marseille Université,
CNRS, UMR 7273, F-13397 Marseille, France
e-mail: frederic.dumur@univ-amu.fr

F. Miomandre and P. Audebert (eds.), *Luminescence in Electrochemistry*,
DOI 10.1007/978-3-319-49137-0_10

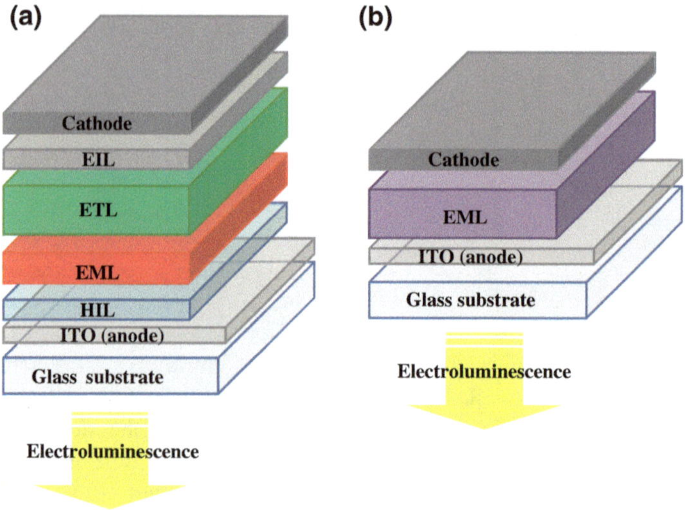

Fig. 1 Structure of an OLED (**a**) and a LEC (**b**) (where *HIL* stands for hole injection layer, *EML* emissive layer, *ETL* electron transport layer, *EIL* electron injection layer)

the two devices differ by their working principle. Notably, LECs are different from OLEDs in that sense that LECs use the same light-emitting materials than OLEDs but the light-emitting materials are mixed with mobile ions [4]. When the device is turned on, mobile ions allow electrochemistry to occur within the emissive layer so that a *p–n* junction structure is created. Once formed, a low resistance pathway for electronic charges is created, allowing electronic current and light emission to turn on sharply. Compared to OLEDs, LECs exhibit several advantages. Notably, OLEDs are vacuum-processed devices formed by coating several layers of different organic materials on top of a transparent electrode (usually indium tin oxide, ITO). Specifically, these devices are extremely sensitive to the thickness of the layers and the chemical properties of the electrode materials.

In contrast, LECs can be fabricated from solely solution-processed materials. These devices require only one or at most two layers of active material to be coated, and these devices are relatively insensitive to the thickness of the layers and the work function of the electrodes, allowing the use of air-stable metals (see Fig. 1). These features make LECs an attractive type of electroluminescent device for low-cost large-scale production. If LECs exhibit appealing features, there are also a number of disadvantages. One of the serious shortcomings of LECs is the slow response time, i.e., the time required for the mobile ions to diffuse during the *p–n* junction formation. As a result of this activation step, ions redistribution implies a delay between turn-on time and light emission. The second major disadvantage of LECs is their relatively short operation lifetime that is currently far from adequate for most commercial applications [5]. More generally, LECs are relatively simple to manufacture yet difficult to understand due to the combination of ionic and electronic charge carriers, rendering the electrochemical analog of OLEDs a richly complex device. In this chapter, recent advances and the main remaining challenges concerning LECs are discussed.

2 Electroluminescence: Origin of the Light Emission

There are two main ways of producing light: incandescence and luminescence. Incandescence is the production of light from heat. When an electrical current is passed through a filament, electricity running through the metal can heat it to temperatures high enough that it can give off light. Light emission directly results from the resistance of the metal to the passage of a current. The greater the resistance of the filament is, the more the light it produces. This technology was typically under use in incandescent lighting bulbs that are phased out in Europe. In contrast, luminescence is the generation of light due to causes other than temperature. Thus, electroluminescence is the response of a luminescent material to an electrical excitation. In other words, electroluminescence is a non-thermal generation of light upon the application of an electric field. But there are a number of other types of luminescence including photoluminescence (resulting from an optical excitation) or chemiluminescence (resulting from a chemical reaction). In electroluminescent devices, electrical excitation results from the radiative recombination of electrons and holes (i.e., a positive charge) within the active layer. Typically, the basic principle of operation of an electroluminescent device is as follows: First, electronic charges are injected and transported into the emissive layer from the electrodes, the electrons from the cathode, and the holes from the anode. In the second step, holes and electrons capture each other through electrostatic interactions, generating excitons. Exciton is a bound state of an electron and a hole which are attracted to each other by electrostatic Coulomb forces. Finally, in a third step, the energy issued of the formation of excitons is transferred to the light-emitting materials. In contact with the photoluminescent material, an electron of the light-emitting material is transferred from its ground state (the valence band) to its excited state (the conduction band) and relaxation of the electron to its ground state leads to light emission (photons) with the wavelength of the emitted light depending on the bandgap of the materials used. Under the application of a constant voltage, the *p–n* junction is constantly fed with a flux of electrons and holes, giving rise to a constant emission of light. In summary, electroluminescence simply originates from the radiative deactivation of electrons transferred to their excited states by the energy formed by the recombination of electrons and holes in the neighborhood of the light-emitting deactivation.

In electroluminescence, injected electrons and holes from electrodes have randomized spin states so that their recombination gives both symmetric (triplet) and antisymmetric (singlet) states in the ratio of 3:1 based on the spin statistics [6]. In fluorescent materials that exhibit a weak spin–orbit coupling, only singlets have an allowed transition to the ground state, meaning that only 25% of the created excitons can contribute to light emission (see Fig. 2). Conversely, in

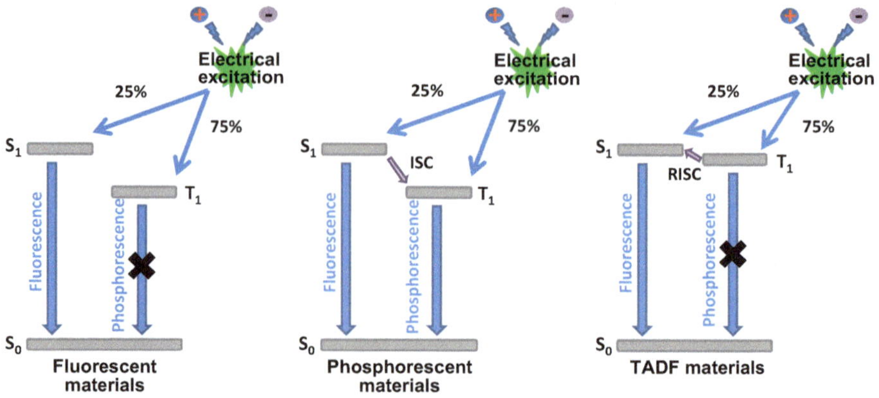

Fig. 2 Radiative deactivation pathways existing in fluorescent, phosphorescent, and TADF materials

phosphorescent materials, owing to the strong spin–orbit coupling in metal complexes, electrons from the singlet state (S_1) are efficiently transferred to the triplet state (T_1) by intersystem crossing (ISC). The radiative decay from the triplet state known as phosphorescence thus harvest both triplet and singlet excitons, enabling to reach the theoretical internal quantum efficiency of 100%. Exciton statistics thus determine the upper limit of internal quantum efficiency (IQE) for fluorescent (25%) and phosphorescent (100%) devices. However, this theoretical limitation to 25% excitons contributing to light emission in fluorescent devices has recently been overcome with materials of specific design allowing the luminescence mechanism of thermally activated delayed fluorescence (TADF) [7]. As specificity, TADF materials exploit both singlet and triplet excitons for fluorescence emission by thermally up-converting T_1 states to emissive S_1 state through reverse ISC (RISC) [8]. These new materials are even reported by Chihaya Adachi, pioneer in developing TADF OLED emitters, as constituting the third-generation organic electroluminescence materials.

Ideally, energy of all excitons should be released under the form of light. In practice, there is always competition between radiative and non-radiative decays. Excitons are mobile and subsequent to their generations they can diffuse within the active materials before decaying. A fraction of them can collide with the dopant molecules, and these collisions cause the excitons to decay non-radiatively. The consequences of this situation are twofold. First, the quenched excitons produce no light, reducing the overall device efficiency, and second, the heat released within the active layer may cause an early light-output degradation by the thermal degradation of the light-emitting material. Also, impurities in the active layer (light-emitting material, polymer electrolyte, electrolyte, solvent) may give rise to further reduction of the radiative fraction. Finally, the driving voltage can give rise to non-radiative processes. Indeed, at high applied voltages, the exciton formation rate is increased, whereas a thinning of the *p–n* junction is simultaneously observed [9]. As a result, the continuous increase in the exciton formation rate is not offset by radiative decay rate of the formed excitons, and a significant fraction of the excitons are diffusing out from the *p–n* junction into the surrounding doped regions. A trade-off between emission intensity and current-to-light conversion has thus to be found.

3 External Quantum Efficiency and Light Extraction

Light-emitting electrochemical cells are really appealing light sources by the possibility to minimize devices to a monolayer sandwiched between two electrodes. To achieve high luminous efficacy, the driving voltage of LECs needs to be as low as possible. By the presence of the two doped interfaces within the emissive layer, LECs achieve driving voltages close to the thermodynamic limit. Another condition to reach high efficiencies is the use of highly emissive materials, and this issue is addressed if phosphors are chosen. Indeed, with these materials, singlet and triplet excitons can be both harvested for light emission, enabling devices to theoretically achieve an IQE of 100%. Beyond the simple production of light, a second parameter of crucial importance is the ability to efficiently extract light out of devices. Even though high IQEs can be realized by using phosphorescent light-emitting materials, EQEs still encounter a severe bottleneck, namely the low light extraction efficiency. Regarding the light extraction, one metric is the external quantum efficiency (EQE) that quantifies the number of photons emitted per injected electrons. More precisely, EQE determines the ratio of outcoupled photons per generated photons. EQE can be expressed as

$$\mathrm{EQE} = \gamma\chi\eta_{\mathrm{PL}}\eta_{\mathrm{OC}}$$

and can be split into four terms where γ is the electrical efficiency (recombination efficiency of injected holes and electrons), χ the radiative efficiency (fraction of excitons that radiatively decays), η_{PL} the intrinsic photoluminescence (PL) efficiency of the material (directly related to the photoluminescence quantum yield (PLQY) of the light-emitting material), and η_{oc} the light outcoupling factor [10]. Due to photon trapping phenomena existing within devices, only a minor fraction of the generated light is emitted out of the device. It directly arises from the fact that light is an electromagnetic wave with a rectilinear propagation. Upon electrical excitation, photons are generated and propagate, albeit in a random direction. Most of the generated photons are not properly oriented and remain confined within devices. More precisely, conventional devices have typically a planar waveguide-like structure, with a surface area of light emission greater than the perimeter edge. As a result of this difference in scale between thickness and surface area of the emissive layer, and due to the refractive index differences between organics, metals, glass substrate, and air, layers adjacent to the emissive layer form waveguides, giving rise to internal reflections occurring at the organic/ITO–glass substrate and glass–air interfaces.

It has to be noticed that the waveguided structure of OLEDs (and to a lower extent of LECs) is directly related to the protocol used to fabricate devices. For OLEDs, thin and perfectly parallel layers are obtained by successive vacuum deposition of organic materials. As a result of this process, extremely smooth surface layers are obtained, greatly favorable to form waveguides (see Fig. 3). On the opposite, for LECs, smooth layers are obtained by spin-coating the light-emitting materials [11]. Considering that most of LECs comprise a polymer within their emissive layers, active layers of low surface roughness are thus obtained as a result of the remarkable film-forming ability of polymers. To come back to light trapping in electroluminescent devices and as explained by the classical ray optics theory (Snell's law), difference in refractive indices between layers results in internal reflections and a low outcoupling efficiency [12]. As a result of these optical modes guiding light in a propagation direction parallel to the waveguides, approximately 40% of the internally generated light is trapped in the organic/ITO layer, and about 30% is trapped in the glass substrate. Therefore, only around 30% of the electrogenerated photons can be extracted from devices. Considering this point, choice of the light-emitting material is of crucial importance. Indeed, EQE is proportional to radiative efficiency (see equation) and therefore strongly depends on the selected light-emitting materials. Thus, for devices comprising phosphorescent emitters, a theoretical IQE of 100% can be achieved and the maximum EQE is thus limited to about 25–30% [13]. Conversely, for devices comprising fluorophores, the upper limit of EQE is extremely low, close to 5%. In these conditions, it also explains why phosphorescent materials have attracted more interest than the fluorescent ones for developing highly emissive devices (see Fig. 3) [14].

If we get a deeper insight into the light confinement, we can determine that photon trap in devices has several origins. As aforementioned, outcoupling losses are mostly due to the refractive indices of the different organic materials, metals, and substrates used to fabricate LECs. As a result, reflected light is trapped within the different materials and absorbed. In fact, several light modes exist and these latter can be distinguished into outcoupled, waveguided, and surface plasmon modes occurring at the organic–metal interfaces (see Fig. 4). Exact contribution of each mode in the loss mechanism is strongly dependent on the device architecture, layer position, and thickness [15]. However, if their exact contribution has to be determined for each investigated device, recent experimental and theoretical investigations on OLEDs assuming a completely isotropic emitter orientation have furnished the following trend [16].

Hence, for an OLED with an EQE of 15%, absorption losses during the transition from the organic layers into the glass substrate and into air were determined around 6%, and the other optical modes to amount to $\sim$24% for the substrate emission, 11% for the waveguide modes, and 44% for the surface plasmon modes (see Fig. 5). Substrate and surface plasmon modes thus constitute the main light loss mechanism. While focusing this time on LECs, other authors demonstrated the detrimental effect of self-absorption in LECs and quantified losses [17]. Thus, if a 4% loss of the emission intensity due to self-absorption was determined of a LEC

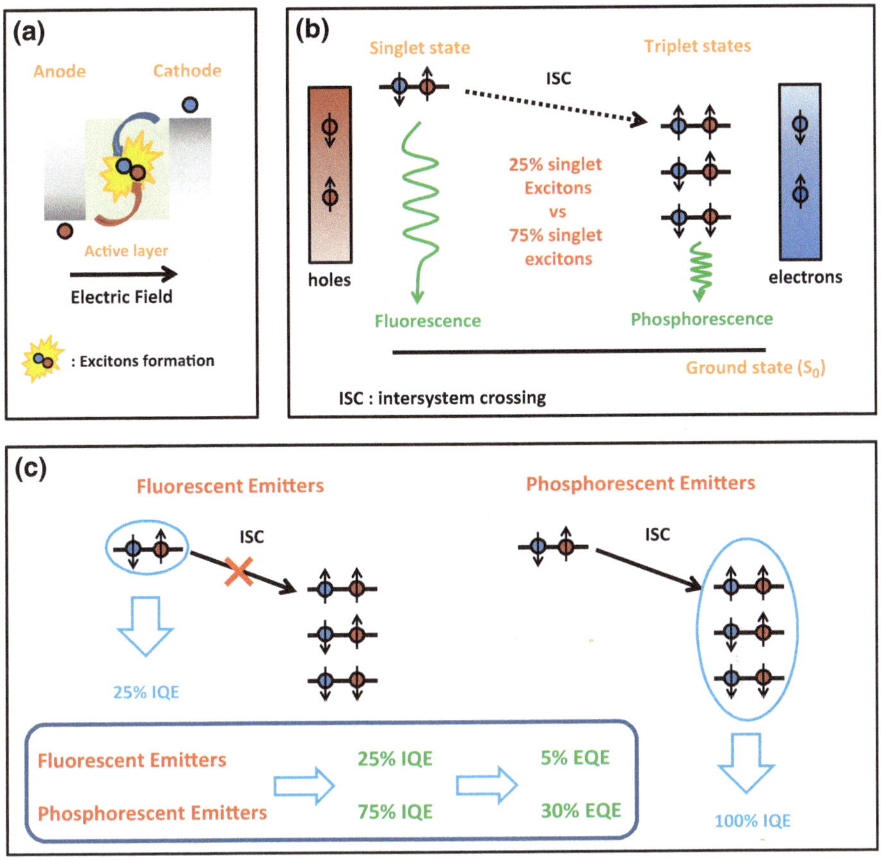

Fig. 3 Internal and external quantum efficiencies of devices comprising a fluorescent or a phosphorescent emitter. **a** Mechanisms of excitons generation, **b** spin statistics of the carrier recombination, and **c** internal and external quantum efficiencies of LECs depending of the light-emitting material

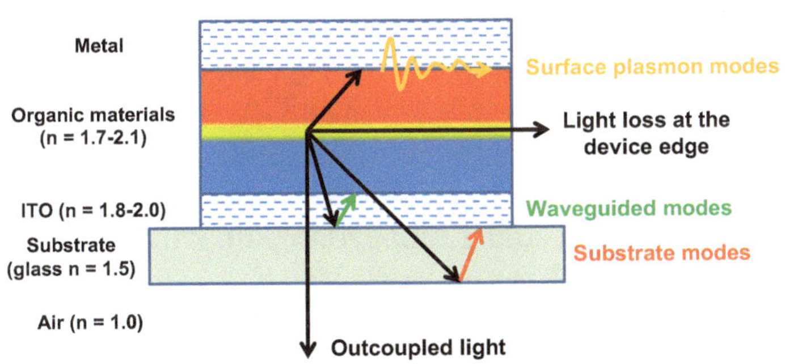

Fig. 4 Light loss mechanisms in electroluminescent devices

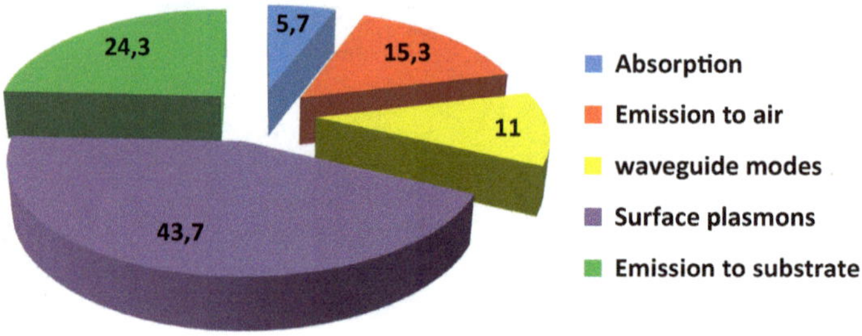

Fig. 5 Amount of power coupled to different optical channels

with an active layer thickness of 95 nm, a staggering 40% was measured for devices with a 1-µm-thick layer. Given the small amount of light extracted from devices, it is therefore not surprising that development of new concepts for improving light extraction has been a major issue over recent years [18]. To overcome the drawback of photon confinement, a large number of methods have been explored, including the use of high-index substrates [19], low-index grids [20], periodic corrugated structure [21], Bragg mirrors [22], photonic crystals [23], and antireflection coatings [24].

It has to be noticed that if a long-term systematic work has been focused toward improving the electrical characteristics of LECs (development of new luminescent materials of ever-improved photoluminescence quantum yield, enhancement of the device lifetime and stability, control of the charge recombination zone), an approach recently developed consists now in taking a closer look into how the light outcoupling can be improved. In this field, first LECs for which performances have been improved by optical design have only been reported in 2014 [25]. An out-coupled light emission improved by a factor of 1.6 was obtained by laminating a light-outcoupling film with a surface structure in the form of a hexagonal array of hemispherical microlenses, onto the front glass of the device. More recently, double photoresist layers doped with TiO_2 nanoparticles inserted between the ITO layer and the glass substrate furnished an easy tool to finely adjust the refractive index of the double photoresist layer and thus scatter light [26]. Compared to reference devices, an enhancement of 90% of the EQE was obtained by tuning the doping concentration of TiO_2 nanoparticles. Considering improvements obtained in these two works, this approach deserves to be more widely studied.

4 Microcavity Effect and Lambertian Repartition

Ordinary LECs are single-layer electroluminescent devices where the emissive layer is positioned between two highly reflective films. Therefore, LECs can be also regarded as a microcavity. A microcavity is an optical structure with dimensions of the order of one wavelength [27]. Typically, the thickness of the emissive layer

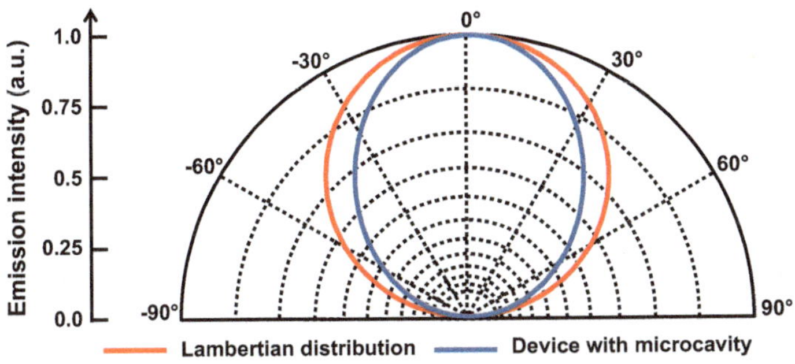

Fig. 6 Emission pattern angular distribution of devices with microcavity (*blue*) compared with the lambertian distribution (*red*)

used in LECs is comparable to the optical wavelength. Microcavity effect is classically observed for devices comprising organic layers sandwiched between two parallel reflectors [28]. Usually, one of the reflectors is a metallic mirror (cathode), while the other is a Bragg reflector. In the case of LECs, devices consist of only one effective reflector, i.e., the metallic cathode, and the reflectance of the cathode is obtained by using a thick layer of metal. Due to the microcavity effect, devices are found to show strong interference effects that alter both the optical mode density and spectrally redistribute the EL spectrum. As main disadvantage of the microcavity effect, interferences introduce an angular dependence of the luminous intensity with the viewing angle, an angular dependence to the LEC's color, and a non-Lambertian angular emission profile, i.e., the emission intensity is not proportional to the cosine of the emission angle (see Fig. 6). As shown in Fig. 6, an electroluminescent device shows a Lambertian-like radiation pattern if the full width at half-maximum is observed at ±60° along the surface normal view. Even if it is necessary to fully characterize the LEC light output, the most common parameters classically used to describe device performances are the luminance (cd/m^2), the power efficiency (lm/W), the current efficiency (cd/A), the external quantum efficiency (%), and the device lifetime.

Determination of the emission profile with the viewing angle is rarely provided as it requires to capture both the total luminous flux and the spectral information of the emitted light. Indeed, the luminous flux is adjusted to reflect the varying sensitivity of the human eye to the different wavelengths, namely that a blue source appears to be brighter than a red source at constant luminance.

5 LECs: Devices Combining Charge and Ions Transport

As already mentioned in the introduction, first LECs were prepared by Heeger et al. in 1995 [29]. In these devices, the luminescent-conjugated polymer was blended with an ion-conducting polymer (a poly(ethylene oxide) POE) and a salt acting as the electrolyte (lithium triflate). It has to be noticed that the operating mechanism of LECs is totally different from OLEDs as the application of a voltage higher than the bandgap of the luminescent polymer is not sufficient to produce light. The nominal difference between a LEC and an OLED is that the former contains mobile ions within the active layer. Therefore, contrarily to OLEDs where only the transport of charges occurs, LECs combine both charge and ion transport [30]. In such devices, upon application of a voltage, the ions are first redistributed to the electrode interfaces with the *p*-doping of the region adjacent to the cathode and the *n*-doping of the region adjacent to the anode, leading to the creation of a *p–n* junction within the active layer (see Fig. 7).

As a result, electrons and holes can move more easily through the high-conductivity doped regions and form excitons at the *p–n* junction. Light emission at the *p–n* junction can thus occur. As a consequence of this transient phenomenon corresponding to the motion of ions and the establishment of the electrochemical doping, the turn-on of LECs is rather slow.

Slow establishment of the *p–n* junction was advantageously used in numerous works to directly image and probe the *p–n* junction formation over time [31]. In fact, ions redistribution within the active layer is highly dependent on the ionic conductivity in the solid state which is often low and the turn-on time classically defined as the time required to reach the luminance of 1 cd/m^2 after application of the driving voltage often exceeds several minutes to hours [32]. However, if a delay is observed between turn-on time and light emission, the presence of the solid electrolyte is beneficial in terms of operating voltage. Thus, the electrolyte is a key component in LECs, as it facilitates in situ electrochemical doping and associated attractive device features [33]. By creating a double doped interface, charge injection is greatly facilitated and LECs can be operated at lower voltages than OLEDs comprising the same light-emitting material. As other consequence of the formation of the *p–n* junction within the emissive layer, the recombination zone of holes and electrons is extremely narrow (see Fig. 7). Figure 8 shows a typical time-dependent *I-L* curve, obtained with $Ir(ppy)_2$(4,4′-dinonyl-2,2′-bipyridine).PF_6 at an applied voltage of 7 V.

From the electrical point of view, two different phases can be distinguished during device operation [34]. The first one consists in the turn-on of the device during the formation of the *p–n* junction [35]. As soon as the two doped interfaces are created, the current increases and simultaneously devices turn on. The increase in the current continues until the *p–n* junction is completely formed. At this moment, the current saturates and the maximum luminance is obtained. Then, a second phase takes place. It corresponds to the progressive degradation of the emitter, resulting in the irreversible decrease in the luminance. At today's state of the art, the origin of the degradation is poorly understood. Various degradation processes have been observed and mechanisms postulated [36]. Notably, a shift of the emissive zone toward the cathode due to a reduction in the electron mobility with the polymer degradation could be demonstrated and imaged [37].

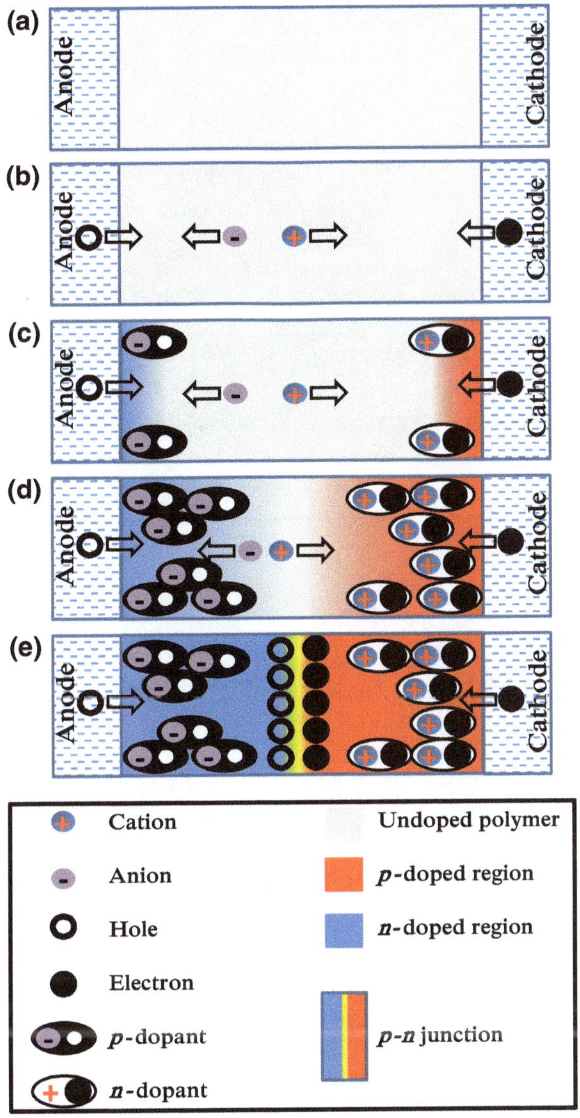

Fig. 7 Schematic operation of LECs. **a** pristine device structure, **b** initial ion transport, **c** electronic charge injection and initial formation of doping next to the electrodes, **d** continuation of the charge separation, and **e** steady state with light emission at the *p*–*n* junction

For LECs comprising metal complexes, ligand exchange with water molecules and subsequent formation of a new complex that functions as a luminance quencher was notably reported for ruthenium complexes [38]. Defluorination of iridium complexes during device operation was also suggested as another source of device degradation [39]. Detrimental effects of oxygen or excimers on the degradation of the EL performances have also been largely demonstrated [40]. Considering that all LECs degrade over the time, a characteristic typically associated with LECs is the

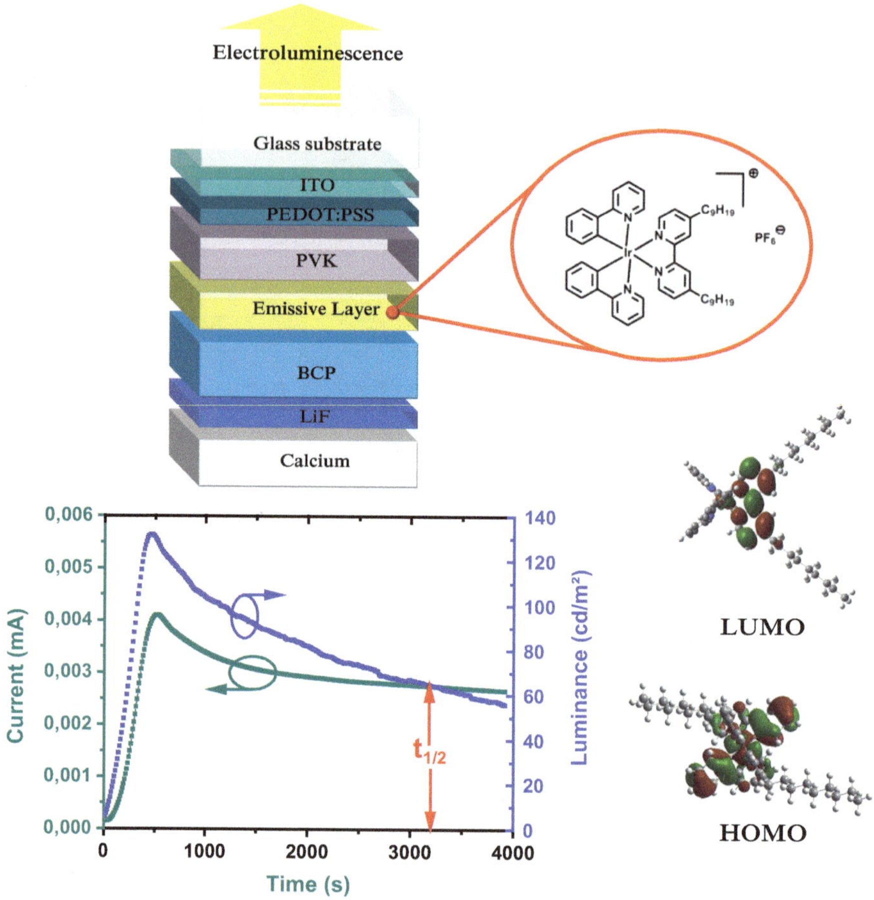

Fig. 8 Typical time-dependent current and brightness curves of LECs obtained in the present case with Ir(ppy)$_2$(4,4′-dinonyl-2,2′-bipyridine).PF$_6$ as the light-emitting material and for devices biased at 7 V

device lifetime. More precisely, this latter is determined by the measurement of the time necessary for the brightness to decay half-maximum under a constant bias.

While comparing polymer-based LECs to metal complex-based LECs, apart marked differences in key properties such as turn-on time, maximum luminance, and lifetime, device characteristics were determined as following the same trend. More precisely, the time required to form the *p–n* junction, i.e., the turn-on time, constitutes the main difference between the two types of LECs, differing by many orders of magnitude (see Fig. 9) [41]. A shorter lifetime as well as a lower maximum luminance is classically determined for polymer-based LECs compared to that measured for metal complex-based LECs. Notably, metal complex-based LECs are intrinsically more favorable candidates for furnishing higher brightness, transition

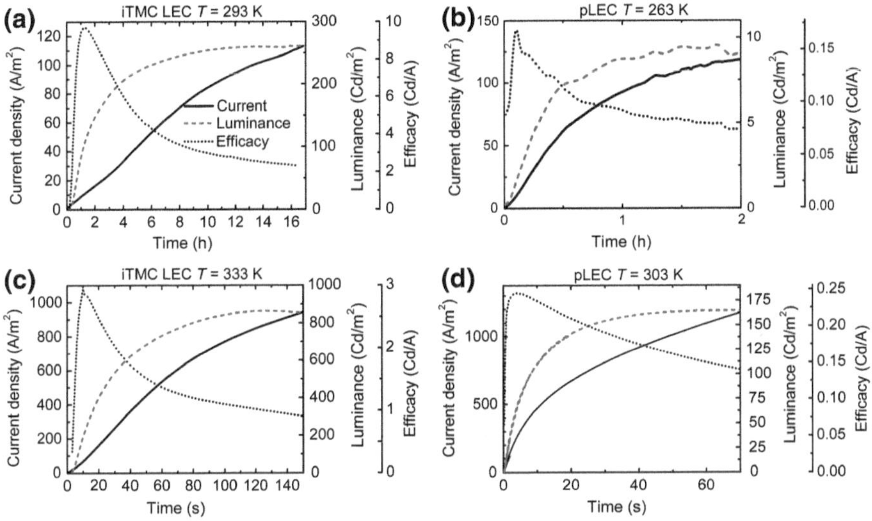

Fig. 9 Variation in the current density, luminance, and current efficiency over time for (b-d) polymer- and (a-c) metal complex-based LECs. (Reprinted with permission from Ref. [41]. Copyright 2013 American Chemical Society)

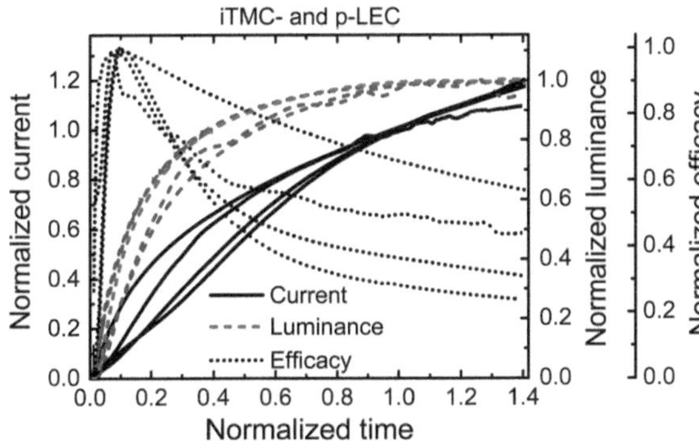

Fig. 10 Normalized current, luminance, and efficacy transients of polymer- and metal complex-based LECs at two temperatures each and biased at 3.5 V. (Reprinted with permission from Ref. [41]. Copyright 2013 American Chemical Society)

metals being triplet emitters. Jointly, metal complexes are often more robust than polymers, elongating the device lifetime.

Besides, as evidenced by Bolink et al. in a comparative study between polymers and metal complexes [41], no fundamental differences exist between the two types of LECs, device characteristics being independent of the semiconductor or the ion type.

In fact, transient behavior of metal complex and polymer-based LECs is only directed by the ionic conductivity of the emissive layer that controls the dynamic properties of devices. This last point was notably rationalized upon normalization of the transients (see Fig. 10), with a good overlap of the transients, irrespective of the semiconductor and the temperature. A strong temperature dependence of the turn-on time with the temperature was evidenced for both types of LECs, indicating the ionic mobility to control the overall device characteristics.

6 Investigations of the Electrical Characteristics of LECs

LECs are different from OLEDs in the sense that LECs contain mobile ions. Therefore, the current–voltage–luminance *(I-V-L)* scans classically used to characterize OLEDs are not suitable for LECs [42]. However, this is a useful tool for the determination of important device characteristics such as the turn-on voltage, the maximum luminance, and the voltage required to get it, or the leakage current. But application of a voltage immediately induces a motion of ions and, therefore, a change in the state of the device. As a result, *I-V-L* measurements are highly sensitive to the scan rate and large hysteresis can be found between reported *I-V-L* scans. For a higher accuracy of the measurements, very slow scan rates were notably used to characterize LECs so that at each voltage, the measure was only acquired after the current saturates [29, 43]. On the opposite, for LECs with frozen junctions, i.e., where the ions are totally immobilized, *I-V-L* curves resemble those of OLEDs [44]. In this context, to reduce ions motion, the use of a pulsed voltage during the *I-V-L* measurements has been reported by several authors [45]. Notably, the pulsed current offers the advantages of lower current, higher stability, faster turn-on, and higher efficiency at higher luminance compared to the traditional constant voltage measurements.

7 Working Principle of LECs: The Two Conflicting Models

For all electroluminescent devices, the operating voltage can be defined as the sum of the voltage corresponding to the photon energy of the emitted light and an over-potential corresponding to the energies required to inject electrons and holes in the emitting layer. Ideally, the driving voltage of devices can be reduced to the photon energy of the emitted light if the charge injection barrier from the electrodes to the EML becomes negligible, so that no energy barrier exits. If these ideal devices have been studied from the theoretical point of view, up-to-date, these devices have never been fabricated (see Fig. 11) [46].

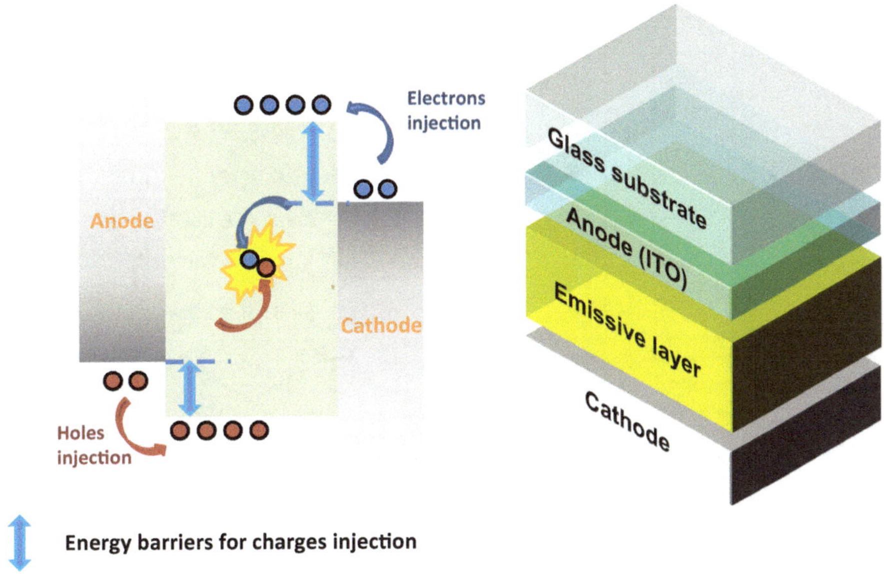

Fig. 11 Energetic diagram of LECs

For LECs, the advantage of having a single active layer also proved to be a disadvantage because of its complexity. Operating mechanism of LECs has been a long-standing controversy, and numerous articles have been dedicated to establish the exact working principle [47]. The most famous pages have been written in 2007 and 2008 in *Nature Materials* with conflicting statements [48]. Indeed, for almost a decade, two contradictory models concerning the device operation in LECs, namely the electrochemical and the electrodynamic model, have coexisted with neither one gaining general acceptance as no definitive experimental evidence was provided to underpin one of the hypotheses. Finally, the rather lively debate was resolved conclusively by Slinker et al. in 2007 [47] by the measurement of the electric field distribution in a LEC.

Finally, more recently, Kemerink et al. demonstrated that these two models are in turn limits of one master model, separated by different rates of carrier injection [49]. The two conflicting models can be briefly recapped. In the electrochemical model, electrons and holes are first injected, respectively, from the cathode and the anode, resulting in the formation of highly conductive *p*- and *n*-doped regions (see Fig. 12). Injection of holes and electrons causes oxidation and reduction in the semiconductor close to the anode and the cathode, respectively. It is from this initial model that light-emitting electrochemical cells get their name. Injection of further charges from these two doped layers is facilitated, and the two doped regions form an ohmic contact with the two electrodes. Finally, the *p*- and *n*-type regions meet in the bulk and form an undoped region placed between the *n*- and *p*-type regions named the *p–n* junction. Electrons and holes meet in this intrinsic region where they

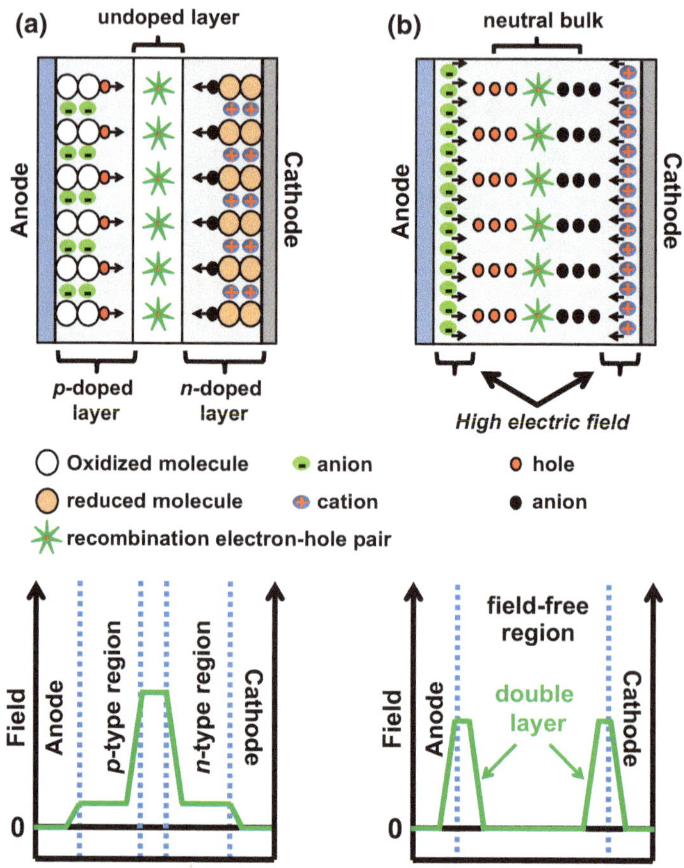

Fig. 12 Electrochemical (**a**) and the electrodynamic (**b**) models suggested as the working principle of LECs. The spatial distribution of the electric field is shown underneath

recombine radiatively. Due to the high conductivity of the two doped layers, the electric field is small in the doped regions but high in the central junction. In the electrodynamic model which was proposed after the statement of the electrochemical model, owing to the applied voltage, anions and cations drift toward the anode and the cathode, respectively. Motion of ions within the active layer causes the formation of a large electric field. Therefore, the specificity of this model is to claim that no electrochemical doping occurs; hence, no *p–n* junction ever forms.

Instead, ions redistribute until the steady state is reached and the electric field is canceled throughout the bulk. Ions' accumulation at both electrodes results in the formation of a high electric field across a thin interfacial layer close to each electrode. Ohmic contact with the electrodes is notably said to be due to the formation of electrical double layers (EDL) near the electrodes [50]. As a result of the presence of these high electric fields at the electrodes, charge injection is facilitated.

Therefore, the two models differ by the internal distribution of the electric field. According to the electrochemical model, the high electric field is located at the central *p–n* junction, whereas the electric field is almost zero everywhere else. On the opposite, according to the electrodynamic model, high electric fields are predicted to be only close to the contacts with a field-free region everywhere else. Finally, by mapping out the electrical field across the device from electrode to electrode in planar cells, no definitive conclusions could be established. Thus, Slinker et al. found high electric fields in the vicinity of the two electrodes, in direct accordance with the second model [47]. But pertinence of their measurements was questioned. Notably, these works were performed on interdigitated planar LECs, whereas best performances are classically obtained with sandwiched LECs. However, the choice of this architecture directly arises from the impossibility of performing scanning probe measurements in stacked devices. In 2009, new attempts to measure the electrostatic potential within a LEC during operation using scanning kelvin probe microscopy (SKPM) presented data providing strong evidences that LECs operate in a manner consistent with the electrochemical doping model [51]. Finally, a third study also using SKPM was inconclusive, the presented potential profile being consistent with both proposed models of operation [52]. To solve these problems, numerical modeling were also carried out to rationalize the experimental observations and find a unifying model for the operation of LECs [53]. In this context, there are still rooms to further debate on the operating mechanism.

8 Frozen Junction LECs

LECs are devices that possess appealing characteristics including insensitivity to the electrodes' work function and the active layer thickness, high efficiency, and low operating voltages, compared to OLEDs comprising the same light-emitting material [54]. Despite the many attractive merits, today's state-of-the-art LECs suffer from some critical drawbacks, including short operating lifetimes and slow response times. These shortcomings are directly related to the dynamic nature of the *p–n* junction formed "in situ" when the *p* and *n* regions meet in the bulk. Indeed, once the applied voltage is removed, the *p–n* junction immediately relaxes and ions are rapidly redistributed within the active layer to reform ion pairs. LECs relax and discharge until the undoped equilibrium state is reestablished. To some extent, this problem of ion migration and relaxation can be overcome with a simple and reliable approach consisting in freezing the *p–n* junction. While forming the *p–n* junction in polymer-based LECs at temperature higher than room temperature and then to quickly cool the device to below the glass transition temperature of the polymer electrolyte, the *p–n* junction is frozen and the ion mobility becomes negligible [55]. Once frozen, the applied voltage used to create the *p–n* junction can be removed and the ionic distribution as well as the *p–n* junction is permanent. Such a strategy is reminiscent of that used for the electric dipole alignment in second-order nonlinear optical materials. As main advantage, the freeze out of ions' motion eliminates the electrochemical reaction, thus enabling devices to be operated as conventional OLEDs by eliminating the delay required to dope interfaces. When operated at temperatures below the glass transition temperature, LECs exhibit typical *p–n* junction LED behaviors including diode rectification, unipolar light emission at low voltages, and fast response. Therefore, aside from the presence of mobile ions within the active layer, nothing in the electrical characteristics differentiates a frozen junction LEC from an OLED. In particular, the two doped regions function as the anode and the cathode for charge injection, and the turn-on voltage of LECs is approximately equal to the bandgap of the light-emitting material. Indeed, for a semiconductor, its respective HOMO and LUMO levels can be determined by electrochemistry from its respective oxidation and reduction potentials. Therefore, considering that in LECs, the semiconductor is oxidized in the *p*-doped region and reduced in the *n*-doped region, the energy gap of the frozen *p–n* junction is thus equal to the difference in electrochemical potentials between the *p*- and *n*-doped layers that correspond to the gap of the material. To create the frozen junction, several strategies have been developed over the years. Notably, the most common ion-solvating polymer classically used in LECs is poly(ethylene oxide) (PEO). However, the glass transition temperature (T_g) of this polymer is low, around 208 K, and devices must be operated at temperatures lower than 200 K to keep the *p–n* junction frozen [56]. Therefore, frozen-junction LECs that can be operated at room temperature were actively pursued. Several strategies were developed to increase the T_g of POE including the copolymerization of POE with various monomers (styrene, acrylate, methacrylate, etc.) [57] or the cross-linkage of PEO oligomers capped with methacrylate end groups [58]. It has to be noticed that parallel to these different works devoted to increase the T_g of POE, replacement of POE by ionic liquids [59] or crown ethers combined with lithium triflate [59] also furnished frozen-junction LECs that could be operated at room temperature.

9 Planar LECs

Classically, LECs are devices manufactured in a sandwich configuration with the emissive layer disposed between a metal cathode and a transparent anode (typically ITO). However, the intense debate around the working principles of LECs has sustained the development of a new type of cell in which the emissive layer can be more easily examined and the dynamic doping process visualized. These devices are planar LECs (see Fig. 13). As specificity, the emissive layer is not deposited on top of an electrode but onto an insulating substrate such as sapphire. The metal electrodes are deposited on top of the emissive layer, and devices are fabricated so that an extremely large inter-electrode spacing is introduced. As a result, in this configuration, the electric field is applied perpendicular to the direction of the light emission, contrarily to that observed in a conventional sandwiched cell where the applied electric field is parallel to the direction of light emission. As a direct

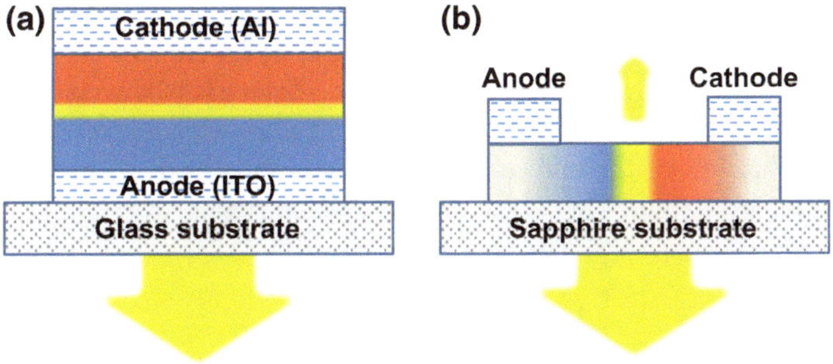

Fig. 13 Two structures of LECs: **a** sandwich structure **b** planar structure

consequence of this specific device configuration, the emission zone does not extend over the entire layer surface as classically observed for sandwich LECs but reduced to an extremely narrow zone. Therefore, the planar cell configuration essentially lends to view inside the LEC and garner a better understanding of the operating mechanism but is not suited for practical applications. For theoretical investigations, the world's largest inter-electrode gap ever used in planar cells was 11 mm, enabling to directly observe the in situ electrochemical doping during device turn-on [60].

10 The Different LECs

On the basis of the nature of the luminescent material, LECs can be distinguished into two different kinds of devices, namely polymer-based LECs (p-LECs) and ionic transition metal complex-based LECs (iTMC-LECs). If these devices vary by the composition of their active layers, marked differences in key properties can also be found such as the operational timescale that differs by many orders of magnitude [41]. In p-LECs, the active layer consists in a polymeric blend comprising a luminescent polymer, a salt, and an ion-conducting polymer in which the salt dissolves. Three components are thus required to prepare a p-LEC. Conversely, iTMC-LECs necessitate only one element since the metal complexes used to fabricate devices are ionic. Organic semiconductors thus have the dual functions of charge transporter and light emitter. To reinforce ions' migration effect, mobile ions are often introduced within the emissive layer. In this aim, ionic liquids are frequently added, especially in the case of iridium-based devices, to accelerate their turn-on [61]. Compared to p-LECs, higher brightness and quantum efficiencies are also expected owing to the phosphorescent nature of the metal complexes. To illustrate this point, best p-LECs exhibit lifetimes of ~5600 h at a brightness of 100 cd/m^2 [62], whereas iTMC-LECs have similar lifetimes for brightness higher than 600 cd/m^2 [63]. Among all measurable characteristics of LECs, the largest difference between p-LECs and iTMC-LECs can be found in the turn-on time.

Thus, for devices biased at 3–4 V, p-LECs typically turn on within a few seconds, whereas iTMC-LECs frequently require hours [64]. If differences of magnitude can be found for current, light output, and efficacy transients between p-LECs and iTMC-LECs, the two types of devices behave the same way, as recently evidenced by the normalization of the different transients and their perfect superimpositions.

11 Mechanisms of Device Failure

LECs possess numerous advantages over conventional OLEDs. However, LECs are also currently plagued by key-adverse factors resulting from the electrochemical process taking place within the emissive layer. One of the main drawbacks is undoubtedly the low device stability. Indeed, electrochemistry is involved in the desired doping of the organic semiconductor and at the origin of light emission. Unfortunately, electrochemistry is also responsible of side reactions that are not predicted by the theory outlined above and thus render the device operation far more complex than theoretically anticipated. Notably, the formation of the electric double layers at the electrode interfaces is the response of the mobile ions to the electric field induced by an externally applied voltage. By accumulating ions at both interfaces, an efficient charge injection is obtained, independently of the electrode material used. Therefore, the electrochemical stability of the electrodes used to fabricate LECs is important. To illustrate this point, aluminum is classically chosen as the cathode material due to its low cost and easy availability. However, aluminum can easily oxidize and is thus unsuitable as an anode [65]. Similarly, redox reactions involving the electrolyte are not considered in the theoretical models. However, an electrolyte that is completely inert over the potential window over which the electrochemistry of the light-emitting material takes place is difficult to find. In this field, the most representative example is PEO. Notably, the emissive layer that is a polymer blend composed of a conjugated polymer (light-emitting semiconductor) and a non-conjugated polymer (POE) makes LECs prone to phase separation [66]. It thus constitutes a major drawback for LECs as high performances are directly related to the stability of the polymeric blend upon device operation. Phase transition and/or phase separation of the polymer blend can locally modify the electrolyte concentration, causing inhomogeneous light emission. In this latter case, light emission is dominated by areas containing the largest amounts of electrolyte as these areas offer the lowest electrical resistance. These regions are also the most doped, and the high current passing throughout these areas leads to rapid degradation. External contaminants can also give rise to unwanted redox reactions. Among all contaminants, the most common species are clearly water and oxygen [67]. If it is easy to get rid from oxygen by working under inert atmosphere, the water is more problematic as the electrolyte (PEO, ionic liquid, ionic metal complex) is ionic and thus intrinsically hydrated. Elemental analyses of metal complexes often indicate the presence of one or more water molecules. These two species seriously harm the function of LECs. Notably, reduction of water at the cathode leads to hydrogen evolution that causes delamination of the cathode by creating bubbles under the cathode surface. It even constitutes the primary causes of appearance of black spots commonly seen within the emissive layer [68]. Therefore, devices must be carefully encapsulated to be operated outside the glove box. Now, encapsulation has become a common practice in the field of OLEDs and LECs. The effect of residual solvent after spin-coating on the performance of LECs was also evidenced [69]. In particular, the role of traces of solvents in the operation of LECs and its impact on the anion mobility and the response time was examined in several studies [70]. To illustrate this point , a complete removal of solvent from the emissive layers resulted in higher driving voltages than that observed for devices still containing traces of solvent within the emissive layers due to a much higher resistance in the dried films and the lower ions mobility. A slower response time as well as a enhanced device stability was observed for devices not containing traces of solvents compared to LECs still containing solvent traces. Purity of the materials and solvents used to fabricate LECs must also be chosen with care as these latter are the main sources of contaminants contributing to the accelerated device degradation. As other factor contributing to decrease, the device lifetime is the temperature. LECs are highly sensitive to temperature, and device stability is improved by operating LECs at low temperature [71]. Therefore, low emissive materials strongly contribute to lower device lifetime by speeding up device degradation due to their poor efficiencies. Indeed, in this last case, most of the excitons are lost in non-emissive processes such as heat, which in turn increases the temperature of the device and contributes to accelerate degradation. As another viable route contributing to reduce the device lifetime, combination of heat with light was even identified as a parameter speeding up the degradation of polymer-based LECs [72]. Conjugated polymers are well known to be sensitive to photochemical degradations [73]. By combining several techniques such as direct optical probing, photoluminescence, and Raman spectroscopy, the main lifetime-limiting process of LECs was determined as occurring in the thin light-emitting *p–n* junction positioned between the doped regions and not at the electrode interfaces. By examining devices after use, a clearly observable "degradation line" exactly located at the previous position of the light-emitting *p–n* junction was evidenced. A close inspection of the material located at the *p–n* junction revealed that the irreversible chemical changes were attributable to a photo-assisted degradation of the vinyl group of the conjugated polymer. By varying the device temperature, this photo-degradation process was determined as being more severe at elevated temperatures. Considering that the *p–n* junction is particularly exposed to self-heating effects (non-radiative decay of excitons, Joule heating), local increase in temperature at the junction contributes to accelerate degradation of the conjugate polymer at this position. It has to be noticed that the elevation of temperature is not negligible. In this field, Yu et al. determined that the temperature of the emitting area (i.e., the *p–n* junction) was 60–80 K higher than the initial device temperature [74a]. Similarly, Zhang et al. fabricated planar LECs of improved lifetime by simply replacing the glass substrate for a sapphire substrate that exhibits an enhanced thermal conductivity [56].

12 The Rise of ITMC-Based LECs

Metal complexes are attractive materials for optimizing light output. This is a consequence of the significantly higher efficiencies obtainable with the phosphorescent materials as compared to the fluorescent ones. Using these triplet emitters and thanks to the internal management of singlet and triplet excitons, it is possible to obtain, in principle, a four times higher electroluminescence efficiency than with classical singlet emitters. Early works on iTMC-based LECs focused on ruthenium (II) complexes [75]. However, the dominant orange-red emission combined with the lack of color tunability rapidly incited researchers to focus on other metal complexes. Parallel to this, their luminescent efficiencies were low and therefore limited their applications in lighting devices [76]. In 2002, the highest EQE reported for a Ru-based LEC was only 5.5% [77]. In this context, iridium complexes were identified as more promising candidates face to the first LEC based on an iridium(III) complex that exhibited a four times higher EQE (23.5%) than the best Ru-based LECs [40a]. The most outstanding characteristic of iridium complexes is without contest the facile variability of their electroluminescent properties. By varying the attached ligands, their emission color can be easily tuned from blue to red [78]. Iridium complexes are able to acquire a broader range of colors than ruthenium complexes favored by an increase in the ligand-field stabilization energy (LFSE), which results in a less thermally accessible MC^3 state [79]. Changing the metal center from a second-row to a third-row transition element also improves the stability of the related complexes, and the strong spin–orbit coupling also leads to enhanced conversion of singlet to triplet state population, resulting in augmented quantum yield of luminescence.

12.1 Ionic Iridium Complexes: Easily Accessible Emitters

To date, all iTMC-based LECs comprising an iridium complex were prepared with an ionic one [80]. From the synthetic point of view, it constitutes a major advantage for LECs as ionic complexes can be synthesized in high yield (most of them can be prepared with reaction yields approaching 100%) [81]. On the opposite, conventional OLEDs require a neutral emitter and these latter are often obtained with yields ranging from 15 to 25% for the most stable ones, namely the *tris*-cyclometalated $Ir(C\^N)_3$ complexes [82]. Typically, ionic complexes are synthesized in two steps according to the strategy depicted in Fig. 14.

It has to be noticed that LECs can be fabricated with cationic and anionic iridium complexes. However, due to the extremely low solubility of the anionic complexes [83] and the high solubility of the cationic ones, these latter have undoubtedly been the most widely studied. By reacting two equivalents of ligands and one equivalent

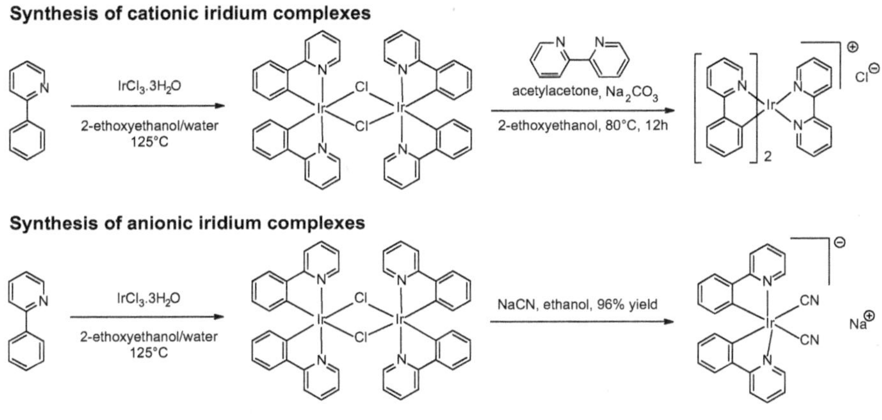

Fig. 14 Typical two-step synthesis of a heteroleptic *bis*-cyclometalated iridium (III) complex

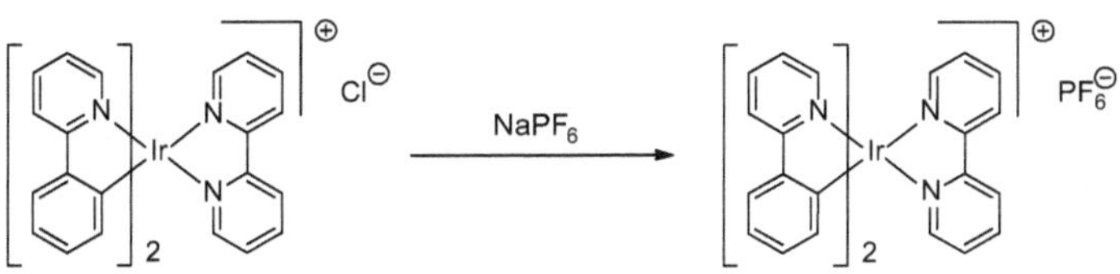

Fig. 15 Metathesis reaction converting a soluble complex to an organic solvent–soluble complex

of iridium chloride hydrate, the Nonoyama reaction yields the corresponding μ-chloride-bridged dinuclear iridium species [84]. In a second time, the cationic *bis*-cyclometalated iridium(III) complex is obtained by a bridge-splitting and substitution reaction of the μ-chloro-bridged iridium dimer with the appropriate ancillary ligand under mild conditions. Splitting the μ-chloro-bridged dimer and introducing bidentate ligands provide access to a wide range of neutral and charged complexes. In this aim, a wide range of 2,2′-bipyridine, 1,10-phenanthroline, and terpyridine derivatives have found use as neutral ligands [85]. As interesting feature, for a given μ-chloro-bridged dimer, change in ancillary ligands provides the additional possibilities of fine-tuning the optical properties of the resulting complexes and the possibility of a further functionalization of the final complex. A library of cationic iridium complexes can be easily obtained. As a result of the bridge-splitting and substitution reaction, all complexes are obtained after ligand exchange under the form of a chloride salt soluble in water. By anion metathesis with tetrabutylammonium hexafluorophosphate, counter-anion exchange occurs, furnishing a new salt soluble in most of the common organic solvent (see Fig. 15) [86].

In combination with the ancillary ligand, emission color can also be tuned by the lateral substitution of the cyclometalated ligands with electron-withdrawing or electron-releasing groups. Combining both approaches, libraries of triplet emitters covering the visible range with excited state lifetimes ranging from nanoseconds to several microseconds and with photoluminescence quantum yield ranging from 0 to almost 100% were reported [87]. Over the years, several trends were determined to control these different parameters (emission wavelengths, PLQY, etc.). Thus, increase in the steric hindrance of the ancillary ligand was determined to enhance the PLQY [88]. By extending the π-conjugation of the ancillary ligands and/or the cyclometalated ligands, emission color of complexes could also be shifted toward longer wavelengths [89].

12.2 The Current Challenge: Designing Blue Emitters with Emission of High Intensity

Up-to-date, considerable research has been invested in developing blue emitters for LECs as this color is still lagging far behind the other colors. In particular, development of white-emitting LECs is still hampered by the current limitations concerning the efficiency and the blue emission color. In electroluminescence, a material is reported as a true-blue emitter if its emission is observed at wavelength below 440 nm [90]. However, a particular issue of this color is the redshift observed between the photoluminescence in solution and in the solid state. Consistent with these observations, the electroluminescence spectrum of blue emitters very often shifts significantly to the red. The exact reason of this redshift is still unclear, and numerous explanations have been given to support this shift. Thus, for the blue-emitting iridium (III) complex [Ir(ppy-F_2)$_2$$Me_4$phen][$PF_6$] **Ir_1** (see Fig. 16), the redshift of 84 nm between the photoluminescence in solution and the electroluminescence was assigned to the high concentration of complex in the thin film, changing the emissive excited state [91]. Supported by theoretical calculations, the concentration-dependent emission was determined as originating from the contiguous presence of three low-energy triplet excited states lying within an energy difference of 0.1 eV. In the same spirit, comparison of the EL spectra of LECs comprising [Ir(dfppy)$_2$(pzpy)]PF_6 **Ir_2** as the emitter and fabricated with/without ionic liquid evidenced a significant redshift of the EL spectrum upon addition of ionic liquid [92]. Instead of the initial blue emission color at 460 nm, LECs using the title complex with 1-butyl-3-methylimidazolium hexafluorophosphate as the ionic liquid emitted a yellowish light at 526 nm. However, in this second case, the PL spectrum of [Ir(dfppy)$_2$(pzpy)]PF_6 in neat film was nearly similar to that measured in solution and only a negligible redshift was observed. In this context, the redshift of the EL spectrum was tentatively assigned to the excitation method, namely the electrical excitation. In particular, a polarization of the molecular orbitals resulting in a decrease in the energy levels was suggested to support the shift. Resulting from a more classical explanation, molecular arrangement of [Ir(4,5-diphenyl-2-methylthiazolo)2(5-methyl-1,10-phenanthroline)]PF_6 **Ir_3** into aggregates gave rise to the formation of excimers that were irreversibly formed once the device was biased [93]. Emission energy of excimers proved to be highly dependent of the applied voltage and the operation time. To limit the intermolecular interactions impacting the emission color, introduction of bulky but small substituents was examined. Indeed, contrarily to neutral emitters used in OLEDs, complexes used in LECs also act as charge transporters in complement to light emitters. Therefore, addition of too large substituents would unfavorably isolate complexes from each other and decrease the charge transport performances. In this respect, *tert*-butyl substituent seemed to be an appropriate choice to provide a balanced isolation of complexes while maintaining suitable charge transportation. Notably, complex **Ir_4** with a photoluminescence at 440 nm was designed and synthesized [94]. Despites its deep blue photoluminescence, LECs fabricated with **Ir_4** showed an EL spectrum centered on 500 nm, evidencing the insufficient steric hindrance generated by this group (see Table 1).

Conversely, a more effective confinement was obtained with benzyl groups, and if this group did not completely suppressed the aggregation in the solid state, a redshift from 468 nm for the photoluminescence in neat film to 487 nm for the electroluminescence was observed for **Ir_5** [95]. The use of carbenes as high-field ligands was more successful since no redshift between the PL and the EL spectrum was observed for **Ir_6** and **Ir_7** (see Fig. 16) [96]. Similarly, a very small spectral shift was observed for **Ir_8** since a redshift of the EL emission maximum of only 3 nm was measured compared to its PL maximum emission wavelength [97]. As specificity, this complex comprises a 1,2,3-triazole derivative as the ancillary ligand. Face to the low availability of pure-blue-emitting iridium complexes, there are still room for new investigations.

13 Conclusions—Perspectives

Light-emitting electrochemical cells are without contest the simplest light-emitting devices ever reported, and these devices have clearly the potential to replace incumbent technologies. By simply mixing mobile ions with the same type of light-emitting material than that used in OLEDs, these latter allow electrochemistry to occur so that a *p–n*-doped structure is created in situ upon application of a driving voltage. To get this sophisticated structure, LECs require only one layer of light-emitting material to be coated so that devices insensitive to the layers' thickness and the electrodes' work function are obtained. If significant improvements in terms of device lifetime, stability, and brightness have been achieved, LECs still suffer from the low availability of blue emitters, impeding the development of highly emissive white light-emitting LECs. Recent progresses as well as

Table 1 Summary of electroluminescent properties of Ir(III) complexes as emitters in LECs

Complex	Device structure	Max η_c (cd.A^{-1})	Max η_p (lm.W^{-1})	CIE coord.	Max brightness (cd. m^{-2})	EQE (%)	λ_{EL} (nm)	Ref.
Ir_1	ITO/Ir_1/Al (80 nm)	5.5	5.8	0.417, 0.533	160 (5 V)	–	560	[91]
Ir_2	ITO/PEDOT: PSS (50 nm)/Ir_2 (75 nm)/Al (120 nm)	0.65	–	0.20, 0.28	39 (6.5 V)	0.28	460, 486 (sh)	[92]
Ir_2	ITO/Ir_2:BMIM.PF_6 (120 nm)/Al (120 nm)	0.51	–	0.33, 0.45	23 (5.0 V)	0.21	526, 490 (sh), 460 (sh)	[92]
Ir_3	ITO/Ir_3 with PMMA (5 wt%) (120 nm)/Al (100 nm)	0.01	–	–	58 (8 V)	–	637	[93]
Ir_4	ITO/PEDOT:PSS (100 nm)/Ir_4:BMIM 3:1 (100 nm)/Al (150 nm)	1.76	0.92	0.28, 0.43	15 (6 V)	–	500	[94]
Ir_5	ITO/PEDOT:PSS (90–100 nm)/Ir_5 (100 nm)/LiF (1 nm)/Al (80 nm)	0.3	–	0.264, 0.363	13 (9 V)	0.14	487	[95]
Ir_6	ITO/PEDOT:PSS (70 nm)/Ir_6.PF_6: TBAOTf 1:1/Al (150–200 nm)	0.73	–	0.27, 0.43	21.6 (5 V)	–	488	[96]
Ir_6	ITO/PEDOT:PSS (70 nm)/Ir_6.BF_4: TBAOTf 1:1/Al (150–200 nm)	0.50	–	0.27, 0.43	25.7 (5 V)	–	488	[96]
Ir_7	ITO/PEDOT:PSS (70 nm)/Ir_7.PF_6: TBAOTf 1:1/Al (150–200 nm)	0.37	–	0.20, 0.34	24.0 (5 V)	–	456	[96]
Ir_7	ITO/PEDOT:PSS (70 nm)/Ir_7.BF_4: TBAOTf 1:1/Al (150–200 nm)	0.85	–	0.22, 0.38	13.4 (5 V)	–	488	[96]
Ir_8	ITO/PEDOT:PSS (100 nm)/Ir_8.BF_4: TBAOTf 1:1/Al (150–200 nm)	–	–	–	14.5 (5 V)	–	460, 488	[97]
Ir_8	ITO/PEDOT:PSS (100 nm)/Ir_8.PF_6: TBAOTf 1:1/Al (150–200 nm)	–	–	–	19.4 (5 V)	–	460, 488	[97]

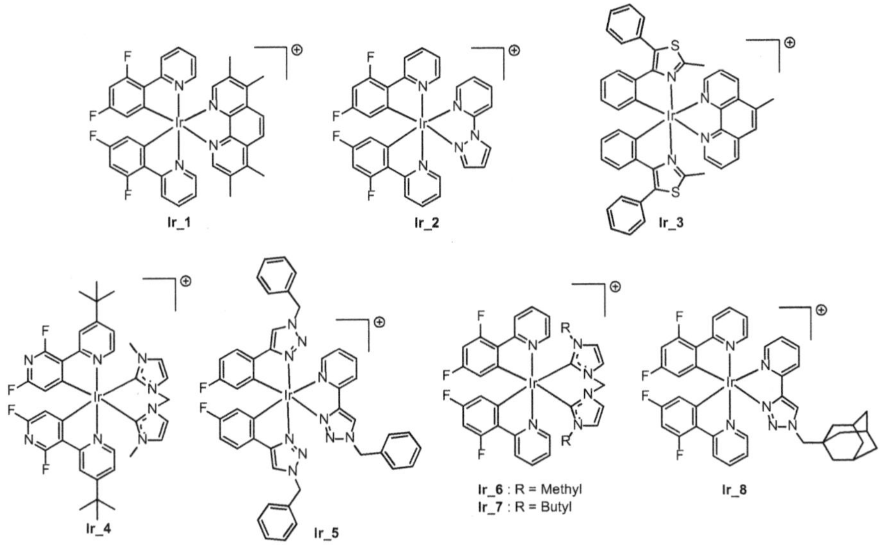

Fig. 16 Structures of blue-emitting iridium(III) complexes reported in the literature

the better understanding of the working principles of LECs make us confident for the future of these devices.

References

1. Pei, Q., Yu, G., Zhang, C., Yang, Y., Heeger, A.J.: Polymer light-emitting electrochemical cells. Science **269**, 1086–1088 (1995)
2. (a) Meier, S.B., Tordera, D., Pertegás, A., Roldán-Carmona, C., Ortí, E., Bolink, H.J.: Light-emitting electrochemical cells: recent progress and future prospects. Mater. Today **17**(5), 217–223 (2014) (b) Su, H.-C., Cheng, C.-Y.: Recent advances in solid-state white light-emitting electrochemical cells. Isr. J. Chem. **54**, 855–866 (2014)
3. Krasnov, A.N.: Electroluminescent displays: history and lessons learned. Displays **24**, 73–79 (2003)
4. van Reenen, S., Janssen, R.A.J., Kemerink, K.: Fundamental tradeoff between emission intensity and efficiency in light-emitting electrochemical cells. Adv. Mater. **25**(20), 3066–3073 (2015)
5. (a) Gao, J., Yu, G., Heeger, A.J.: Polymer light-emitting electrochemical cells with frozen *p-i-n* junction. Appl. Phys. Lett. **71**, 1293–1295 (1997) (b) Kervella, Y., Armand, M., Stephan, O.: Organic light-emitting electrochemical cells based on polyfluorene. Investigation of the failure modes. J. Electrochem. Soc. **148**, H155–H160 (2001) (c) Edman, L., Moses, D., Heeger, A.J.: Influence of the anion on the kinetics and stability of a light-emitting electrochemical cell. Synth. Met. **138**, 441–446 (2003) (d) Shin, J.H., Xiao, S., Edman, L.: Polymer light-emitting electrochemical cells: the formation and effects of doping-induced micro shorts. Adv. Funct. Mater. **16**, 949–956 (2006) (e) Habrard, F., Ouisse, T., Stephan, O., Aubouy, L., Gerbier, P., Hirsch, L., Huby, N., Van der Lee, A.: Organic light-emitting diodes

and organic light-emitting electrochemical cells based on silole-fluorene derivatives. Synth. Met. **156**, 1262–1270 (2006)

6. (a) Baldo, M.A., O'Brien, D.F., Thompson, M.E., Forrest, S.R.: Excitonic singlet-triplet ratio in a semiconducting organic thin film. Phys. Rev. B **66**, 14422–14428 (1999) (b) Sun, Y., Giebink, N.C., Kanno, H., Ma, B., Thompson, M.E., Forrest, S.R.: Management of singlet and triplet excitons for efficient white organic light-emitting devices. Nature **440**, 908–912 (2006)
7. Adachi, C.: Third-generation organic electroluminescence materials. Jap. J. Appl. Phys. **53**, 060101-1–060101-11 (2014)
8. Ishimatsu, R., Matsunami, S., Shizu, K., Adachi, C., Nakano, K., Imato, T.: Solvent effect on thermally activated delayed fluorescence by 1,2,3,5-tetrakis(carbazol-9-yl)-4,6-dicyanobenzene. J. Phys. Chem. A **117**, 5607–5612 (2013)
9. Munar, A., Sandström, A., Tang, S., Edman, L.: Shedding light on the operation of polymer light-emitting electrochemical cells using impedance spectroscopy. Adv. Funct. Mater. **22**(7), 1511–1517 (2012)
10. Shukla, M., Brahme, N.: Analytical measurements for quantum efficiency of organic light emitting diodes. J. Int. Acad. Phys. Sci. **15**(2), 231–238 (2011)
11. Sandström, A., Edman, L.: Towards high-throughput coating and printing of light-emitting electrochemical cells: a review and cost analysis of current and future methods. Energy Tech. **3**(4), 329–339 (2015)
12. (a) Benisty, H., Neve, H.D., Weisbuch, C.: Impact of planar microcavity effects on light extraction—Part I: basic concepts and analytical trends. IEEE J. Quantum Electron. **34**, 1612–1631 (1998) (b) Benisty, H., Neve, H.D., Weisbuch, C.: Impact of planar microcavity effects on light extraction—Part II: selected exact simulations and role of photon recycling. IEEE J. Quantum Electron. **34**, 1632–1643 (1998)
13. (a) Zhen, C.-G., Dai, Y.-F., Zeng, W.-J., Ma, Z., Chen, Z.-K., Kieffer, J.: Achieving highly efficient fluorescent blue organic light-emitting diodes through optimizing molecular structures and device configuration. Adv. Funct. Mater. **21**, 699–707 (2011) (b) Hofmann, S., Thomschke, M., Freitag, P., Furno, M., Lussem, B., Leo, K.: Top-emitting organic light-emitting diodes: influence of cavity design. Appl. Phys. Lett. **97**, 253308-1–253308-3 (2010) (c) Furno, M., Meerheim, R., Hofmann, S., Lussem, B., Leo, K.: Efficiency and rate of spontaneous emission in organic electroluminescent devices. Phys. Rev. B **85**, 115205-1–115205-21 (2012)
14. Chiang, C.-J., Kimyonok, A., Etherington, M.K., Griffiths, G.C., Jankus, V., Turksoy, F., Monkman, A.P.: Ultrahigh efficiency fluorescent single and bilayer organic light emitting diodes: the key role of triplet fusion. Adv. Funct. Mater. **23**, 739–746 (2013)
15. Bruckner, R., Lyssenko, V.G., Hofmann, S., Leo, K.: Lasing of tamm states in highly efficient organic devices based on small-molecule organic semiconductors. Faraday Disc **174**, 183–201 (2014)
16. Brütting, W., Frischeisen, J., Schmidt, T.D., Scholz, B.J., Mayr, C.: Device efficiency of organic light-emitting diodes: progress by improved light outcoupling. Phys. Status Solidi A **210**(1), 44–65 (2013)
17. Kaihovirta, N., Longo, G., Gil-Escrig, L., Bolink, H.J., Edman, L.: Appl. Phys. Lett. **106**, 103502-1–103502-4 (2015)
18. Saxena, K., Jain, V.K., Singh Mehtam, D.: A review on the light extraction techniques in organic electroluminescent devices. Opt. Mater. **32**(1), 221–233 (2009)
19. Peng, H.J., Ho, Y.L., Qiu, C.F., Wong, M., Kwok, H.S.: Coupling efficiency enhancement of organic light emitting devices with refractive microlens array on high index glass substrate. SID Symp. Dig. Tech. Pap. **35**(1), 158–161 (2004)
20. (a) Sun, Y., Forrest, S.R.: Enhanced light outcoupling of organic light-emitting devices using embedded low-index grids. Nature Photon **2**(8), 483–487 (2008) (b) Koh, T.W., Choi, J.M., Lee, S., Yoo, S.: Optical outcoupling enhancement in organic light-emitting diodes: highly conductive polymer as a low-index layer on microstructured ITO electrodes. Adv. Mater. **22**(16), 849–1853

21. Bi, Y.G., Feng, J., Li, Y.F., Jin, Y., Liu, Y.-F., Chen, Q.-D., Sun, H.-D.: Enhanced efficiency of organic light-emitting devices with metallic electrodes by integrating periodically corrugated structure. Appl. Phys. Lett. **100**, 053304–1–053304-4 (2012)
22. (a) Cho, S.-H., Song, Y.-W., Lee, J-g., Kim, Y.-H., Lee, J.H., Ha, J., Oh, J.-S., Lee, S.Y., Lee, S.Y., Hwang, K.H., Zang, D.S., Lee, Y.-H.: Weak-microcavity organic light-emitting diodes with improved light out-coupling. Opt. Express **16**(17), 12632–12639 (2008) (b) Lin, C.-L., Cho, T.-Y., Chang, C.-H., Wu, C.-C.: Enhancing light outcoupling of organic light-emitting devices by locating emitters around the second antinode of the reflective metal electrode. Appl. Phys. Lett. **88**:081114-1–081114-3 (2006)
23. (a) Do, Y.R., Kim, Y.C., Song, Y.W., Cho, C.-O., Jeon, H., Lee, Y.-J., Kim, S.H., Lee, Y.H.: Enhanced light extraction from organic light-emitting diodes with 2D $SiO_2/SiNx$ photonic crystals. Adv. Mater. **15**(14), 1214–1218 (2003) (b) Xu, W., Li, Y.: The effect of anisotropy on light extraction of organic light-emitting diodes with photonic crystal structure. J. Nanomater. 969120 (2013)
24. Saxena, K., Mehta, D.S., Rai, V.K., Srivastava, R., Chauhan, G., Kamalasanan, M.N.: Implementation of anti-reflection coating to enhance light outcoupling in organic light-emitting devices. J. Lumin. **128**(3), 525–530 (2008)
25. Kaihovirta, N., Larsen, C., Edman, L.: Improving the performance of light-emitting electrochemical cells by optical design. ACS Appl. Mater. Interf. **6**, 2940–2947 (2014)
26. Cheng, C.-Y., Wang, C.-W., Cheng, J.-R., Chen, H.-F., Yeh, Y.-S., Su, H.-C., Chang, C.-H., Wong, K.-T.: Enhancing device efficiencies of solid-state white light-emitting electrochemical cells by employing waveguide coupling. J. Mater. Chem. C **3**, 5665–5673 (2015)
27. So, S.K., Choi, W.K., Leung, L.M., Neyts, K.: Interference effects in bilayer organic light-emitting diodes. Appl. Phys. Lett. **74**(14), 1939–1941 (1999)
28. (a) Jordan, R.H., Rothberg, L.J., Dodabalapur, A., Slusher, R.E.: Efficiency enhancement of microcavity organic light emitting diodes. Appl. Phys. Lett. **69**, 1997–1999 (1996) (b) Fisher, T.A., Lidzey, D.G., Pate, M.A., Weaver, M.S., Whittaker, D.M., Skolnick, M.S., Bradley, D. D.C.: Electroluminescence from a conjugated polymer microcavity structure. Appl. Phys. Lett. **67**, 1355–1357 (1995) (c) Tsutsui, T., Takada, N., Saito, S., Ogino, E.: Sharply directed emission in organic electroluminescent diodes with an optical-microcavity structure. Appl. Phys. Lett. **65**, 1868–1870 (1994)
29. Pei, Q., Yu, G., Zhang, C., Heeger, A.J.: Polymer light-emitting electrochemical cells: in situ formation of a light-emitting *p–n* junction. J. Am. Chem. Soc. **118**(16), 3922–3929 (1996)
30. (a) Slinker, J.D., Rivnay, J., Moskowitz, J.S., Parker, J.B., Bernhard, S., Abruna, H.D., Malliaras, G.G.: Electroluminescent devices from ionic transition metal complexes. J. Mater. Chem. **17**, 2976–2988 (2007) (b) Edman, L.: Bringing light to solid-state electrolytes: the polymer light-emitting electrochemical cell. Electrochim. Acta. **50**, 3878–3885 (2005) (c) Marcilla, R., Mecerreyes, D., Winroth, G., Brovelli, S., del Mar Rodriguez Yebra, M., Cacialli, F.: Light-emitting electrochemical cells using polymeric ionic liquid/polyfluorene blends as luminescent material. Appl. Phys. Lett. **96**, 043308-1–043308-3 (2010) (d) Mindemark, J., Edman, L.: Illuminating the electrolyte in light-emitting electrochemical cells. J. Mater. Chem. C **4**, 420–432 (2016)
31. Hu, Y., Gao, J.: Direct imaging and probing of the *p-n* junction in a planar polymer light-emitting electrochemical cell. J. Am. Chem. Soc. **133**(7), 2227–2231 (2011)
32. Costa, R.D., Pertegas, A., Orti, E., Bolink, H.J.: Improving the turn-on time of light-emitting electrochemical cells without sacrificing their stability. Chem. Mater. **22**, 1288–1290 (2010)
33. Tang, S., Mindemark, J., Moyses Graca Araujo, C., Brandell, D., Edman, L.: Identifying key properties of electrolytes for light-emitting electrochemical cells. Chem. Mater. **26**(17), 5083–5088 (2014)
34. Tordera, D., Lenes, M., Bolink, H.J.: Dynamic doping in bright and stable light-emitting electrochemical cells. J. Nanosci. Nanotechnol. **13**(7), 5170–5174 (2013)
35. Dumur, F., Bertin, D., Mayer, C.R., Guerlin, A., Wantz, G., Nasr, G., Dumas, E., Miomandre, F., Clavier, G., Gigmes, D.: Design of blue or yellow emitting devices controlled by the deposition process of a cationic iridium (III) complex. Synth. Met. **161**, 1934–1939 (2011)

36. (a) Scott, J.C., Kaufman, J.H., Brock, P.J., Di Pietro, R., Salem, J., Goitia, J.A.: Degradation and failure of MEH-PPV light-emitting diodes. J. Appl. Phys. **79**, 2745–2751 (1996) (b) Cumpston, B.H., Jensen, K.F.: Photooxidation of electroluminescent polymers. Trends Polym. Sci. **4**, 151–157 (1996) (c) Sutherland, D.G.J., Carlisle, J.A., Elliker, P., Fox, G., Hagler, T.W., Jimenez, I., Lee, H.W., Pakbaz, K., Terminello, L.J., Williams, S.C., Himpsel, F.J., Shuh, D.K., Tong, W.M., Jia, J.J., Callcott, T.A., Ederer, D.L.: Photo-oxidation of electroluminescent polymers studied by core-level photoabsorption spectroscopy. Appl. Phys. Lett. **68**, 2046–2048 (1996) (d) Cumpston, B.H., Parker, I.D., Jensen, K.F.: In situ characterization of the oxidative degradation of a polymeric light emitting device. J. Appl. Phys. **81**, 3716–3720 (1997) (e) Parker, I.D., Cao, Y., Yang, C.Y.: Lifetime and degradation effects in polymer light-emitting diodes, J. Appl. Phys. **85**, 2441–2447 (1999) (f) Bliznyuk, V. N., Carter, S.A., Scott, J.C., Klarner, G., Miller, R.D., Miller, D.C.: Electrical and photoinduced degradation of polyfluorene based films and light-emitting devices, Macromolecules **32**, 361–369 (1999) (g) Giebeler, C., Whitelegg, S.A., Lidzey, D.G., Lane, P.A., Bradley, D.D.C.: Device degradation of polymer light emitting diodes studied by electro-absorption measurements. Appl. Phys. Lett. **75**, 2144–2146 (1999) (h) Silvestre, G.C. M., Johnson, M.T., Giraldo, A., Shannon, J.M.: Light degradation and voltage drift in polymer light-emitting diodes. Appl. Phys. Lett. **78**, 1619–1621 (2001)
37. (a) Dane, J., Gao, J.: Imaging the degradation of polymer light-emitting devices. Appl. Phys. Lett. **85**, 3905–3907 (2004) (b) Su, H.-C., Hsu, J.-H.: Improving the carrier balance of light-emitting electrochemical cells based on ionic transition metal complexes. Dalton Trans. **44**(18), 8330–8345 (2015)
38. (a) Kalyuzhny, G., Buda, M., McNeill, J., Barbara, P., Bard, A.J.: Stability of thin-film solid-state electroluminescent devices based on *tris*(2,2′-bipyridine)ruthenium(II) complexes. J. Am. Chem. Soc. **125**, 6272–6283 (2003) (b) Soltzberg, L.J., Slinker, J., Flores-Torres, S., Bernards, D., Malliaras, G.G., Abruna, H.D., Kim, J.S., Friend, R.H., Kaplan, M.D., Goldberg, V.: Identification of a quenching species in ruthenium *tris*-bipyridine electroluminescent devices. J. Am. Chem. Soc. **128**, 7761–7764 (2006)
39. (a) Sivasubramaniam, V., Brodkorb, F., Hanning, S., Loebl, H.P., van Elsbergen, V., Boerner, H., Scherf, U., Kreyenschmidt, M.: Fluorine cleavage of the light blue heteroleptic triplet emitter FIrpic. J. Fluorine Chem. **130**, 640–649 (2009) (b) Sivasubramaniam, V., Brodkorb, F., Hanning, S., Loebl, H.P., van Elsbergen, V., Boerner, H., Scherf, U., Kreyenschmidt, M.: Investigation of FIrpic in PhOLEDs via LC/MS technique. Cent. Eur. J. Chem. **7**(4), 836–845 (2009)
40. (a) Slinker, J.D., Gorodetsky, A.A., Lowry, M.S., Wang, J., Parker, S., Rohl, R., Bernhard, S., Malliaras, G.G.: Efficient yellow electroluminescence from a single layer of a cyclometalated iridium complex. J. Am. Chem. Soc. **126**, 2763–2767 (2004) (b) Graber, S., Doyle, K., Neuburger, M., Housecroft, C.E., Constable, E.C., Costa, R.D., Orti, E., Repetto, D., Bolink, H.J.: A supramolecularly-caged ionic iridium(III) complex yielding bright and very stable solid-state light-emitting electrochemical cells. J. Am. Chem. Soc. **130**, 14944–14945 (2008) (c) Costa, R.D., Orti, E., Bolink, H.J., Graber, S., Housecroft, C.E., Constable, E.C.: Intramolecular π-stacking in a phenylpyrazole-based iridium complex and its use in light-emitting electrochemical cells. J. Am. Chem. Soc. **132**, 5978–5980 (2010) (d) Bolink, H. J., Coronado, E., Costa, R.D., Orti, E., Sessolo, M., Graber, S., Doyle, K., Neuburger, M., Housecroft, C.E., Constable, E.C.: Long-living light-emitting electrochemical cells – control through supramolecular interactions. Adv. Mater. **20**, 3910–3913 (2008)
41. van Reenen, S., Akatsuka, T., Tordera, D., Kemerink, M., Bolink, H.J.: Universal transients in polymer and ionic transition metal complex light-emitting electrochemical cells. J. Am. Chem. Soc. **135**, 886–891 (2013)
42. (a) Walzer, K., Maennig, B., Pfeiffer, M., Leo, K.: Highly efficient organic devices based on electrically doped transport layers. Chem. Rev. **107**(4), 1233–1271 (2007) (b) Blom, P.W.M., Vissenberg, M.: Charge transport in poly(*p*-phenylenevinylene) light-emitting diodes. Mater. Sci. Eng. R-Rep. **27**, 53–94 (2000)

43. (a) Rudmann, H., Shimada, S., Rubner, M.F.: Operational mechanism of light-emitting devices based on Ru(II) complexes: Evidence for electrochemical junction formation. J. Appl. Phys. **94**(1), 115–122 (2003) (b) Wu, A., Yoo, D., Lee, J.K., Rubner, M.F.: Solid-state light-emitting devices based on the *tris*-chelated ruthenium(II) complex: 3. High efficiency devices via a layer-by-layer molecular-level blending approach. J. Am. Chem. Soc. **121**(20), 4883–4891 (1999)
44. (a) Shao, Y., Gong, X., Heeger, A.J., Liu, M., Jen, A.K.Y.: Long-lifetime polymer light-emitting electrochemical cells fabricated with crosslinked hole-transport layers. Adv. Mater. **21**(19), 1972 (2009) (b) Leger, J.M., Rodovsky, D.B., Bartholomew, G.P.: Self-assembled, chemically fixed homojunctions in semiconducting polymers. Adv. Mater. **18**(23), 3130–3134 (2006)
45. (a) Garcia, A., Bakus Ii, R.C., Zalar, P., Hoven, C.V., Brzezinski, J.Z., Nguyen, T.-Q.: Controlling ion motion in polymer light-emitting diodes containing conjugated polyelectrolyte electron injection layers. J. Am. Chem. Soc. **133**(8), 2492–2498 (2011) (b) Elbing, M., Garcia, A., Urban, S., Nguyen, T.-Q., Bazan, G.C.: In Situ conjugated polyelectrolyte formation. Macromolecules **41**(23), 9146–9155 (2008) (c) Shavaleev, N.M., Scopelliti, R., Grätzel, M., Nazeeruddin, M.K., Pertegás, A., Roldán-Carmona, C., Tordera, D., Bolink, H.J.: Pulsed-current versus constant-voltage light-emitting electrochemical cells with trifluoromethyl-substituted cationic iridium(III) complexes. J. Mater. Chem. C **1**(11), 2241–2248 (2013)
46. Manzanares, J.A., Reiss, H., Heeger, A.J.: Polymer light-emitting electrochemical cells: a theoretical study of junction formation under steady-state conditions. J. Phys. Chem. B **102** (22), 4327–4336 (1998)
47. Slinker, J.D., Defranco, J.A., Jaquith, M.J., Silveira, W.R., Zhong, Y.-W., Moran-Mirabal, J. M., Craighead, H.G., Abruna, H.D., Maron, J.A., Malliaras, G.G.: Direct measurement of the electric-field distribution in a light-emitting electrochemical cell. Nature Mater. **6**(11), 894–899 (2007)
48. (a) Pei, Q., Heeger, A.J.: Nature Mater. **7**, 167–167 (2008) (b) Malliaras, G.G., Slinker, J.D., DeFranco, J.A., Jaquith, M.J., Silveira, W.R., Zhong, Y.-W., Moran-Mirabal, J.M., Craighead, H.G., Abruña, H.D., Marohn, J.A.: Operating mechanism of light-emitting electrochemical cells. Nature Mater. **7**, 168–168 (2008) (c) deMello, J.C.: Organic Electronics: What's in a name? Nature Mater. **6**, 796–797 (2007)
49. van Reenen, S., Janssen, R.A.J., Kemerink, M.: Doping dynamics in light-emitting electrochemical cells. Org. Electron. **12**, 1746–1753 (2011)
50. deMello, J.C., Tessler, N., Graham, S., Friend, F.: Ionic space-charge effects in polymer light-emitting diodes. Phys. Rev. B **57**(20), 12951–12963 (1998)
51. Matyba, P., Maturova, K., Kemerink, M., Robinson, N.D., Edman, L.: The dynamic organic *p–n* junction. Nat. Mater. **8**, 672–676 (2009)
52. Pingree, L.S.C., Rodovsky, D.B., Coffey, D.C., Bartholomew, G.P., Ginger, D.S.: Scanning kelvin probe imaging of the potential profiles in fixed and dynamic planar LECs. J. Am. Chem. Soc. **129**, 15903–15910 (2007)
53. van Reenen, S., Matyba, P., Dzwilewski, A., Janssen, R.A.J., Edman, L., Kemerink, M.: A unifying model for the operation of light-emitting electrochemical cells. J. Am. Chem. Soc. **132**(39), 13776–13781 (2010)
54. Hohertz, D., Gao, J.: How electrode work function affects doping and electroluminescence of polymer light-emitting electrochemical cells. Adv. Mater. **20**(17), 3298–3302 (2008)
55. (a) Gao, J., Li, Y., Yu, G., Heeger, A.J.: Polymer light-emitting electrochemical cells with frozen junctions. J. Appl. Phys. **86**(8), 4594–4599 (1999) (b) Edman, L.: Planar polymer light-emitting device with fast kinetics at a low voltage. J. Appl. Phys. **95**(8), 4357–4361 (2004)
56. Zhang, Y., Gao, J.: Lifetime study of polymer light-emitting electrochemical cells. J. Appl. Phys. **100**(8), 084501-1–084501-8 (2006)

57. Wantz, G., Gautier, B., Dumur, F., Phan, T.N.T., Gigmes, D., Hirsch, L., Gao, J.: Towards frozen organic PN junctions at room temperature using high-T_g polymeric electrolytes. Org. Electron. **13**, 1859–1864 (2012)
58. Yu, Z., Wang, M., Lei, G., Liu, J., Li, L., Pei, Q.: Stabilizing the dynamic *p–i–n* junction in polymer light-emitting electrochemical cells. J. Phys. Chem. Lett. **2**, 367–372 (2011)
59. (a) Shao, Y., Bazan, G.C., Heeger, A.J.: Long-lifetime polymer light-emitting electrochemical cells. Adv. Mater. **19**, 365–370 (2007) (b) Shin, J.H., Xiao, S., Fransson, A., Edman, L. Polymer light-emitting electrochemical cells: frozen-junction operation of an "ionic liquid" device. Appl. Phys. Lett. **87**, 043506-1–043506-3 (2005) (c) Yang, C., Sun, Q., Qiao, J., Li, Y.: Ionic liquid doped polymer light-emitting electrochemical cells. J. Phys. Chem. B **107**:12981–12988 (2003)
60. Hu, Y., Tracy, C., Gao, J.: High-resolution imaging of electrochemical doping and dedoping processes in luminescent conjugated polymers. Appl. Phys. Lett. **88**(12), 123507-1–123507-3 (2006)
61. Parker, S.T., Slinker, J.D., Lowry, M.S., Cox, M.P., Bernhard, S., Malliaras, G.G.: Improved turn-on times of iridium electroluminescent devices by use of ionic liquids. Chem. Mater. **17**, 3187–3190 (2005)
62. (a) Asadpoordarvish, A., Sandstrom, A., Tang, S., Granstrom, J., Edman, L.: Encapsulating light-emitting electrochemical cells for improved performance. Appl. Phys. Lett. **100**(19), 193508-1–193508-4 (2012) (b) Fang, J.F., Matyba, P., Edman, L.: The design and realization of flexible, long-lived light-emitting electrochemical cells. Adv. Funct. Mater. **19**, 2671–2676 (2009) (c) Tang, S., Edman, L.: Quest for an appropriate electrolyte for high-performance light-emitting electrochemical cells. J. Phys. Chem. Lett. **1**, 2727–2732 (2010) (d) Sandstrom, A., Matyba, P., Edman, L.: Yellow-green light-emitting electrochemical cells with long lifetime and high efficiency. Appl. Phys. Lett. **96**, 053303-1–053303-3 (2010)
63. (a) Tordera, D., Meier, S., Lenes, M., Costa, R.D., Orti, E., Sarfert, W., Bolink, H.J.: Simple, fast, bright, and stable light sources. Adv. Mater. **24**(7), 897–900 (2012) (b) Tordera, D., Delgado, M., Orti, E., Bolink, H.J., Frey, J., Nazeeruddin, M.K., Baranoff, E.: Stable green electroluminescence from an iridium *tris*-heteroleptic ionic complex. Chem. Mater. **24**(10), 1896–1903 (2012)
64. (a) Costa, R.D., Orti, E., Bolink, H.J., Monti, F., Accorsi, G., Armaroli, N.: Luminescent ionic transition-metal complexes for light-emitting electrochemical cells. Angew. Chem. Int. Ed. **51** (33), 8178–8211 (2012) (b) Lenes, M., Garcia-Belmonte, G., Tordera, D., Pertegas, A., Bisquert, J., Bolink, H.J.: Operating modes of sandwiched light-emitting electrochemical cells. Adv. Funct. Mater. **21**(9), 1581–1586 (2011) (c) Gautier, B., Gao, J.: A light emitting transistor based on a hybrid metal oxide-organic semiconductor lateral heterostructure. Appl. Phys. Lett. **101**, 093302-1–093302-3 (2012) (d) Hoven, C.V., Wang, H.P., Elbing, M., Garner, L., Winkelhaus, D., Bazan, G.C.: Chemically fixed *p–n* heterojunctions for polymer electronics by means of covalent B–F bond formation. Nat. Mater. **9**(3), 249–252 (2010) (e) Kosilkin, I.V., Martens, M.S., Murphy, M.P., Leger, J.M.: Polymerizable ionic liquids for fixed-junction polymer light-emitting electrochemical cells. Chem. Mater. **22**(17), 4838–4840 (2010)
65. (a) Shin, J.H., Shin, J.H., Matyba, P., Robinson, N.D., Edman, L.: The influence of electrodes on the performance of light-emitting electrochemical cells. Electrochim. Acta **52**(23), 6456–6462 (2007) (b) Leger, J.M., Carter, S.A., Ruhstaller, B.: Recombination profiles in poly 2-methoxy-5-(2-ethylhexyloxy)-1,4-phenylenevinylene light-emitting electrochemical cells. J. Appl. Phys. **98**(12), 124907-1–124907-7 (2005)
66. Hernandez-Sosa, G., Eckstein, R., Tekoglu, S., Becker, T., Mathies, F., Lemmer, U., Mechau, N.: The role of the polymer solid electrolyte molecular weight in light-emitting electrochemical cells. Org. Electron. **14**(9), 2223–2227 (2013) (b) Summers, M.A., Buratto, S.K., Edman, L.: Morphology and environment-dependent fluorescence in blends containing a phenylenevinylene-conjugated polymer. Thin Solid Films **515**(23), 8412–8418 (2007)

67. Edman, L., Summers, M.A., Buratto, S.K., Heeger, A.J.: Polymer light-emitting electrochemical cells: doping, luminescence, and mobility. Phys. Rev. B **70**(11), 115212-1–115212-7 (2004)
68. Schaer, M., Nüesch, F., Berner, D., Leo, W., Zuppiroli, L.: Water vapor and oxygen degradation mechanisms in organic light emitting diodes. Adv. Funct. Mater. **11**(2), 116–121 (2001)
69. Zhao, W., Liu, C.-Y., Wang, Q., White, J.M., Bard, A.J.: Effect of residual solvent on Ru $(bpy)_3(ClO_4)_2$-based light-emitting electrochemical cells. Chem. Mater. **17**, 6403–6406 (2005)
70. (a) Buda, M., Kalyuzhny, G., Bard, A.J.: Thin-film solid-state electroluminescent devices based on *tris*(2,2′-bipyridine)ruthenium(II) complexes. J. Am. Chem. Soc. **124**(21), 6090–6098 (2002) (b) Kalyuzhny, G., Buda, M., McNeil, J., Barbara, P., Bard, A.J.: Stability of thin-film solid-state electroluminescent devices based on *tris*(2,2′-bipyridine)ruthenium(II) complexes. J. Am. Chem. Soc. **125**, 6272–6283 (2002)
71. (a) Sun, Q.J., Li, Y.F., Pei, Q.B.: Polymer light-emitting electrochemical cells for high-efficiency low-voltage electroluminescent devices. J. Disp. Technol. **3**, 211–224 (2007) (b) Chung, S., Lee, J.-H., Jeong, J., Kim, J.J., Hong, Y.: Substrate thermal conductivity effect on heat dissipation and lifetime improvement of organic light-emitting diodes. Appl. Phys. Lett. **94**(25), 253302-1–253302-3 (2009)
72. Wågberg, T., Ralph Hania, P., Robinson, N.D., Shin, J.-H., Matyba, P., Edman, L.: On the limited operational lifetime of light-emitting electrochemical cells. Adv. Mater. **20**(9), 1744–1749 (2008)
73. (a) Alem, S., Wakim, S., Lu, J., Robertson, G., Ding, J., Tao, Y.: Degradation mechanism of benzodithiophene-based conjugated polymers when exposed to light in air. ACS Appl. Mater. Interf. **4**(6), 2993–2998 (2012) (b) Chambon, S., Rivaton, A., Gardette, J.-L., Firon, M., Lutsen, L.: Aging of a donor conjugated polymer: photochemical studies of the degradation of poly[2-methoxy-5-(3′,7′-dimethyloctyloxy)-1,4-phenylenevinylene]. J. Polym. Sci. A Polym. Chem. **45**(2), 317–331 (2007)
74. (a) Yu, G., Cao, T., Andersson, M., Gao, J., Heeger, A.J.: Polymer light-emitting electrochemical cells with frozen *p-i-n* junction at room temperature. Adv. Mat. **10**, 385–388 (1998) (b) Edman, L., Pauchard, M., Moses, D., Heeger, A.J.: Planar polymer light-emitting device with fast kinetics at a low voltage. J. Appl. Phys. **95**, 4357–4361 (2004)
75. Elliott, C.M., Pichot, F., Bloom, C.J., Rider, L.S.: Highly efficient solid-state electrochemically generated chemiluminescence from ester-substituted *tris*-bipyridine ruthenium(II)-based polymers. J. Am. Chem. Soc. **120**(27), 6781–6784 (1998)
76. Bernanose, A., Comte, M., Vouaux, P.: A new method of emission of light by certain organic compounds. J. Chim. Phys. **50**, 64–68 (1953)
77. Rudmann, H., Shimada, S., Rubner, M.F.: Solid-state light-emitting devices based on the tris-chelated ruthenium(II) complex. 4. High-efficiency light-emitting devices based on derivatives of the *tris*(2,2′-bipyridyl)ruthenium(II) complex. J. Am. Chem. Soc. **124**(17), 4918–4921 (2002)
78. (a) Lowry, M.S., Bernhard, S.: Synthetically tailored excited states: phosphorescent, cyclometalated iridium(III) complexes and their applications. Chem. Eur. J. **12**(31), 7970–7977 (2006) (b) Lamansky, S., Djurovich, P., Murphy, D., Abdel-Razzaq, F., Lee, H.-E., Adachi, C., Burrows, P.E., Forrest, S.R., Thompson, M.E.: Highly phosphorescent *bis*-cyclometalated iridium complexes: synthesis, photophysical characterization, and use in organic light emitting diodes. J. Am. Chem. Soc. **123**(18), 4304–4312 (2001) (c) You, Y., Park, Y.: Inter-ligand energy transfer and related emission change in the cyclometalated heteroleptic iridium complex: facile and efficient color tuning over the whole visible range by the ancillary ligand structure. J. Am. Chem. Soc. **127**(36), 12438–12439 (2005)
79. Juris, A., Balzani, V., Barigelletti, F., Campagna, S., Belser, P., von Zelewsky, A.: Ru(II) polypyridine complexes: photophysics, photochemistry, electrochemistry, and chemiluminescence. Coord. Chem. Rev. **84**, 85–277 (1988)

80. (a) Hu, T., He, L., Duan, L., Qiu, Y.: Solid-state light-emitting electrochemical cells based on ionic iridium(III) complexes. J. Mater. Chem. **22**(10), 4206–4215 (2012) (b) Dumur, F., Bertin, D., Gigmes, D.: Iridium (III) complexes as promising emitters for solid-state light-emitting electrochemical cells (LECs). Int. J. Nanotechnol. **9**, 377–395 (2011)
81. (a) Lepeltier, M., Dumur, F., Graff, B., Xiao, P., Gigmes, D., Lalevée, J., Mayer, C.R.: Triple heteroleptic *tris*-cyclometalated iridium (III) complexes with three different ligands: new example with 2-(2,4-difluorophenyl)pyridine-based complex. Helv. Chim. Acta. **97**, 939–956 (2014) (b) Lepeltier, M., Dumur, F., Marrot, J., Contal, E., Bertin, D., Gigmes, D., Mayer, C. R.: Unprecedented combination of regioselective hydrodefluorination and ligand exchange reaction during the syntheses of *tris*-cyclometalated iridium (III) complexes. Dalton Trans. **42**, 4479–4486 (2013)
82. (a) Zhuang, J., Li, W., Wu, W., Song, M., Su, W., Zhou, M., Cu, Z.: Homoleptic *tris*-cyclometalated iridium(III) complexes with phenylimidazole ligands for highly efficient sky-blue OLEDs. New J. Chem. **39**(1), 246–253 (2015) (b) Tavasli, M., Moore, T.N., Zheng, Y., Bryce, M.R., Fox, M.A., Griffiths, G.C., Jankus, V., Al-Attarc, H.A., Monkman, A.P.: Colour tuning from green to red by substituent effects in phosphorescent *tris*-cyclometalated iridium(III) complexes of carbazole-based ligands: synthetic, photophysical, computational and high efficiency OLED studies. J. Mater. Chem. **22**(13), 6419–6428 (2012) (c) Beeby, A., Bettington, S., Samuel, I.D.W., Wan, Z.: Tuning the emission of cyclometalated iridium complexes by simple ligand modification. J. Mater. Chem. **13**(1), 80–83 (2003) (d) Kappaun, S., Slugovc, C., List, E.J.W.: Phosphorescent organic light-emitting devices: working principle and iridium based emitter materials. Int. J. Mol. Sci. **9**(8), 1527–1547 (2008)
83. (a) Dumur, F., Yuskevitch, Y., Wantz, G., Mayer, C.R., Bertin, D., Gigmes, D.: Light-emitting electrochemical cells based on a solution-processed multilayered device and an anionic iridium (III) complex. Synth. Met. **177**, 100–104 (2013) (b) Chen, H.-F., Wu, C., Kuo, M.-C., Thompson, M.E., Wong, K.-T.: Anionic iridium complexes for solid state light-emitting electrochemical cells. J. Mater. Chem. **22**, 9956–9561 (2012)
84. Nonoyama, M.: Benzo[*h*]quinolin-10-yl-*N* iridium(III) complexes. Bull. Chem. Soc. Jpn **47** (3), 767–768 (1974)
85. (a) Ulbricht, C., Beyer, B., Friebe, C., Winter, A., Schubert, U.S.: Recent developments in the application of phosphorescent iridium(III) complex systems. Adv. Mater. **21**(44), 4418–4441 (2009) (b) Lo, K.K.-W., Hui, W.-K., Chung, C.-K., Tsang, K.H.-K., Ng, D.C.-M., Zhu, N., Cheung, K.-C.: Biological labelling reagents and probes derived from luminescent transition metal polypyridine complexes. Coord. Chem. Rev. **249**(13), 1434–1450 (2005) (c) Mi, B.X., Wang, P.F., Gao, Z.Q., Lee, C.S., Lee, S.T., Hong, H.L., Chen, X.M., Wong, M.S., Xia, P.F., Cheah, K.W., Chen, C.H., Huang, W.: Strong luminescent iridium complexes with C^N=N structure in ligands and their potential in efficient and thermally stable phosphorescent OLEDs. Adv. Mater. **21**(3), 339–343 (2009) (d) Whittle, B., Everest, N.S., Howard, C., Ward, M.D.: Synthesis and electrochemical and spectroscopic properties of a series of binuclear and trinuclear ruthenium and palladium complexes based on a new bridging ligand containing terpyridyl and catechol binding sites. Inorg. Chem. **34**(8), 2025–2032 (1995)
86. Denisov, S.A., Cudré, Y., Verwilst, P., Jonusauskas, G., Marín-Suarez, M., Fernandez-Sanchez, J.F., Baranoff, E., McClenaghan, N.D.: Direct observation of reversible electronic energy transfer involving an iridium center. Inorg. Chem. **53**(5), 2677–2682 (2014)
87. (a) Lowry, M.S., Goldsmith, J.I., Slinker, J.D., Pohl, R., Pascal, R.A., Jr, Malliaras, G.G., Bernhard, S.: Single-layer electroluminescent devices and photoinduced hydrogen production from an ionic iridium(III) complex. Chem. Mater. **17**(23), 5712–5719 (2005) (b) Lowry, M. S., Hudson, W.R., Pascal, R.A., Jr, Bernhard, S.: Accelerated luminophore discovery through combinatorial synthesis. J. Am. Chem. Soc. **126**(43), 14129–14135 (2004)
88. Su, H.-C., Fang, F.-C., Hwu, T.-Y., Hsieh, H.-H., Chen, H.-F., Lee, G.-S., Peng, S.-M., Wong, K.-T., Wu, C.-C.: Highly efficient orange and green solid-state light-emitting electrochemical cells based on cationic IrIII complexes with enhanced steric hindrance. Adv. Mater. **17**(6), 1019–1027 (2007)

89. Zhao, Q., Lin, S., Shi, M., Wang, C., Yu, M., Li, L., Li, F., Yi, T., Huang, C.: Series of new cationic iridium(III) complexes with tunable emission wavelength and excited state properties: structures, theoretical calculations, and photophysical and electrochemical properties. Inorg. Chem. **45**(16), 6152–6160 (2006)
90. Xu, H., Chen, R., Sun, Q., Lai, W., Su, Q., Huang, W., Liu, X.: Recent progress in metal–organic complexes for optoelectronic applications. Chem. Soc. Rev. **43**(10), 3259–3302 (2014)
91. Bolink, H.J., Cappelli, L., Cheylan, S., Coronado, E., Costa, R.D., Lardies, N., Nazeeruddin, M.K., Orti, E.: Origin of the large spectral shift in electroluminescence in a blue light emitting cationic iridium (III) complex. J. Mater. Chem. **17**(48), 5032–5041 (2007)
92. He, L., Duan, L., Qiao, J., Wang, R., Wei, P., Wang, L., Qiu, Y.: Blue-emitting cationic iridium complexes with 2-(1*H*-pyrazol-1-yl)pyridine as the ancillary ligand for efficient light-emitting electrochemical cells. Adv. Mater. **18**(14), 2123–2131 (2008)
93. Margapoti, E., Shukla, V., Valore, A., Sharma, A., Dragonetti, C., Kitts, C.C., Roberto, D., Murgia, M., Ugo, R., Muccini, M.: Excimer emission in single layer electroluminescent devices based on [Ir(4,5-diphenyl-2-methylthiazolo)2(5-methyl-1,10-phenanthroline)]$^+$ [PF6]$^-$. J. Phys. Chem. C **113**(28), 12517–12522 (2009)
94. Meier, S.B., Sarfert, W., Junquera-Hernández, J.M., Delgado, M., Tordera, D., Orti, E., Bolink, H.J., Kessler, F., Scopelliti, R., Grätzel, M., Nazeeruddin, M.K., Baranoff, E.: A deep-blue emitting charged *bis*-cyclometalated iridium(III) complex for light-emitting electrochemical cells. J. Mater. Chem. **1**, 58–68 (2013)
95. Fernández-Hernández, J.M., Ladouceur, S., Shen, Y., Iordache, A., Wang, X., Donato, L., Gallagher-Duval, S., de Anda, Villa M., Slinker, J.D., De Cola, L., Zysman-Colman, E.: Blue light emitting electrochemical cells incorporating triazole-based luminophores. J. Mater. Chem. **1**(44), 7440–7452 (2013)
96. Yang, C.-H., Beltran, J., Lemaur, V., Cornil, J., Hartmann, D., Sarfert, W., Fröhlich, R., Bizzarri, C., De Cola, L.: Iridium metal complexes containing N-heterocyclic carbene ligands for blue-light-emitting electrochemical cells. Inorg. Chem. **49**(21), 9891–9901 (2010)
97. Mydlak, M., Bizzarri, C., Hartmann, D., Sarfert, W., Schmid, G., De Cola, L.: Positively charged iridium(III) triazole derivatives as blue emitters for light-emitting electrochemical cells. Adv. Funct. Mater. **20**(11), 1812–1820 (2010)

FSC
www.fsc.org
MIX
Papier | Fördert gute Waldnutzung
FSC® C083411

Kolibri 360 GmbH
Ferdinand-Jühlke-Straße 7
99095 Erfurt, Deutschland
produktsicherheit@kolibri360.de